I0055098

Plant Genomics: Methods and Functions

Plant Genomics: Methods and Functions

Editor: Isabelle Nickel

R CALLISTO REFERENCE

www.callistoreference.com

Callisto Reference,
118-35 Queens Blvd., Suite 400,
Forest Hills, NY 11375, USA

Visit us on the World Wide Web at:
www.callistoreference.com

© Callisto Reference, 2019

This book contains information obtained from authentic and highly regarded sources. Copyright for all individual chapters remain with the respective authors as indicated. All chapters are published with permission under the Creative Commons Attribution License or equivalent. A wide variety of references are listed. Permission and sources are indicated; for detailed attributions, please refer to the permissions page and list of contributors. Reasonable efforts have been made to publish reliable data and information, but the authors, editors and publisher cannot assume any responsibility for the validity of all materials or the consequences of their use.

ISBN: 978-1-64116-159-6 (Hardback)

Trademark Notice: Registered trademark of products or corporate names are used only for explanation and identification without intent to infringe.

Cataloging-in-Publication Data

Plant genomics : methods and functions / edited by Isabelle Nickel.
    p. cm.
Includes bibliographical references and index.
ISBN 978-1-64116-159-6
1. Plant genomes. 2. Plant genetic engineering. 3. Genomics.
4. Agricultural genome mapping. I. Nickel, Isabelle.
SB123.57 .P53 2019
631.523 3--dc23

Table of Contents

Preface

The purpose of the book is to provide a glimpse into the dynamics and to present opinions and studies of some of the scientists engaged in the development of new ideas in the field from very different standpoints. This book will prove useful to students and researchers owing to its high content quality.

Plant genomics is a vast field of study. It studies the structure, mapping, function, evolution and editing of genomes. It is instrumental in the splicing and deletion of specific genes to produce desired phenotypes as a response to the changed genotype. Such genetic modifications can be generated by the "Gene Gun" method and by transformation via Agrobacterium. Introduction of new traits and control over existing traits in plants is possible due to research in plant genetics and genomics. Most of the topics introduced in this book cover new techniques and applications of plant genomics. While understanding the long-term perspectives of the topics, the book makes an effort in highlighting their impact as a modern tool for the growth of the discipline. Those with an interest in this field would find this book helpful.

At the end, I would like to appreciate all the efforts made by the authors in completing their chapters professionally. I express my deepest gratitude to all of them for contributing to this book by sharing their valuable works. A special thanks to my family and friends for their constant support in this journey.

Editor

Tos17 rice element: incomplete but effective

Francois Sabot

Abstract

Background: *Tos17* was the first LTR retrotransposon (*Copia*) described as active in cultivated rice, and is present in two copies in the genome of the sequenced Nipponbare variety. Only the chromosome 7 copy is active and able to retrotranspose, at least during *in vitro* culture, and this ability was widely used in insertional mutagenesis assays.

Results: Here the structure of the active *Tos17* was thoroughly annotated using a set of bioinformatic analyses.

Conclusions: Unexpectedly, *Tos17* appears to be a non-autonomous LTR retrotransposon, lacking the *gag* sequence and thus unable to transpose by itself.

Background

The long terminal repeats (LTR) retrotransposon life cycle involves a cytosolic reverse-transcription step within a multiproteic core called virus-like particle (VLP), formed by the polymerization of the Group-specific antigen (GAG) proteins, normally encoded in the element itself; for a recent review, see [1]. This GAG protein classically harbors three domains, from external to internal:

1) the matrix domain (MA), for membrane targeting and capsid assembly;
2) the capsid hydrophobic region (CA) and the most conserved part of GAG, in charge of polymerization, and the
3) nucleocapsid (NC), targeting the specific mRNA through the PSI region [1].

In addition, a CCHC zinc-finger motif is located at the C-terminus of the protein, single or twice repeated (or even thrice), and is in charge of the protein-nucleic acid interactions [1]. This protein is theoretically specific of its own RNA, and is an essential and mandatory component of the retrotransposition of LTR retrotransposons. A second open reading frame (ORF), *pol*, encodes the reverse transcriptase-RNaseH (RT-RNaseH), which drives the synthesis of a double-stranded cDNA from two RNA matrices and the integrase (INT) which allows the insertion of the new cDNA copy. However, in some cases,

some non-autonomous elements have been shown capable of hijacking the GAG from other elements [2].

In cultivated Asian rice (*Oryza sativa* L.), LTR retrotransposons compose at least 20% of the genome (MSUv7.0 reference genome [3], http://rice.plantbiology.msu.edu/index.shtml). The *Copia Tos17* element (for Transposon of *Oryza sativa* 17) was the first identified as active [4] and able to transpose in this genome. Moreover, *Tos17* seems to be the most transpositionally competent one in regenerated plants [5].

Two almost identical genomic copies of *Tos17* reside in the reference genome (on chromosomes 7 and 10; Figure 1). Only the chromosome-7 copy is transpositionally active (during *in vitro* culture at least), whereas the other, located on chromosome 10, is inactive, heavily methylated and contains several stop codons and indels in its predicted coding region [6]. This last copy can, however, be reactivated (transcriptionally) in methylation-defective mutants [6]. In the whole *Oryza* genus, the copy number as well as the location of active copies (if there are any) may differ [7].

The *Tos17* activation during *in vitro* culture was widely used in mutagenesis assays, which allowed reverse genetics analyses through the generation of insertional mutants without transformation [8-10]. In the present study, a detailed functional analysis of *Tos17* was performed, showing that both genomic *Tos17* copies lack a *gag* ORF, making *Tos17* a non-autonomous element requiring an active one in order to ensure its transposition.

Results and discussion

The two *Tos17* genomic copies were extracted from their respective location in the rice MSUv7.0 genome,

Correspondence: francois.sabot@ird.fr
UMR DIADE IRD/UM2, 911 Avenue Agropolis BP64503, F-34394 Montpellier Cedex 5, France

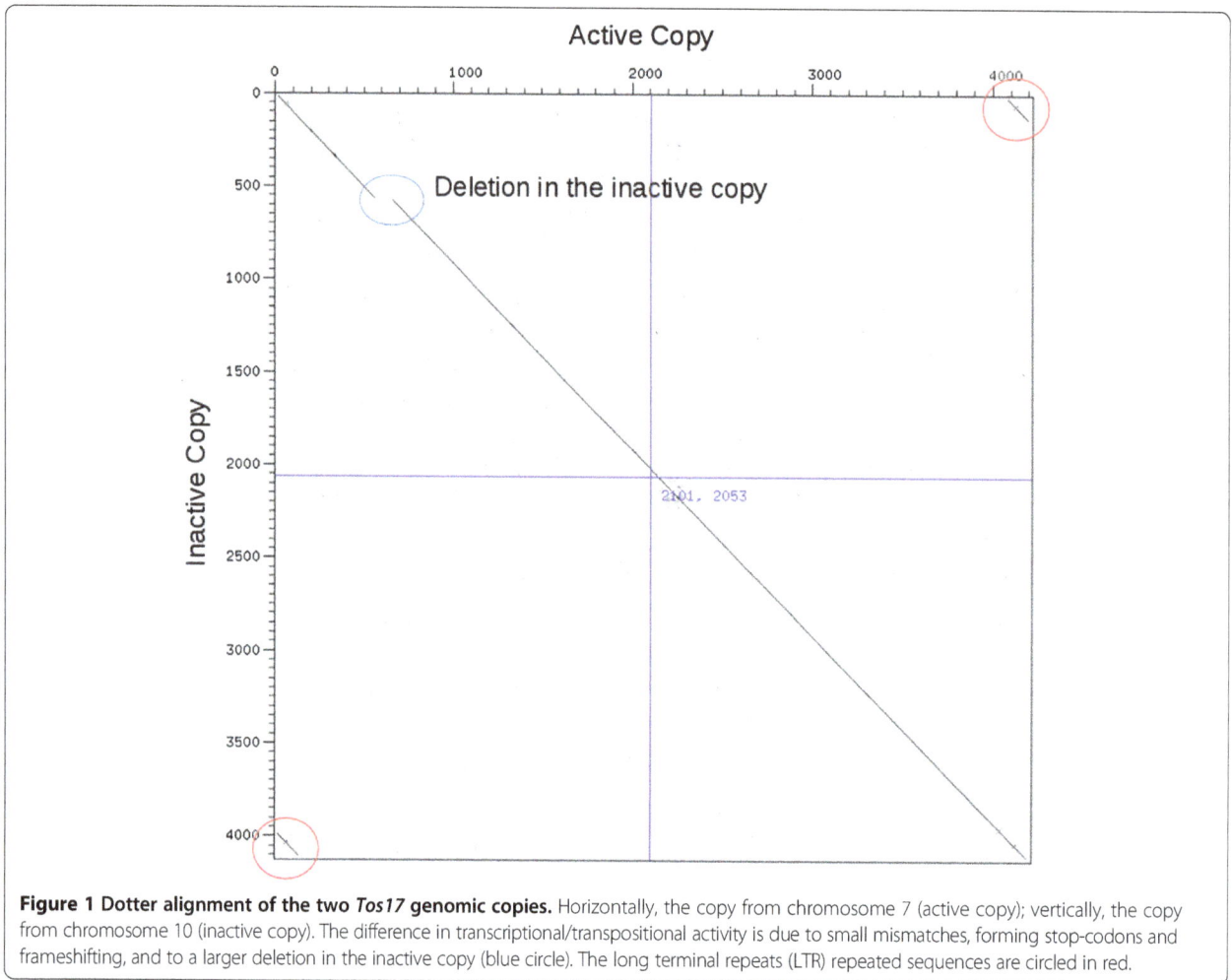

Figure 1 Dotter alignment of the two *Tos17* genomic copies. Horizontally, the copy from chromosome 7 (active copy); vertically, the copy from chromosome 10 (inactive copy). The difference in transcriptional/transpositional activity is due to small mismatches, forming stop-codons and frameshifting, and to a larger deletion in the inactive copy (blue circle). The long terminal repeats (LTR) repeated sequences are circled in red.

and manually annotated using a series of basic local alignment search tool (BLAST), ProSite and Protein families (Pfam) analyses. A predicted long ORF (from position 659 to 3835, Figure 2A; annotated as the *gag-pol* ORF [4]) of 1,058 residues can be detected on the active copy (chromosome 7), whereas no apparent ORFs (that is, more than 100 residues starting with Met) exist on the inactive copy. On this long ORF, INT (*gag_pre-integrase* and *rve*) and RT (*RVT_2*) *Pfam*-A motifs can be easily identified (see Table 1), which suggests that this ORF is the polyprotein (POL) one. However, none of the truly GAG-related

motifs, such as CCHC zinc-finger (18 residues) or the *UBN2* group (100 to 150 residues), could be identified, and the first confidently identified motif related to the INT (and thus to the pol ORF) in the *Pfam* database starts at residue 79 (Figure 2A; *gag_pre-integrase* motif) of this ORF (base 757 of the internal sequence).

The *Pfam* analysis was performed on the largest *Tos17* ORF.

This ORF was then compared to ORFs from those of the active *Copia* elements, *RIRE1* from *Oryza. australiensis* [11,12] [BAA22288; EMBL/GB] (Figure 2B), and *Houba*

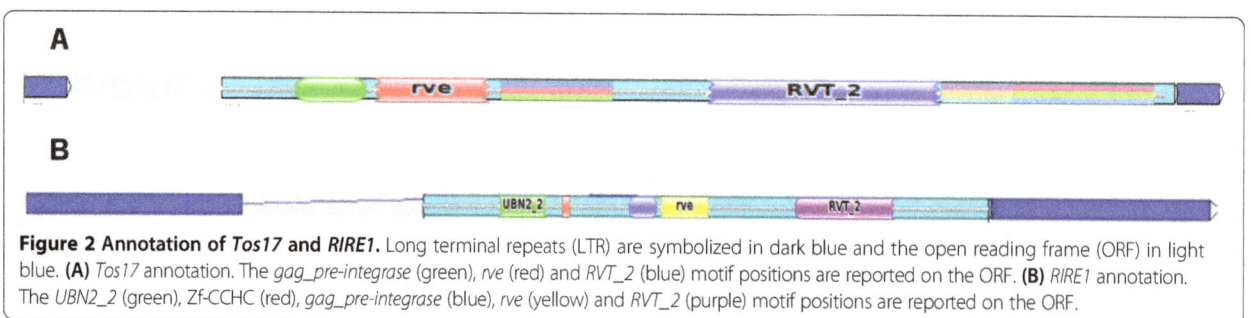

Figure 2 Annotation of *Tos17* and *RIRE1*. Long terminal repeats (LTR) are symbolized in dark blue and the open reading frame (ORF) in light blue. **(A)** *Tos17* annotation. The *gag_pre-integrase* (green), *rve* (red) and *RVT_2* (blue) motif positions are reported on the ORF. **(B)** *RIRE1* annotation. The *UBN2_2* (green), Zf-CCHC (red), *gag_pre-integrase* (blue), *rve* (yellow) and *RVT_2* (purple) motif positions are reported on the ORF.

Table 1 Tos17 Open reading frame (ORF)2 *Pfam* motifs

Motif	Start	Stop	e-value
gag_pre-integrase PF13976	79	153	6.50e-016
rve PF00665	164	284	3.60e-026
RVT_2 PF07727	519	762	1.10e-095

from *O. sativa* (known to be one of the most recently retrotransposed *Copia* in rice; [13]). As shown on Figure 3, the ORFs aligned on the whole POL part the elements that are compared two by two; the *Tos17* ORF, however, lacked the GAG region, while the ORFs from *RIRE1* and *Houba* are also aligned on the GAG part. No GAG-related region can be detected on the whole *Tos17* genomic sequences in *BLASTx* against *nr* and protein databases (data not shown). Various *tBLASTn* (protein query versus nucleic database) analyses against the rice EST databases from NCBI were performed, and no ESTs resembling a larger ORF than the ones known were detected. Finally, no other *Tos17 gag*-like sequence can be amplified in PCR on the NipponBare genomic DNA (data not shown).

RT-sequence phylogenetic analysis showed that only *RN304* and *Lullaby* are closely related to *Tos17* [14]. Interestingly, *RN304*, the closest element to *Tos17*, is itself also a non-autonomous element also lacking the *gag* sequence, similar to *Tos17*, but no information about its transpositional activity is available. The closest complete element to *Tos17* (that is, one that harbors a complete *gag-pol* ORF) is the *Lullaby* element, recently shown as transitionally active in only some of the regenerated lines in which its expression was detected [14]. *Lullaby* is a 5′142-long element, and has two copies in the Nipponbare genome, on chromosomes 6 and 9, with only the chromosome-6 copy active [14]. The DNA similarity between *Tos17* and *Lullaby* is 57% at the DNA level (whole element sequence), and 64% at the protein level (*gag-pol* region sequence). At the DNA level, the similarity

is limited to the internal sequence, whereas at the protein level the two POL sequences aligned well. Moreover, the primer binding site (PBS) region, located immediately after the 5′ LTR, and involved in RNA-GAG recognition [1], is almost identical between the two elements (5′-TGGTATCAGAGC(a/t)A(t/-)GGT-3′), starting at positions 126 and 139 for *Lullaby* and *Tos17* respectively. However, no common INT signal (at the 3′-end of the 3′ LTR [1]) is shared between *Lullaby* and *Tos17*, highlighting the use of *Lullaby* GAG by *Tos17* only.

Tos17, the most active LTR retrotransposon in cultivated rice, and the most commonly used element as an insertional mutation tool [8], is thus a non-autonomous element, because no *gag* sequence exists in the *Oryza sativa* genome, even if *Tos17* is able to retrotranspose in this species. The simplest explanation is that *Tos17* is coupled with an active LTR retrotransposon for its mobility, and that the former is able to use the *gag* (and VLP) from the latter. Such hitchhiking implies a structural (same GAG-recognition signals) as well as translational (same time of expression) relationship between *Tos17* and its autonomous partner. This association is probably a long-term association, as the structural annotation of the *Tos17* elements (Figure 2A) reveals a complete removal of the *gag* region, without any identifiable remnants, but without damaging any other structural features of the element (LTR, PBS or polypurine tract (PPT)). Indeed, such clean elimination might have occurred during *Tos17* evolution, with only elements within this correct deletion selected (able to be correctly expressed and mobilized by its partner), as no other *Tos17*-like element with *gag* remnants has been detected.

The use of *Tos17* as an insertional tool for reverse genetics is not affected by this non-autonomous state, as long as requested functional and complementation analyses are performed to validate or invalidate the insertion as the real cause of the observed phenotype.

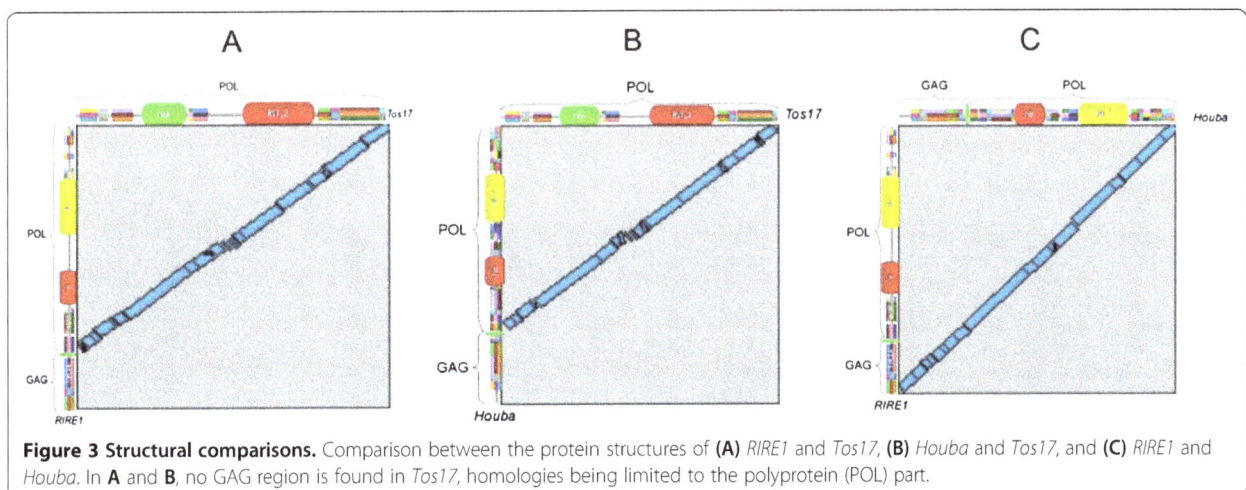

Figure 3 Structural comparisons. Comparison between the protein structures of **(A)** *RIRE1* and *Tos17*, **(B)** *Houba* and *Tos17*, and **(C)** *RIRE1* and *Houba*. In **A** and **B**, no GAG region is found in *Tos17*, homologies being limited to the polyprotein (POL) part.

The fact that *Tos17* is not able to retrotranspose by itself may help to explain the high rate (almost 90%) of morpho-physiological variations untagged by *Tos17* (or the transferred T-DNA) observed among regenerated lines ([9]; M Lorieux, unpublished data; B Hsingh, personal communication), which is probably also due to transposition of other elements, as shown previously [5].

Conclusion

Analyses, such as the one described here, highlight the need for a better knowledge of transposable elements (TEs), in order to ensure a better understanding of their effects upon the host genome. In particular, it may be of interest to further study the details of the relationships between the non-autonomous elements and their autonomous counterparts, because existing data suggest that the former are more active than the latter, as shown for *BARE2* and *Tos17*.

Methods

The nucleotidic sequences from genomic copies of each element were launched in Artemis [15], and the ORFs longer than 100 residues were automatically extracted from the element sequences. The ORFs were then scanned online using a combination of Pfam, ProSite and BLASTp analyses [16] with standard parameters. The results were then reported on Artemis, in order to manually reconstruct the complete structure of each element. The LTRs were identified using Dotter [17], and the PBS and PPT were manually determined. The comparison between putative GAG-POL sequences was performed using the Align2Sequence graphical tool from the NCBI, through a BLASTp analysis, for a better presentation. The identity/similarity levels were calculated using the Stretcher program from the EMBOSS suite.

Abbreviations

BLAST: basic local alignment search tool; GAG: group-specific antigene; INT: integrase; LTR: long terminal repeats; ORF: open reading frame; PBS: primer binding site; POL: Polyprotein; PPT: Polypurine tract; TE: transposable element; Tos: transposon of *Oryza sativa*; VLP: virus-like particle.

Competing interests

The author declares having no competing interests.

Acknowledgements

The author thanks Cristian Chaparro and Benoit Piegu for their comments on the analyses, and Dr Timothy Tranberger for his help with English corrections.

References

1. Sabot F, Schulman AH: **Parasitism and the retrotransposon life cycle in plants: a hitchhiker's guide to the genome.** *Heredity* 2006, **97**:381–388.
2. Tanskanen JA, Sabot F, Vicient C, Schulman AH: **Life without GAG: the BARE-2 retrotransposon as a parasite's parasite.** *Gene* 2007, **390**:166–174.
3. Kawahara Y, de la Bastide M, Hamilton JP, Kanamori H, McCombie WR, Ouyang S, Schwartz DC, Tanaka T, Wu J, Zhou S, Childs KL, Davidson RM, Lin H, Quesada-Ocampo L, Vaillancourt B, Sakai H, Lee SS, Kim J, Numa H, Itoh T, Buell CR, Matsumoto T: **Improvement of the *Oryza sativa* Nipponbare reference genome using next generation sequence and optical map data.** *Rice* 2013, **6**:4.
4. Hirochika H, Sugimoto K, Otsuki Y, Tsugawa H, Kanda M: **Retrotransposons of rice involved in mutations induced by tissue culture.** *Proc Natl Acad Sci USA* 1996, **93**:7783–7788.
5. Sabot F, Picault N, ElBaidouri M, Llauro C, Chaparro C, Piegu B, Roulin A, Guiderdoni E, Delabastide M, McCOMBIE R, Panaud O: **Transpositional landscape of the rice genome revealed by paired-end mapping of high-throughput re-sequencing data.** *Plant J* 2011, **66**:241–246.
6. Hirochika H: **Contribution of the *Tos17* retrotransposon to rice functional genomics.** *Curr Opin Plant Biol* 2001, **4**:118–122.
7. Petit J, Bourgeois E, Stenger W, Bès M, Droc G, Meynard D, Courtois B, Ghesquière A, Sabot F, Panaud O, Guiderdoni E: **Diversity of the *Ty-1 copia* retrotransposon *Tos17* in rice (*Oryza sativa* L.) and the AA genome of the *Oryza* genus.** *Mol Genet Genom* 2009, **282**:633–652.
8. Hirochika H, Guiderdoni E, An G, Hsing Y-I, Eun MY, Han C-D, Upadhyaya N, Ramachandran S, Zhang Q, Pereira A, Sundaresan V, Leung H: **Rice mutant resources for gene discovery.** *Plant Mol Biol* 2004, **54**:325–334.
9. Miyao A, Iwasaki Y, Kitano H, Itoh J-I, Maekawa M, Murata K, Yatou O, Nagato Y, Hirochika H: **A large-scale collection of phenotypic data describing an insertional mutant population to facilitate functional analysis of rice genes.** *Plant Mol Biol* 2007, **63**:625–635.
10. Piffanelli P, Droc G, Mieulet D, Lanau N, Bès M, Bourgeois E, Rouvière C, Gavory F, Cruaud C, Ghesquière A, Guiderdoni E: **Large-scale characterization of *Tos17* insertion sites in a rice T-DNA mutant library.** *Plant Mol Biol* 2007, **65**:587–601.
11. Noma K, Nakajima R, Ohtsubo H, Ohtsubo E: **RIRE1, a retrotransposon from wild rice *Oryza australiensis*.** *Genes Genet Syst* 1997, **72**:131–140.
12. Piegu B, Guyot R, Picault N, Roulin A, Sanyal A, Saniyal A, Kim H, Collura K, Brar DS, Jackson S, Wing RA, Panaud O: **Doubling genome size without polyploidization: dynamics of retrotransposition-driven genomic expansions in *Oryza australiensis*, a wild relative of rice.** *Genome Res* 2006, **16**:1262–1269.
13. Vitte C, Ishii T, Lamy F, Brar D, Panaud O: **Genomic paleontology provides evidence for two distinct origins of Asian rice (*Oryza sativa* L.).** *Mol Genet Genom* 2004, **272**:504–511.
14. Picault N, Chaparro C, Piegu B, Stenger W, Formey D, Llauro C, Descombin J, Sabot F, Lasserre E, Meynard D, Guiderdoni E, Panaud O: **Identification of an active LTR retrotransposon in rice.** *Plant J* 2009, **58**:754–765.
15. Rutherford K, Parkhill J, Crook J, Horsnell T, Barrell B, Rice P: **Artemis: sequence visualization and annotation.** *Bioinformatics* 2000, **16**:944–945.
16. Altschul SF, Gish W, Miller W, Myers EW, Lipman DJ: **Basic local alignment search tool.** *J Mol Biol* 1990, **215**:403–410.
17. Sonnhammer ELL, Durbin R: **A dot-matrix program with dynamic threshold control suited for genomic DNA and protein sequence analysis.** *Gene* 1996, **167**:1–10.

2

The diversification of PHIS transposon superfamily in eukaryotes

Min-Jin Han[1], Chu-Lin Xiong[1], Hong-Bo Zhang[1], Meng-Qiang Zhang[1], Hua-Hao Zhang[2] and Ze Zhang[1]*

Abstract

Background: PHIS transposon superfamily belongs to DNA transposons and includes *PIF/Harbinger*, *ISL2EU*, and *Spy* transposon groups. These three groups have similar DDE domain-containing transposases; however, their coding capacity, species distribution, and target site duplications (TSDs) are significantly different.

Results: In this study, we systematically identified and analyzed PHIS transposons in 836 sequenced eukaryotic genomes using transposase homology search and structure approach. In total, 380 PHIS families were identified in 112 genomes and 168 of 380 families were firstly reported in this study. Besides previous identified *PIF/Harbinger*, *ISL2EU*, and *Spy* groups, three new types (called *Pangu*, *Nuwal*, and *Nuwall*) of PHIS superfamily were identified; each has its own distinctive characteristics, especially in TSDs. *Pangu* and *Nuwall* transposons are characterized by 5'-ANT-3' and 5'-C|TNA|G-3' TSDs, respectively. Both transposons are widely distributed in plants, fungi, and animals; the *Nuwal* transposons are characterized by 5'-CWG-3' TSDs and mainly distributed in animals.

Conclusions: Here, in total, 380 PHIS families were identified in eukaryotes. Among these 380 families, 168 were firstly reported in this study. Furthermore, three new types of PHIS superfamily were identified. Our results not only enrich the transposon diversity but also have extensive significance for improving genome sequence assembly and annotation of higher organisms.

Keywords: Transposable elements, PHIS, Diversification, Identification

Background

Transposable elements (TEs) are fragments of DNA that can move from one site to another in a genome [1, 2]. TEs are classified into two classes (class 1 and class 2) according to their mechanism of transposition. The transposition mechanism of class 1 elements can be described as copy-and-paste mode, whereas class 2 transposons can be transposed by cut-and-paste mechanism. Recently, more and more genome sequencing revealed that TEs constitute the largest components of most eukaryotic genomes [2–13]. TEs not only have significant impact on the evolution of the host genomes and biological complexity but also are challenges for host genome sequencing, assembly, and annotation due to their repeatability. Thus, the knowledge about TEs characteristics and categories will promote the development of genomics.

In the past decade, many studies focused on identification, annotation, and function of TEs. So far, huge amounts of TEs have been identified and annotated. For example, 42 class 1 superfamilies and 19 class 2 superfamilies were annotated and cataloged in the RepBase database. However, the number of reported TEs could be just the tip of the iceberg. There are a larger number of TEs to be annotated due to their great diversification. For instance, 658 families were classified into unknown TEs in the silkworm; 163 unknown TE families in the maize and about 0.38 % of mouse genome sequences are unknown TEs [12–14]. Thus, the work of identification and annotation of TEs is far from finished.

Recently, we have identified a new group of cut-and-paste transposons designated as *Spy* [15]. *Spy* transposons are distinct from all other groups of DNA transposons by their strong insertion preference within the AAATTT motif and the lack of target site duplications (TSDs) upon insertion. In addition, we showed that *PIF/Harbinger*, *ISL2EU*, and *Spy* are evolutionarily related and share a preference for insertion into AT-rich target

* Correspondence: zezhang@cqu.edu.cn
[1]School of Life Sciences, Chongqing University, Chongqing 400044, China
Full list of author information is available at the end of the article

sequences [15]. For instance, the *ISL2EU* transposons are characterized by 5′-AT-3′ TSDs and the *PIF/Harbinger* transposons by 5′-TWA-3′ [16, 17]. Thus, these three groups *PIF/Harbinger*, *ISL2EU*, and *Spy* were classified into the same superfamily that is designated as "PHIS". The PHIS transposon superfamily is high polymorphism in the target sequences, coding capacity, and conserved motifs of transposase [15]. It is common to find some distinct groups within a given superfamily. Previously, variable nucleotide composition and length of TSDs were found in some superfamilies [16–18]. However, the detailed diversification of PHIS transposon superfamily still remains unclear.

Here, we systematically identified and analyzed PHIS transposons in 836 sequenced eukaryotic genomes using transposase homology search combined with structure approach. Totally, 380 PHIS families including 212 previously reported families and 168 unpublished families were identified in this study. The 380 PHIS families are classified into six groups including three previously reported groups (*PIF/Harbinger*, *ISL2EU*, and *Spy*) and three new groups, called *Pangu*, *NuwaI*, and *NuwaII*. Each new group has its own particular characteristics, especially in TSDs.

Results
The landscape of PHIS transposons in eukaryotic genomes

To investigate the detailed diversification and evolution of PHIS superfamily in eukaryotes, we systematically identified and analyzed the characteristics and distribution of PHIS transposons in 836 eukaryotic genomes using transposase homology search and structure approach. Finally, we identified 380 PHIS transposon families. Furthermore, each of the PHIS consensus sequence defined in this study was subject to homology search against RepBase (as of October 20, 2014) and National Center for Biotechnology Information (NCBI) non-redundant (nr) nucleotide database using Censor and BlastN program. The results of these searches showed that 168 of 380 PHIS families were not reported, and other TEs (212) had been released and cataloged in RepBase, NCBI, or published papers [15].

Based on the characteristics (TSDs, coding capacity, and secondary structure of transposase, etc.) of these 380 families, we found that 214 families belong to the *PIF/Harbinger* transposon group (Additional file 1: Table S1). Among the 214 families, 80 families had been previously identified and cataloged in RepBase, and 134 families were firstly identified in this study. These 214 families shared the following characteristics. (1) The TSD sequence is 5′-TWA-3′ tri-nucleotide ('W' represents A or T nucleotide) (Fig. 1). (2) Most candidate autonomous elements contain two open reading frames

(ORFs), one ORF encoding the DDE and helix-turn-helix (HTH) motif-containing transposase and the other ORF encoding a DNA-binding protein with a Myb/SANT domain. The potential active families of PIF/Harbinger group were defined as those including both two intact ORFs. Finally, we identified 88 potential active families in the eukaryotic genomes (Additional file 1: Table S1 and Fig. 2b). (3) The TIR (terminal inverted repeat) lengths of different PIF/Harbinger families are highly variable (5–1042 bp), but the lengths of most TIRs (~93 %) are less than 60 bp, and the first nucleotide of TIRs is usually A or G (Fig. 1). (4) The average length of consensus sequences of candidate autonomous is ~4124 bp. (5) These families are distributed in 75 species including plants, fungi, and animals. The above-described characteristics of *PIF/Harbinger* transposons are consistent with previous reports [15, 16, 19].

Meanwhile, 25 families belong to *ISL2EU* group. Among these 25 families, 8 families were firstly identified in this study. The others had been cataloged in RepBase (Additional file 1: Table S2). These families shared the following characteristics. (1) The TSDs are 5′-AT-3′ di-nucleotide; however, there is a conserved single A nucleotide in the flank of 5′ terminal of TSDs and a conserved single T nucleotide in the flank of 3′ terminal of TSDs (Additional file 2: Figure S1). Thus, we speculated that the target site sequence of *ISL2EU* transposons is A|AT|T (where '|' marks the cut site), the analysis of paralogous empty sites further confirmed the target site sequence of *ISL2EU*. Additional file 2: Figure S2 shows the possible generation mechanism of this TSDs. (2) Most autonomous candidate transposons of *ISL2EU* contain two ORFs, one ORF encoding the DDE, HTH, and THAP domain-containing transposase, the other ORF encoding a DNA-binding protein with a YqaJ exonuclease domain. Similar to a standard mentioned before, TEs with two intact ORFs are defined as the potential active transposons. Thus, 12 potential active families of *ISL2EU* group were identified in the eukaryotic genomes (Additional file 1: Table S2 and Fig. 2b) (3). The TIR length ranges from 6 to 259 bp, and the first two nucleotides of TIRs are usually "GG" di-nucleotide (Fig. 1). (4) The average length of consensus sequences of autonomous elements is ~4840 bp. (5) These families are distributed in 14 species. All these species belong to animals.

In this study, we found 54 families that belong to the *Spy* transposons; however, we did not identify any new *Spy* transposon family. All these families have been identified in previous study, and the characteristics of *Spy* transposons were also shown previously [15]. Besides the above three identified PHIS groups (*PIF/Harbinger*, *ISL2EU*, and *Spy*), we also found three new types of PHIS transposons distinct from the previous PHIS

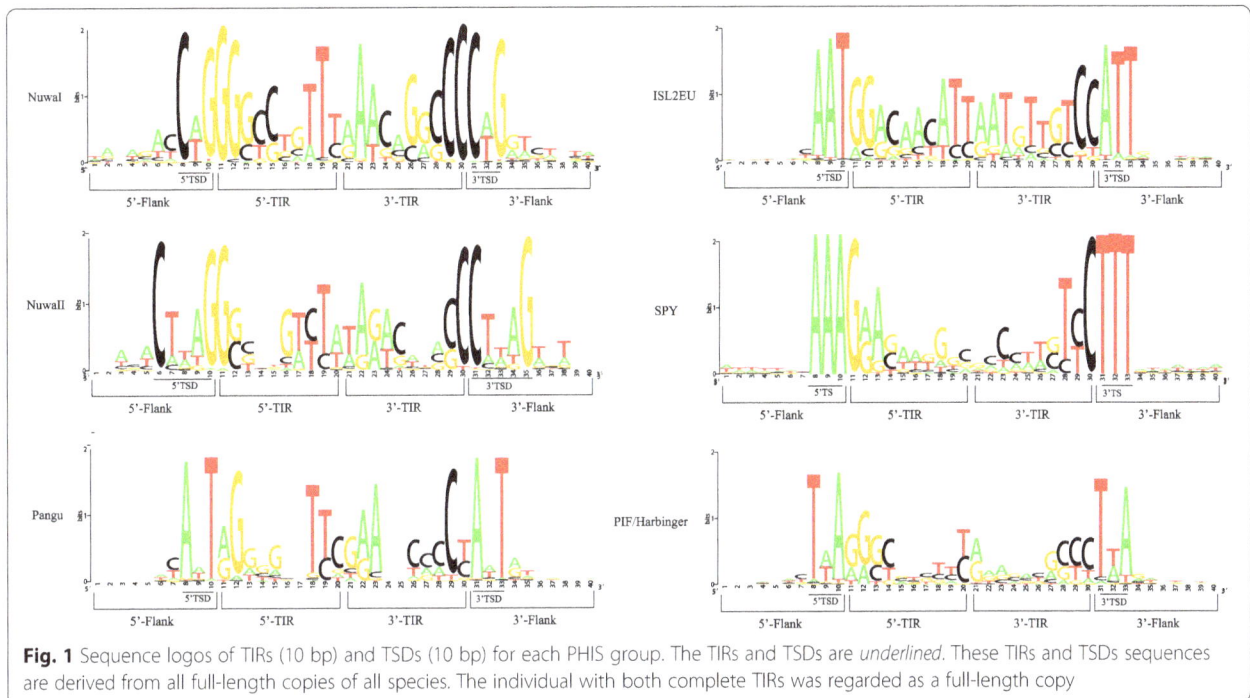

Fig. 1 Sequence logos of TIRs (10 bp) and TSDs (10 bp) for each PHIS group. The TIRs and TSDs are *underlined*. These TIRs and TSDs sequences are derived from all full-length copies of all species. The individual with both complete TIRs was regarded as a full-length copy

transposons in TSDs, and these new types transposons are called *Pangu*, *NuwaI*, and *NuwaII*, respectively.

Characterization and distribution of *Pangu* transposons

Thirty four *Pangu* families were identified in this study (Additional file 1: Table S3). The length of TIRs in these families varies from 11 to 40 bp, and the first two nucleotides of TIRs are usually "AG" and "GG" di-nucleotide (Fig. 1). The average consensus sequence length of autonomous candidates is ~3487 bp. Most autonomous candidates of *Pangu* transposon contain two ORFs, one ORF encoding the DDE motif-containing transposase and without any other domains. Meanwhile, we did not detect any known motifs in the other ORF. Given that the potential active families should contain the two intact ORFs, we identified two potential active families of *Pangu* group in the eukaryotic genomes (Additional file 1: Table S3 and Fig. 2b). Secondary structure prediction of *Pangu* DDE-containing transposases suggests that the first D is located between two beta-sheets, the second D is located between a beta-sheet and an alpha-helix, and the last E is present within an alpha-helix (Fig. 3). This result is consistent with the eukaryotic *PIF/Harbinger* and *ISL2EU* transposons [15]. The results of paralogous empty site confirmed that the TSDs of these families are 5′-ANT-3′ ('N' represents A, T, C, or G nucleotide) (Fig. 3). This characteristic of TSDs is significantly different from eukaryotic *PIF/Harbinger*, *ISL2EU*, and *Spy* transposons but consistent with the bacterial *IS5* transposons. Thus, both *Pangu* and *IS5* transposons

could belong to the same group or were derived from the same ancient element.

These 34 *Pangu* transposons are distributed in 15 eukaryotic genomes. These species include two coleopterans, one dipteran, one arachnidan, one molluscan, one hydrozoan, one anthozoan, two ascomycetes, three basidiomycetes, one heterokontophyta, and two algae (Fig. 2). And these species are widely distributed in plants, fungi, and animals. Thus, the *Pangu* transposons could be ancient elements in the eukaryotic genomes. To estimate the abundance of *Pangu* transposons in the eukaryotic genomes, the consensus sequence of each family of *Pangu* was used as query in BlastN ($e < 10^{-5}$) search against the corresponding genome. A copy for the same family was defined by e value less than e^{-5}, length larger than 50 bp, and nucleotide identity larger than 80 %. Finally, we identified 3270 copies of *Pangu* group in the eukaryotic genomes (Additional file 1: Table S3, Additional file 3: Table S4, and Fig. 2c).

Characterization and distribution of *Nuwal* transposons

Twenty-three *NuwaI* families were identified in this study (Additional file 1: Table S5). The results of paralogous empty site confirmed that the TSDs of these families are 5′-CWG-3′ ('W' represents A or T nucleotide) (Fig. 4). This characteristic is significantly different from previously the identified *PIF/Harbinger*, *ISL2EU*, and *Spy* transposons (AT-rich TSDs). Most autonomous candidates of *NuwaI* transposons contain two ORFs, one ORF encoding the DDE motif-containing transposase and

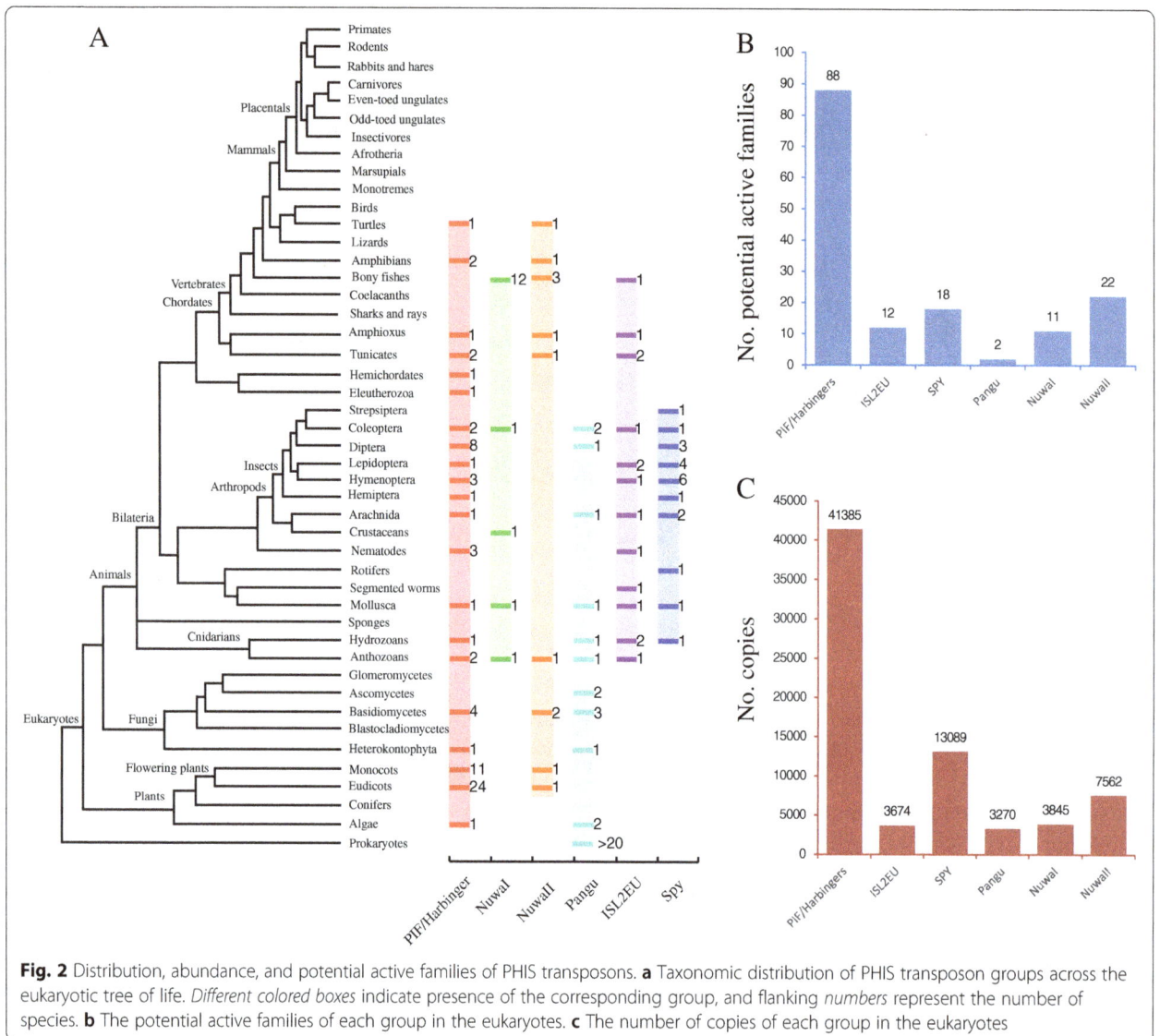

Fig. 2 Distribution, abundance, and potential active families of PHIS transposons. **a** Taxonomic distribution of PHIS transposon groups across the eukaryotic tree of life. *Different colored boxes* indicate presence of the corresponding group, and flanking *numbers* represent the number of species. **b** The potential active families of each group in the eukaryotes. **c** The number of copies of each group in the eukaryotes

without any other domain, the other ORF encoding a DNA-binding protein with a Myb/SANT domain. We identified 11 potential active families in the eukaryotic genomes because these TEs contain the two intact ORFs (Additional file 1: Table S5 and Fig. 2b). The secondary structure of *NuwaI* transposase is very similar to the *PIF/Harbinger*, *ISL2EU*, and *Pangu* transposases. For instance, the first D is located between two beta-sheets, the second D is typically between a beta-sheet and an alpha-helix, and the last E occurs within an alpha-helix (Fig. 4). The TIR lengths of *NuwaI* families range from 12 to 61 bp, and the first three nucleotides of TIRs are usually 'GGG' tri-nucleotide (Fig. 1). The average length of consensus sequences of autonomous candidates is ~4462 bp. These *NuwaI* transposons are distributed in 16 animal genomes. These species include 12 bony fish, 1 coleopteran, 1 crustacean, 1 molluscan,

and 1 anthozoan (Fig. 2a). However, these species are distributed only in the kingdom of animals. Thus, the *NuwaI* transposons could be relatively younger elements in the eukaryotes. Finally, 3845 copies of *NuwaI* group were identified in the eukaryotic genomes. The genomic abundance and copy number of each *NuwaI* family in each species were shown in Fig. 2c, Additional file 1: Table S5, and Additional file 3: Table S6.

Characterization and distribution of *NuwaII* transposons

There are 30 out of 380 families which belong to the *NuwaII* families (Additional file 1: Table S7). According to the paralogous empty site, we cannot judge that the TSDs of NuwaII group are 3 bp (TNA) or 5 bp (CTNAG) (Fig. 5). However, most PHIS elements are typically associated with 3-bp TSD. Thus, the TSDs of NuwaII elements are most likely 3-bp TSDs. Meanwhile,

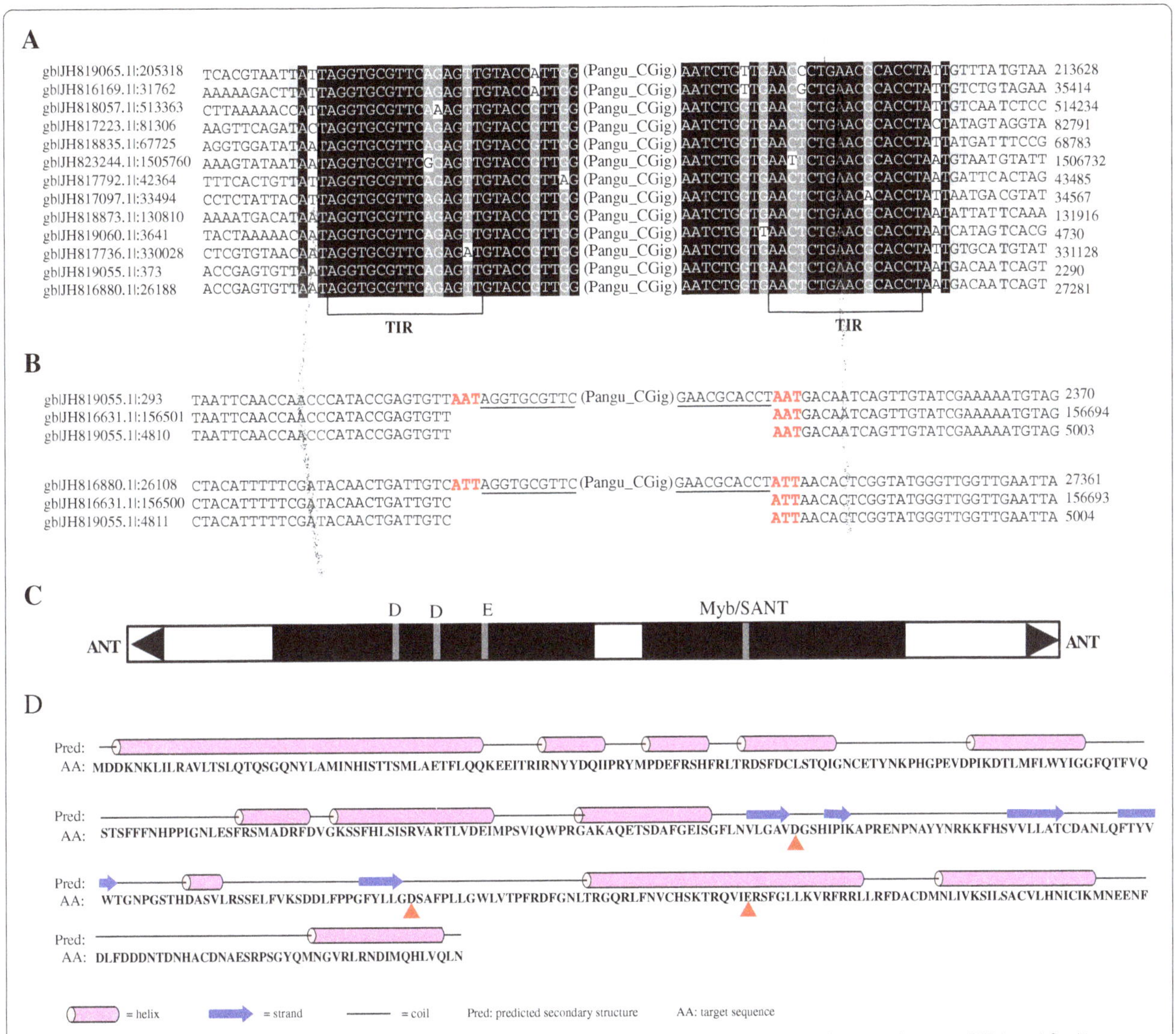

Fig. 3 Characteristics of *Pangu* transposons. **a** Sequence alignments for *Pangu_CGig* family. The terminal inverted repeats (TIRs) and flanking sequences (10 bp) are shown. **b** Two examples of alignments of the flanking sequences of *Pangu_CGig* insertions with a paralogous sequences found within the same genome but devoid of the transposon. The TIRs of the element are *underlined*. **c** Structure of *Pangu_CGig*. *Black triangles* and *solid black boxes* represent the TIRs and ORFs, respectively, and the position of the DDE triad is shown. **d** Predicted secondary structure of the DDE motif-containing transposase of the *Pangu_CGig*. The DDE triads is marked with *red triangles* below the sequence

there is a conserved single C nucleotide in the flank of 5′ terminal of TSDs and a conserved single G nucleotide in the flank of 3′ terminal of TSDs. Thus, the target of NuwaII is preferentially C|TNA|G ('N' represents A, T, C, or G nucleotide, '|' represents the cut site).

The transposase of *NuwaII* is very similar to that of *NuwaI* in the coding capacity, conserved motifs, and second enzyme structure. For instance, the most autonomous elements of *NuwaII* transposons contain two ORFs, one ORF encoding the DDE motif-containing transposase (Additional file 2: Figure S3), and the other ORF encoding a Myb/SANT domain-containing protein. Twenty-two potential active *NuwaII* families with the two intact ORFs were identified in the eukaryotic

genomes (Additional file 1: Table S7 and Fig. 2b). In the secondary structure of *NuwaII* transposase, the first D is located between two beta-sheets, the second D is typically between a beta-sheet and an alpha-helix, and the last E occurs within an alpha-helix (Fig. 5). The average length of consensus sequences of autonomous candidates is ~4685 bp; TIRs length of each family ranges from 13 to 46 bp, and the first two nucleotides of most TIRs are conserved GG. These *NuwaII* transposons are distributed in 12 species, including 1 turtle, 1 amphibian, 3 bony fishes, 1 amphioxus, 1 tunicate, 1 anthozoan, 2 basidiomycetes, 1 monocot, and 1 eudicot (Fig. 2a). Meanwhile, these species are also distributed in the kingdoms of plants, fungi, and animals. Thus, the

Fig. 4 Characteristics of *Nuwal* transposons. **a** Sequence alignments for *Nuwal-4_DRer* family. The terminal inverted repeats (TIRs) and flanking sequences (10 bp) are shown. **b** Two examples of alignments of the flanking sequences of *Nuwal-4_DRer* insertions with a paralogous sequences found within the same genome but devoid of the transposon. The TIRs of the element are *underlined*. **c** Structure of *Nuwal-4_DRer*. *Black triangles* and *solid black boxes* represent the TIRs and ORFs, respectively, and the position of the DDE triad is shown. **d** Predicted secondary structure of the DDE motif-containing transposase of the *Nuwal-4_DRer*. The DDE triads is marked with *red triangles* below the sequence

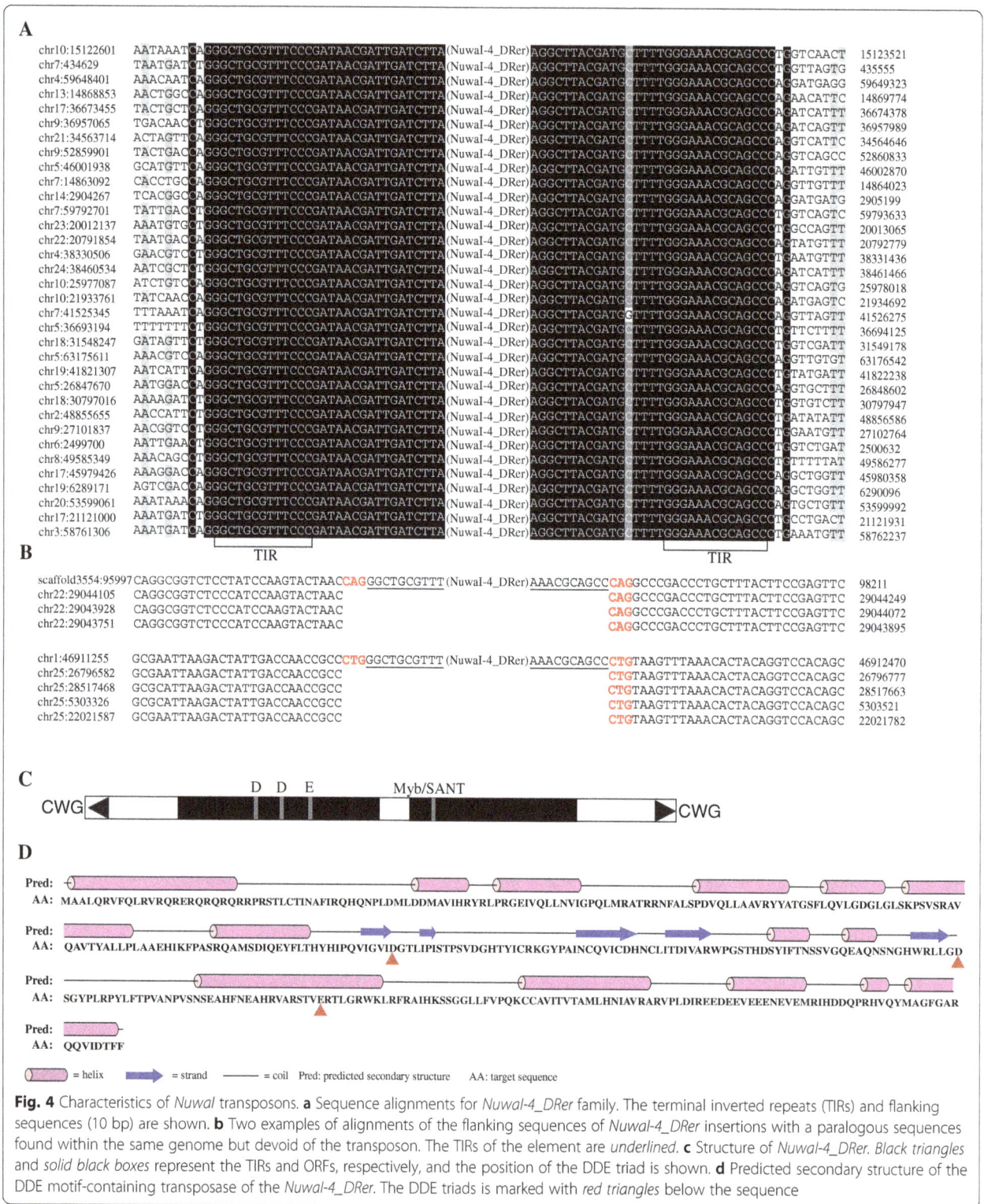

NuwaII transposons could be also relatively old elements. Finally, we found 7564 copies of *NuwaII* group. The genomic abundance and copy number of each *NuwaII* family in each species are shown in Fig. 2c and Additional file 3: Table S8.

Evolutionary relationships of PHIS transposons

To investigate the evolutionary relationships of six PHIS transposon groups (*PIF/Harbinger*, *ISL2EU*, *Spy*, *Pangu*, *NuwaI*, and *NuwaII*), the core catalytic DDE domain of 16 representative transposases (include intact DDE

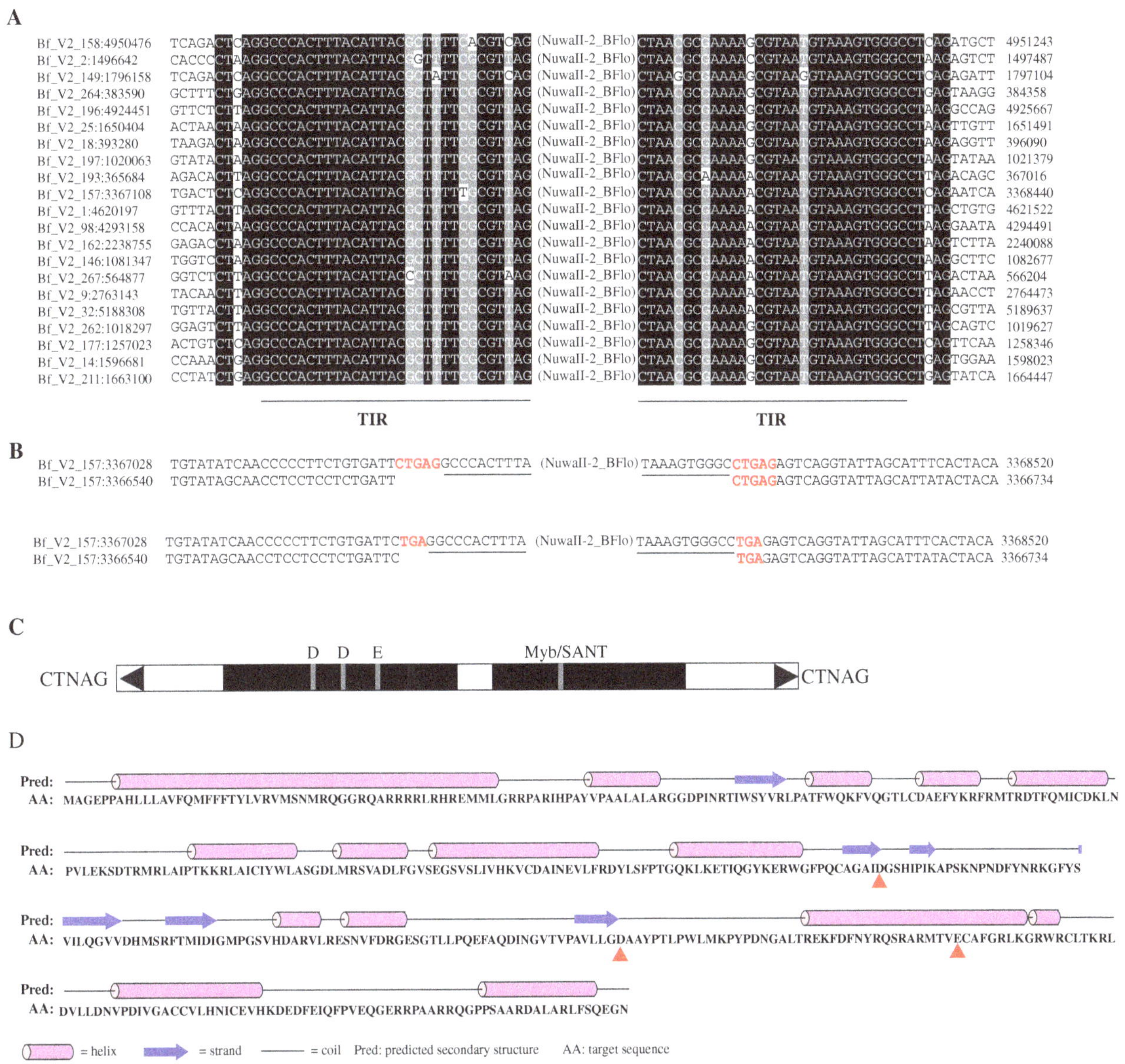

Fig. 5 Characteristics of *NuwaII* transposons. **a** Sequence alignments for *NuwaII-2_BFlo* family. The terminal inverted repeats (TIRs) and flanking sequences (10 bp) are shown. **b** Examples of alignments of the flanking sequences of *NuwaII-2_BFlo* insertions with a paralogous sequences found within the same genome but devoid of the transposon. The TIRs of the element are *underlined*. **c** Structure of *NuwaII-2_BFlo*. *Black triangles* and *solid black boxes* represent the TIRs and ORFs, respectively, and the position of the DDE triad is shown. **d** Predicted secondary structure of the DDE motif-containing transposase of the *NuwaII-2_BFlo*. The DDE triads is marked with *red triangles* below the sequence

domain) of *Pangu*, 12 *NuwaI*, 21 *NuwaII*, 33 *PIF/Harbinger*, 18 *ISL2EU*, 11 *Spy*, and 11 bacterial *IS5* were used to perform a Bayesian phylogenetic analysis. The resulting tree (Fig. 6) showed that the eukaryotic PHIS transposases formed five distinct highly supported monophyletic clades beside the individual clade of bacterial *IS5* transposons. In the phylogenetic tree, *Pangu*, *PIF/Harbinger*, *ISL2EU*, and *SPY* transposons formed four separate clades. Meanwhile, *NuwaI* and *NuwaII* transposons formed a single clade in the phylogenetic tree.

Discussion

Identification and characterization of PHIS transposons

Previous study suggested that the PHIS is a DNA transposon superfamily with a great diversity in the eukaryotic genomes [15]. However, the detailed diversification and evolution of PHIS superfamily are still unknown. In this study, we systematically identified PHIS transposons in the eukaryotic genomes. A total of 380 families of PHIS superfamily were identified in 112 sequenced eukaryotic genomes. These families were classified into six groups based on the characteristic of each family's TSDs. Among

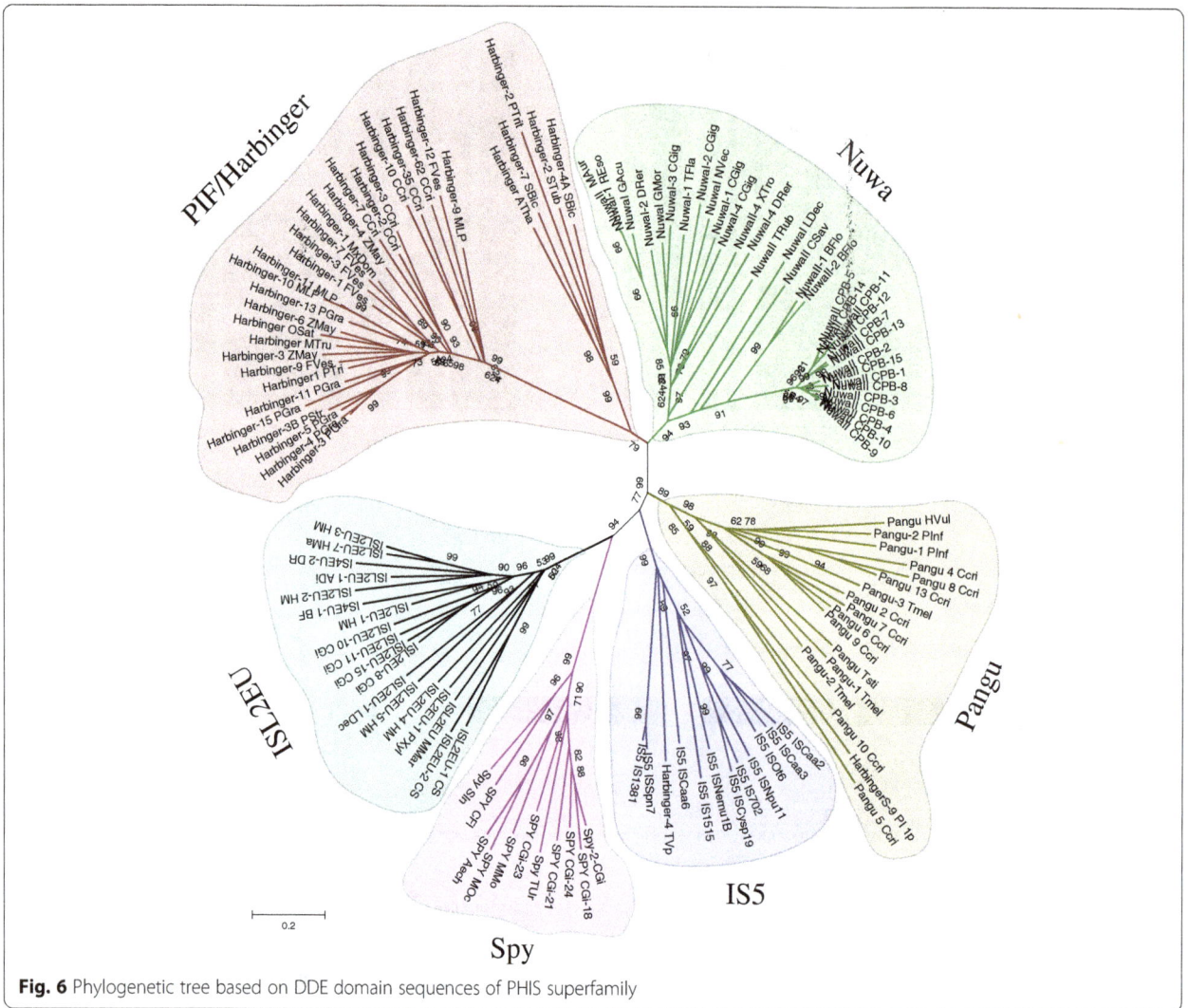

Fig. 6 Phylogenetic tree based on DDE domain sequences of PHIS superfamily

these groups, three (*PIF/Harbinger*, *ISL2EU*, and *Spy*) have been reported in the previous studies [15, 20, 21]. Beside the above three groups, we found three new transposon groups, called *Pangu*, *NuwaI*, and *NuwaII*.

These types shared similar transposases with DDE motif. However, each group has unique TSDs distinguished from others (Additional file 2: Figure S2). According to the criteria of previous TE classification [16], the transposases can be aligned over their entire catalytic regions (*e* value less than e^{-4}), then they belong to the same superfamily. The same group of a superfamily was defined by the same TSD composition. In addition, previous studies showed that variable length or composition of TSDs have been identified in some superfamilies, such as 8–9 bp TSDs in *Merlin* superfamily, 5–8 bp in *hAT*, 2–4 bp in *CMC*, and 4–5 bp in *Ginger* [16, 22, 23]. Thus, it may be better to define *Spy*, *PIF/Harbinger*, and *ISL2EU* and *Pangu*, *NuwaI*, and *NuwaII* as different groups (at the same level) of the same superfamily (PHIS).

To estimate the abundance of each group in the eukaryotic genomes, the consensus sequence of each family of each group was used as a query in BlastN ($e < 10^{-5}$) search against corresponding genome. Finally, we found that the abundances of these transposon groups varied in the eukaryotic genomes. For instance, there were 41,385 copies of *PIF/Harbinger* group, 3647 copies of *ISL2EU*, 13,089 copies of *SPY*, 3270 copies of *Pangu*, 3845 copies of *NuwaI*, and 7562 copies of *NuwaII* in the eukaryotic genomes (Additional files 1 and 3: Table S1–S8 and Fig. 2c). However, it should be noted that PHIS transposons were investigated using transposase homology search. Thus, some nonautonomous PHIS transposons (such as MITEs) might be missed in this study. In addition, we found that the number of potential active families varied. For example, there were 88 potential active families of *PIF/harbinger*, 12 families of *ISL2EU*, 18 families of *SPY*, 2 families of *Pangu*, 11 families of *NuwaI*, and 22 families of *NuwaII* in the eukaryotes (Fig. 2b). Furthermore, the abundance of each group was significantly positively correlated with the number of potential active families (Pearson's product-moment correlation, $r = 0.9816605$, $P = 0.0005$). This phenomenon is easy to understand, and the more potential active families will have more copies for a group of PHIS transposon superfamily.

Most groups of PHIS superfamily include two ORFs, one coding for transposase containing DDE motif and the other ORF encoding a DNA-binding protein. However, *SPY* transposons include only one transposase containing DDE motif [15]. In addition, the additional ORFs of the four groups (including *Pangu*, *PIF/Harbinger*, *NuwaI*, and *NuwaII*) encode a protein with Myb/SANT domain except that of the *ISL2EU* transposon that encodes a protein with the Yqaj domain. At present, the functions of the additional ORFs are still unknown, and whether these ORFs are related to the transposition mechanisms also remains unclear [24]. This question could be answered using biochemical studies in the future.

The results of species distribution of PHIS transposons showed that the PHIS elements are completely absent in mammals, birds, sponges, sharks, and coelacanths. This is consistent with a previous study [16]. In addition, it is interesting to see that in some lineages, there is only one of the six groups of PHIS superfamily or only one of the six groups is absent. To our knowledge, the above results could be caused by two reasons. First, some PHIS transposons were lost or degenerated in some species by drift or selection in their original lineages. Second, some species gain different families from other species through horizontal transfer (HT). In addition, almost all of the DNA transposons have the ability of HT, and more and more HT of DNA transposons have been reported in the eukaryotic genomes [25–29]. Furthermore, previous studies suggested that PIF/Harbinger experienced HT events between *Drosophila* species [30]. However, HT of PHIS transposons remains to be studied in the future.

Evolutionary relationships of PHIS transposons
The result of phylogenetic analysis showed that *Pangu* elements formed a single clade and were adjacent to *IS5* group in the phylogenetic tree. In addition, both *Pangu* and *IS5* transposons shared the same target site sequence (5′-ANT-3′). Furthermore, *Pangu* elements were widely distributed in plants, fungi, and animals. Thus, we proposed that *Pangu* is a relatively old PHIS group in the eukaryotic genomes.

Meanwhile, *NuwaI* and *NuwaII* transposons formed a single clade in the phylogenetic tree, and they shared the same coding capacity (encoding two ORFs) and the conserved domains (DDE motif and Myb/SANT domain). However, the TSDs of *NuwaI* are significantly different

from the *NuwaII* transposons. *NuwaI* and *NuwaII* transposons should belong to two different groups of PHIS superfamily. Nevertheless, these two types might diverge recently. Thus, the two types cannot be distinguished from each other in the phylogenetic tree.

HarbingerS-9_PI and *Harbinger-4_TV* had been released as *PIF/Harbinger* families cataloged in RepBase. However, our phylogenetic analysis indicated that *HarbingerS-9_PI* was grouped into the clade of *Pangu* group. Meanwhile, *Harbinger-4_TV* was grouped into the *IS5* clade (Fig. 6). However, we could not find distinct target site duplications (TSDs) in the flank of *HarbingerS-9_PI* and *Harbinger-4_TV* families. Right now, we cannot judge if both families should belong to which group of PHIS superfamily.

Conclusions
In the present study, 380 PHIS transposon families were identified in 112 of 836 sequenced eukaryotic genomes using transposase homology search and structure approach. Among these families, 168 families are firstly identified in this study. We systematically analyzed their characteristics including TSDs, TIRs, coding capacity, conserved transposase domain and species distribution, etc. The phylogenetic analysis based on the core catalytic DDE domain of these identified transposases showed that these PHIS transposon families were divided into five clusters including three previous reported clusters (*PIF/Harbinger*, *ISL2EU*, and *Spy*) and two new clusters (*Pangu* and *Nuwa*). *Nuwa* cluster includes two groups called *NuwaI* and *NuwaII*. Furthermore, each new group has its own distinctive characteristics, especially in target site sequences. For instance, the *Pangu* transposons are characterized by 5′-ANT-3′ TSDs, the *NuwaI* transposons by 5′-CWG-3′, and the *NuwaII* transposons by 5′-C|TNA|G-3′. Our results reveal the diversification and evolution of PHIS transposons in the eukaryotic genomes and imply that further study on the generation mechanism of varied target sequences of PHIS superfamily will promote the development of new transgenic vectors.

Methods
Identification of PHIS superfamily
Eukaryotic genomes including animals (295 species), plants (105 species), fungi (315 species), and protists (121 species) were downloaded from NCBI (http://www.ncbi.nlm.nih.gov/) (as of January 16, 2014), and the information of each species is listed in Additional file 3: Table S9. All published autonomous PHIS elements were downloaded from RepBase (v19.07) [31]. PHIS elements of eukaryotic genomes were identified using the transposase homology search that includes three steps (Additional file 2: Figure S4): (1) the transposase sequences of published

PHIS elements were used as a query to do TblastN and TESeeker searches against each genome [32], where a hit with e value less than 10^{-4} was considered as candidate PHIS sequence; (2) each candidate PHIS nucleotide sequence was used as a query to BlastN search (e value $< e^{-5}$, sequence length >50 bp, and nucleotide identity >80 %) against the corresponding genome; (3) the sequences of each cluster were extended in both directions using a Perl script and aligned using MUltiple Sequence Comparison by Log-Expectation (MUSCLE) [33], then the boundaries of each cluster were manually defined.

Characterization and phylogenetic analysis of PHIS superfamily

To estimate the abundance of each PHIS family in the corresponding genome, the consensus sequence of each family was used as a query in BlastN search against the corresponding genome. Finally, the sequences with the e value less than e^{-5}, length larger than 50 bp, and a minimum nucleotide identity of 80 % were classified as members of the same family. Transposase coding sequences, transposase domains, secondary structures of representative transposases, and the paralogous empty sites were analyzed as described previously [15]. Sequence logos of TIRs and TSDs were created by WebLogo (http://weblogo.berkeley.edu/logo.cgi) [34]. Multiple sequences alignments were performed using MUSCLE software with default parameters. The phylogenetic tree was constructed based on the DDE domains of transposases using MrBayes software (v3.1.2) [35] with the Blosum model and other parameters with default. The Blosum model was estimated by protest-3.2 software [36]. Meanwhile 3,000,000 generations of Bayesian inference were performed.

Additional files

Additional file 1: Table S1. Distribution and characteristics of all identified *PIF/Harbinger* transposons. **Table S2.** Distribution and characteristics of all identified *ISL2EU* transposons. **Table S3.** Distribution and characteristics of all identified *Pangu* transposons. **Table S5.** Distribution and characteristics of all identified *Nuwal* transposons. **Table S7.** Distribution and characteristics of all identified *Nuwall* transposons. **Table S9.** The eukaryotes used in this study.

Additional file 2: Figure S1. (A) Sequence alignments for *ISL2EU-2_Pxyl* family. The terminal inverted repeats (TIRs) and flanking sequences (10 bp) are shown. (B) An example of alignments of the flanking sequences of *ISL2EU-2_Pxyl* insertion with a paralogous sequences found within the same genome but devoid of the transposon. The TIRs of the element are underlined. **Figure S2.** Speculated transposition mechanism of each PHIS groups. **Figure S3.** The alignment of DDE domain of *Pangu* and *Nuwa* groups after redundancy elimination. Distances between the conserved blocks are indicated in the number of amino acid residues. Conserved residues within each superfamily are highlighted in the same color. The DDE triad identified here is marked with asterisks below alignments. **Figure S4.** Pipeline for PHIS transposons identification. Where a hit with e value less than 10^{-4} was considered as a homology sequence. The ones with an e value less than e^{-5}, sequence length larger

than 50 bp, and nucleotide sequence identity larger than 80 % were classified as member of the same family. Target site duplications (TSDs) were identified using the paralogous empty sites.

Additional file 3: Table S4. Positions of *Pangu* transposons in the corresponding genome. **Table S6.** Positions of *Nuwal* transposons in the corresponding genome. **Table S8.** Positions of *Nuwall* transposons in the corresponding genome.

Abbreviations
MITEs: Miniature inverted-repeat transposable elements; TEs: Transposable elements; TIRs: Terminal inverted repeats; TSDs: Target site duplications.

Competing interests
The authors declare that they have no competing interests.

Authors' contributions
MJH designed the study, carried out the analyses, and drafted the manuscript. CLX, HBZ, MQZ, and HHZ did the data analyses and revised the manuscript. ZZ designed the study, supervised the study, and revised the manuscript. All authors read and approved the final manuscript.

Acknowledgements
This work was supported by the National Natural Science Foundation of China (No. 31471197 to ZZ and No. 31401106 to MJH), Postdoctoral Science Foundation of Chongqing (No. Xm2014080 to MJH), and Chongqing Graduate Student Research Innovation Project (CYB14041).

Author details
[1]School of Life Sciences, Chongqing University, Chongqing 400044, China. [2]College of Pharmacy and Life Science, Jiujiang University, Jiujiang 332000, China.

References
1. Feschotte C, Jiang N, Wessler SR. Plant transposable elements: where genetics meets genomics. Nat Rev Genet. 2002;3:329–41.
2. Finnegan DJ. Eukaryotic transposable elements and genome evolution. Trends Genet. 1989;5:103–7.
3. de Koning AP, Gu W, Castoe TA, Batzer MA, Pollock DD. Repetitive elements may comprise over two-thirds of the human genome. PLoS Genet. 2011;7:e1002384.
4. Holt RA, Subramanian GM, Halpern A, Sutton GG, Charlab R, Nusskern DR, et al. The genome sequence of the malaria mosquito *Anopheles gambiae*. Science. 2002;298:129–49.
5. Kapitonov VV, Jurka J. Molecular paleontology of transposable elements in the Drosophila melanogaster genome. Proc Natl Acad Sci U S A. 2003;100:6569–74.
6. Kidwell MG. Transposable elements and the evolution of genome size in eukaryotes. Genetica. 2002;115:49–63.
7. Lander ES, Linton LM, Birren B, Nusbaum C, Zody MC, Baldwin J, et al. Initial sequencing and analysis of the human genome. Nature. 2001;409:860–921.
8. Meyers BC, Tingey SV, Morgante M. Abundance, distribution, and transcriptional activity of repetitive elements in the maize genome. Genome Res. 2001;11:1660–76.
9. Nene V, Wortman JR, Lawson D, Haas B, Kodira C, Tu ZJ, et al. Genome sequence of *Aedes aegypti*, a major arbovirus vector. Science. 2007;316:1718–23.
10. Sanmiguel P, Bennetzen JL. Evidence that a recent increase in maize genome size was caused by the massive amplification of intergene retrotransposons. Ann Bot. 1998;81:37–44.
11. Vicient CM, Suoniemi A, Anamthawat-Jónsson K, Tanskanen J, Beharav A, Nevo E, et al. Retrotransposon BARE-1 and its role in genome evolution in the genus hordeum. Plant Cell. 1999;11:1769–84.
12. Waterston RH, Lindblad-Toh K, Birney E, Rogers J, Abril JF, Agarwal P, et al. Initial sequencing and comparative analysis of the mouse genome. Nature. 2002;420:520–62.

13. Xu HE, Zhang HH, Xia T, Han MJ, Shen YH, Zhang Z. BmTEdb: a collective database of transposable elements in the silkworm genome. Database (Oxford). 2013;2013:bat055.
14. Schnable PS, Ware D, Fulton RS, Stein JC, Wei F, Pasternak S, et al. The B73 maize genome: complexity, diversity, and dynamics. Science. 2009;326:1112–5.
15. Han MJ, Xu HE, Zhang HH, Feschotte C, Zhang Z. Spy: a new group of eukaryotic DNA transposons without target site duplications. Gonome Bio Evol. 2014;6:1748–57.
16. Yuan YW, Wessler SR. The catalytic domain of all eukaryotic cut-and-paste transposase superfamilies. Proc Natl Acad Sci U S A. 2011;108:7884–9.
17. Bao W, Jurka MG, Kapitonov VV, Jurka J. New superfamilies of eukaryotic DNA transposons and their internal divisions. Mol Biol Evol. 2009;26(5):983–93.
18. Kapitonov VV, Jurka J. A universal classification of eukaryotic transposable elements implemented in Repbase. Nat Rev Genet. 2008;9:411–2.
19. Zhang X, Feschotte C, Zhang Q, Jiang N, Eggleston WB, Wessler SR. P instability factor: an active maize transposon system associated with the amplification of Tourist-like MITEs and a new superfamily of transposases. Proc Natl Acad Sci U S A. 2001;98:12572–7.
20. Kapitonov VV, Jurka J. Molecular paleontology of transposable elements from Arabidopsis thaliana. Genetica. 1999;107:27–37.
21. Walker EL, Eggleston WB, Demopulos D, Kermicle J, Dellaporta SL. Insertions of a novel class of transposable elements with a strong target site preference at the r locus of maize. Genetics. 1997;146:681–93.
22. Feschotte C. Merlin, a new superfamily of DNA transposons identified in diverse animal genomes and related to bacterial IS1016 insertion sequences. Mol Biol Evol. 2004;21:1769–80.
23. Bao W, Kapitonov VV, Jurka J. Ginger DNA transposons in eukaryotes and their evolutionary relationships with long terminal repeat retrotransposons. Mob DNA. 2010;1:3.
24. Jiang N, Bao Z, Zhang X, Hirochika H, Eddy SR, McCouch SR, et al. An active DNA transposon family in rice. Nature. 2003;421:163–7.
25. Schaack S, Gilbert C, Feschotte C. Promiscuous DNA: horizontal transfer of transposable elements and why it matters for eukaryotic evolution. Trends Ecol Evol. 2010;25:537–46.
26. Daniels SB, Peterson KR, Strausbaugh LD, Kidwell MG, Chovnick A. Evidence for horizontal transmission of the P transposable element between Drosophila species. Genetics. 1990;124:339–55.
27. Maruyama K, Hartl DL. Evidence for interspecific transfer of the transposable element mariner between Drosophila and Zaprionus. J Mol Evol. 1991;33:514–24.
28. Zhang HH, Xu HE, Shen YH, Han MJ, Zhang Z. The origin and evolution of six miniature inverted-repeat transposable elements in Bombyx mori and Rhodnius prolixus. Genome Biol Evol. 2013;5:2020–31.
29. Gilbert C, Hernandez SS, Flores-Benabib J, Smith EN, Feschotte C. Rampant horizontal transfer of SPIN transposons in squamate reptiles. Mol Biol Evol. 2012;29:503–15.
30. Casola C, Lawing AM, Betrán E, Feschotte C. PIF-like transposons are common in drosophila and have been repeatedly domesticated to generate new host genes. Mol Biol Evol. 2007;24:1872–88.
31. Jurka J, Kapitonov VV, Pavlicek A, Klonowski P, Kohany O, Walichiewicz J. Repbase Update, a database of eukaryotic repetitive elements. Cytogenet Genome Res. 2005;110:462–7.
32. Kennedy RC, Unger MF, Christley S, Collins FH, Madey GR. An automated homology-based approach for identifying transposable elements. BMC Bioinformatics. 2011;12:130.
33. Edgar RC. MUSCLE: multiple sequence alignment with high accuracy and high throughput. Nucleic Acids Res. 2004;32:1792–7.
34. Schneider TD, Stephens RM. Sequence logos: a new way to display consensus sequences. Nucleic Acids Res. 1990;18:6097–100.
35. Ronquist F, Huelsenbeck JP. MrBayes 3: Bayesian phylogenetic inference under mixed models. Bioinformatics. 2003;19:1572–4.
36. Darriba D, Taboada GL, Doallo R, Posada D. ProtTest 3: fast selection of best-fit models of protein evolution. Bioinformatics. 2011;27:1164–5.

3

A host factor supports retrotransposition of the TRE5-A population in *Dictyostelium* cells by suppressing an Argonaute protein

Anika Schmith[1], Thomas Spaller[1], Friedemann Gaube[1], Åsa Fransson[2,6], Benjamin Boesler[3], Sandeep Ojha[4], Wolfgang Nellen[3,7], Christian Hammann[4], Fredrik Söderbom[5] and Thomas Winckler[1]*

Abstract

Background: In the compact and haploid genome of *Dictyostelium discoideum* control of transposon activity is of particular importance to maintain viability. The non-long terminal repeat retrotransposon TRE5-A amplifies continuously in *D. discoideum* cells even though it produces considerable amounts of minus-strand (antisense) RNA in the presence of an active RNA interference machinery. Removal of the host-encoded C-module-binding factor (CbfA) from *D. discoideum* cells resulted in a more than 90 % reduction of both plus- and minus-strand RNA of TRE5-A and a strong decrease of the retrotransposition activity of the cellular TRE5-A population. Transcriptome analysis revealed an approximately 230-fold overexpression of the gene coding for the Argonaute-like protein AgnC in a CbfA-depleted mutant.

Results: The *D. discoideum* genome contains orthologs of RNA-dependent RNA polymerases, Dicer-like proteins, and Argonaute proteins that are supposed to represent RNA interference pathways. We analyzed available mutants in these genes for altered expression of TRE5-A. We found that the retrotransposon was overexpressed in mutants lacking the Argonaute proteins AgnC and AgnE. Because the *agnC* gene is barely expressed in wild-type cells, probably due to repression by CbfA, we employed a new method of promoter-swapping to overexpress *agnC* in a CbfA-independent manner. In these strains we established an in vivo retrotransposition assay that determines the retrotransposition frequency of the cellular TRE5-A population. We observed that both the TRE5-A steady-state RNA level and retrotransposition rate dropped to less than 10 % of wild-type in the *agnC* overexpressor strains.

Conclusions: The data suggest that TRE5-A amplification is controlled by a distinct pathway of the *Dictyostelium* RNA interference machinery that does not require RNA-dependent RNA polymerases but involves AgnC. This control is at least partially overcome by the activity of CbfA, a factor derived from the retrotransposon's host. This unusual regulation of mobile element activity most likely had a profound effect on genome evolution in *D. discoideum*.

Keywords: *Dictyostelium*, Retrotransposition, siRNA, RNAi, Argonaute

* Correspondence: t.winckler@uni-jena.de
[1]Department of Pharmaceutical Biology, Institute of Pharmacy, University of Jena, Semmelweisstrasse 10, 07743 Jena, Germany
Full list of author information is available at the end of the article

Background

Transposable elements are found in virtually all organisms and play central roles in shaping their host's genomes. The amplification of these genomic parasites is a constant threat to host fitness due to the intrinsic process of integration into the genomic DNA that can cause mutagenesis of genes and force illegitimate recombinations between distant transposon copies [1–4]. Eukaryotic cells have evolved several pathways of RNA interference (RNAi) to restrain the amplification of transposons at the posttranscriptional level [5–8]. In this process, long RNA duplexes (dsRNA), which may occur in cells as intermediates of transposon or RNA virus replication, are typically processed into 20–30 nucleotide double-stranded small interfering RNAs (siRNAs) by ribonuclease III-type enzymes such as Dicer. The siRNAs are loaded onto RNA-induced silencing complexes (RISCs), which are minimally composed of an Argonaute protein and a small RNA [9, 10]. Argonaute proteins are characterized by an RNA-binding PAZ (Piwi-Argonaut-Zwille) domain and a catalytic, ribonuclease H-like PIWI (P-element-induced wimpy testis) domain. Argonaute proteins bind siRNAs via their PAZ domains, unwind the siRNA duplex and use one of the single-stranded RNA molecules as guides to bind mRNAs in a sequence-specific manner [9]. If the guide RNA is fully complementary to the target RNA across the active site of the Argonaute protein, the enzyme is able to degrade the target RNA by a single endonucleolytic cut executed by the PIWI domain, a function termed slicing. If slicing is precluded by mismatches between the annealing guide RNA and cellular mRNA, translation is repressed and mRNA can be degraded by deadenylation and decapping.

The social amoeba *Dictyostelium discoideum* has a haploid genome in which nearly two thirds of DNA are protein-coding genes [11]. Despite the remarkable compactness of its genome, *D. discoideum* accommodates a large number of mobile elements that add up to approximately 10 % of the entire genomic DNA [12]. Most likely for the purpose of suppressing transposition, the organism has evolved a sophisticated RNAi machinery that includes, for example, three RNA-dependent RNA polymerases (RdRPs), two Dicer-like proteins, and five Argonaute-like proteins [13–17]. Intriguingly, the non-long terminal repeat retrotransposon TRE5-A has established a fairly high amplification rate in growing *D. discoideum* cells [18, 19] despite the constitutive production of minus-strand RNA from an element-internal promoter [20, 21]. Thus, how TRE5-A manipulates the cellular RNAi machinery to maintain its remarkable retrotransposition activity is of interest.

Clearly, *D. discoideum* cells could take advantage of TRE5-A's minus-strand RNA production to downregulate TRE5-A plus-strand RNA, the substrate for retrotransposition, using an RNAi pathway. This strategy is actually realized in the silencing of the tyrosine recombinase retrotransposon DIRS-1 in *D. discoideum* cells [22]. To suppress TRE5-A amplification, promoter activity of the C-module, the distinguished minus-strand RNA promoter at the 3′ end of the TRE5-A element, could be positively regulated by a host-encoded transcription factor. This could elevate the level of TRE5-A-derived dsRNA, which could be processed into small RNAs that guide Argonaute proteins to degrade TRE5-A plus-strand RNA and prevent retrotransposition. Consistent with this idea, we previously isolated the C-module-binding factor (CbfA), a host-encoded DNA-binding protein that interacts with the C-module of TRE5-A in vitro [23–25].

The gene CbfA-coding could not be inactivated by conventional homologous recombination (knockout) and may be essential for the growth of *D. discoideum* cells. We constructed a knock-in mutant, JH.D, in which the *cbfA* gene was replaced by a *cbfA* variant containing an *amber* stop codon at amino acid position 455 [25]. The expression of an *amber* suppressor tRNA gene in *D. discoideum* cells allows read-through translation without causing an inherent phenotype [26]. Due to the low efficacy of this *amber* suppression, JH.D cells produce less than 5 % of full-length CbfA protein from the expressed *cbfA*(*amber*) mRNA [25].

JH.D cells have an aberrant developmental phenotype that can be fully rescued by ectopic expression of CbfA in the mutant [27]. Transcriptome analyses revealed that CbfA has general gene regulatory functions in *D. discoideum* cells [28], making this protein an attractive candidate as a host protein that could limit TRE5-A expression and retrotransposition by elevating TRE5-A-derived minus-strand RNA. Interestingly, we observed that both plus- and minus-strand RNA of TRE5-A were reduced concurrently in the CbfA mutant by more than 90 %, and this reduction of transcript levels was accompanied by a sharp drop in TRE5-A's retrotransposition activity in vivo [21]. Remarkably, the promoter activity of neither the A-module (TRE5-A's plus-strand RNA promoter) nor the C-module was altered in reporter gene assays in the CbfA mutant compared to wild-type cells [21]. Thus, we hypothesized that CbfA supports TRE5-A amplification indirectly by down-regulating one or several components of the cellular RNAi machinery. In support of this assumption, a previous transcriptome analysis revealed an approximately 230-fold and 3-fold overexpression of the genes encoding *D. discoideum* Argonaute-like proteins AgnC and AgnE, respectively, in the CbfA-depleted mutant [28].

Here, we found that TRE5-A expression was elevated in knockout strains of *agnC* and *agnE*, suggesting that CbfA may support the accumulation of TRE5-A transcripts by suppressing an RNAi pathway that involves these Argonaute proteins. To determine whether control of TRE5-A expression by AgnC and/or AgnE leads to a reduction in TRE5-A retrotransposition in vivo, we first developed a new gene activation (GA) strategy to construct strains that overexpress *agnC* in the absence of any residual plasmid sequences inserted in their genomes. We found that the accumulation of TRE5-A RNA was reduced in both *agnC^{GA}* and *agnE^{GA}* strains. Next, we employed the previously developed "TRE trap" retrotransposition assay [18, 19] to determine the retrotransposition activity of the cellular TRE5-A population in *agnC^{GA}* cells. The retrotransposition frequency of the cellular TRE5-A population was determined to be less than 10 % of the wild-type level, suggesting that TRE5-A amplification in *D. discoideum* cells is under surveillance of a distinct RNAi pathway that requires AgnC function and that this control of mobile element expansion is at least in part overcome by CbfA, a factor derived from the retrotransposon's host cell.

Results

CbfA regulates the expression of the Argonaute-like protein AgnC

Even though the accumulation of TRE5-A RNA in *D. discoideum* cells strictly depends on CbfA and this factor binds to the C-module of TRE5-A in vitro, it does not regulate the C-module's promoter activity in vivo [21]. A probable explanation for this paradox could be that CbfA exerts an indirect effect by regulating an RNAi pathway that is involved in the control of TRE5-A expression. In concordance with a rather indirect and probably broader function of CbfA in the control of mobile elements, including TRE5-A, the re-evaluation of previously obtained mRNA-seq data [28] suggested a considerable amount of deregulation of transposable elements in the CbfA-depleted mutant JH.D compared to the parental strain AX2 (Fig. 1). Typically the differential expression of mobile elements between AX2 and JH.D cells was highly variable among different cell cultures and could not be unequivocally verified by quantitative RT-PCR. Nevertheless, this observation strengthened the hypothesis that the regulation of steady-state levels of TRE5-A RNA may not be directly regulated by CbfA by binding to the C-module, but rather be the result of CbfA regulating an RNAi pathway involved in the regulation of TRE5-A.

Following this hypothesis we used previously obtained mRNA-seq data [28] to determine differential expression of putative RNAi components between the CbfA mutant JH.D and parent AX2 cells. The Argonaute genes *agnC* and *agnE* were 228-fold and a 2.7-fold, respectively, overexpressed in JH.D cells (Fig. 2). The genes coding for Argonaute proteins AgnA and AgnB and the RdRP RrpC were slightly underexpressed in the JH.D mutant cells, whereas expression of the genes coding for the

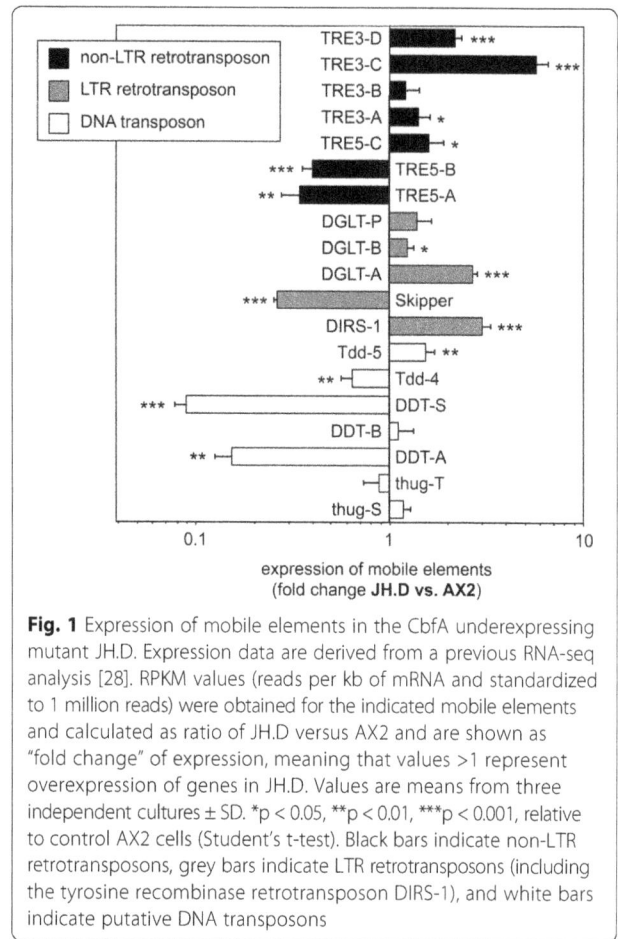

Fig. 1 Expression of mobile elements in the CbfA underexpressing mutant JH.D. Expression data are derived from a previous RNA-seq analysis [28]. RPKM values (reads per kb of mRNA and standardized to 1 million reads) were obtained for the indicated mobile elements and calculated as ratio of JH.D versus AX2 and are shown as "fold change" of expression, meaning that values >1 represent overexpression of genes in JH.D. Values are means from three independent cultures ± SD. *p < 0.05, **p < 0.01, ***p < 0.001, relative to control AX2 cells (Student's t-test). Black bars indicate non-LTR retrotransposons, grey bars indicate LTR retrotransposons (including the tyrosine recombinase retrotransposon DIRS-1), and white bars indicate putative DNA transposons

RdRPs *rrpA* and *rrpB* were unaffected by CbfA depletion (Fig. 2). RNA-seq also revealed normal expression of the genes coding for the two Dicer-like proteins of *D. discoideum*, *drnA* and *drnB*, in the CbfA mutant (Fig. 2). To confirm the RNA-seq data, we determined the expression the Dicer genes, the three RdRPs, and the five Argonaute genes by qRT-PCR. For these measurements we combined three RNA samples used in the previous RNA-seq experiment with RNA preparations from three additional independent cultures. The data were consistent in all six biological replicates and are presented in Fig. 2. The strong overexpression of *agnC* in the CbfA mutant was confirmed (257-fold, p < 0.01, Student's t-test). The weak overexpression of *agnE* seen in RNA-seq could not be verified by qRT-PCR at a statistically significant level (4.3-fold overexpression; p = 0.17), although the trend to *agnE* overexpression in JH.D was reproduced (Fig. 2). The weak but highly significant underexpression of *agnA* in the JH.D mutant observed by RNA-seq (2.1-fold; p < 0.001) was confirmed by qRT-PCR (3.6-fold; p < 0.01), whereas results for *agnB* were inconclusive (Fig. 2).

CbfA can be divided into an amino-terminal part containing a "carboxy-terminal jumonji domain" and two zinc

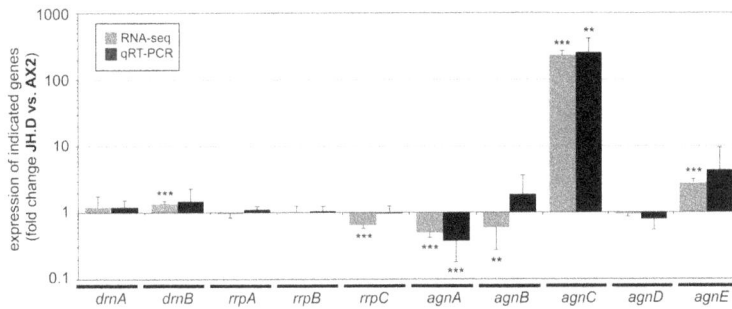

Fig. 2 Expression of RNAi components in CbfA mutant JH.D cells. Expression of Dicer-like proteins (*drnA*, *drnB*), RdRPs (*rrpA-C*) and Argonaute genes (*agnA-E*) was analyzed by RNA-seq (gray bars, n = 3) in wild-type AX2 and CbfA-mutant JH.D cells from three independent cultures [28]. These three RNA samples, and RNA from three additional independent cultures, were analyzed by qRT-PCR (black bars, n = 6). Data are shown as fold change of expression with values >1 meaning that expression in the mutant cells was higher than in the control cells. Values are means ± SD from three and six independent cultures, respectively. **p < 0.01, ***p < 0.001, relative to wild-type AX2 cells (Student's t-test)

finger-like motifs, as well as a carboxy-terminal domain (CbfA-CTD) that contains a DNA-binding AT hook motif (Fig. 3a). We have previously determined that the CbfA-CTD is able to mediate most of CbfA's gene-regulatory activity without requiring the rest of the CbfA protein [28] and it also completely restores TRE5-A expression in JH.D cells [21]. We wanted to evaluate whether the aberrant expression of *agnC* and *agnE* in JH.D cells was reversed by expression of CbfA in the mutant. To this end, we expressed either full-length CbfA or a GFP fusion to CbfA-CTD in JH.D cells using multicopy expression plasmids as shown in Fig. 3b. To confirm functional expression of CbfA or CbfA-CTD in JH.D, we first determined TRE5-A expression (Additional file 1: Figure S1). TRE5-A underexpression was rescued to wild-type level by full-length CbfA and was even "over-complemented" by CbfA-CTD, which was likely due to overexpression of this protein relative to normal CbfA amounts present in AX2 cells (compare Fig. 3b). The overexpression of *agnC* in JH.D was reduced by full-length CbfA by 80 % (p < 0.05) and by CbfA-CTD by 92 % (p < 0.05) (Fig. 3c). This result was similar to previous RNA-seq data [28], which revealed 97 % reduction (p < 0.001) of *agnC* overexpression in response to the presence of CbfA-CTD [28]. Taken together, the data indicate that *agnC* is a genuine CbfA-regulated gene that requires CbfA-CTD for proper expression. The data argued for a role of AgnC in CbfA-dependent TRE5-A regulation because the accumulation of TRE5-A transcripts is also regulated by CbfA-CTD [21].

Expression of full-length CbfA in JH.D cells had only a minor effect on the observed overexpression of *agnE* in JH.D cells. Likewise, expression of CbfA-CTD in JH.D cells did not affect *agnE* overexpression in the CbfA mutant (Additional file 1: Figure S1). The latter results were consistent with previous RNA-seq data, which did not indicate an effect of CbfA-CTD on *agnE* expression [28]. Therefore, we cannot definitely conclude from our data that *agnE* is a genuine CbfA-regulated gene.

AgnC and AgnE downregulate TRE5-A expression

We performed qRT-PCR measurements to determine whether putative components of the *D. discoideum* RNAi machinery are involved in the silencing of TRE5-A expression. No significant changes of TRE5-A expression were determined when the genes coding for the Dicer-like protein DrnB or the RdRP proteins RrpA and RrpB were inactivated; however, we detected a mild but significant underexpression of TRE5-A in *rrpC* knockout cells (Fig. 4). Inactivation of *agnA* or *agnB* had no effect on TRE5-A expression (Fig. 4), whereas a 4.3- and 5.9-fold overexpression of TRE5-A was observed in knockout mutants of *agnC* and *agnE*, respectively (Fig. 4). Overexpression of TRE5-A in *agnC* and *agnE* knockout strains was completely reversed when AgnC or AgnE was expressed from a multicopy plasmid in the respective knockout mutant (Fig. 4), suggesting that both AgnC and AgnE contribute to TRE5-A regulation.

The role of AgnC and AgnE in the suppression of the TRE5-A population could be analyzed in more detail if the corresponding genes would be overexpressed in a wild-type background, i.e., in a strain with normal CbfA activity. This was an important consideration because previous data indicated that CbfA may have functions in TRE5-A retrotransposition beyond the regulation of TRE5-A RNA levels in supporting the integration process upstream of tRNA genes [21]. Usually, overexpression of proteins in wild-type cells is facilitated by transforming cells with expression plasmids. We assumed that this would be a suboptimal strategy for our experiments because transformants would contain insertions of multicopy plasmids at random genomic positions that could compromise the subsequent determination of retrotransposition activity of the TRE5-A population using the TRE trap assay (see below). With this consideration in mind, we decided to generate gene activation (GA) strains in which the endogenous promoter of either *agnC* or *agnE* was replaced by the strong *actin15* promoter (*act15P*) by

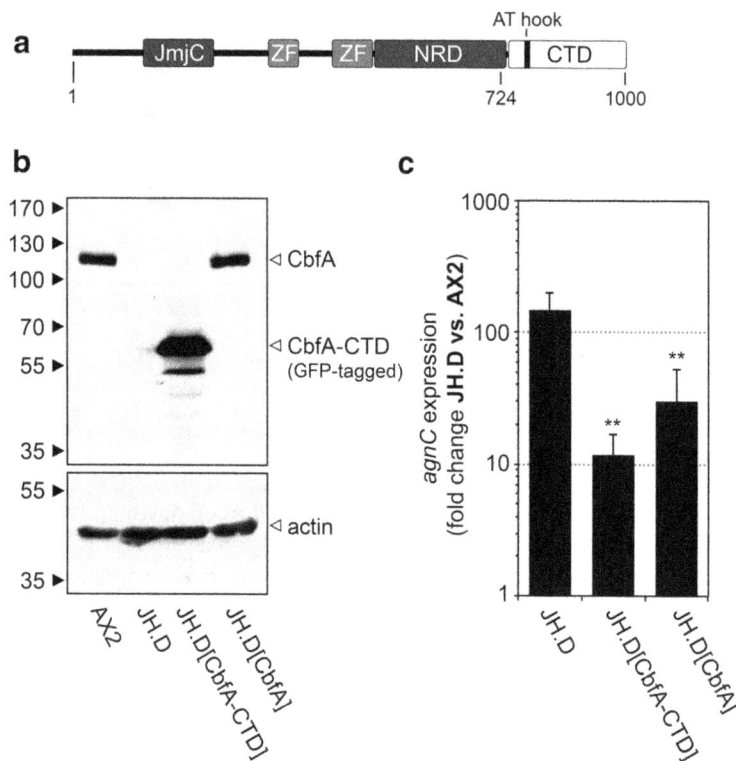

Fig. 3 CbfA controls AgnC expression. **a** Scheme of the CbfA protein. The 1000-amino acid protein can be roughly divided into an amino-terminal part that may have chromatin-remodeling activity and a carboxy-terminal part that may facilitate DNA-binding [28, 39]. JmjC: carboxy-terminal jumonji domain", ZF: zinc finger-like motif; NRD: asparagine-rich domain; CTD: carboxy-terminal domain. The CbfA proteins expressed in mutant JH.D comprised either the full length protein (amino acids 2–1000) or the CbfA-CTD (amino acids 724–1000). **b** Expression of CbfA in JH.D cells. CbfA mutant JH.D cells were transformed with plasmids allowing for the expression of either untagged, full-length CbfA [21] or the GFP-tagged, carboxy-terminal domain of CbfA (CbfA-CTD) [28]. Shown is a western blot of whole-cell extracts prepared from the indicated strains. CbfA was visualized with the monoclonal antibody 7 F3 that detects CbfA-CTD. Numbers to the left indicate the sizes of the protein standards in kDa. **c** Complementation of the *agnC* overexpression phenotype in JH.D cells. Expression of *agnC* in the indicated strains was determined by qRT-PCR. Expression levels in JH.D cells and JH.D transformants were compared to AX2 wild-type cells and are expressed as "fold change" of expression, meaning that values >1 represent overexpression of genes in the JH.D strains and a value of 1 would indicate complete reversion of the overexpression in JH.D cells. Values are means from six independent cultures ± SD. **p < 0.01, relative to control AX2 cells (Student's t-test)

homologous recombination. The advantage of this approach would be that the resulting overexpressor strains had stable genetic modifications at known genomic locations with an absence of remaining plasmid sequences, rather than random (multicopy) plasmid insertions, which have been shown to generate somewhat aberrant expression [29].

The *agnC* locus on chromosome 2 is indicated in Fig. 5a. The *agnC* gene shares its upstream sequence with gene DDB_G0271884, which is transcribed in the opposite direction. The gene activation construct consisted of a DNA fragment containing a fused *actin6/ actin15* promoter element, which allowed for the expression of a blasticidin resistance gene under the control of *act6P* and the *agnC* gene under the control of *act15P* in opposite directions (Fig. 5a). A 1200 bp fragment of the *agnC* gene, including its authentic translation start, was inserted downstream of the *act15* promoter, whereas

another 1200 bp fragment covering the complete coding sequence of gene DDB_G0271884, including 273 bp of upstream sequence, was cloned downstream of the blasticidin resistance cassette to generate a classical two-armed knockout plasmid. After double homologous recombination in the transformants, 749 bp of upstream *agnC* sequence were replaced by the *act15* promoter. PCR on genomic DNA from several independent *agnC*^GA mutants confirmed the promoter exchange and ensured that the organization at the locus was otherwise unaffected, particularly with respect to the upstream gene DDB_G0271884.

Semi-quantitative RT-PCR of *agnC*^GA mutants revealed strong overexpression of *agnC* in growing cells, whereas the transcript was barely detectable in wild-type cells (Fig. 5b). Whereas a 23-fold overexpression of *agnC* in JH.D cells relative to wild-type AX2 cells was determined by qRT-PCR, AX2[*agnC*^GA] strains contains between 995- to

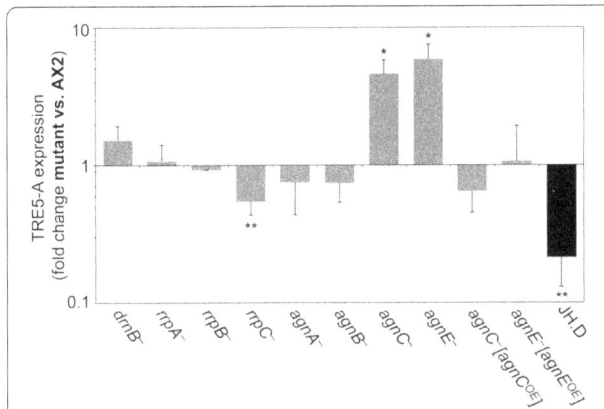

Fig. 4 Expression of TRE5-A in knockout mutants of RNAi components. Expression of TRE5-A ORF1 was analyzed by qRT-PCR in the indicated knockout mutants of Argonaute genes and the Dicer-like protein DrnB. Phenotype reversion in *agnC* and *agnE* knockouts was accomplished using TAP-tagged *agnC* and *agnE* overexpressed in the respective mutants (*agnC⁻[agnCᴼᴱ]* and *agnE⁻[agnEᴼᴱ]*). TRE5-A expression in JH.D cells is shown for comparison. Expression levels were compared to AX2 wild-type cells and are expressed as fold change of expression, meaning that values >1 represent overexpression of genes in the mutants. Data represent means from three independent cultures ± SD. *p < 0.05, **p < 0.01 relative to AX2 cells (Student's t-test)

2016-fold excess *agnC* mRNA compared to AX2 cells. Yet the *agnCᴳᴬ* mutants had no obvious phenotype during growth and multicellular development. Expression of the *agnC*-upstream gene DDB_G0271884 was not affected by the homologous recombination yielding the *agnCᴳᴬ* strains (Fig. 5b). Semi-quantitative RT-PCR revealed that the TRE5-A steady-state transcript level in the *agnCᴳᴬ* mutants dropped sharply and was even more lower than in JH.D cells (Fig. 5b). This was confirmed by qRT-PCR, which suggested 24- to 45-fold lower expression of the TRE5-A ORF1 transcript in *agnCᴳᴬ* mutants than in wild-type cells, compared to a 4-fold decrease seen in JH.D cells (Fig. 5c). The data suggested that AgnC is directly involved in suppressing TRE5-A transcripts in *D. discoideum* cells.

Strains overexpressing the *agnE* gene in the AX2 background were constructed by employing the gene activation strategy as outlined in Additional file 1: Figure S2. Apparently overexpression of AgnE in the recovered *agnEᴳᴬ* strains resulted in an average of 2.3-fold underexpression of TRE5-A in four strains tested, thus confirming that AgnE may be involved in TRE5-A suppression.

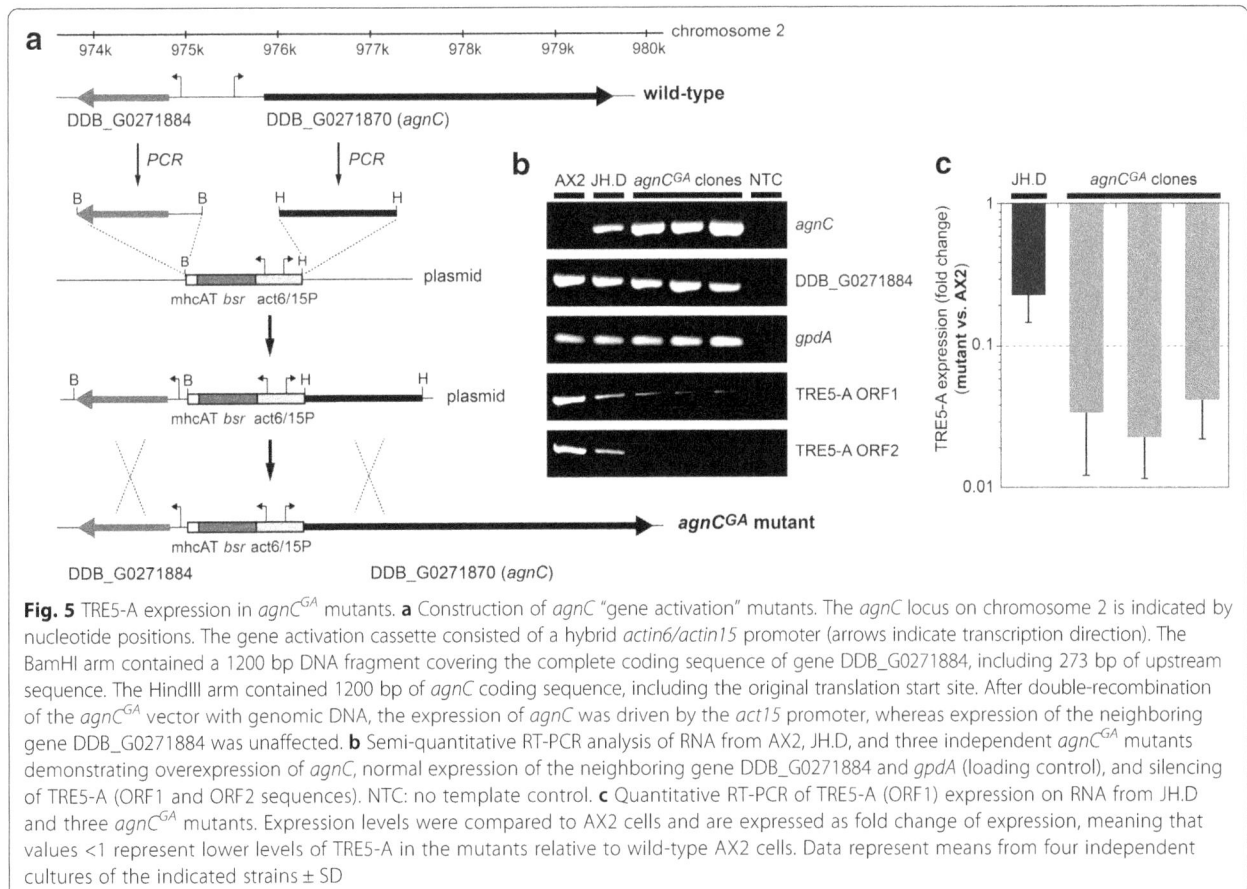

Fig. 5 TRE5-A expression in *agnCᴳᴬ* mutants. **a** Construction of *agnC* "gene activation" mutants. The *agnC* locus on chromosome 2 is indicated by nucleotide positions. The gene activation cassette consisted of a hybrid *actin6/actin15* promoter (arrows indicate transcription direction). The BamHI arm contained a 1200 bp DNA fragment covering the complete coding sequence of gene DDB_G0271884, including 273 bp of upstream sequence. The HindIII arm contained 1200 bp of *agnC* coding sequence, including the original translation start site. After double-recombination of the *agnCᴳᴬ* vector with genomic DNA, the expression of *agnC* was driven by the neighboring gene DDB_G0271884 was unaffected. **b** Semi-quantitative RT-PCR analysis of RNA from AX2, JH.D, and three independent *agnCᴳᴬ* mutants demonstrating overexpression of *agnC*, normal expression of the neighboring gene DDB_G0271884 and *gpdA* (loading control), and silencing of TRE5-A (ORF1 and ORF2 sequences). NTC: no template control. **c** Quantitative RT-PCR of TRE5-A (ORF1) expression on RNA from JH.D and three *agnCᴳᴬ* mutants. Expression levels were compared to AX2 cells and are expressed as fold change of expression, meaning that values <1 represent lower levels of TRE5-A in the mutants relative to wild-type AX2 cells. Data represent means from four independent cultures of the indicated strains ± SD

AgnC is a suppressor of TRE5-A retrotransposition

It seemed plausible that silencing of TRE5-A expression would diminish retrotransposition of the TRE5-A population in the *agnC* and *agnE* overexpressor strains. To directly measure the retrotransposition activity of the cellular TRE5-A population, we set up a previously described "TRE trap" in vivo retrotransposition assay [18, 19] in *agnC*^{GA} cells. The TRE trap is based on a modified *pyr56* gene (*TRE*^{trap}) that codes for UMP synthase (Fig. 6). The *TRE*^{trap} gene contains an intron into which a *Val*^{UAC} tRNA gene was inserted as target for TRE5-A integrations. After transformation of the *TRE*^{trap} gene into ura$^-$ cells, the transformants present a ura$^+$ phenotype due to the expression of functional UMP synthase from the *TRE*^{trap} gene (with the intron including the tRNA gene being spliced out); however, the cells are prone to mutations in the *TRE*^{trap} gene by integration of cellular TRE5-A elements upstream of the embedded tRNA gene. Cells affected by TRE5-A integration into the *TRE*^{trap} gene can no longer splice out the intron and are converted to the ura$^-$ phenotype; they can be recovered after clonal growth in medium complemented with 5-fluoroorotic

acid (5-FOA) and uracil. In previous studies, approximately 100 insertions into the TRE trap were analyzed for integration by mobile elements [18, 19]. The data revealed that ~1 % of recovered ura$^-$ clones had spontaneous loss-of-function mutations of the *TRE*^{trap} gene, whereas ~99 % of the clones carried a TRE5-A element. No insertions of other members of the tRNA gene-specific TRE retrotransposon family were detected in this assay, suggesting that they amplify at a very low rate. Thus, the number of clones obtained in the TRE trap assay is an estimate of the TRE5-A retrotransposition activity in *D. discoideum* cells [18, 19].

In the experiments described above, AX2 was the parent strain for the generation of *agnC*^{GA} and *agnE*^{GA} mutants because we wanted to be able to directly compare them to JH.D cells, which were also derived from AX2. Because no suitable uracil-auxotrophic AX2 mutant was available, we reproduced *agnC*^{GA} mutants in the ura$^-$ strain DH1, which is an AX3 derivative [30]. As shown in Fig. 7a, overexpression of *agnC* in the recovered DH1[*agnC*^{GA}] strains was comparable to AX2[*agnC*^{GA}] cells. Likewise, TRE5-A ORF1 expression was suppressed in DH1[*agnC*^{GA}] strains, albeit at somewhat lower efficacy as in AX2[*agnC*^{GA}] cells. qRT-PCR revealed that TRE5-A expression in the particular DH1[*agnC*^{GA}] strain that we subsequently used to determine TRE5-A retrotransposition was 7.7-fold lower than in the parental DH1 strain. This DH1[*agnC*^{GA}] strain was transformed with plasmids carrying either the empty TRE trap (i.e., no tRNA gene inserted in the trap) or the *TRE*^{trap} gene, which contained a *Val*^{UAC} tRNA gene as bait for TRE5-A integrations. Five plates, each containing 10^7 cells, were cultured in minimal medium supplemented with 5-FOA and uracil until clones appeared. As the positive control, the TRE5-A retrotransposition frequency in DH1[*TRE*^{trap}] cells was determined at 2.03×10^{-5}, whereas it was $<0.01 \times 10^{-5}$ in DH1[*TRE*^{trap}] cells in which the *Val*^{UAC} tRNA gene was omitted as the negative control. In two independently recovered DH1[*agnC*^{GA}/*TRE*^{trap}] strains, TRE5-A retrotransposition activity was determined at 0.14×10^{-5} and 0.05×10^{-5}, representing a more than 90 % drop retrotransposition in the *agnC* overexpressing cells compared to control cells ($p < 0.001$, Student's t-test) (Fig. 7b). These data indicate that AgnC controls the amplification of TRE5-A elements in *D. discoideum* cells by limiting the accumulation of retrotransposon-derived RNA.

Although we were able to establish *agnE*^{GA} strains in the DH1 background, we could not recover viable cells after transformation of the *TRE*^{trap} gene into these cells. Therefore, we were unable to determine the retrotransposition activity of the TRE5-A population under these conditions. Thus, it remains elusive at this point whether the moderate downregulation of

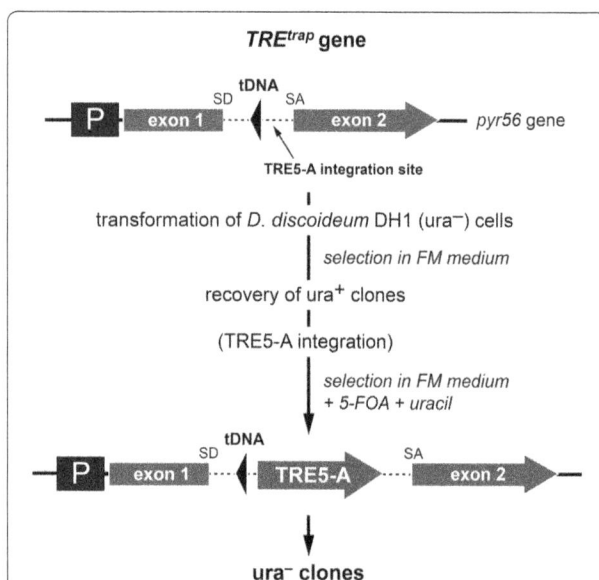

Fig. 6 Outline of the TRE trap retrotransposition assay. The *TRE*^{trap} gene is a modified version of the *pyr56* gene, which codes for the *D. discoideum* UMP synthase. The *pyr56* gene contains an intron (dashed line; SD: splice donor site, SA: splice acceptor site) into which a tRNA gene is inserted as bait for the integration of TRE5-A. The *TRE*^{trap} gene is transformed into the ura$^-$ strain DH1 that has a complete deletion of the *pyr56* gene. The *TRE*^{trap} gene converts transformants to ura$^+$ because the intron is functionally spliced. If an element of the endogenous TRE5-A population targets the tRNA gene in the *TRE*^{trap} gene for integration, the *TRE*^{trap} gene is disrupted even if the integration actually occurs in the intronic sequence. The resulting ura$^-$ cells gain resistance to the drug 5-fluoroorotic acid (5-FOA) and grow out clonally if uracil is added to the medium [18]

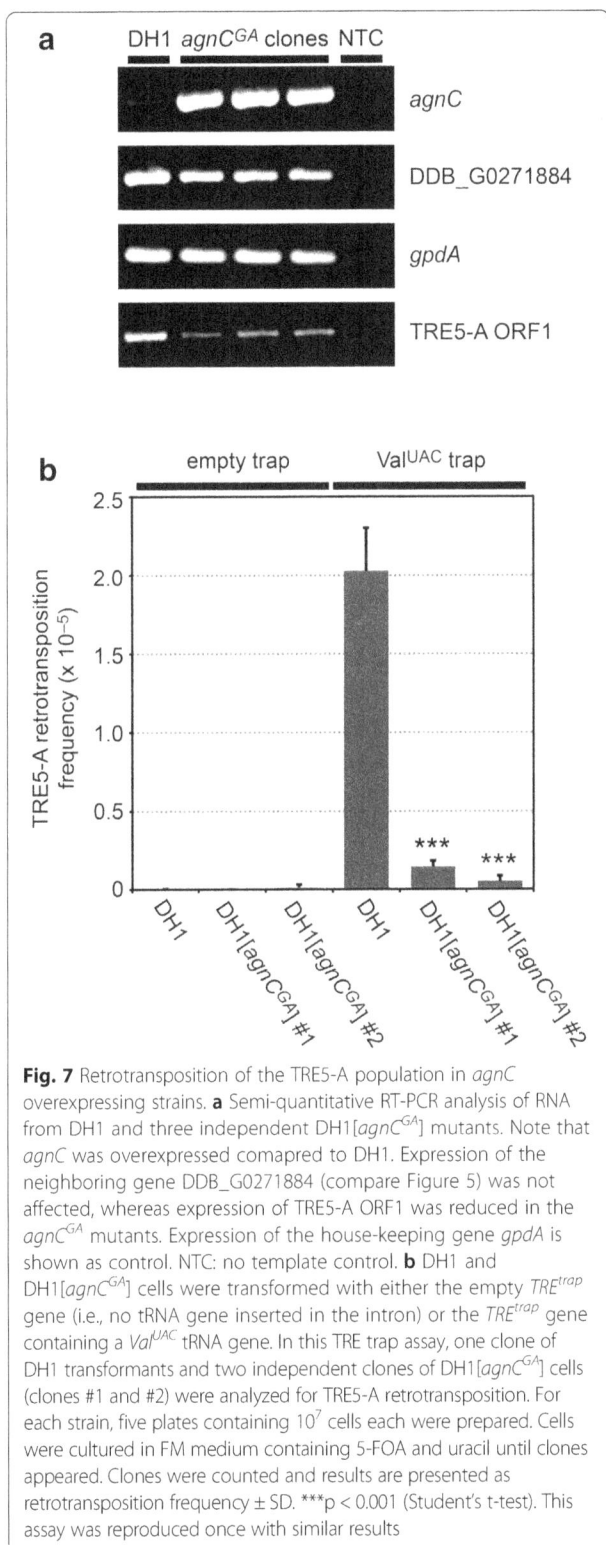

Fig. 7 Retrotransposition of the TRE5-A population in *agnC* overexpressing strains. **a** Semi-quantitative RT-PCR analysis of RNA from DH1 and three independent DH1[*agnC^GA*] mutants. Note that *agnC* was overexpressed comapred to DH1. Expression of the neighboring gene DDB_G0271884 (compare Figure 5) was not affected, whereas expression of TRE5-A ORF1 was reduced in the *agnC^GA* mutants. Expression of the house-keeping gene *gpdA* is shown as control. NTC: no template control. **b** DH1 and DH1[*agnC^GA*] cells were transformed with either the empty *TRE^trap* gene (i.e., no tRNA gene inserted in the intron) or the *TRE^trap* gene containing a Val^UAC tRNA gene. In this TRE trap assay, one clone of DH1 transformants and two independent clones of DH1[*agnC^GA*] cells (clones #1 and #2) were analyzed for TRE5-A retrotransposition. For each strain, five plates containing 10^7 cells each were prepared. Cells were cultured in FM medium containing 5-FOA and uracil until clones appeared. Clones were counted and results are presented as retrotransposition frequency ± SD. ***p < 0.001 (Student's t-test). This assay was reproduced once with similar results

TRE5-A transcripts in *agnE^GA* strains correlates with an appreciable suppression of the retrotransposition activity of the TRE5-A population.

Discussion

AgnC and AgnE act in RNAi pathways to suppress TRE5-A retrotransposition

Because the TRE5-A element produces both plus- and minus-strand RNA, we assumed that the retrotransposition frequency of the TRE5-A population may be under surveillance by the cellular RNAi machinery. In this study, we provide evidence to support this assumption. Genetic inactivation of either AgnC or AgnE resulted in overexpression of TRE5-A, suggesting that both proteins have functions in TRE5-A regulation in which the loss of one cannot be compensated for by expression of the other. The *D. discoideum* Argonaute proteins are a part of the PIWI subfamily of Argonaute proteins. They all have divergent amino-terminal domains, but possess conserved PAZ and PIWI domains including an intact DEDH catalytic tetrad and probably possess slicer activity. One model to explain TRE5-A silencing is the generation of PIWI-interacting RNAs (piRNAs) by AgnC and AgnE in a ping-pong piRNA replication mechanism typical for PIWI proteins [31]. This model is intuitive given that piRNAs are often generated from long single-stranded RNA precursors produced from transposable elements, such as the minus strand RNA of TRE5-A. However, piRNA are usually 23–30 nt long [31, 32], which contradicts results from a previous deep sequencing of small RNA libraries which revealed the formation of ~21 nt siRNAs from TRE5-A elements at a very low level (0.05 % of total small RNAs) in growing *D. discoideum* cells [15].

The silencing of the retrotransposon DIRS-1 is a model to study RNAi pathways in *D. discoideum* [14, 15, 17, 22]. Previous deep sequencing of small RNAs revealed high levels of DIRS-1-derived ~21 nt siRNAs [15] that add up to 20 % of all small RNAs detected in *D. discoideum* cells [17]. The difference in the amount of ~21 nt siRNAs derived from DIRS-1 and TRE5-A may be interpreted as the less efficient dsRNA formation from TRE5-A RNA compared to DIRS-1 RNA. DIRS-1 silencing is enhanced by the RdRP RrpC that synthesizes new DIRS-1 dsRNA that can be diced into secondary siRNAs [16]. This amplification step may be missing in the RNAi pathway that controls TRE5-A expression, because TRE5-A transcripts were not stabilized in RdRP-deficient mutants. The Dicer homolog DrnB, which is mainly required for miRNA formation in *D. discoideum* [15], is apparently not involved in the regulation of DIRS-1 RNA levels [17] and it seems to be also dispensable in the process of TRE5-A regulation (Fig. 4). Thus, the RNAi pathways that regulate DIRS-1 and TRE5-A may overlap at the stage of primary siRNA formation, presumably involving the Dicer homolog DrnA, but use different RISCs that contain either AgnA for DIRS-1 silencing [22] or AgnC/AgnE for TRE5-A suppression.

CbfA abrogates TRE5-A suppression by repressing AgnC

The C-module at the 3′ end of the TRE5-A element has promoter activity that is responsible for the production of minus-strand RNA by the element [20]. CbfA was originally identified as a "C-module-binding factor" because it binds to the C-module of TRE5-A in vitro [24], but it does not regulate the C-module promoter activity in vivo [21]. Considering that CbfA regulates more than 1000 genes of the *D. discoideum* genome, the present data suggest that the observations of in vitro binding of CbfA to the C-module and in vivo regulation of steady-state levels of TRE5-A transcripts by CbfA are purely coincidental. Together with the data obtained in this study, we propose instead that the accumulation of TRE5-A transcripts in *D. discoideum* cells is indirect and a result of the CbfA-mediated suppression of a post-transcriptional pathway involving AgnC. This assumption is supported by the observation that both the plus- and minus-strand RNA of TRE5-A simultaneously vanish upon removal of CbfA from cells, but reappear when CbfA is re-introduced into CbfA-underexpressing cells [21]. In a previous mRNA-seq experiment comparing gene expression in JH.D with wild-type cells [28] we detected underexpression of putative DNA transposons such as DDT-A and DDT-S, the long-terminal repeat (LTR) retrotransposon Skipper, and the non-LTR retrotransposon TRE5-B (see Fig. 1). RNA-seq also predicted overexpression of some mobile elements in the absence of CbfA such as the tyrosine recombinase retrotransposon DIRS-1, the LTR retrotransposon DGLT-A, and the non-LTR retrotransposons TRE3-C and TRE3-D. However, differential expression of the mentioned mobile elements in JH.D cells could not be unequivocally confirmed by qRT-PCR. This was obviously due to high biological variation between independent cultures that was never observed when analyzing the expression of coding genes (i.e., Fig. 2) and the reason for this phenomenon remains elusive. At least the overexpression of DIRS-1 in JH.D cells could be explained by the weak, but reproducible overexpression of the genes *rrpC* and *agnA* (Fig. 2), which were both shown to be involved in the downregulation of DIRS-1 [22]. Thus, DIRS-1 and TRE5-A may be suppressed by different RNAi pathways and are affected indirectly by CbfA's broad-ranging gene-regulatory activity. Unfortunately, a high variability of mobile element expression among biological replicates was also observed when analyzing either *agnC* knockout or *agnC* overexpressor cells. Therefore, we are unable at this point to predict whether other mobile elements are regulated by the same AgnC-involving RNAi pathway that controls TRE5-A retrotransposition.

TRE5-A belongs to a family comprising seven tRNA gene-targeting retrotransposons in *D. discoideum* cells. The TRE5-A and TRE5-B elements are closely related and share a common ancestor, but only TRE5-A was amplified to a high copy number rather late in the evolution of this species [12]. It is puzzling that TRE5-A amplification in *D. discoideum* was apparently accelerated after the acquisition of the C-module (i.e., an antisense promoter) and after the split from its common ancestor with the TRE5-B element that lacks a C-module. Intuitively, the incorporation of an antisense promoter into a mobile element should make it vulnerable to silencing by RNAi-related mechanisms and thus prevent its amplification. Because the C-module was most likely acquired by 3′-transduction, a process not uncommon in this class of retrotransposons [33], the question of what advantage the element may have gained by incorporating an antisense promoter at its 3′ end remains. Did *D. discoideum* cells gain a selective advantage from TRE5-A expansion? The release of TRE5-A from RNAi surveillance by a regulated process involving a host-encoded factor such as CbfA may have evolved because it could be used for cellular purposes such as enhancing genome flexibility. Alternatively, TRE5-A release from suppression by RNAi may have been incidentally caused by adaptation to evolutionary pressure forcing alterations in AgnC-mediated posttranscriptional regulation that are unrelated to transposon suppression. It is unknown under which conditions the repression of *agnC* by CbfA would be released or for which functions AgnC would be required; at least, it seems to be unrelated to the multicellular development of *D. discoideum* because *agnC* is barely upregulated during development [34] and neither *agnC* knockouts nor *agnC* overexpressor display a developmental phenotype. Repression of AgnC by CbfA may provide an efficient way to respond to changes in particular environmental conditions that require specialized functions of this Argonaute protein. Even if this mode of gene regulation by CbfA would come at the cost of TRE5-A amplification, it is reasonably tolerable because TRE5-A's targeted integration to regions upstream of tRNA genes would largely prevent insertion mutagenesis of the genome.

Whereas RNAi may have been developed to restrict mobile element expansion in *D. discoideum* as in other eukaryotes, as exemplified by DIRS-1 silencing, our study shows an intriguing example of a transposable element that is under surveillance by the cellular RNAi machinery, but the control of which can be overcome by suppression of a distinct RNAi pathway by a host factor.

Conclusions

The social amoeba *D. discoideum* has a compact and haploid genome that requires tight control of mobile element activity to maintain genome stability. The non-long terminal repeat retrotransposon TRE5-A actively amplifies in the genome of *D. discoideum* even though the element should be vulnerable to posttranscriptional silencing due to the production of antisense RNA from an element-internal promoter. The host-encoded factor CbfA has global gene-regulatory functions in *D. discoideum* that include the suppression of the Argonaute-like proteins AgnC and AgnE. Whereas TRE5-A transicts were found to accumulate in mutants lacking AgnC or AgnE, expression and retrotransposition of the element vanished in AgnC and AgnE overexpressing cells. These observations suggest that TRE5-A amplification is under surveillance by an RNAi pathway that involves AgnC and AgnE and that this control is at least partially overcome by the activity of CbfA. This unusual regulation of mobile element activity by a host factor most likely had a profound effect on genome evolution in *D. discoideum*.

Methods
Strains and plasmids
The CbfA-depleted mutant JH.D and the plasmids used for the expression of full-length CbfA and the GFP-tagged carboxy-terminal domain of CbfA (CbfA-CTD) have been previously described [21, 25]. *D. discoideum* strains harboring knockouts of RdRP genes *rrpA*, *rrpB*, and *rrpC* were described by Wiegand & Hammann [16]. Knockout strains of *agnA* and *agnB* were described elsewhere [22]. The *drnB⁻* strain was described in Avesson et al., 2012 [35]. Knockout mutants of *agnC* and *agnE* as well as plasmids allowing for the expression of TAP-tagged AgnC and AgnE will be described in a separate publication (F.S. et al., manuscript in preparation).

Construction of gene activation mutants
The *D. discoideum* expression vector pDM326 [36] contains a blasticidin resistance cassette driven by the *act6* promoter and an upstream *act15* promoter in opposite direction for the expression of transgenes. A DNA fragment containing both the blasticidin cassette and the *act15* promoter was isolated from pDM326 by digestion with BamHI and BglII. The DNA fragment was inserted into the BamHI site of pGEM7Zf(−) (Promega), such that the former BglII site was placed next to the HindIII site of the pGEM vector to generate pGEM-GA. To generate the *agnC^GA* vector, the "BamHI arm" covering the entire coding sequence of the gene DDB_G0271884, which shares its upstream region with *agnC*, was amplified including 273 bp of residual upstream sequence and inserted into the BamHI site of pGEM-GA. The "HindIII arm" was generated by amplification of nucleotides 1–1166 of the *agnC* gene including its authentic translation start codon. The pGEM-agnC-GA plasmid was linearized and transformed into *D. discoideum* AX2 or DH1 cells and transformants were selected in HL5 medium (Formedium, Hunstanton, UK) containing 6 μg/ml blasticidin (Life Technologies, Carlsbad,

USA) [37]. From such clones genomic DNA was isolated and screened by PCR for insertion of the GA cassette at the targeted locus using one primer specific for the blasticidin resistance gene and a second primer that hybridized outside of the DNA sequences covered by the HindIII arm. RT-PCR was used to confirm that the expression of the *agnC*-upstream gene DDB_G0271884 was not affected by insertion of the GA cassette.

Reverse transcription-PCR
Total RNA was prepared from frozen cell pellets and RT-PCR was done as described previously [21]. In quantitative RT-PCR gene expression levels were standardized to the gene coding for catalase (*catA*). The following qRT-PCR primers were used: rrpA-01, GAACGTCAAGAACTTGGTAAATTGTATC; rrpA-02, TAACCTACAGTTTGTAAC CGAATGTTTAC; rrpB-01, GAACGTCAAGAACTTGGTAAAATGTATAA; rrpB-02, GTGGATAACCTTTAGTTTTTAACCAAAC; rrpC-01, GGTGTTTATAGTAAAAAAGAATCATTC; rrpC-02, CAACTATCCAAGAATTTATGAACATTTAC; agnA-01, GCCGAAACTCCTTCTTCTTGGGGTAC; agnA-02, GTTCATCCAATAAGACATGGTAATGAG; agnB-01, GTGATGGTGTTGGTGATGGTATGTTAG; agnB-02, CTTGGTAATCCTGATCAAGGTGTTGTTG; agnC-03, GTGCACTTTTATGAGAGTATTGGCATAC; agnC-04, GTACATGATAATGAGTTGGATTTGTAG; agnD-01, CATCATATTAATAGTCGTTTACCAGAG; agnD-02, GTACCAATCCACCCAATGGTACAATGG; agnE-03, GAGCATAATTACAAGGAGCAGGTGTTC; agnE-04, CAGTGCTAACCATTGTACCATTGGGTG; catA-01, GTTTCGCTGCTCGTCAACCATACAATC; catA-02, GCACGAACTTGAATTTCTTTGATGGTG; gpdA-01, GGTTGTCCCAATTGGTATTAATGG; gpdA-02, CCGTGGGTTGAATCATATTTGAAC; TRE5-A ORF1 Rep-108, GTCATAAACATCAATCCGAACCAGAC; TRE5-A ORF1 Rep-109, GTTAGATTGTCTAGTTCAATGATAGTGTC; TRE5-A ORF2 Rep-75, GACTGTTCAGTGGATAATAACC; TRE5-A ORF2 Rep-176, CTCGAGTTAAAGGAAGATTGCTCTTGAATC; DDB_G0271884-01, GAGTTGGCCAAATTAGTTAAGCAATTG; DDB_G0271884-02, CCTTGTTCAACCCAAGAGAAAATTTCTG.

TRE trap retrotransposition assay
The TRE trap is an in vivo retrotransposition assay that measures the activity of the cellular TRE5-A population. It was essentially performed as described previously [21]. The TRE trap consists of the complete *pyr56* gene modified to contain a functional intron into which a *Val^UAC* bait tRNA gene was inserted. This gene is referred to as the *TRE^trap* gene. After transformation into *D. discoideum* DH1 cells, ura⁺ cells harboring chromosomal integrations of the *TRE^trap* gene were recovered by cultivation in FM medium without supplements. After integration of a

TRE5-A element into the trap, the TRE^{trap} gene is disrupted and no functional UMP synthase is expressed. Thus, affected cells were converted to the ura⁻ phenotype and gained resistance to the drug 5-fluoroorotic acid (5-FOA). In a typical retrotransposition assay, 5 plates each containing 10^7 cells were prepared, and cells were cultured in FM medium containing 150 μg/ml 5-FOA and 20 μg/ml uracil. Clones that arose were counted, and the data presented are the means from 5 plates ± SD.

Western blots

D. discoideum cells were washed in phosphate buffer and stored as frozen pellets of 2×10^7 cells at –80 °C. SDS/polyacrylamide gel electrophoresis of whole-cell extract proteins and western blotting were done as described [38]. We used monoclonal antibody 7 F3 to detect CbfA and a polyclonal antiserum to detect actin8 [38].

Additional data files

The following additional data are available with the online version of this paper. Additional file 1: Figure S1 shows the functional complementation of strain JH.D with CbfA or its carboxy-terminal domain with respect to TRE5-A and *agnE* expression. Additional file 1: Figure S2 illustrates the construction of $agnE^{GA}$ mutants and shows the expression of TRE5-A in these mutants.

Additional file

Additional file 1: Figure S1. Functional complementation of strain JH.D with CbfA or its isolated carboxy-terminal domain. RNA levels of retrotransposon TRE5-A (A) and *agnE* (B) were determined by qRT-PCR in JH.D cells, JH.D cells expressing full-length CbfA, and JH.D cells expressing CbfA-CTD. Expression levels in JH.D cells and JH.D transformants were compared to AX2 wild-type cells and are expressed as "fold change" of expression, meaning that values >1 represent overexpression of genes in the JH.D strains and a value of 1 would indicate complete reversion of the overexpression in JH.D cells. Values are means from six independent cultures ± SD. **p < 0.01, relative to control AX2 cells (Student's t-test). Note that TRE5-A is actually overexpressed in the JH.D[CbfA-CTD] transformant, which is an effect of overexpression of CbfA-CTD (see Fig. 2, main text). **Figure S2.** TRE5-A expression in $agnE^{GA}$ mutants. (A) Construction of *agnE* "gene activation" mutants. The *agnE* locus on chromosome 5 is indicated by nucleotide positions. The gene activation cassette consisted of a hybrid *actin6/actin15* promoter (arrows indicate transcription direction). The BamHI arm contained a 1070 bp DNA fragment covering part of the coding sequence of gene DDB_G0289385. The HindIII arm contained 1080 bp of *agnE* coding sequence, including the original translation start site. After double-recombination of the $agnE^{GA}$ vector with genomic DNA, the expression of *agnE* was driven by the *act15* promoter, whereas expression of the neighboring gene DDB_G0289385 was unaffected. (B) Semi-quantitative RT-PCR analysis of RNA from AX2, JH.D, and three independent $agnE^{GA}$ mutants demonstrating overexpression of *agnE*, normal expression of the neighboring gene DDB_G0289385 and *gpdA* (loading control), and silencing of TRE5-A (ORF1 and ORF2 sequences). NTC: no template control. (C) Quantitative RT-PCR of TRE5-A (ORF1) expression on RNA from JH.D and four $agnE^{GA}$ mutants. Expression levels were compared to AX2 cells and are expressed as fold change of expression, meaning that values <1 represent lower levels of TRE5-A in the mutants relative to wild-type AX2 cells. Data represent means from four independent cultures of the indicated strains ± SD. (PDF 1033 kb)

Abbreviations

CbfA: C-module-binding factor; CTD: carboxy-terminal domain (of CbfA); GA: gene activation (by promoter swapping); LTR: long terminal repeat; qRT-PCR: quantitative reverse-transcription PCR; RdRP: RNA-dependent RNA polymerase; RISC: RNA-induced silencing complex; RNAi: RNA interference; siRNA: small interfering RNA; TAP: tandem affinity purification; TRE5-A: tRNA gene-targeted retroelement 5-A.

Competing interests

The authors declare that they have no competing interests.

Authors' contributions

TW conceived the study. AS, TS, FG, SO, ÅF, and BB designed and performed the experiments and revised the manuscript critically for important intellectual content. TW, WN, FS, and CH analyzed the data and drafted the manuscript. All authors read and approved the final manuscript.

Acknowledgments

This work was supported by grants to TW and HC (WI 1142/7-1, HA3459/7-1) from the German Research Foundation (Deutsche Forschungsgemeinschaft, DFG) and by support from the Swedish Research Council (to FS and to Uppsala RNA Research Center) and The Swedish Research Council for Environment, Agricultural Sciences and Spatial Planning (FORMAS) to FS. SO acknowledges a stipend by the Fritz Thyssen Foundation.

Author details

[1]Department of Pharmaceutical Biology, Institute of Pharmacy, University of Jena, Semmelweissstrasse 10, 07743 Jena, Germany. [2]Department of Molecular Biology, Biomedical Center, Swedish University of Agricultural Sciences, Uppsala, Sweden. [3]Institute of Biology – Genetics, University of Kassel, Kassel, Germany. [4]Ribogenetics@Biochemistry Lab, Department of Life Sciences and Chemistry, Molecular Life Sciences Research Center, Jacobs University Bremen, Bremen, Germany. [5]Department of Cell and Molecular Biology, Biomedical Center, Uppsala University, Uppsala, Sweden. [6]Present address: Aprea AB, Karolinska Institutet Science Park, Nobels väg 3, 17175 Solna, Sweden. [7]Present address: Department of Biology, Brawijaya University, Jl. Veteran, Malang, East Java, Indonesia.

References

1. Cordaux R, Batzer MA. The impact of retrotransposons on human genome evolution. Nature Rev Genet. 2009;10:691–703.
2. Kazazian HH. Mobile elements: drivers of genome evolution. Science. 2004;303:1626–32.
3. Levin HL, Moran JV. Dynamic interactions between transposable elements and their hosts. Nat Rev Genet. 2011;12:615–27.
4. Scheifele LZ, Cost GJ, Zupancic ML, Caputo EM, Boeke JD. Retrotransposon overdose and genome integrity. Proc Natl Acad Sci USA. 2009;106:13927–32.
5. Siomi H, Siomi MC. On the road to reading the RNA-interference code. Nature. 2009;457:396–404.
6. Ghildiyal M, Zamore PD. Small silencing RNAs: an expanding universe. Nat Rev Genet. 2009;10:94–108.
7. Kim VN, Han J, Siomi MC. Biogenesis of small RNAs in animals. Nat Rev Mol Cell Biol. 2009;10:126–39.
8. Castel SE, Martienssen RA. RNA interference in the nucleus: roles for small RNAs in transcription, epigenetics and beyond. Nat Rev Genet. 2013;14:100–12.
9. Kawamata T, Tomari Y. Making RISC. Trends Biochem Sci. 2010;35:368–76.
10. Kuhn C-D, Joshua-Tor L. Eukaryotic Argonautes come into focus. Trends Biochem Sci. 2013;38:263–71.
11. Eichinger L, Pachebat JA, Glöckner G, Rajandream M-A, Sucgang R, Berriman M, et al. The genome of the social amoeba *Dictyostelium discoideum*. Nature. 2005;435:43–57.
12. Glöckner G, Szafranski K, Winckler T, Dingermann T, Quail M, Cox E, et al. The complex repeats of *Dictyostelium discoideum*. Genome Res. 2001;11:585–94.
13. Martens H, Novotny J, Oberstrass J, Steck TL, Postlethwait P, Nellen W. RNAi in *Dictyostelium*: the role of RNA-directed RNA polymerases and double-stranded RNase. Mol Biol Cell. 2002;13:445–53.

14. Kuhlmann M, Borisova BE, Kaller M, Larsson P, Stach D, Na JB, et al. Silencing of retrotransposons in *Dictyostelium* by DNA methylation and RNAi. Nucleic Acids Res. 2005;33:6405–17.

15. Hinas A, Reimegard J, Wagner EG, Nellen W, Ambros VR, Söderbom F. The small RNA repertoire of *Dictyostelium discoideum* and its regulation by components of the RNAi pathway. Nucleic Acids Res. 2007;35:6714–26.

16. Wiegand S, Hammann C. The 5′ spreading of small RNAs in *Dictyostelium discoideum* depends on the RNA-dependent RNA polymerase RrpC and on the Dicer-related nuclease DrnB. PLOS ONE. 2013;8:e64804.

17. Wiegand S, Meier D, Seehafer C, Malicki M, Hofmann P, Schmith A, et al. The *Dictyostelium discoideum* RNA-dependent RNA polymerase RrpC silences the centromeric retrotransposon DIRS-1 post-transcriptionally and is required for the spreading of RNA silencing signals. Nucleic Acids Res. 2014;42:3330–45.

18. Beck P, Dingermann T, Winckler T. Transfer RNA gene-targeted retrotransposition of *Dictyostelium* TRE5-A into a chromosomal UMP synthase gene trap. J Mol Biol. 2002;318:273–85.

19. Siol O, Boutliliss M, Chung T, Glöckner G, Dingermann T, Winckler T. Role of RNA polymerase III transcription factors in the selection of integration sites by the *Dictyostelium* non-long terminal repeat retrotransposon TRE5-A. Mol Cell Biol. 2006;26:8242–51.

20. Schumann G, Zündorf I, Hofmann J, Marschalek R, Dingermann T. Internally located and oppositely oriented polymerase II promoters direct convergent transcription of a LINE-like retroelement, the *Dictyostelium* Repetitive Element, from *Dictyostelium discoideum*. Mol Cell Biol. 1994;14:3074–84.

21. Bilzer A, Dölz H, Reinhardt A, Schmith A, Siol O, Winckler T. The C-module-binding factor supports amplification of TRE5-A retrotransposons in the *Dictyostelium discoideum* genome. Eukaryot Cell. 2011;10:81–6.

22. Boesler B, Meier D, Förstner KU, Friedrich M, Hammann C, Sharma CM, et al. Argonaute proteins affect siRNA levels and accumulation of a novel extrachromosomal DNA from the *Dictyostelium* retrotransposon DIRS-1. J Biol Chem. 2014;289:35124–38.

23. Horn J, Dietz-Schmidt A, Zündorf I, Garin J, Dingermann T, Winckler T. A *Dictyostelium* protein binds to distinct oligo(dA)•oligo(dT) DNA sequences in the C-module of the retrotransposable element DRE. Eur J Biochem. 1999;265:441–8.

24. Geier A, Horn J, Dingermann T, Winckler T. A nuclear protein factor binds specifically to the 3′-regulatory module of the long-interspersed-nuclear-element-like *Dictyostelium* repetitive element. Eur J Biochem. 1996;241:70–6.

25. Winckler T, Trautwein C, Tschepke C, Neuhauser C, Zündorf I, Beck P, et al. Gene function analysis by amber stop codon suppression: CMBF is a nuclear protein that supports growth and development of *Dictyostelium* amoebae. J Mol Biol. 2001;305:703–14.

26. Dingermann T, Reindl N, Brechner T, Werner H, Nerke K. Nonsense suppression in *Dictyostelium discoideum*. Dev Genetics. 1990;11:410–7.

27. Winckler T, Iranfar N, Beck P, Jennes I, Siol O, Baik U, et al. CbfA, the C-module DNA-binding factor, plays an essential role in the initiation of *Dictyostelium discoideum* development. Eukaryot Cell. 2004;3:1349–58.

28. Schmith A, Groth M, Ratka J, Gatz S, Spaller T, Siol O, et al. Conserved gene-regulatory function of the carboxy-terminal domain of dictyostelid C-module-binding factor. Eukaryot Cell. 2013;12:460–8.

29. Windhof IM, Dubin MJ, Nellen W. Chromatin organisation of transgenes in *Dictyostelium*. Pharmazie. 2013;68:595–600.

30. Caterina MJ, Milne JLS, Devreotes PN. Mutation of the third intracellular loop of the cAMP receptor, cAR1, of *Dictyostelium* yields mutants impaired in multiple signaling pathways. J Biol Chem. 1994;269:1523–32.

31. Thomson T, Lin H. The biogenesis and function of PIWI proteins and piRNAs: Progress and prospect. Annu Rev Cell Dev Biol. 2009;25:355–76.

32. Saito K, Siomi MC. Small RNA-mediated quiescence of transposable elements in animals. Dev Cell. 2010;19:687–97.

33. Moran JV, DeBerardinis RJ, Kazazian Jr HH. Exon shuffling by L1 retrotransposon. Science. 1999;283:1530–4.

34. Stajdohar M, Jeran L, Kokosar J, Blenkus D, Janez T, Kuspa A et al. dictyExpress: visual analytics of NGS gene expression in Dictyostelium. 2015. https://www.dictyexpress.org.

35. Avesson L, Reimegård J, Wagner EG, Söderbom F. MicroRNAs in Amoebozoa: deep sequencing of the small RNA population in the social amoeba *Dictyostelium discoideum* reveals developmentally regulated microRNAs. RNA. 2012;18:1771–82.

36. Veltman DM, Akar G, Bosgraaf L, van Haastert PJM. A new set of small, extrachromosomal expression vectors for *Dictyostelium discoideum*. Plasmid. 2009;61:110–8.

37. Siol O, Spaller T, Schiefner J, Winckler T. Genetically tagged TRE5-A retrotransposons reveal high amplification rates and authentic target site preference in the *Dictyostelium discoideum* genome. Nucleic Acids Res. 2011;39:6608–19.

38. Hentschel U, Zündorf I, Dingermann T, Winckler T. On the problem of establishing the subcellular localization of *Dictyostelium* retrotransposon TRE5-A proteins by biochemical analysis of nuclear extracts. Anal Biochem. 2001;296:83–91.

39. Lucas J, Bilzer A, Moll L, Zündorf I, Dingermann T, Eichinger L, et al. The carboxy-terminal domain of *Dictyostelium* C-module-binding factor is an independent gene regulatory entity. PLoS One. 2009;4(4):e5012.

4

Precise repair of *mPing* excision sites is facilitated by target site duplication derived microhomology

David M. Gilbert[1], M. Catherine Bridges[2], Ashley E. Strother[1], Courtney E. Burckhalter[1], James M. Burnette III[3] and C. Nathan Hancock[1*]

Abstract

Background: A key difference between the *Tourist* and *Stowaway* families of miniature inverted repeat transposable elements (MITEs) is the manner in which their excision alters the genome. Upon excision, *Stowaway*-like MITEs and the associated *Mariner* elements usually leave behind a small duplication and short sequences from the end of the element. These small insertions or deletions known as "footprints" can potentially disrupt coding or regulatory sequences. In contrast, *Tourist*-like MITEs and the associated *PIF/Pong/Harbinger* elements generally excise precisely, returning the genome to its original state. The purpose of this study was to determine the mechanisms underlying these excision differences, including the role of the host DNA repair mechanisms.

Results: The transposition of the *Tourist*-like element, *mPing*, and the *Stowaway*-like element, *14T32*, were evaluated using yeast transposition assays. Assays performed in yeast strains lacking non-homologous end joining (NHEJ) enzymes indicated that the excision sites of both elements were primarily repaired by NHEJ. Altering the target site duplication (TSD) sequences that flank these elements reduced the transposition frequency. Using yeast strains with the ability to repair the excision site by homologous repair showed that some TSD changes disrupt excision of the element. Changing the ends of *mPing* to produce non-matching TSDs drastically reduced repair of the excision site and resulted in increased generation of footprints.

Conclusions: Together these results indicate that the difference in *Tourist* and *Stowaway* excision sites results from transposition mechanism characteristics. The TSDs of both elements play a role in element excision, but only the *mPing* TSDs actively participate in excision site repair. Our data suggests that *Tourist*-like elements excise with staggered cleavage of the TSDs, which provides microhomology that facilitates precise repair. This slight modification in the transposition mechanism results in more efficient repair of the double stranded break, and thus, may be less harmful to host genomes by disrupting fewer genes.

Keywords: *mPing*, Excision site repair, Target site duplication

Background

Type II DNA transposable elements (TE) are present in most, if not all, eukaryotic genomes, but are especially abundant in plants where they play a role in genome evolution [1]. Plant DNA TEs have been classified into superfamilies including *hAT*, *MuDR/MU*, *CACTA*, *Mariner*, and *Harbinger/Pong* [2]. Each of these superfamilies is composed of autonomous elements that encode the proteins required for mobilization and non-autonomous elements that can only be mobilized *in trans* [3, 4]. Of special interest are the small (<500 bp) non-autonomous miniature inverted repeat TEs (MITEs). These are the most abundant TEs in the genome, often reaching thousands of copies, due to their ability for rapid proliferation [5–7]. The two best characterized MITE families, *Stowaway* and *Tourist*, have unique characteristics stemming from differences in their transposition mechanisms. *Stowaway*-like MITEs are mobilized by transposase proteins encoded

* Correspondence: nathanh@usca.edu
[1]Department of Biology and Geology, University of South Carolina Aiken, 471 University Parkway, Aiken, SC 29801, USA
Full list of author information is available at the end of the article

by autonomous *Mariner*-like elements, produce a 2 bp target site duplication (TSD) upon insertion, and commonly leave small insertions or deletions (footprints) at their excision site [8]. *Tourist*-like MITEs are mobilized by transposase proteins encoded by the autonomous *PIF/Pong*-like elements, produce a 3 bp TSD, and generally excise precisely leaving no footprints at their excision site [9].

DNA TEs and their associated MITEs are mobilized by a "cut and paste" mechanism in which transposase proteins bind to the terminal inverted repeats (TIRs), effectively positioning the catalytic domain for the DNA cleavage that is required for both excision and insertion [10]. Staggered cleavage of the genomic DNA at the insertion site results in either a 5′ or 3′ overhang, both of which create small TSDs that flank the inserted elements. Based on the fact that *Mariner*-like and *Stowaway*-like elements have 2 bp TSDs, the transposase proteins likely produce a 2 bp overhang upon cleavage of the DNA [8]. The *PIF/Pong*-like and *Tourist*-like elements have 3 bp TSDs, indicating cleavage by their encoded transposases produce a 3 bp overhang [11]. Analysis of the excision sites of the elements can elucidate differences in the catalytic mechanism of their specific transposases. For example, the excision sites of the *Ac* and *Ds* elements in both plants and yeast demonstrate that their footprints are palindromic sequences from the flanking DNA, as opposed to pieces of the TE itself [12–14]. This suggests that this transposase cleaves at the end of the element, causing hairpin formation at the ends of the double stranded break. In contrast, the excision sites of *Mariner*/*Stowaway*-like elements contain footprints that often include some of the sequences of the element in addition to retaining the TSDs [15]. This indicates that that the *Mariner*-like transposase cleaves with a staggered cut at the end of the TIR for excision, leaving behind the TSD and a short region of single stranded TIR [15].

Excision of these DNA TEs produces double stranded breaks that are repaired by the host DNA repair mechanisms. This can be accomplished using a complementary template for homologous recombination (HR) or by the non-homologous end joining (NHEJ) pathway [16]. In plants, excision site analysis indicates that many of the repaired sites include insertions or deletions consistent with NHEJ [17–19]. In addition, yeast transposition experiments with the *Ac* element superfamily showed that repair of the double stranded break after excision required NHEJ proteins [12]. This study also showed that microhomology (<6 bp) exposed by end processing between the two strands flanking the element is often used to facilitate repair [12]. Differences in the proteins required for the repair of these ends may hint at the nature of the DNA breaks produced by the different transposases.

The focus of this study was to further characterize the transposition mechanism of the best studied *Tourist*-like MITE, *mPing*. In contrast to the previously mentioned elements, *mPing* excision sites are repaired precisely (leaving no element or TSD sequences). Based on these unique excision sites, we hypothesize that the *mPing* transposase proteins may cut at the TSD sequences adjacent to the element instead of within the element as seen for *Mariner*-like elements. Because the TSD sequences are identical, staggered cleavage at this location produces compatible sticky ends, providing microhomology for NHEJ that would easily restore the genome back to its original state before insertion of the element. Based on this hypothesis, we predict that alteration of *mPing*'s TSDs would alter the microhomology and reduce the effectiveness of NHEJ repair. Using a previously developed yeast transposition assay [20, 21], we tested the result of changing the TSDs for a *Tourist*-like MITE (*mPing*) and *Stowaway*-like MITE (*OsMar 14T32* or the hyperactive *OsMar 14T32-T7*). By performing these assays in yeast strains with a defective NHEJ DNA repair pathway, we were able to distinguish between impaired element excision and DNA repair.

Results and discussion
NHEJ is used for excision site repair
The yeast transposition assay used for these experiments measures the rate at which the *ADE2* gene is repaired in-frame following excision of the TE (Additional file 1) [14, 15, 20, 21]. Traditionally, these assays have been performed in haploid yeast lacking an *ADE2* homologous template for HR repair of the excision site. Under these conditions the excision site should be repaired only by NHEJ. Performing transposition assays with *mPing* and *14T32* in haploid yeast strains lacking the NHEJ pathway proteins *KU70*, *MRE11*, or *RAD50* showed that these proteins are required for efficient repair of the excision sites of both elements (Fig. 1a). Almost no *ADE2* revertant colonies were obtained in the *ku70* strain, as *KU70* is a highly conserved protein involved in the initial binding of the double stranded breaks [22]. For both elements, the *rad50* strain showed a higher DNA repair rate than the *mre11* strain. This is consistent with a previous study indicating that *MRE11* function is more important for repair than *RAD50* even though these two proteins function together in the MRX complex to process double stranded breaks before ligation [22, 23]. These results also indicated *RAD50* plays a more important role in excision site repair for the *14T32* element than the *mPing* element [92 % vs. 56 % decrease in repair efficiency (Fig. 1a)]. However, some of this change could be due to a difference in the amount of repair products that result in reading frame disruption. Analysis of excision sites produced in the *rad50* background showed that the *mPing* excision

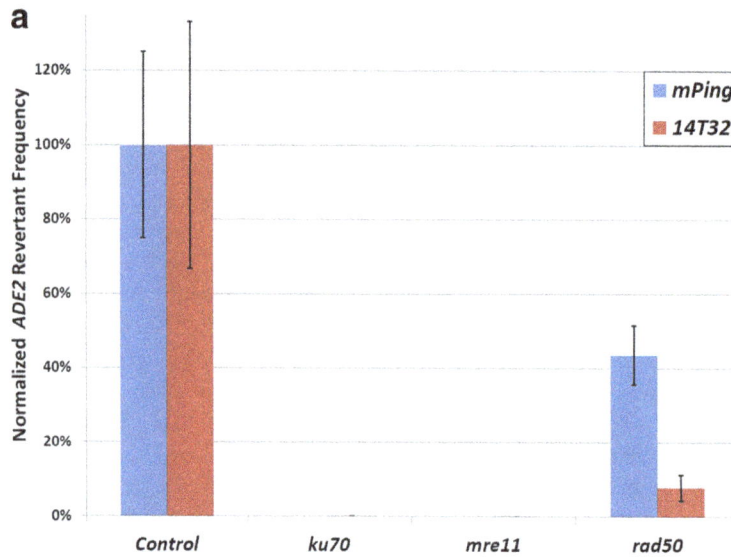

Fig. 1 Transposition assays in NHEJ deficient yeast. Normalized *ADE2* revertant frequency for the *mPing* (*blue*) and *14T32* (*red*) elements in control (JIM17) and NHEJ mutant yeast strains (**a**). Error bars indicate the standard error for 6 replicates. Repaired excision sites from control and *rad50* yeast strains (**b**). Lowercase letters indicate the bases derived from the TSD (*mPing*) or TIRs and TSDs (*14T32*)

sites were still repaired precisely, while the *14T32* excision sites had more bases deleted (less precise repair) compared to the control (Fig. 1b). This difference in repair efficiency and quality observed for the two elements in the *rad50* strain provides evidence that there are important differences in the nature of the double strand breaks produced by these two elements.

Performing yeast transposition assays in a yeast strain that provides a partial *ADE2* template is an effective strategy to evaluate whether HR can be used for excision site repair. This approach has been used to study the *Ac* element (*hAT* superfamily, also creates footprints upon excision) where it was reported that when a template is available, about half of the excision sites are repaired by

HR [12]. In this study, we employed similar methodology to determine if *mPing* and the hyperactive *OsMar 14T32-T7* excision sites are repaired by HR; we used the CB101 yeast strain that contains a partial *ADE2* template called *ADE2**. This experiment showed that while no significant difference in the rate of *ADE2* revertant colonies is observed with or without the *ADE2** template, CB101 seems to show slightly lower average *ADE2* revertants (Fig. 2). This may be due to competition between the two repair pathways or some unknown genetic change present in CB101. This slight difference did not affect our experiments because we were able to normalize within strains. To determine if HR was

occurring in this strain, we analyzed 96 *mPing* excision sites and fifteen *14T32-T7* excision sites by PCR and digestion with *Hae*III (present in *ADE2** but not in the original *ADE2*). Under these conditions, none of the excision sites in either element contained the *Hae*III site, and thus, were not repaired by HR at detectable levels (Fig. 2b). This indicates that even when a homologous template is present, the predominant repair pathway for these excision sites is NHEJ.

In order to allow separate analysis of element excision and repair, we developed a yeast strain (DG21B9) that was capable of performing HR at the excision site (contained the *ADE2** template), but also had an impaired NHEJ pathway (*ku70*). In this strain, the number of

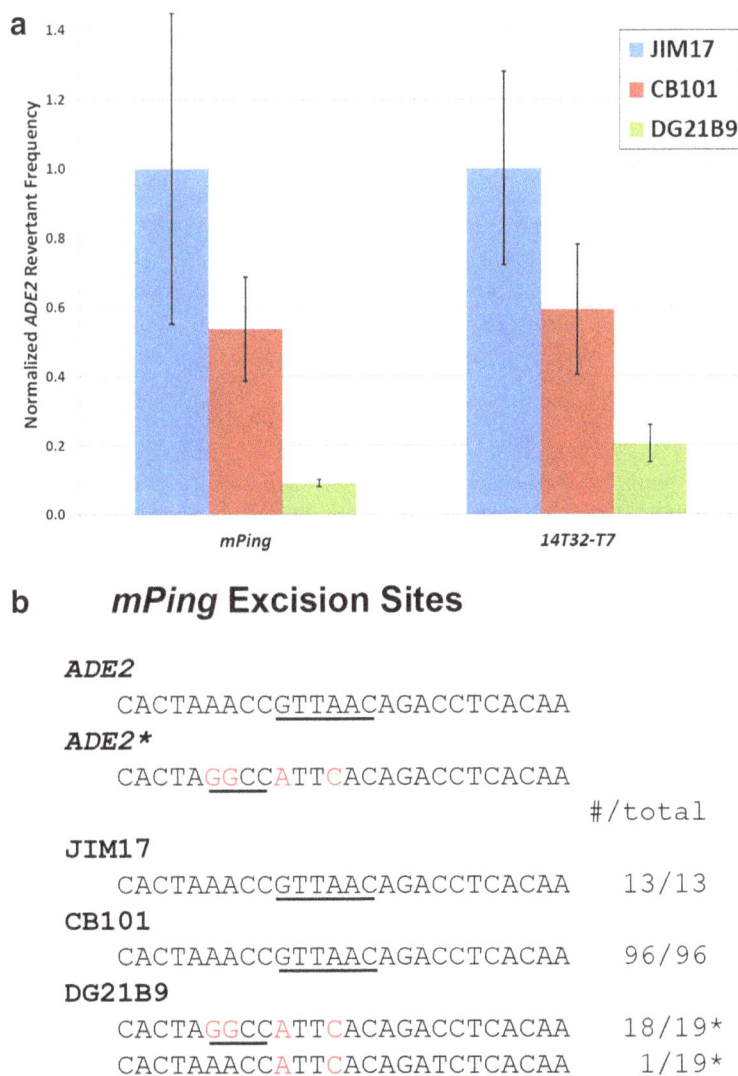

Fig. 2 Transposition assays in yeast with altered DNA repair potentials. *ADE2* revertant frequencies for the *mPing* and *14T32-T7* elements in yeast strains with different DNA repair mechanisms available for excision site repair (**a**). JIM17 repairs by NHEJ, CB101 is capable of both HR and NHEJ, and DG21B9 can only repair by HR. Frequencies were normalized to the activity of each transposable element in JIM17. Error bars represent standard error. Sequences identified at the *mPing* (5′ TAA/3′ TAA TSDs) excision sites by restriction site analysis and sequencing (**b**). Underlined sequences indicate the *Hpa*I and *Hae*III sites used for analysis. Red bases are unique to the *ADE2** template. *indicates the excision site was repaired by HR using the *ADE2** template

ADE2 revertant colonies was drastically reduced for both *mPing* and *14T32-T7* (Fig. 2), but was still higher than observed in the absence of a homologous template (Fig. 1). This drop in activity in this NHEJ deficient strain was consistent with the finding that NHEJ is the dominant pathway for repair of the excision sites. This together with the results for CB101 suggest that HR repair of these breaks functions as a backup to NHEJ and only occurs at about 10–20 % of the rate of NHEJ repair. Analysis of the DG219B *ADE2* revertant *mPing* excision sites by digestion and sequencing showed that 100 % were repaired by HR (Fig. 2). Most of these excision sites (18/19) contained the *ADE2** specific *Hae*III site and the remaining site showed that *ADE2** was used in such a way as to only remove the *Hpa*I site and not add the *Hae*III site (Fig. 2).

The ability to perform transposition assays in this NHEJ deficient strain (DG21B9) makes it possible to exclude the effects that the quality (i.e. blunt, staggered cut, presence or absence of microhomology) of the DNA break has on repair efficiency. This is because HR is less dependent on the immediate sequence at the end of the double stranded break, instead using sequences farther away from the cleavage site. Thus, this strain provides a method to differentiate whether a mutation affects the rate of NHEJ repair or the rate of excision.

TSD alteration disrupts element excision

Previous studies have shown that *Mariner*-like elements require the TSD (TA on both ends) for transposition in vitro [24]. In this study, we confirmed the importance of the conserved TSD for the *14T32-T7* element by changing the TSDs and performing yeast transposition assays. In CB101, changing both bases of the TSDs from TA/TA (5′/3′) to AT/AT almost completely inhibited transposition, while changing just one base (TT/TT or AA/AA) allowed transposition, but at highly reduced rates (Fig. 3a). This experiment was also performed in DG21B9 (HR competent, NHEJ deficient) to confirm that this decrease in activity was due to inhibited excision and not inhibited excision site repair. Figure 3a shows that in DG21B9 alteration of the *14T32-T7* TSDs produced a comparable decrease in activity to the one observed in CB101 yeast. Thus, the drop in activity upon changing the TSDs is likely due to a decrease in excision, and not due to changes in the efficiency of NHEJ. Other researchers have shown that the *Mariner*-like transposase proteins bind to the TIRs and not the TSDs [25, 26]. Therefore, the TSDs do not likely play a role in binding, but instead play a role in the catalytic mechanism that cleaves the element from the genome.

To determine what role the TSDs play in *mPing* transposition, we performed yeast assays with *mPing* elements with altered TSDs. These experiments indicate that

alteration of *mPing*'s TSDs also inhibits its transposition (Fig. 3b, Additional file 2a). Based on insertion site analysis, it was already known that T or A was acceptable at the middle position of the TSD [7]. Changing the middle base to C or G (i.e. from TAA/TAA (5′/3′) to TCA/TCA) had a small effect with TGA/TGA TSDs producing more colonies than TCA/TCA TSDs (Additional file 2a). Changing the first base (i.e. GAA/GAA) or third base (i.e. TAC/TAC) caused a more severe drop in the number of *ADE2* revertants. Changing all three bases completely disrupted the transposition of the element (Fig. 3b, Additional file 3a). To determine if this decrease in *ADE2* revertants was caused by a drop in excision or from a decreased rate of repair, a subset of these altered elements were tested in the DG21B9 strain (HR only). If altering the TSDs to this extent only affects repair of the excision site and not excision itself, all of these altered TSDs would have the same *ADE2* revertant rate as the control in DG21B9. However, almost no *ADE2* revertant colonies were detected in the TAC/TAC or GCC/GCC TSD (5′/3′) combinations (Fig. 3b), indicating that these base changes inhibit the ability of the transposase proteins to catalyze excision. It is not clear if this is due to altered enzyme binding or if these bases are directly involved in the catalytic mechanism.

In addition to reducing the number of *ADE2* revertant colonies by decreasing excision, sequencing the excision sites indicated that altering the TSDs can result in imprecise repair (Additional file 3b). The production of footprints was especially pronounced for the TAC/TAC TSDs, with 10 of 16 excision sites having indels. The inefficient excision of these altered elements may have resulted in strand cleavage in a non-standard position, creating double stranded breaks that were not as easily repaired.

mPing excision site repair is facilitated by TSD homology

Based on these initial experiments, we hypothesized that a difference in the double stranded breaks created by the *mPing* and *14T32* elements results in their excision site differences. Analysis of repaired excision sites shows that *Mariner*-like transposase proteins produce staggered DNA cleavage within the element, leaving behind some of the TIR sequences (Fig. 1b, Fig. 4) [15]. In contrast, our model for *PIF/Pong/Harbinger* transposition is that they are mobilized by staggered cleavage of the TSDs, producing three bases of microhomology that facilitates NHEJ (Fig. 4). Based on this, we predicted that changing the TSDs in such a way as to disrupt the microhomology would affect the quality and efficiency of *mPing*'s excision site repair.

The fact that multiple bases are equally acceptable in the middle position of *mPing*'s TSDs allowed experiments to determine if homology between the two TSDs facilitates repair of *mPing* excision sites. Yeast transposition assays comparing *mPing* constructs with matching

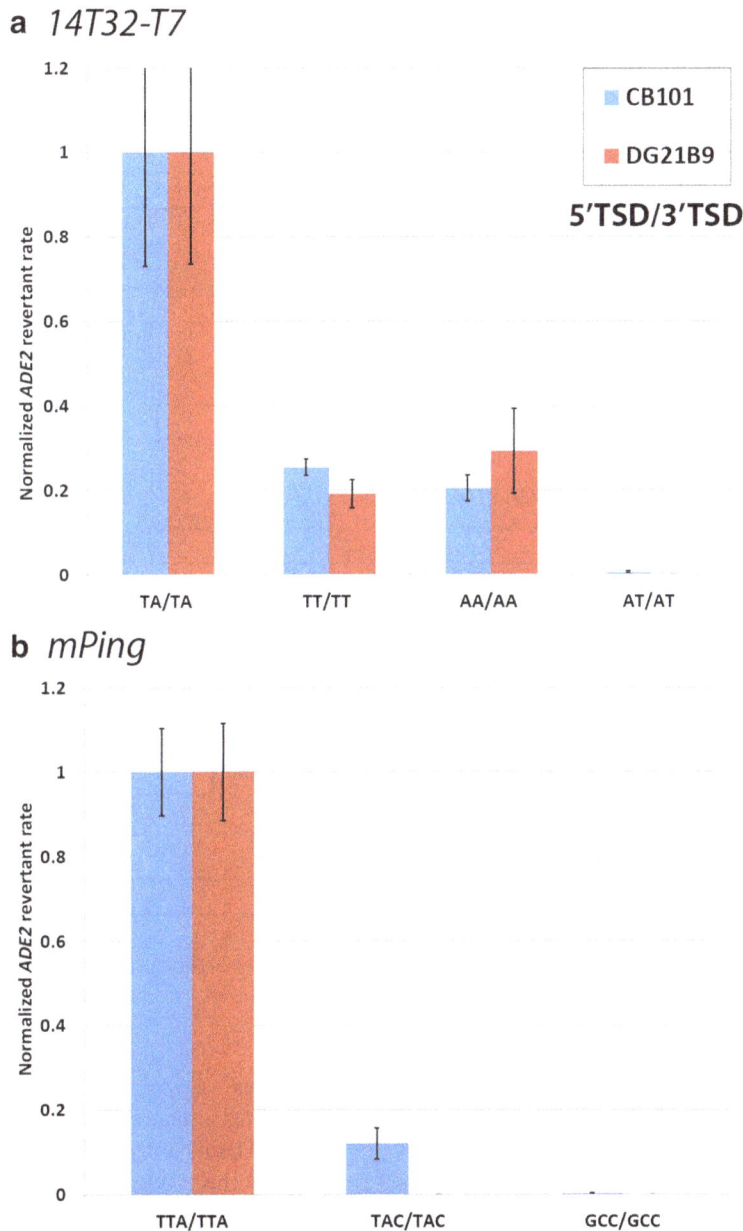

Fig. 3 Transposition assays with altered but matching TSDs. *ADE2* revertant rates for *14T32-T7* (**a**) and *mPing* (**b**) elements with altered but matching TSDs. Blue bars indicate the rate in CB101 (capable of both NHEJ and HR), while red bars indicate the rate in DG21B9 (only capable of HR). Values were normalized to the control TSDs (TA/TA for *14T32-T7* and TAA/TAA for *mPing*) for each yeast strain separately. Error bars represent standard error

TSDs (TTA/TTA and TAA/TAA) and non-matching TSDs (TTA/TAA and TAA/TTA) were performed (Fig. 5, Additional file 3). As shown in CB101 (Fig. 5b) or JIM17 (Additional file 3) yeast strains, the *mPing* elements with non-matching TSDs showed significantly lower transposition than those with matching TSDs. Performing this assay in the DG21B9 strain, which is only capable of repair by HR did not show this effect, with all TSD combinations showing a similar number of *ADE2* revertant colonies (Fig. 5b). Together these results indicate that

the reduction in *ADE2* revertant colonies for non-matching TSDs is caused by reduced or inaccurate NHEJ repair efficiency. For comparison, similar experiments using the *14T32-T7* element showed that non-matching TSDs produced a similar effect in both NHEJ competent (CB101) and NHEJ deficient (DG21B9) strains (Fig. 5a). This indicates that changing the TSDs of *14T32-T7* only affected its excision and not the repair of the excision site.

Analysis of the excision sites from non-matching TSDs by restriction digest and sequencing was performed to

a *mPing* (*Tourist*-like MITE)

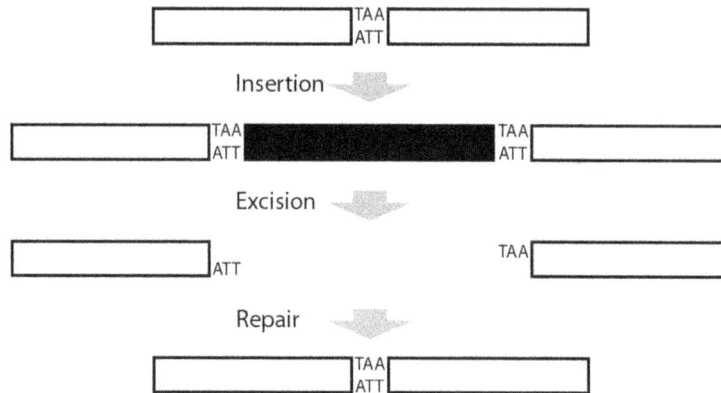

b *14T32* (*Stowaway*-like MITE)

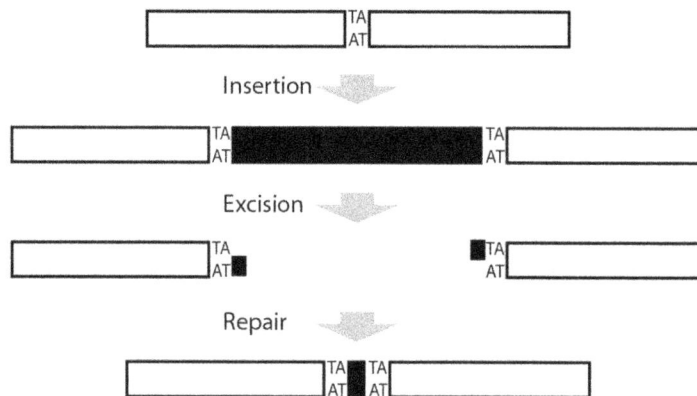

Fig. 4 Model of *Tourist*-like and *Stowaway*-like MITE transposition. *mPing* (**a**) and *14T32-T7* (**b**) elements are represented by black boxes, with the TSDs (3 bp and 2 bp respectively) created upon insertion shown as letters. Excision of the *mPing* element produces TSD derived 5′ overhangs that result in precise repair, while *14T32* excision leaves element derived overhangs that results in footprint production

determine how these sites were repaired. Figure 6 shows that in JIM17 and CB101 the excision site produced by an element with TAA/TTA TSDs was repaired fairly precisely with most excision sites only showing one of the TSDs. However, two of the excision sites retained both TSDs, consistent with staggered cleavage at the TSDs that was repaired by NHEJ without microhomology. In contrast, we found that repair of the excision site produced by an element with TTA/TAA TSDs was repaired less precisely in JIM17, or exclusively by HR repair using the *ADE2** template in CB101 (Fig. 6). This result suggests that the staggered ends created by the TTA/TAA combination were not as easily joined by NHEJ pathway as the TAA/TTA combination. Since a 5′ overhang would create a different set of mismatched bases than a 3′ overhang at the excision site (Table 1), we compared our results to the expected base pairing for each non-matching TSD. Based on this result, we propose that *mPing*'s TSDs cleavage produces a 5′ overhang (Fig. 4, Table 1). A three base 5′ overhang would result in the TAA/TTA TSDs forming a

T:T (pyrimidine:pyrimidine) pairing at the middle base of the overhang, a more compatible pairing than the A:A (purine:purine) base paring created by the TTA/TAA TSDs.

Based on this model, we should see that some TSD combinations are more detrimental to excision site repair than others. In fact, analysis of additional combinations of *mPing* TSDs (TCA and TGA) showed that non-matching TSDs, that according to our model would result in T:C (pyrimidine:pyrimidine) or A:G (purine:purine) mismatches, produced fewer *ADE2* revertants than TSD combinations that produce C:A (pyrimidine:purine) and T:G mismatches (pyrimidine:purine) (Additional file 4). Sequence analysis of the excision sites produced by selected TCA and TGA mismatched TSDs (Supplemental 4c) indicates that, for the most part, only one of the TSD sequences is left behind, as is expected of precise repair. However, about 14 % of the time both of the TSDs remained, leaving a footprint. This is in stark contrast to *mPing* elements with matching TSDs, which have never been observed to leave behind both TSDs upon excision (Fig. 2b) [20, 27].

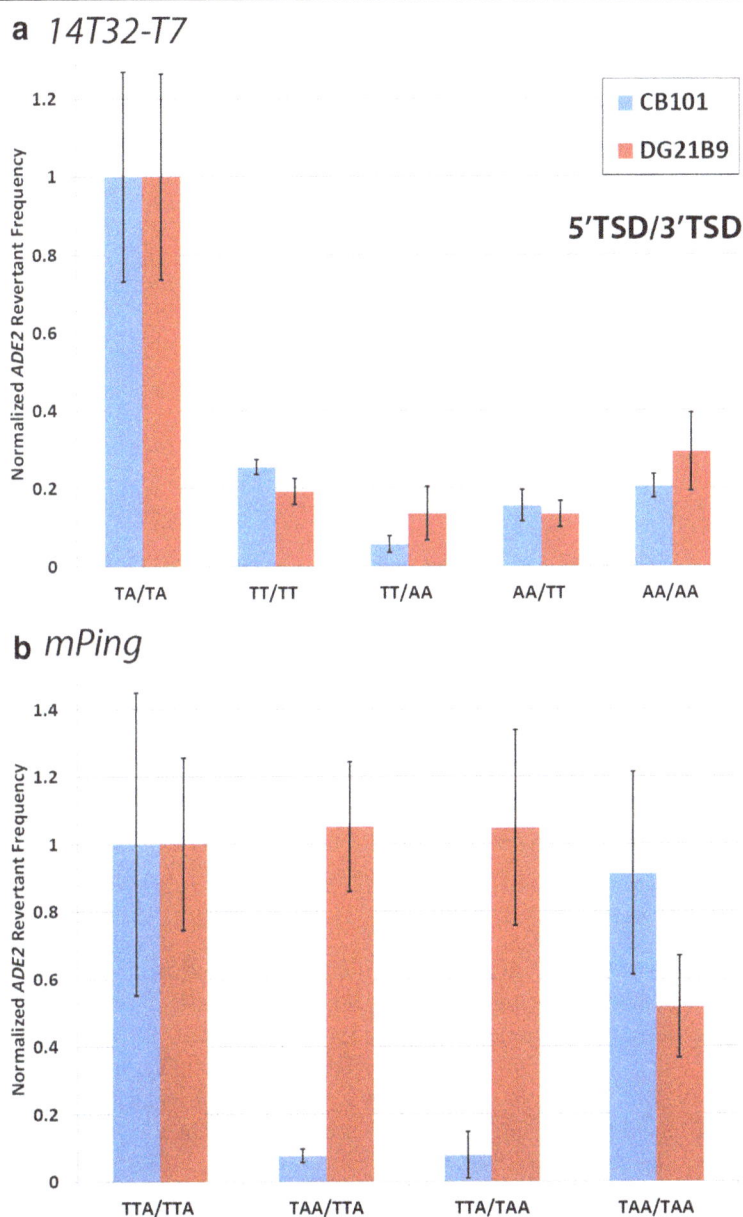

Fig. 5 Transposition assays with non-matching TSDs. Normalized *ADE2* revertant frequencies for *14T32-T7* (**a**) and *mPing* (**b**) elements with altered TSDs. Blue bars indicate the rate in CB101 (capable of both NHEJ and HR), while red bars indicate the rate in DG21B9 (only capable of HR). Values were normalized to the wild-type TSD (left column). Error bars indicate the standard error

It is not clear how common the excision site creation and repair mechanisms observed for *mPing* are present in other transposon superfamilies. Interestingly, alteration of the *P-element* TSDs from *Drosophila* showed a reduction in transposition activity [28]. Also, a recent study with the *Os3378* element (*Mutator* superfamily from rice) that also excises precisely, indicated that alteration of its TSDs reduces the rate of precise excision in yeast [29]. Analysis of these elements in the CB101 and DG21B9 yeast strains would be able to determine if this is due to disruption of excision or excision site repair.

mPing TSDs do not influence target site insertion

Previous research has shown that *mPing* exhibits a strong preference for insertion into TAA or TTA sequences in the genome [20, 27, 30]. This is consistent with the findings of this study indicating that these sequences are required for efficient excision of the element. However, it was not known if the TSD sequences might play a role in the insertion preference of the element. To address this, 46 insertions of an *mPing* element with TCA/TCA TSDs were analyzed by sequencing transposon display PCR products [31]. We observed that 45 of the insertions were in TTA or TAA, and only

```
                    ADE2   AAACACTAAACCGT      TTA      CAGACCTCACAATCA
                    ADE2*  AAACACTAGGCCAT      TCA      CAGACCTCACAATCA

         5'TSD/3'TSD                                                      #/total

         JIM17
         TAA/TAA    AAACACTAAACCGT      TAA      CAGACCTCACAATCA          13/13

         TAA/TTA    AAACACTAAACCGT      TTA      CAGATCTCACAATCA          7/8
                    AAACACTAAACCGT      TAA      CAGACCTCACAATCA          1/8

         TTA/TAA    AAACACT gctaattctttag TAA  CAGACCTCACAATCA          4/10
                    AAACACTAAACCGT      TTA      CAGACCTCACAATCA          2/10
                    AAACACTAAACCGT      TAA      CAGACCTCACAATCA          1/10
                    AAACACTAAACCGT      TTATAA   CAGACCTCACAATCA          1/10
                    AAACA tgactgctaattctttagTAACAGACCTCACAATCA           1/10
                    AAACACTAAACCGTTA aggccagtcaCAGACCTCACAATCA           1/10

         CB101
         TAA/TAA    AAACACTAAACCGT      TAA      CAGACCTCACAATCA          96/96

         TAA/TTA    AAACACTAAACCGT      TAA      CAGATCTCACAATCA          9/20
                    AAACACTAAACCGT      TTA      CAGACCTCACAATCA          5/20
                    AAACACTAAACGGT      TTA      CAGACCTCACAATCA          2/20
                    AAACACTAAACCGT      TAATTA   CAGACCTCACAATCA          4/20

         TTA /TAA   AAACACTAGGCCAT      TCA      CAGACCTCACAATCA          14/20*
                    AAACACTAAACCAT      TCA      CAGATCTCACAATCA          6/20*
```

Fig. 6 *mPing* excision sites for non-matching TIRs. Sequences identified at the *mPing* excision sites by restriction site analysis and sequencing in JIM17 (NHEJ only) and CB101 (HR and NHEJ). Lowercase letters indicate inserted sequences and a base change is in red. * indicates that the site was repaired by HR using the *ADE2** template

one was in TCA. This is consistent with the results observed for the wild type *mPing* element [20], suggesting that the TSDs do not play a large role in target site selection.

Conclusions

These results demonstrate a key difference in the transposition mechanisms used by the *Tourist*-like and *Stowaway*-like MITEs. While the excision sites of both *mPing* and *14T32* elements are primarily repaired by the NHEJ pathway in yeast, the *14T32* element appears to be more sensitive to alteration of NHEJ pathway genes. Our study suggests that the TSDs flanking both elements are required for their efficient excision. On the other hand, complementarity of the two TSDs was found only to be

Table 1 Base pairing that results after 5' or 3' staggered cleavage of the *mPing* TSDs

mPing target site duplications	Proposed middle base pairing	
	5' overhang	3' overhang
TTA/TTA	A:T	T:A
TTA/TAA	A:A	T:T
TAA/TTA	T:T	A:A
TAA/TAA	T:A	A:T

critical to the efficiency and precision of *mPing's* excision site repair. Based on this finding, we conclude that the transposases that excise *mPing*, and presumably other *Tourist*-like MITEs, produce a staggered cut at the TSDs that provides microhomology that facilitates precise repair of the excision site.

Methods
Yeast strains and vectors

Strain name	Genotype
JIM17	MATa ade2Δ::hphMX4 his3Δ1 leu2Δ0 met15Δ0 ura3Δ0
CB101	MATa ade2Δ::hphMX4 his3Δ1 leu2Δ0 met15Δ0 ura3Δ0 lys2Δ::ADE2*
JIM16	MATa rad50Δ::kanMX4 ade2Δ::hphMX4 his3Δ1 leu2Δ0 met15Δ0 ura3Δ0
JIM22	MATa mre11Δ::kanMX4 ade2Δ::hphMX4 his3Δ1 leu2Δ0 met15Δ0 ura3Δ0
JIM21	MATa ku70Δ::kanMX4 ade2Δ::hphMX4 his3Δ1 leu2Δ0 met15Δ0 ura3Δ0
DG21B9	MATa ku70Δ::kanMX4 ade2Δ::hphMX4 his3Δ1 leu2Δ0 met15Δ0 ura3Δ0 lys2Δ::ADE2*

Saccharomyces cerevisiae strains, BY4741 (JIM17) or Yeast Deletion Project strains [32, 33] in the BY4741 background (JIM16, JIM22, JIM21), were adapted for the study by deleting the *ADE2* gene using the *hphMX4 (pAG32)* cassette replacement technique [34] using the following primers: *ADE2hphMX* For-CAATCAAGAAAA ACAAGAAAATCGGACAAAACAATCAAGTCCTTGA CAGTCTTGACGTGC, *ADE2hphMX* Rev-ATAATTATT TGCTGTACAAGTATATCAATAAACTTATATACGCAC TTAACTTCGCATCTG.

The partial *ADE2* template (*ADE2**) was synthesized with the following sequence 5′-TTTGGCATACGATGG AAGAGGTAACTTCGTTGTAAAGAATAAGGAAATG ATTCCGGAAGCTTTGGAAGTACTGAAGGATCGTC CTTTGTACGCCGAAAAATGGGCACCATTTACTAA AGAATTAGCAGTCATGATTGTGAGATCTGTGAAT GGCCTAGTGTTTTCTTACCCAATTGTAGAGACTA TCCACAAGGACAATATTTGTGACTTATGTTATGC GCCTGCTAGAGTTCCGGACTCCGTTCAACTTAAG GCGAAGTTGTTGGCAGAAAATGCAATCAAATCTTT T-3′ and cloned between the *Bgl*II and *Hind*III sites of the pIS 385 disintegrator plasmid [35]. To make the CB101 and DG21B9 yeast strains, this plasmid was then linearized with *Nru*I (New England Biolabs, Massachusetts, USA) and transformed into the *LYS2* locus of JIM17 and JIM21, respectively. Selection and screening were performed as described [35] to remove the *URA3* selectable marker and identify transformants that maintained the genomic copy of the *ADE2** template.

The pAG413 *Pong* ORF1, pAG415 *Pong* transposase L418A, L420A and pWL89A *mPing* plasmids were described previously [20]. The pAG415 *Osmar14* transposase was made by PCR amplification of the open reading frame from a previously described *Osmar14* transposase plasmid [21] with the following primers *Osmar 14* For – GGGGACAAGTTTGTACAAAAAAGCAGGCTTCATG CAAGAGTACGGCGTGTATGC, *Osmar 14* Rev- GGGG ACCACTTTGTACAAGAAAGCTGGGTCTTAAACT GCACTTGGTTGGCTAATGCT. The PCR product was inserted into the Gateway® pDONR™/Zeo vector using a BP clonase reaction (Life Technologies, Carlsbad, CA), then transferred into pAG415 GAL ccdb using an LR clonase Reaction (Life Technologies, Carlsbad, CA). The reporter plasmids pwL89A *14T32* and *14T32-T7* were described previously [14, 21]. TSD mutations were made to the MITEs *mPing* and *Osmar 14T32-T7* by PCR amplification using primers altered at the TSD (underlined positions indicate TSD), for example:

mPing TGA For – AGTCTCTACAATTGGGTAAGA AAACACTAAACCGT**TGA**GGCCAGTCACAATGGGG GTTTC

mPing TGA Rev – ACTAAAGAATTAGCAGTCATG ATTGTGAGGTCTGT**TCA**GGCCAGTCACAATGGCTA GTGTC

14T32 AT For –CTAAAGAATTAGCAGTCATGATT GTGAGGTCTGTT**AT**CTCCCTCCGTCCCAGAAAGAA GG, and

14T32 AT Rev – GTCTCTACAATTGGGTAAGAAAA CACTAAACCGTT**AT**CTCCCTCCGTCCCAGAAAGAAGC

The resulting PCR products were purified using a clean and concentrate kit (Zymo Research, Irvine, CA) and then transformed together with *Hpa*I digested pWL89A using the LiAc method [36]. Mutations were verified by sequencing PCR products or purified plasmids with following primers that flank the *ADE2 Hpa*I site: *ADE2*-CF-GG GTTTTCCATTCGTCTTGAAGTCGAGGAC and *ADE2*-CR-CATTTCCACACCAAATATACCACAACCGGGA.

Yeast transposition assay

Transposition assays were performed using two techniques depending on the relative transposition rates. For low activity combinations (i.e. Figs. 1, 3 and 4b, and Additional files 3 and 4) transformed yeast were grown in 5 ml of selective media (2 % dextrose) at 30 °C for 48 h, centrifuged to concentrate the culture, plated on selective 2 % galactose plates (150 mm) lacking adenine, and incubated at 30 °C for 15 days as described [20]. For experiments with higher rates of transposition (i.e. Fig. 2, 5a and Additional file 1), a 3 ml liquid (2 % dextrose) culture was grown for 24 h at 30 °C and 100 μl was plated on selective 2 % galactose plates (100 mm) and incubated at 30 °C for 10 days. A time course of this procedure showed that the number of *ADE2* revertant colonies had a linear rate of appearance (Additional file 1). Dilution series of the liquid cultures plated on complete YPD media were used to determine the total number of cells plated. Transposition rate was calculated by dividing the number of *ADE2* revertant colonies by the total number of yeast plated.

Excision site analysis

ADE2 revertant colonies were suspended in 20 μl of 1 unit/μl Zymolyase (Zymo Research, Irvine, CA) and incubated for 15 min at 37 °C to lyse the yeast cells. PCR amplification of the excision site was performed using the ADE2-CF and ADE2-CR primers in a 20 μl reaction with 2 μl of lysed yeast as the template. PCR products were diluted and digested with *Hpa*I or *Hae*III (New England Biolabs, Massachusetts, USA) and then analyzed by agarose gel electrophoresis. PCR products were treated with ExoSAP-IT (USB Corporation, Ohio, USA) per instruction of the manufacturer prior to sequencing.

Insertion site analysis

Transposon display analysis of *mPing* insertion sites were performed as described previously [20, 30, 31]. Individual bands were sequenced after cutting them from the gel and performing PCR amplification with the transposon display primers.

Additional files

Additional file 1: Time course of a yeast transposition assay. *ADE2* revertant frequencies for *mPing* in the JIM17 strain of yeast. (PDF 305 kb)

Additional file 2: Additional altered matching TSDs. Additional examples of the of *ADE2* revertant frequency for *mPing* elements with alternative but matching TSDs and their associated excision site sequences. (PDF 360 kb)

Additional file 3: Non-matching *mPing* TIRs in JIM17. Chart comparing the frequency of *ADE2* revertant colonies produced for *mPing* elements with matching (i.e. TAA/TAA) and non-matching TSD sequences (i.e. TAA/TTA) in the JIM17 strain. (PDF 403 kb)

Additional file 4: Additional non-matching *mPing* TIRs. Charts comparing the frequency of *ADE2* revertant colonies produced for *mPing* elements with additional combinations of matching (i.e. TCA/TCA) and non-matching TSD sequences (i.e. TAA/TCA) in the JIM17 strain. (PDF 593 kb)

Abbreviations
TE: Transposable element; MITE: Miniature inverted repeat transposable element; TSD: Target site duplication; TIR: Terminal inverted repeat; NHEJ: Non-homologous end joining; HR: Homologous recombination.

Competing interests
The authors declare that they have no competing interests.

Authors' contributions
CNH and JMB developed the experimental design, trained the undergraduates, and drafted the manuscript. DMG, MCB, AES, CEB helped draft and edit the manuscript. Yeast strains were produced by CNH, JMB, MCB, CEB, and DMG. Plasmids were constructed by DMG, CNH, and MCB. Yeast assays and excision site analysis were performed by MCB, AES, and DMG. All authors read and approved the final manuscript.

Authors' information
The research described in this manuscript was performed almost exclusively as undergraduate research projects. The majority of the work was performed at the University of South Carolina Aiken, a primarily undergraduate institution.

Acknowledgements
We thank Dr. Guojun Yang for providing the *14T32* and *14T32-T7* constructs. We also thank Dr. Clifford Weil for advice and Christie Bradshaw for technical assistance. We also thank the many undergraduates including Tyler Shealy, Wesley Tindal, Lee B. Sharpe, Lucy Fu, and the spring 2009 HHMI Dynamic Genome class (Justin Brown, Charles B. Allen Jr., Krelin Naidu, Ashley Turner) that helped with aspects of the project. Portions of this research were funded by a grant from the Howard Hughes Medical Institute to Susan R. Wessler. David Gilbert was funded by USCA Connections and USC Magellan grant.

Author details
[1]Department of Biology and Geology, University of South Carolina Aiken, 471 University Parkway, Aiken, SC 29801, USA. [2]Present Address: Department of Pathology and Laboratory Medicine, Medical University of South Carolina, Charleston, SC 29425, USA. [3]Present Address: College of Natural and Agricultural Sciences, University of California Riverside, Riverside, CA 92521, USA.

References
1. Feschotte C, Jiang N, Wessler SR. Plant transposable elements: Where genetics meets genomics. Nat Rev Genet. 2002;3(5):329–41.
2. Wicker T, Sabot F, Hua-Van A, Bennetzen JL, Capy P, Chalhoub B, et al. A unified classification system for eukaryotic transposable elements. Nat Rev Genet. 2007;8(12):973–82.
3. Wessler SR, Bureau TE, White SE. LTR-retrotransposons and MITEs—Important players in the evolution of plant genomes. Curr Opin Genet Dev. 1995;5(6):814–21.
4. Casacuberta J, Santiago N. Plant LTR-retrotransposons and MITEs: control of transposition and impact on the evolution of plant genes and genomes. Gene. 2003;311:1–11.
5. Casa AM, Brouwer C, Nagel A, Wang LJ, Zhang Q, Kresovich S, et al. The MITE family *Heartbreaker* (*Hbr*): molecular markers in maize. Proc Natl Acad Sci U S A. 2000;97(18):10083–9.
6. Feschotte C, Zhang XY, Wessler SR. Miniature Inverted-repeat Transposable Elements (MITEs) and their relationship with established DNA transposons. In: Craig NL, Craige R, Gellert M, Lambowitz A, editors. Mobile DNA II. Washington D.C.: American Society of Microbiology Press; 2002.
7. Naito K, Cho E, Yang GJ, Campbell MA, Yano K, Okumoto Y, et al. Dramatic amplification of a rice transposable element during recent domestication. Proc Natl Acad Sci U S A. 2006;103(47):17620–5.
8. Plasterk R, Izsvak Z, Ivics Z. Resident aliens—the *Tc1/Mariner* superfamily of transposable elements. Trends Genet. 1999;15(8):326–32.
9. Zhang XY, Jiang N, Feschotte C, Wessler SR. *PIF*- and *Pong*-like transposable elements: distribution, evolution and relationship with *Tourist*-like miniature inverted-repeat transposable elements. Genetics. 2004;166(2):971–86.
10. Yuan YW, Wessler SR. The catalytic domain of all eukaryotic cut-and-paste transposase superfamilies. Proc Natl Acad Sci U S A. 2011;108(19):7884–9.
11. Zhang XY, Feschotte C, Zhang Q, Jiang N, Eggleston WB, Wessler SR. *P instability factor*: an active maize transposon system associated with the amplification of *Tourist*-like MITEs and a new superfamily of transposases. Proc Natl Acad Sci U S A. 2001;98(22):12572–7.
12. Yu JH, Marshall K, Yamaguchi M, Haber JE, Weil CF. Microhomology-dependent end joining and repair of transposon-induced DNA hairpins by host factors in *Saccharomyces cerevisiae*. Mol Cell Biol. 2004;24(3):1351–64.
13. Rinehart TA, Dean C, Weil CF. Comparative analysis of non-random DNA repair following *Ac* transposon excision in maize and Arabidopsis. Plant J. 1997;12(6):1419–27.
14. Weil CF, Kunze R. Transposition of maize *Ac/Ds* transposable elements in the yeast *Saccharomyces cerevisiae*. Nat Genet. 2000;26(2):187–90.
15. Yang GJ, Weil CF, Wessler SR. A rice *TC1/Mariner*-like element transposes in yeast. Plant Cell. 2006;18(10):2469–78.
16. Aylon Y, Kupiec M. DSB repair: the yeast paradigm. DNA Repair. 2004;3(8–9):797–815.
17. Wessler SR. Phenotypic diversity mediated by the maize transposable elements *Ac* and *Spm*. Science. 1988;242(4877):399–405.
18. Scott L, LaFoe D, Weil CF. Adjacent sequences influence DNA repair accompanying transposon excision in maize. Genetics. 1996;142(1):237–46.
19. Doseff A, Martienssen R, Sundaresan V. Somatic excision of the *Mu1* transposable element of maize. Nucleic Acids Res. 1991;19(3):579–84.
20. Hancock C, Zhang F, Wessler S. Transposition of the *Tourist*-MITE *mPing* in yeast: an assay that retains key features of catalysis by the Class 2 *PIF/Harbinger* superfamily. Mobile DNA. 2010;1(5):5.
21. Yang GJ, Nagel DH, Feschotte C, Hancock CN, Wessler SR. Tuned for transposition: molecular determinants underlying the hyperactivity of a *Stowaway* MITE. Science. 2009;325(5946):1391–4.
22. Daley JM, Palmbos PL, Wu D, Wilson TE. Nonhomologous end joining in yeast. Annu Rev Genet. 2005;39:431–51.
23. Tsukamoto Y, Kato J, Ikeda H. Budding yeast *Rad50*, *Mre11*, *Xrs2*, and *Hdf1*, but not *Rad52*, are involved in the formation of deletions on a dicentric plasmid. Mol Gen Genet. 1997;255(5):543–7.
24. Vos JC, DeBaere I, Plasterk RHA. Transposase is the only nematode protein required for in vitro transposition of *Tc1*. Genes Dev. 1996;10(6):755–61.
25. Zhang L, Dawson A, Finnegan DJ. DNA-binding activity and subunit interaction of the *Mariner* transposase. Nucleic Acids Res. 2001;29(17):3566–75.
26. Auge-Gouillou C, Hamelin MH, Demattei MV, Periquet G, Bigot Y. The ITR binding domain of the mariner *Mos-1* transposase. Mol Genet Genomics. 2001;265(1):58–65.
27. Yang GJ, Zhang F, Hancock CN, Wessler SR. Transposition of the rice miniature inverted repeat transposable element *mPing* in *Arabidopsis thaliana*. Proc Natl Acad Sci U S A. 2007;104(26):10962–7.
28. Mullins MC, Rio DC, Rubin GM. *cis*-acting DNA-sequence requirements for *P-element* transposition. Genes Dev. 1989;3(5):729–38.
29. Zhao D, Ferguson A, Jiang N. Transposition of a rice *Mutator*-like element in the yeast *Saccharomyces cerevisiae*. Plant Cell. 2015;27(1):132–48.
30. Hancock C, Zhang F, Floyd K, Richardson A, LaFayette P, Tucker D, et al. The rice miniature inverted repeat transposable element *mPing* is an effective insertional mutagen in soybean. Plant Physiol. 2011;157(2):552–62.

31. Jiang N, Bao ZR, Zhang XY, Hirochika H, Eddy SR, McCouch SR, et al. An active DNA transposon family in rice. Nature. 2003;421(6919):163–7.

32. Winzeler EA, Shoemaker DD, Astromoff A, Liang H, Anderson K, Andre B, et al. Functional characterization of the *S. cerevisiae* genome by gene deletion and parallel analysis. Science. 1999;285(5429):901–6.

33. Giaever G, Shoemaker DD, Jones TW, Liang H, Winzeler EA, Astromoff A, et al. Genomic profiling of drug sensitivities via induced haploinsufficiency. Nat Genet. 1999;21(3):278–83.

34. Goldstein AL, McCusker JH. Three new dominant drug resistance cassettes for gene disruption in *Saccharomyces cerevisiae*. Yeast. 1999;15(14):1541–53.

35. Sadowski I, Su TC, Parent J. Disintegrator vectors for single-copy yeast chromosomal integration. Yeast. 2007;24(5):447–55.

36. Gietz RD, Woods RA. Transformation of yeast by lithium acetate/single-stranded carrier DNA/polyethylene glycol method. Method Enzymol. 2002;350:87–96.

5

Identification and characterization of a minisatellite contained within a novel miniature inverted-repeat transposable element (MITE) of *Porphyromonas gingivalis*

Brian A. Klein[1,2], Tsute Chen[2], Jodie C. Scott[2], Andrea L. Koenigsberg[1], Margaret J. Duncan[2] and Linden T. Hu[1]*

Abstract

Background: Repetitive regions of DNA and transposable elements have been found to constitute large percentages of eukaryotic and prokaryotic genomes. Such elements are known to be involved in transcriptional regulation, host-pathogen interactions and genome evolution.

Results: We identified a minisatellite contained within a miniature inverted-repeat transposable element (MITE) in *Porphyromonas gingivalis*. The *P. gingivalis* minisatellite and associated MITE, named 'BrickBuilt', comprises a tandemly repeating twenty-three nucleotide DNA sequence lacking spacer regions between repeats, and with flanking 'leader' and 'tail' subunits that include small inverted-repeat ends. Forms of the BrickBuilt MITE are found 19 times in the genome of *P. gingivalis* strain ATCC 33277, and also multiple times within the strains W83, TDC60, HG66 and JCVI SC001. BrickBuilt is always located intergenically ranging between 49 and 591 nucleotides from the nearest upstream and downstream coding sequences. Segments of BrickBuilt contain promoter elements with bidirectional transcription capabilities.

Conclusions: We performed a bioinformatic analysis of BrickBuilt utilizing existing whole genome sequencing, microarray and RNAseq data, as well as performing *in vitro* promoter probe assays to determine potential roles, mechanisms and regulation of the expression of these elements and their affect on surrounding loci. The multiplicity, localization and limited host range nature of MITEs and MITE-like elements in *P. gingivalis* suggest that these elements may play an important role in facilitating genome evolution as well as modulating the transcriptional regulatory system.

Keywords: Species-specific repeat, DNA structure, Miniature Inverted-repeat Transposable Element, BrickBuilt, Transcriptional regulation, *Porphyromonas*

Background

Porphyromonas gingivalis, a gram-negative, anaerobic, asaccharolytic, black-pigmenting bacterium, is a keystone pathogen in the development and progression of periodontal disease [1, 2]. Multiple repetitive and transposable elements were previously identified in the *P. gingivalis* genomes [3–12]. Genome sequences are now available for multiple strains of *P. gingivalis* which has greatly facilitated

genetic and genomic analyses of the species [9–16]. Each of the sequenced *P. gingivalis* genomes has contained multiple repetitive and transposable elements, an aspect that makes sequencing and alignment difficult.

Repetitive Elements (REs) are DNA sequences present in multiple copies throughout a genome, chromosome or vector. They are broadly classified into 'terminal', 'tandem' and 'interspersed' repeats, however, each of these classifications encompasses several sub-types of REs. Tandem repeats are classified as either identical or non-identical based on the level of nucleic acid matching. They are then further classified as either micro, mini or macro

* Correspondence: linden.hu@tufts.edu
[1]Department of Molecular Biology and Microbiology, Tufts University Sackler School of Biomedical Sciences, Boston, MA 02111, USA
Full list of author information is available at the end of the article

satellites based on size of the repeat. Repetitive elements can either be localized at a single site where a motif is recurrent sequentially adjacent to each other or at many loci as reiteration [17–19].

Transposable Elements (TEs) are 'mobile' DNA sequences that can change locus or multiply and insert into new loci within a genome or between genomes via excision/replication and insertion. They can insert into chromosomes, plasmids and bacteriophages. Class I TEs are retrotransposons, which require reverse-transcriptase activity to transpose. Class II TEs are DNA transposons, which unlike reverse transcriptase-utilizing Class I elements, require a transposase or a replicase to transpose [19–21]. Class II elements can either be autonomous or non-autonomous, the latter [canonically] having undergone mutations involving the transposase such that they can no longer duplicate or excise without the assistance of a parent element that utilizes a similar transposase. Within the non-autonomous element sub-class are miniature inverted-repeat transposable elements, or MITEs [22–25].

MITEs have a distinct structure relative to other TEs. They are between 50–1000 bp in length and are often present in high copy numbers per genome. MITEs are typically AT-nucleotide (nt) rich and frequently contain terminal inverted repeats (TIRs) and target site duplications (TSDs), but they lack the capacity to code for functional transposases [22–25]. Transposable elements, in particular MITEs, can be found in all taxa, varying in number and type between species and can account for greater than half of a genome. Bacteria typically carry between 10–20 copies of a MITE per genome, while plants may have up to 20,000 copies of a given MITE. Copy numbers are suggested to depend on non-coding region availability, polyploidy, the presence of a fully-functional autonomous version of a transposase, evolutionary 'burst' opportunities and regulatory potential of the given element [26–29]. Eukaryotic MITEs are frequently found in or closely associated with the coding region while prokaryotic MITEs are almost exclusively found intergenically [26, 30–36]. Intergenically located MITEs in prokaryotes have been shown to be able to affect gene expression [23, 25].

Several studies have demonstrated potential interactions of repetitive elements with transposable elements, which are generally thought to work independently and be mutually exclusive. In the wedge clam (*Donax trunculu*) genome as well as the butterfly and moth (*Lepidoptera*) genomes, 'hitchhiking' microsatellites were found within transposable elements [37, 38]. Microsatellites and simple sequence repeats have also been found closely associated with transposable elements in *Neisseria meningitidis* [39].

Here we describe 'BrickBuilt', a miniature inverted-repeat transposable element containing a minisatellite, in *P. gingivalis*. The sequences, location, copy number, prevalence throughout the species, as well as implications on genome (in)stability and transcriptional regulation are described. Similarities to other autonomous and non-autonomous *P. gingivalis* transposable elements are addressed with the goal of defining a potential network for the biogenesis of these elements in *P. gingivalis* and their effects on the *P. gingivalis* genome.

Results and discussion
Identification of a repetitive element in *Porphyromonas gingivalis*

We identified a DNA element, 'BrickBuilt', in the genome of *P. gingivalis* strain ATCC 33277. The element was initially identified as a tandemly-repeated sequence of 23 nt located intergenically at a single site (Additional file 1: Figure S1). A more thorough investigation of the genome revealed 19 independent, non-identical segments of the element scattered throughout the genome of strain ATCC 33277 (Table 1). The smallest number of 23 nt direct repeats is 1 (BrickBuilt_1) and the largest 22.8 (BrickBuilt_12). The 23 nt direct repeats are imperfect within a given element, imperfect bases vary from one element to another and imperfections do not correlate with length or total number of repeats within a given element (Fig. 1). The percent of mismatches within a given element varies from 0 to 11, and the percent of insertions and deletions within an element varies from 0 to 6. Within the 23 nt repeats there are conserved and non-conserved nucleotide sites, with the latter half of the element containing the majority of non-conserved sites (Fig. 1). Although similar in length to CRISPR element spacers and microRNAs, BrickBuilt elements are seemingly unrelated to these other entities.

After determining the length and locations of each independent direct repeat element we performed alignments of the sequences flanking the direct repeats to determine whether specific DNA sequences or motifs were necessary for the presence of the element. Alignments of the sequences flanking the direct repeats revealed regions of homology, different for the two flanks of the repeat, which were determined to be 'leader' and 'tail' regions that encompassed the direct repeats (Fig. 2). Of the 19 elements, 11 are flanked by portions of both a leader and a tail, 3 by just leader, 2 by just tail, and 3 by neither. When considered as a single whole element, all BrickBuilt elements are intergenic, although some are within regions where annotation pipelines predicted hypothetical genes that do not appear to be expressed based on proteomic data [40–42]. Total length of the complete elements ranges from 991 nt (BrickBuilt_5) to 84 nt (BrickBuilt_14), which is determined by number of internal 23 nt repeats as well as the specific element may contain full, partial or no leader and tail segments. The longest leader segment is 285 nucleotides (BrickBuilt_17) and the longest tail

Table 1 Genes/Coding Sequences located 5' and 3' to BrickBuilt elements. Gene numbers and characterizations correspond to strain ATCC 33277. Loci of BrickBuilt elements across four sequenced and annotated *P. gingivalis* strains. BrickBuilt elements are situated intergenically between the genes noted. Grayed-out boxes represent loci at which BrickBuilt is aberrant

MITE	33277 Locus	W83 Locus	TDC60 Locus	HG66 Locus	Gene Characterization	Gene	Strand
BrickBuilt_1	PGN_0031	PG0033	PGTDC60_0032	EG14_02395	RmuC domain (DUF 805)		-
	PGN_0033	PG0034	PGTDC60_0034	EG14_02400	Thioredoxin	trx	-
BrickBuilt_2	PGN_0204	PG2159	PGTDC60_1262	EG14_03005	Protoporphyrinogen oxidase	hemG	+
	PGN_0205	PG2161	PGTDC60_1265	EG14_03010	AraC family transcriptional regulator		-
BrickBuilt_3	PGN_0303	PG0196	PGTDC60_0466	EG14_04795	Zinc protease (Peptidase M16)		+
	PGN_0306	PG0198	PGTDC60_0471	EG14_04805	PF05656 family protein (DUF 805)		+
BrickBuilt_4	PGN_0336	**NP**	PGTDC60_1661	EG14_04940	Immunoreactive antigen/PorSS CTD		-
	PGN_0340	**NP**	PGTDC60_1665	EG14_04960	Peptidase S41/PorSS CTD		+
BrickBuilt_5	PGN_0361	PG0264	PGTDC60_0543	EG14_05065	Glycosyl transferase family 2		+
	PGN_0365	PG0267	PGTDC60_0547	EG14_05075	Arginyl-tRNA synthetase	argS	+
BrickBuilt_6	PGN_0400	PG1715	PGTDC60_0586	EG14_05255	TonB-dependent receptor Cna protein		+
	PGN_0403	PG1714	PGTDC60_0590	EG14_05265	Pyridoxamine-phosphate oxidase	pdxH	+
BrickBuilt_7	PGN_0455	PG0549	PGTDC60_0639	EG14_03135	Partial ISPg5		+
	PGN_0456	PG0553	PGTDC60_0641	EG14_03150	Methylmalonyl-CoA mutase	scpA	-
BrickBuilt_8	PGN_0550	PG1559	PGTDC60_0739	EG14_03610	Glycine cleavage system subunit T	gcvT	+
	PGN_0553	PG1556	PGTDC60_0743	EG14_03615	Conserved hypothetical (DUF2149)		-
BrickBuilt_9	PGN_0558	PG1548	PGTDC60_0748	EG14_03640	Haem-binding protein	hmuY	-
	PGN_0559	PG1550	PGTDC60_0751	EG14_03655	Serine protease (Peptidase C10)	prtT	-
BrickBuilt_10	PGN_0632	PG0585	PGTDC60_1709	EG14_09225	Aspartyl-tRNA amidotransferase B		+
	PGN_0633	PG0587	PGTDC60_1713	EG14_09235	Membrane protein putative ion channel	btuF	-
BrickBuilt_11	PGN_0667	PG0625	PGTDC60_1753	EG14_09060	GTP cyclohydrolase I/PorSS CTD	folE	+
	PGN_0668	PG0627	PGTDC60_1756	EG14_09045	RNA-binding protein/PorSS CTD		-
BrickBuilt_12	PGN_0819	PG0796	PGTDC60_1912	EG14_08255	Leucyl-tRNA synthetase	leuS	-
	PGN_0823	PG0800	PGTDC60_1917	EG14_08240	NAD-utilizing dehydrogenase		-
BrickBuilt_13	PGN_0831	PG0807	PGTDC60_1926	EG14_08205	N utilization substance/PorSS CTD		-
	PGN_0832	PG0809	PGTDC60_1927	EG14_08200	Gliding motility protein/PorSS CTD	sprA	-
BrickBuilt_14	PGN_0871	PG1389	PGTDC60_2074	EG14_07995	Membrane protein		-
	PGN_0872	PG1391	PGTDC60_2073	EG14_08000	DNA-binding protein (PF00216)		+
BrickBuilt_15	PGN_0898	PG1424	PGTDC60_2039	EG14_07870	Peptidylarginine deiminase/PorSS CTD	PAD	-
	PGN_0900	PG1427	PGTDC60_2036	EG14_07865	Peptidase C10/PorSS CTD		-
BrickBuilt_16	PGN_1207	PG1117	PGTDC60_1098	EG14_06320	Transport multidrug efflux		+
	PGN_1208	PG1118	PGTDC60_1096	EG14_06310	ClpB chaperone and protease	clpB	-
BrickBuilt_17	PGN_1476	PG0494	PGTDC60_1611	EG14_09640	PorSS C-terminal sorting domain		-
	PGN_1479	PG0491	PGTDC60_1606	EG14_09650	Peptidase S10/PorSS CTD	dppVII	-
BrickBuilt_18	PGN_1777	PG1784	PGTDC60_0106	EG14_00235	Cysteine protease (Peptidase C1)		-
	PGN_1780	PG1786	PGTDC60_0110	EG14_00245	Endoribonuclease L-PSP		-
BrickBuilt_19	PGN_2035	PG0088	PGTDC60_0367	EG14_01925	Peptidase M16		+
	PGN_2037	PG0090	PGTDC60_0370	EG14_01930	DNA-binding protein from starved cells	dps	+

'NP' stands for 'Not Present'

segment is 318 nucleotides (BrickBuilt_4). No BrickBuilt element is bisected by a full-length autonomous transposable element. Thus, although repetitive intergenic sequences may be targets for insertion of exogenous or duplicated endogenous genes, such events have yet to be detected. Additionally, no leader-to-tail versions are present without a full 23 nt repeat and no TIR-containing individual leader or tail segments are present without the

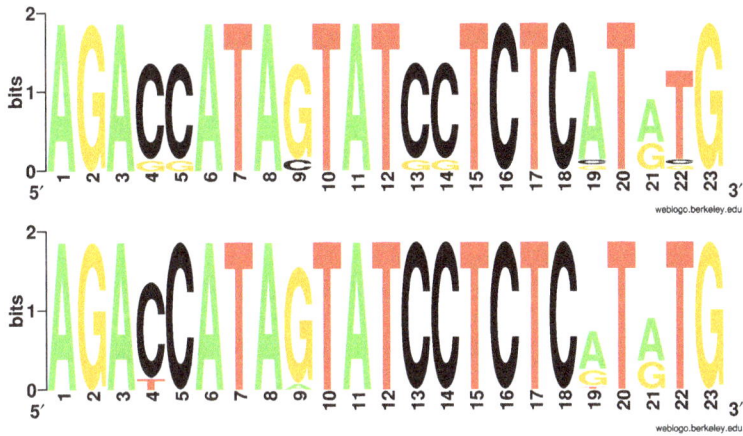

Fig. 1 Sequence logo representation of the 23 nucleotide repeat region from *P. gingivalis* strain ATCC 33277 multiple alignments. Generated with Weblogo software. Total height of a nucleic acid stack represents sequence conservation at a given position. Height of symbols at a stack represents relative frequency of a given nucleic acid at that position. Top sequence logo corresponds to compilation of the consensus of the 19 BrickBuilt elements; consensus for each made prior to combining for sequence logo. Bottom sequence logo corresponds to BrickBuilt_5 alone, constructed from its 18 full repeats

23 nt repeats. A single site adjacent to the Hmu operon contains a partial tail-only version that lacks the terminal 20 nt that would include the TIR; no partial leader-only versions of the element are present.

The genome sequence of strain ATCC 33277 contains 2,345,886 bases. When complied together all 19 BrickBuilt elements in strain ATCC 33277 make up 10,276 bases, or 0.44 percent of the overall genome; the equivalent of 9 protein coding sequences in this strain on average.

Conservation of BrickBuilt elements in other strains of *P. gingivalis*

Of the 19 versions of BrickBuilt found within strain ATCC 33277, 16 are conserved between the analogous coding sequences within strains W83, TDC60 and HG66 (Table 1 and Additional file 2: Table S1). Strains HG66 and TDC60 contain 19 versions of BrickBuilt, equivalent to the number in strain ATCC 33277. However, strain W83 only contains 18 versions of the element. Strains ATCC 33277 and HG66 share the exact same 19 loci for

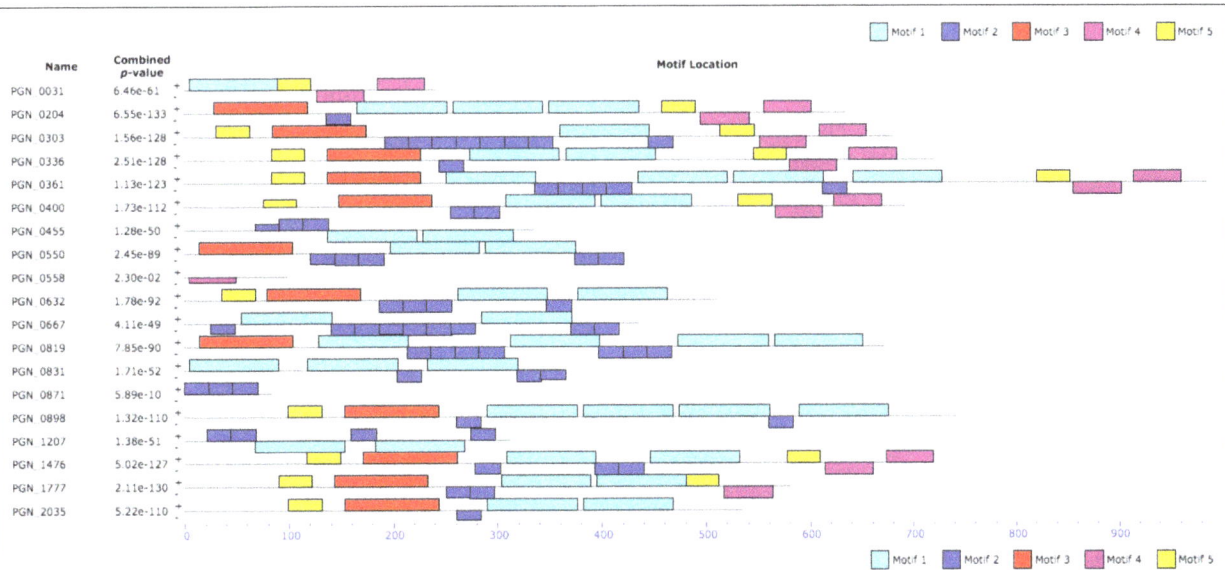

Fig. 2 MEME motif analysis block output of the 19 BrickBuilt elements in *P. gingivalis* strain ATCC 33277. Entire element FASTA sequences were used as the input. Settings for analysis were: Distribution of motif occurrence as 'any number of repetitions', number of different motifs as '5', minimum motif with as '23', and maximum motif width as '200'. Dark and light blue blocks correspond to 23 nucleotide repeat regions, red blocks to leader regions and purple blocks to tail regions. Yellow blocks, potentially representing a 5th motif, were only found on the positive strand and had the lowest e-values associated with their significance scores

BrickBuilt elements. One locus in strain TDC60 (Brick-Built_4) is deviant and is encompassed by two IS*Pg*1 elements. Strain W83 has three sites that differ from the other strains, all which are located adjacent to other types of IS or repetitive elements. In this strain Brick-Built_4 is completely lacking, while BrickBuilt_7 and BrickBuilt_18 are aberrant with respect to size having only maintained 23 nt repeats.

BrickBuilt elements can be identified in the genome of *P. gingivalis* strain JCVI SC001, which is not yet included in the default NCBI BLAST nucleotide database settings. Genome searches of the FASTA files from the JCVI SC001 revealed that most BrickBuilt loci in the JCVI SC001 genome contained strings of undetermined bases, which can be attributed to the manner of isolation, sequencing and assembly. Eight other *P. gingivalis* genomes have since been sequenced and deposited in NCBI, yet they are not completely assembled. Assembly gaps are located at sites where the corresponding surrounding CDS from ATCC 33277 contain BrickBuilt elements, suggesting BrickBuilt is present in those genomes as well and potentially capable of causing assembly difficulties (Additional file 3: Figure S2). Additionally, a degenerate version of the 23 nt repeat consensus sequence (AGAYCATARTATCCTCTCRTRTG) was searched against all 13 (8 unfinished) *P. gingivalis* genomes, each giving positive hits (data not shown). Because of assembly gaps and undetermined bases around BrickBuilt sites in the unfinished sequencing projects they were not included in multiple alignments.

Multiple alignments of BrickBuilt elements using the sequences from strains ATCC 33277, W83, TDC60 and HG66 revealed that sequences from strain ATCC 33277 align most closely with HG66, and those from strain W83 with TDC60 (Fig. 3). Similarly, a phylogenetic analysis with PHYML showed similar clustering of ATCC33277 with HG66 and W83 with TDC60 (Fig. 3). The matching of the sequences between the strains in the above pairings is consistent throughout 18 of 19 elements. Branching of BrickBuilt elements is congruent with the dendrogram generated based on genomic BLAST for all 13 *P. gingivalis*

genomes. Interestingly, strain HG66 was deposited as 'being closely related to strain W83', yet based on the results of dendrogram from the full genome BLAST and from alignments of the BrickBuilt, this seems to be incorrect. The BrickBuilt_5, BrickBuilt_8, BrickBuilt_10, BrickBuilt_11, BrickBuilt_12, BrickBuilt_13, BrickBuilt_14 and BrickBuilt_15 sites all lie between the same two CDS within the respective genomes, with strains ATCC 33277 and HG66 usually having more 23 nucleotide repeats than W83 and TDC60. BrickBuilt_6 is the only site at which strains W83 and TDC60 have more 23 nt repeats than ATCC 33277 and HG66. BrickBuilt_9, the shortest element which is also the only element without a 23 nt repeat has only one SNP across the 100 nucleotides. Unlike all the other BrickBuilts, that single SNP would align strain ATCC 33277 with W83 and strain TDC60 with HG66.

The repetitive nature of the BrickBuilt elements, both the internal repeats and that they are found multiple times through the genome, can lead to sequencing, assembly and annotation issues. Because the strains W83, ATCC 33277, TDC60, HG66 and JCVI SC001 are unique strains, were sequenced independently, and were *de novo* assembled, placement of BrickBuilt elements at the same locus across genomes is unlikely to be due to use of a shared scaffold. However, care should be taken when aligning newly-sequenced *P. gingivalis* genomes to a scaffold.

Homology to MITEs and other repetitive elements

The 23 nt repeats and the leader and tail segments of the element were analyzed using BLAST (NCBI server) to determine whether the element is present in genomes other than the species *P. gingivalis* [43]. With default BLAST nucleotide settings, a full-length BrickBuilt and each of the three distinct parts of the element match solely to *P. gingivalis*. All four sequenced and annotated strains of *P. gingivalis* available for BLAST searching harbored hits for BrickBuilt. Through querying discontiguous megablast as well as using less stringent search constraints within megablast with 'max target sequences', 'expect threshold', 'word size' and 'filter low complexity regions', low-homology hits were obtained with the

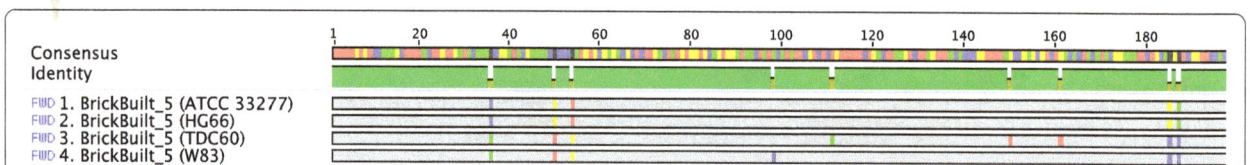

Fig. 3 BrickBuilt_5 multiple alignment generated using MAFFT on the Geneious R8 platform with strain ATCC 33277, W83, TDC60 and HG66 inputs. The multiple alignment is focused on the first 200 nt of the leader region. Grey within the tracks denotes complete conservation at a site. Color within the tracks denotes variable sites. At all five sites where more than one strain is variable strains ATCC 33277 and HG66 cluster together, as do strains TDC60 and W83. Within this region only strains TDC60 and W83 have sites where they alone differ from the other three strains; this is consistent throughout for this specific element

terminal inverted repeat regions. However, matches identified in this manner were only homologous in the TIR sections. Of note, when BLASTx searches (protein database search using a translated nucleotide query) were performed with the leader and tail sequences under default settings several *Bacteroidetes* species contained tail hits and one species contained a leader hit. *Porphyromonas gulae* contained strong hits with both leader and tail, while *Prevotella tannerae* and *P. dentalis* contained weak tail hits only. All of the BLASTx hits were either part of a predicted transposase/partial transposase or a hypothetical protein. If BrickBuilt were indeed a non-autonomous transposable element, homology to sections of related transposons through BLASTx would not be unexpected. As such, low BLASTx homology within *Prevotella tannerae* and *P. dentalis* does not point to these hits being potential parent or identical elements of BrickBuilt.

Genome analysis of recently-uploaded *Porphyromonas gulae* strains was carried out using the NCBI-deposited WGS shotgun sequencing data, from which we determined that *P. gulae* strains do in fact carry BrickBuilt homologues (Additional file 4: Figure S3) [44, 45]. Some of the *P. gingivalis* BrickBuilt element locations are conserved within *P. gulae* strains. However, greater strain variation seems evident in *P. gulae* at certain BrickBuilt loci than between *P. gingivalis* strains (Additioal file 5: Figure S3). Of note, the original *P. gulae* genome was obtained from a wolf and the subsequent strains were obtained from domesticated dogs. The original strain (DSM 15663) only contains 4 BrickBuilt homologues within the genome, and importantly lacks the BrickBuilt_5 homologue that was used for the majority of BLAST database queries.

Within *P. gingivalis* there have been three previously identified groups of MITEs or non-autonomous transposable elements; named the 239, 464 (PgRS) and 700 groups [10, 13]. These numbers are the names of three types of MITEs already noted in *P. gingivalis* genomics publications. The numbers were initially related to the overall length of the elements, however, the 464 type was renamed in subsequent publications and in NCBI genome graphics annotations. General copy numbers of the four MITE versions are similar, holding around 10–20. The number of full copies and partial or fragment copies of each element differs slightly between genomes within the species. During our examination of BrickBuilt we analyzed whether any sequence overlap was apparent between the elements and found that the terminal inverted repeats are similar, yet the rest of the elements do not bear similarity. 464/PgRS elements were previously identified as containing 41 nucleotide tandem direct repeats [10, 13]. The 23 and 41 nucleotide internal tandem direct repeats of the elements do not share homology with each other and neither have non-*P. gingivalis* BLAST matches within

the NCBI database. The segments of the 464/PgRS elements flanking the 41 nt tandem direct repeats are themselves repetitive, which is unlike the non-repetitive leader and tail segments of BrickBuilt. Although not related by sequence, similarities in copy number between 464/PgRS and BrickBuilt elements are evident. With *P. gingivalis* harboring four types of MITE-like elements it is interesting that two types, BrickBuilt and 464/PgRS, contain microsatellite repeats. Although several 236 and 700-type elements are located near repeats or other repetitive elements, they seem not to have encompassed any mini- or microsatellites from analyses of the currently available genomes.

In addition to Tn and IS elements, multiple groups have described repetitive sequences within *P. gingivalis* genomes ranging from single nucleotide tracts to mini- and microsatellites [13, 46]. Several 41, 23 and 22 nucleotide tandem direct repeats were described in *P. gingivalis* strain W83, yet the exact locations of such repeats were not identified, nor were comparative genomics an option at the time of the report [13]. Some of the 23 nt tandem direct repeats noted are presumably the direct repeat portions of BrickBuilt.

Within *P. gingivalis* there are 11 recognized IS elements and 2 different composite transposon (Ctn) elements [13, 47, 48]. Ten of the terminal inverted repeats for the 13 IS and Ctn have been previously characterized. The TIRs of BrickBuilt were identified by first determining where non-repetitive DNA sequences immediately flanked repetitive ones (i.e. leader and tail segments). Next, all sequences were compared in multiple alignments, and only versions that maintained intact leaders or tails were then used for determination of consensus sequences (Fig. 4). The TIRs of BrickBuilt and MITEPgRS elements are almost identical, as are the TIRs of MITE293 and MITE700 elements. The MITE-like elements in *P. gingivalis* share either identical or within one nucleotide TIRs with those of full-length IS elements within the *P. gingivalis* genomes; ISPg1, ISPg3, ISPg4 and ISPg9 (Table 2). The matching full-length ISPg elements are all categorized within the IS5 family. Brick-Built's TIRs are most similar to those of ISPg1 and ISPg9 (which share identical TIRs); ISPg4 is the next closest match with 2 nucleotides different (Table 2). MITE293 and MITE700 TIRs match with ISPg3. Although the TIRs are similar, no remnant of a transposase from any *P. gingivalis* IS or Tn element remains within any of the BrickBuilt copies.

Only 4 of the 19 BrickBuilt copies contain both an intact leader and tail associated TIR (Table 2). In order to determine whether BrickBuilt makes target site duplications (TSD), the DNA sequence immediately adjacent to the proposed TIRs were examined of these 4 copies. Three of the four copies of BrickBuilt do carry TSDs, however, they are not the same length or sequence for

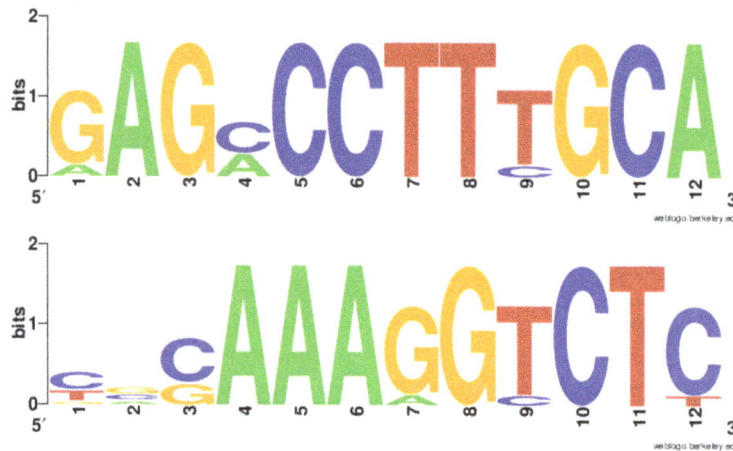

Fig. 4 Consensus Terminal Inverted Repeats (TIRs) of BrickBuilt elements visualized with Weblogo software. Only elements that contained intact leader or tail segments, which carry the TIRs, were considered and used for consensus construction

each element. BrickBuilt_3 has a 'CT' dinucleotide flanking its TIRs; BrickBuilt_4 has a 'GAAA' tetranucleotide flanking its TIRs; BrickBuilt_5 has an 'AAAAA' heptanucleotide; and BrickBuilt_18 does not contain a putative TSD because one of its two TIRs is shared by a MITE293 element. These duplications on either side of the elements may not reflect canonical TSDs, however, if these elements are mobilized by multiple transposases that each make different restrictions to the target DNA this could potentially occur. Within *P. gingvialis*, ISPg elements generate TSDs varying from 2–9 bp; some can lack TSDs completely and may frequently nest into other mobile elements and thus eliminate TSD identification. Additional TSD data related to element mobility will be presented below.

After determining the IS5 family-like TIRs of BrickBuilt other IS5 elements were scanned for potential similarity. Identical TIRs to that of BrickBuilt were found in the *Neisseria meningitidis* ISNmeI and its derivatives (Table 2). ISNmeI is the proposed (based on TIRs) parent element for the type II MITE ATR (_AT_-rich _R_epeat) in *Neisseria meningitidis* genomes [39]. ATR elements are found 19 times within *N. meningitidis* genomes, which is similar to that of BrickBuilt's distribution. Additionally, ATR elements are frequently associated with direct repeat elements of *N. meningitidis* known as REP2.

From initial characterizations of the configuration and locations of BrickBuilt elements, they can be classified within the large group of non-autonomous transposable elements, potentially best fitting within the MITE subcategory. A caveat must be placed, however, given that MITE elements are typically described as being comprised of two homologous flanking regions, and we have determined that BrickBuilt elements contain distinct

'leader' and 'tail' segments. Since all accessible genome sequences of *P. gingivalis* strains contain BrickBuilt elements, the parent element or first version of BrickBuilt probably occurred early within the phylogeny of the *P. gingivalis* species. Insertion of the 23 nt repeat(s) into the original parent element may be the event that catalyzed the inactivation of an autonomous transposable element. Alternatively, a version of BrickBuilt already containing the 23 nt repeats could have been laterally-transferred via plasmid or horizontally-transferred via phage. Given that no full-length (TIR-containing) leader or tail regions are present without a 23 nt repeat it may be deleterious to maintain a full leader or tail region on the chromosome, or the 23 nt repeat is required by the autonomous element.

The limited host range nature of BrickBuilt identified through NCBI BLAST is intriguing yet not uncommon for non-autonomous transposable elements [49–52]. Once a non-autonomous element occurs within a genome, potentially by deletions of an autonomous transposable element, as well as via conjugation-based horizontal transfer of a plasmid or transduction via a bacteriophage, movement between species will become less likely. Additionally, few bacterial species have multiple genome assemblies available for intraspecies comparisons, which could lead to missed elements due to strain variation. Furthermore, it is possible that repetitive sequences could be mis-sequenced or left out of genome assemblies due to repeat region sequencing difficulties or unassigned bases.

Predicted secondary structure of BrickBuilt

The direct repeats within BrickBuilt are predicted to form long stem loop structures (Figs. 5 and 6). Three

Table 2 Terminal Inverted Repeats (TIRs) of BrickBuilt elements from strain ATCC 33277. Terminal Inverted Repeats and family of selected IS and MITE-like elements in *Porphyromonas gingivalis* as well as ISNme1 from *Neisseria meningitidis*

Locus (33277)	TIR 5' (nt)	TIR 3' (nt)	Both TIRs
BrickBuilt_1		CCGAAAGGTCTC	
BrickBuilt_2		CCGAAAGGTCTC	
BrickBuilt_3	AAGACCTTTGCA	CCGAAAGGTCTC	YES
BrickBuilt_4	GAGACCTTTGCA	TGCAAAGGTCTC	YES
BrickBuilt_5	GAGACCTTTGCA	CGCAAAGGTCTC	YES
BrickBuilt_6		CACAAAGGTCTT	
BrickBuilt_7			
BrickBuilt_8		TGCAAAAGTCTC	
BrickBuilt_9			
BrickBuilt_10			
BrickBuilt_11			
BrickBuilt_12		GACAAAGGTCTC	
BrickBuilt_13			
BrickBuilt_14			
BrickBuilt_15	GAGCCCTTTGCA		
BrickBuilt_16			
BrickBuilt_17	GAGCCCTTCGCA		
BrickBuilt_18	GAGCCCTTTGCA	TGCAAAGGCCTC	YES
BrickBuilt_19	GAGCCCTTTGCA		
Element	Left TIR	Right TIR	Family
ISPg1	GAGACCATTGCA	TTCAAAGGTCTC	IS5
ISPg3	ACGTCAGTTCGA	TCGAACTGACGT	IS5
ISPg4	GAGACTGTTGCA	CGCAACAGTCTC	IS5
ISPg9	GAGACCATTGCA		IS5
MITE239	ACGTGAGTTCGATATAAAGGAA	TTCGCTTAAATCGAACTGGCGT	
MITEPgRS/MITE464	GAGACTGTTGCA	TGCAACGGTCTC	
MITE700	ACGTCATTCGA	TCGAACTCACGT	
BrickBuilt	GAGACCTTCGCA	TGAAAGGTCTC	
Nsm ISNme1	GAGACCTTTGCAAAA	TTTTGCAAAGGTCTC	IS5

Blank spaces for TIRs indicate situations in which the region is degenerate or not present

DNA/RNA structure prediction programs, Mfold, RNAstructure and RegRNA2.0, independently predicted long stem loops to form from/within the element [53–55]. The length of the version of BrickBuilt affects the size of the predicted stem loop structure and the associated entropy. BrickBuilt_1, BrickBuilt_9 and BrickBuilt_14 are not predicted to form long stem loops by the RegRNA2.0 program due to the length of the internal 23 nt repeats, however, shorter stem loops due to dyad symmetry may occur. BrickBuilt elements are predicted to be surrounded/flanked by Rho-independent terminators and/or polyadenyltaion sites in 10 of 19 instances. No portion of BrickBuilt matched to any structures in Rfam [56].

Predicted structures of BrickBuilt vary slightly between strains at a given conserved locus. The BrickBuilt_5

Mfold entropy predictions for strains ATCC 33277 and W83 are −127.96 and −102.50, respectively (Fig. 6). BrickBuilt_5 in ATCC 33277 is 991 nucleotides long and the analogous W83 version is 807 nucleotides. In this case the length difference is due to W83 BrickBuilt_5 having fewer 23 nt internal direct repeats; the leader and tails are of the same length. Within a multiple alignment of the four *P. gingivalis* strains at the BrickBuilt_5 locus there are 18 single nucleotide polymorphisms (SNPs) that separate the lineages (Fig. 6). Substituting SNPs between strains at the BrickBuilt_5 locus into the ATCC 33277 model changes the predicted entropy of the element by −7.25, or 5.7 %, to −135.17.

Although no BLASTn matches for BrickBuilt were found outside of the *P. gingivalis* species, MITEs

▶ Map View

▶ Table View

Motif Type	Motif Name	Position	Length	Sequence
open reading frame (ORF)				
rho-independent terminator	Rho-independent Terminator	51 ~ 86	36	aggccgtgtgtgatcgtattcggatcgccatggccg
Riboswitches				
cis-regulatory elements of ERPIN				
cis-regulatory elements of Rfam				
long stems	Stem_1_5prime	427 ~ 638	212	atagaccatagtatcctctcatgtgagaccatagtatcctctcgtatgag accatagtatcctctcatgtgagatcatagtatcctctcatgtgagatca tagtatcctctcttgtgagaccatagtatcctctcatgtgagaccatagt atcctctcatgtgagaccatagtatcctctcgtatgagaccatagtatcc tctcatatgaga
	Stem_1_3prime	650 ~ 862	213	tctcatgtgagaccatagtatcctctcatatgagaccatagtatcctctc atatgagaccatagtatcctctcatatgagaccatagtatcctctcgtat gagaccatagtatcctctcatatgagaccataatatcctctcatatgaga ccatagtatcctctcgtatgagaccatagtatcctctcgtatgagaccat agtatcctctcat
	Stem_2_5prime	345 ~ 684	340	atctatcggtttaacagctggtacggaacctcccactccccgaatctgcc atccggcagagagcagatagcatcttccatcaatagaccatagtatcctc tcatgtgagaccatagtatcctctcgtatgagaccatagtatcctctcat gtgagatcatagtatcctctcatgtgagatcatagtatcctctcttgtga gaccatagtatcctctcatgtgagaccatagtatcctctcatgtgagacc atagtatcctctcgtatgagaccatagtatcctctcatatgagaccatag tatcctctcatgtgagaccatagtatcctctcatatgaga
	Stem_2_3prime	696 ~ 1053	358	tctcatatgagaccatagtatcctctcatatgagaccatagtatcctctc gtatgagaccatagtatcctctcatatgagaccataatatcctctcatat gagaccatagtatcctctcgtatgagaccatagtatcctctcgtatgaga ccatagtatcctctcatgtaagttcattctttgtatctcagggaaaaacc aatgagcaggtgtatggctgccaagtaaacagaaagcagtgtccgaatac gtcgccaatcgataagcatatcgacttggttcgaaaatacattccgtgct ttcctgtatcgattcgttgacatgacgtccctgaatgagacctcttcgtc ggatgaat

Fig. 5 RegRNA2.0 analysis output of *P. gingivalis* strain ATCC 33277 BrickBuilt_5 and surrounding CDS-to-CDS area. The immediate 5'PGN_0361 and the immediate 3'Rho-independent terminators, transcriptional regulatory motifs, riboswitches, cis-regulatory elements, ERPINs, Rfam database matches, long stems and functional RNA sequences were queried. The Rho-independent terminate identified occurs prior to the BrickBuilt element

from other species have been shown or are predicted to be of similar modular makeup and form long stem loops [23, 35, 38, 57–59]. In addition, repetitive sequences that are not MITE-associated also frequently form stem loop structures [58, 60–62]. Stem loop structures, especially long stem loops, are capable of modulating transcript half-life, modulating translational efficiency as well as serving as docking/receptor sites for proteins [12, 58, 61]. The Rho-independent terminator upstream of BrickBuilt_5 is located 111–148 nt from the 5′ CDS, with the leader region of BrickBuilt_5 located 182 nt from the 5′ CDS (Fig. 6). Thus, in this case, the BrickBuilt element has not disrupted the 'natural' terminator for the 5′ CDS. However, given the proximity to the Rho-independent terminator, this BrickBuilt element may be able to modulate the stability or accessibility of the terminator. The long stem loop structures of BrickBuilt_5 start 257 nt from the Rho-independent terminator and end 965 nt away.

Genome locations and surroundings

All BrickBuilt elements are located intergenically; no direct overlap or interruptions of genuine protein coding sequences are apparent in the complete genomes available to date (Table 1 and Additional file 2: Table S1). Several 'hypothetical proteins' are annotated to be within BrickBuilt elements, however expression of these proteins has not been confirmed experimentally [40–42]. Several of the predicted hypothetical proteins are part of repeated/overlapping probes on *P. gingivalis* microarrays. Additionally, the 23 nt repeats within BrickBuilt elements are predicted to cause frequent translational stops (data not shown). Lack of experimental confirmation of protein products, nonunique microarray probes and abundant translational stops suggests that translation of these regions is unlikely, and even if translation were to occur it would probably be truncated versions of a repetitive or mobile element.

Of the 38 genes surrounding the BrickBuilt elements in the ATCC 33277 genome there are several functional clusters (Table 1). Six genes encode proteases of the C1, (2) C10, (2) M16 and S41 families. Five genes are predicted to encode DNA/RNA-binding proteins, and another four are involved in tRNA metabolism. Noticeably, of the 19 BrickBuilt elements in strain ATCC 33277, 5

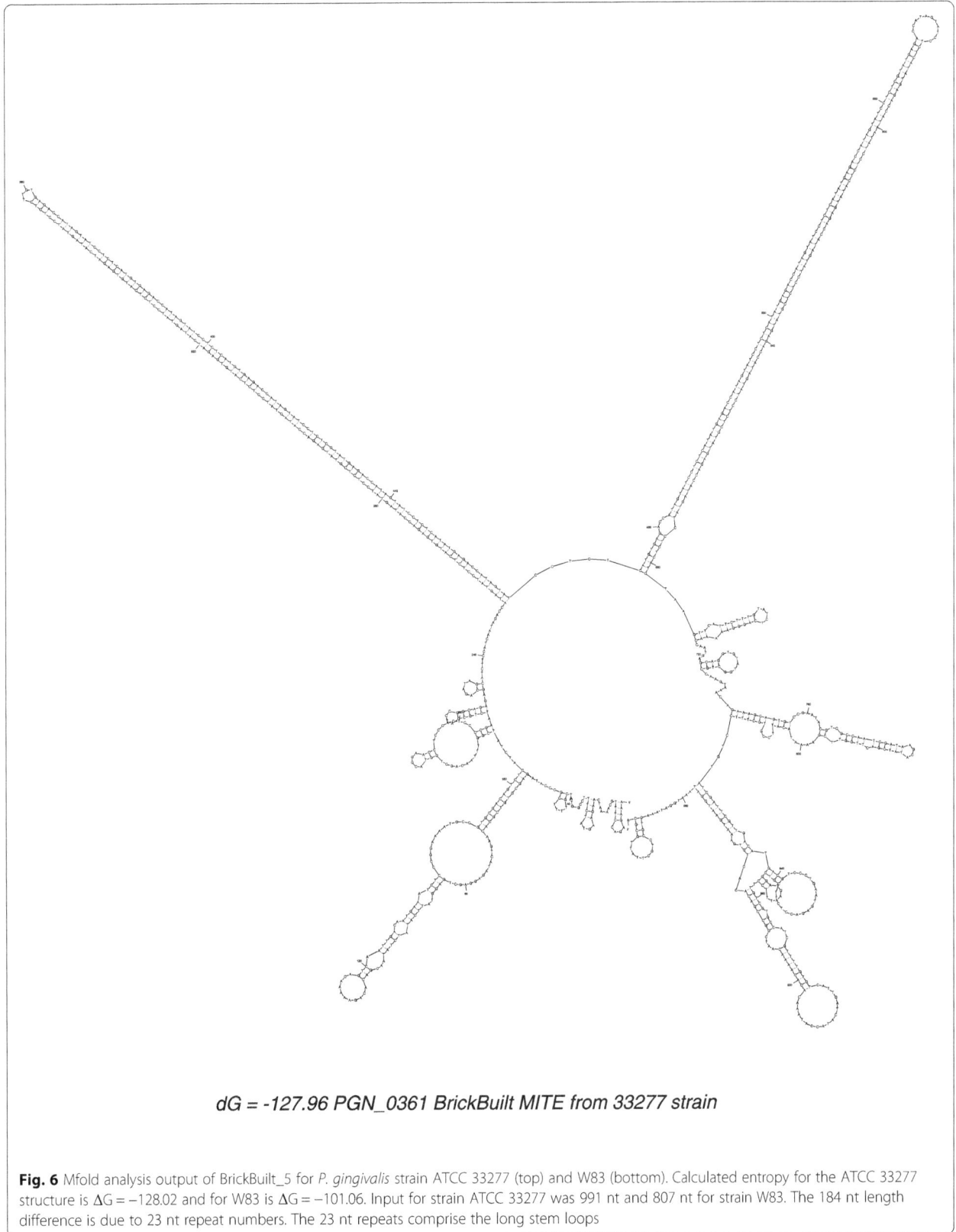

dG = -127.96 PGN_0361 BrickBuilt MITE from 33277 strain

Fig. 6 Mfold analysis output of BrickBuilt_5 for *P. gingivalis* strain ATCC 33277 (top) and W83 (bottom). Calculated entropy for the ATCC 33277 structure is $\Delta G = -128.02$ and for W83 is $\Delta G = -101.06$. Input for strain ATCC 33277 was 991 nt and 807 nt for strain W83. The 184 nt length difference is due to 23 nt repeat numbers. The 23 nt repeats comprise the long stem loops

are located adjacent to a gene/protein containing a Por Secretion System C-Terminal Domain (PorSS CTD) [63] (Table 1). Likewise, 5 of the previously identified *P. gingivalis* MITEs within the W83 genome are also located next to PorSS CTDs. A total of 34 PorSS CTDs have been predicted within the W83 genome (only 22 annotated on NCBI with TIGR); 29 % of PorSS CTDs are associated with MITEs [64]. The PorSS is connected to pigmentation and haem acquisition in *P. gingivalis*. Apart from those associated with PorSS, other genes surrounding BrickBuilt elements are *hemG, dps, trx, and hmuY*, which are involved in haem biosynthesis, acquisition and detoxification, respectively. Additionally, two separate DUF 805 motifs are found in genes surrounding BrickBuilt elements, which are associated with phage integrases. The locations relative to CDS raise the possibility that BrickBuilt could be acting as or similar to a Putative Mobile Promoter (PMP); a secondary regulatory circuit or mechanism to canonical transcription and translation modulation [65].

Expansion and contraction of the 23 nt repeats between strains is evident at the conserved BrickBuilt loci. Entire 23 nt repeat segments have been removed or added. Full and/or partial deletions of the leader and tail regions are also apparent. Deletions of the leader and tail regions occur from the distal ends of each segment with respect to the 23 nt internal repeats.

Pairwise and multiple alignments of a respective Brick-Built locus across the four strains of *P. gingivalis* revealed SNPs that potentially suggest lineages or selected and compensatory mutations (Fig. 3). Whether the SNPs are

generated *de novo* at each site, occur in stages and are distributed, or occur through site-to-site recombination cannot be determined definitively from currently published genome assemblies alone. However, multiple alignments of the conserved BrickBuilt loci within a given strain show patterns of non-random mutation. For sites at which SNPs have occurred, the SNP is frequently distributed at several positions within the element, yet this occurs at intervals. Additionally, SNPs appear localized around a 2–4 nt site when compared in multiple alignments. Long Term Evolution Experiments (LTEE) and plasmid-based recombination systems could be employed to determine mutation rates within BrickBuilt in comparison to the rest of the genome, whether 23 nt repeats expand and contract at a given locus, and how recombinogenic the elements are.

BrickBuilt_4 may best demonstrate the lingering 'mobility' of BrickBuilt elements within *P. gingivalis*. Strain HG66 shares the same locus with strain ATCC 33277. However, the TDC60 BrickBuilt_4 is not located between the same two genes (Fig. 7). No other mis-located Brick-Built elements occur in strain TDC60 and the sequence of this element aligns closely with the ATCC 33277 and HG66 versions at this locus. Thus, it is probable that the BrickBuilt_4 homologue has been induced to transpose by or transposed with the surrounding ISPg1 elements. Additionally, no BrickBuilt_4 homologue is present in strain W83, adding to a mobility pattern of BrickBuilt_4 (Fig. 7). Importantly, BrickBuilt_4 is the only of these elements that has maintained perfect 12 bp TIR matches, increasing the likelihood that a surrogate transposase could act on the element. Further evidence of TSDs can

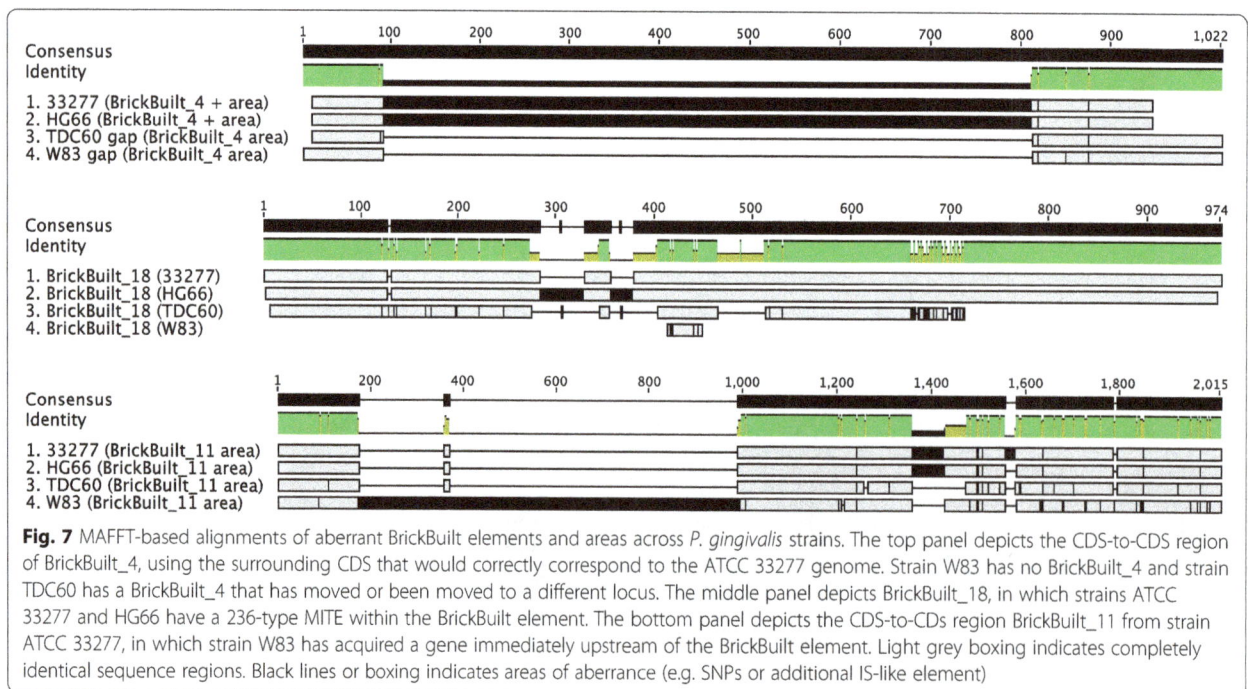

Fig. 7 MAFFT-based alignments of aberrant BrickBuilt elements and areas across *P. gingivalis* strains. The top panel depicts the CDS-to-CDS region of BrickBuilt_4, using the surrounding CDS that would correctly correspond to the ATCC 33277 genome. Strain W83 has no BrickBuilt_4 and strain TDC60 has a BrickBuilt_4 that has moved or been moved to a different locus. The middle panel depicts BrickBuilt_18, in which strains ATCC 33277 and HG66 have a 236-type MITE within the BrickBuilt element. The bottom panel depicts the CDS-to-CDs region BrickBuilt_11 from strain ATCC 33277, in which strain W83 has acquired a gene immediately upstream of the BrickBuilt element. Light grey boxing indicates completely identical sequence regions. Black lines or boxing indicates areas of aberrance (e.g. SNPs or additional IS-like element)

be gleaned from this specific element by comparing the 'filled' and 'empty' sites between the strains. Strains ATCC 3377 and HG66, which contain the element at this locus, have a 'GAAA' tetranucleotide on each side of the intact TIRs. However, strains TDC60 and W83, which lack the element at this locus, only have a single 'GAAA' tetranucleotide copy.

With respect to *P. gulae* strains, BrickBuilt_5 demonstrates the possibility or history of mobility. At this site, 4 of the 11 *P. gulae* strains contain (and 7 lack) BrickBuilt copies. In the strains that contain an element a 'AAAA' TSD can be seen ('AAAAA' in *P. gingivalis* at that site). The *P. gulae* strains lacking an element at that site only have one 'AAAA' tetranucelotide. This site is completely conserved in the published *P. gingivalis* strains.

BrickBuilt_18 in the strains ATCC 33277 and HG66 contain/encompass MITE239_11 between nucleotides 659–904 of the sequence. Strain TDC60 has a gap where MITE239_11 occurs in the other two strains, while the flanking portions of the BrickBuilt match (Fig. 7). The strain W83 version at this site is diminutive, having been reduced to 1.5 copies of the 23 nt internal repeat. While strain W83 doesn't harbor a MITE-within-a-MITE configuration at any locus, BrickBuilt_11 in strain W83 contains an 'extra' gene adjacent to the 5′ region of the element unlike any other strain (Fig. 7).

Transcriptional expression of BrickBuilt

Repetitive and transposable elements are capable of modulating the genome stability and evolution of species [17, 19, 58, 66]. Interestingly, no endogenous plasmids have been found for *P. gingivalis* to date. The presence of many copies and types of repetitive and transposable elements could serve a quick way by which *P. gingivalis* could recombine/adapt to external stimuli beyond traditional host-directed transcriptional and translational controls [17, 18, 67–69].

Analysis of previously published data

Previous microarray and RNAseq studies have shown transcripts originating from within BrickBuilt elements, yet none characterized these regions in detail [9, 70]. Several of the microarray probes are themselves repetitive and many of the oligos/~20 mers used for identifying transcripts could map to multiple sites within the genome. Although BrickBuilt elements are highly conserved and repetitive, small variations due to SNPs, size of leader and tail regions, and the surrounding intergenic context make it possible to map at least some transcripts to the correct sites (Fig. 8 and Additional file 5: Figure S4). For situations where completely identical regions could produce the same transcript, the mapping programs and settings used will determine whether the transcripts are placed at one of the matching loci exclusively, distributed amongst the loci evenly, or left out of the results entirely. Importantly, the placement of any transcript at one or distributed across all of a given repetitive oligo/~20 mer sites suggests that at least one of the sites contributes active transcription.

Within the transcriptome transcript levels of individual BrickBuilt elements vary markedly and also vary according to growth medium, e.g. in/on minimal, tryptic soy, and blood media [9, 70]. Generally, transcripts from BrickBuilt regions are lowest on the blood-containing media. Transcript levels and distribution of transcripts of BrickBuilt elements, using strain W83 RNAseq data, can be grouped generally into three categories. Group one, displaying relatively high transcript levels throughout the element on only one strand bridging the entire CDS-to-CDS gap, includes BrickBuilt_1, 3, and 13. BrickBuilt_13, comprised only of the internal 23 nt repeats. The element's expression correlates directly with that of the upstream gene, thus, the expression of Brick-Built_13 could be completely due to transcriptional read-through from adjacent genes. Consistent with this, BrickBuilt_13-associated transcripts are all on the negative strand. Group two, displaying low to medium intermittent transcript in tryptic soy and minimal media but none on blood agar, includes BrickBuilt_2, 5, 6, 8, 10, 11, 12, 14, 15, 16, 17, and 19. Group three, displaying no transcript yet adjacent to upstream transcript that is well beyond an annotated CDS, includes only BrickBuilt_9.

Additional information about BrickBuilt elements and their surrounding regions can be garnered from the above microarrays and RNAseq studies as well as additional studies that have been carried out with *P. gingivalis* under defined conditions. High-density tiling microarray of *P. gingivalis* strain W83 by Chen *et al.* showed differential expression of BrickBuilt elements at several loci [9]. Using a W83 strain based microarray, genes PG0626 and PG0089 were found to be aberrant in strain ATCC 33277, corresponding to BrickBuilt_ 11 and BrickBuilt_ 19 loci. The area in and around BrickBuilt_10 was identified as a potential sRNA (sRNA35) by Philips et al. [71]. The highest expression of the putative sRNA35 occurred during mid-log cultures grown under hemin excess conditions after an initial period of hemin starvation. Under the experimental methods used by Philips et al., no other Brick-Built loci were determined to be or be part of putative sRNAs expressed in response to hemin-variable growth conditions. BrickBuilt elements are not directly affected by FimR or LuxS regulation [72, 73]. However, genes surrounding BrickBuilt elements are regulated by LuxS. Lack of expression as well as partial expression of annotated genes surrounding BrickBuilt elements is evident from *P. gingivalis* strain W83 transcriptomic analyses by Hovik et al. [70]. Five of the conserved 30 genes flanking BrickBuilt elements in the W83 genome are predicted to

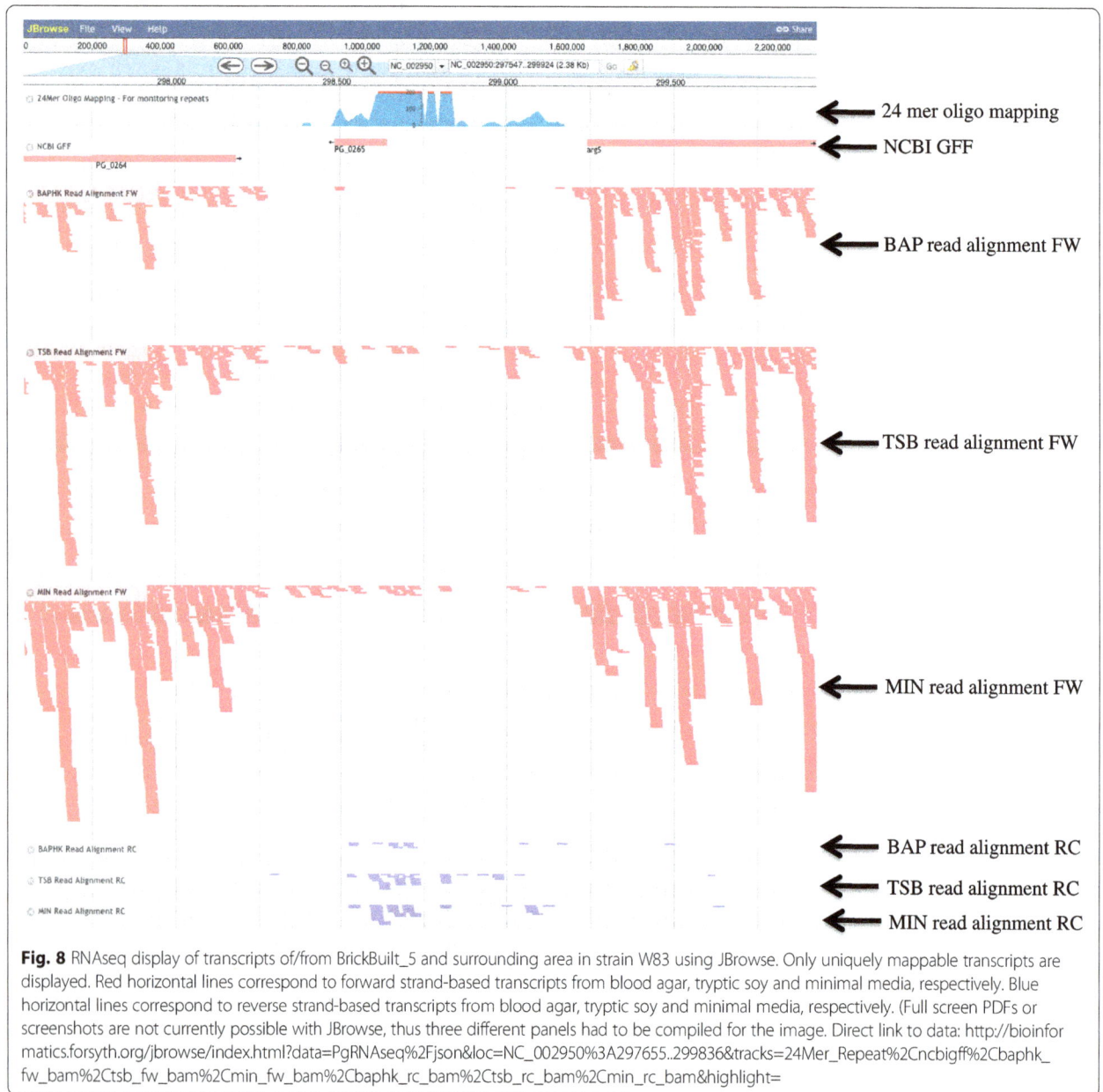

Fig. 8 RNAseq display of transcripts of/from BrickBuilt_5 and surrounding area in strain W83 using JBrowse. Only uniquely mappable transcripts are displayed. Red horizontal lines correspond to forward strand-based transcripts from blood agar, tryptic soy and minimal media, respectively. Blue horizontal lines correspond to reverse strand-based transcripts from blood agar, tryptic soy and minimal media, respectively. (Full screen PDFs or screenshots are not currently possible with JBrowse, thus three different panels had to be compiled for the image. Direct link to data: http://bioinformatics.forsyth.org/jbrowse/index.html?data=PgRNAseq%2Fjson&loc=NC_002950%3A297655..299836&tracks=24Mer_Repeat%2Cncbigff%2Cbaphk_fw_bam%2Ctsb_fw_bam%2Cmin_fw_bam%2Cbaphk_rc_bam%2Ctsb_rc_bam%2Cmin_rc_bam&highlight=

not be expressed and 11 (including 3 of the 5 'not expressed') give partial or abortive transcripts in blood, tryptic soy or minimal media.

The genomic association with haem biosynthesis and pigmentation-associated genes in conjunction with transcriptional data from RNAseq and microarray studies may point to regulation of BrickBuilt regions by haem or iron. DNA tandem repeats have been shown previously to affect transcription of iron and haem-associated genes [18, 74].

E. coli expression vectors

Promoter probe vectors pCB182 and pCB192 were used to determine the potential for transcription and

transcriptional regulation of the full BrickBuilt element and segments. Four potential promoter sites were hypothesized based on previous RNAseq and microarray data (Fig. 9). Four configurations of the leader, tail and element were constructed using BrickBuilt_5 as a template: full element in leader-to-tail orientation ('normal', with tail abutting *lacZ*); full element in tail-to-leader orientation ('reverse', with flipped leader abutting *lacZ*); tail-only in reverse orientation; and leader-only in forward orientation. The reverse orientation of the full element, with the beginning of the leader abutting the promoter-less *lacZ*, displayed the greatest promoter activity of the four constructs (Fig. 10). All four constructs displayed statistically

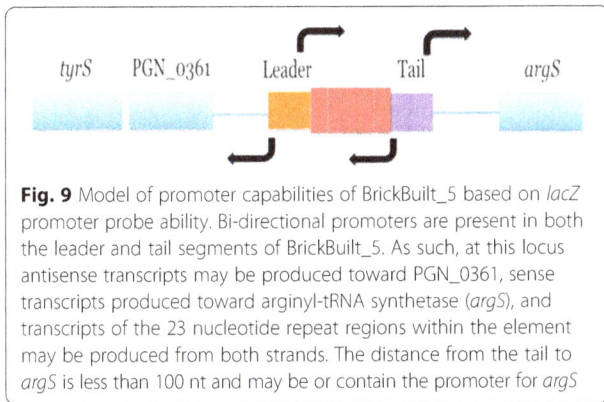

Fig. 9 Model of promoter capabilities of BrickBuilt_5 based on *lacZ* promoter probe ability. Bi-directional promoters are present in both the leader and tail segments of BrickBuilt_5. As such, at this locus antisense transcripts may be produced toward PGN_0361, sense transcripts produced toward arginyl-tRNA synthetase (*argS*), and transcripts of the 23 nucleotide repeat regions within the element may be produced from both strands. The distance from the tail to *argS* is less than 100 nt and may be or contain the promoter for *argS*

significant expression under heterologous expression in *E. coli*. No expression from the vectors lacking inserts was seen on plates with X-gal, while each insert-containing construct showed blue colonies due to expression by 24 h (Additional file 6: Figure S5). Expression from these constructs demonstrates bi-directional promoter ability (when tested in an *E. coli* system) within the tail segment as well as in the leader segment facing out of the element.

Conclusions

We identified and provided preliminary characterization of a genetic element, 'BrickBuilt', in the genome of *Porphyromonas gingivalis*. BrickBuilt appears to be a MITE-like element that has trapped a 23 nt direct repeat; propagating itself and the direct repeat throughout the genome. From promoter-less *lacZ* assays and analyses of previous microarray and RNAseq data we determined certain BrickBuilt elements contain promoter elements capable of bi-directional transcription. Given the element's exclusively intergenic locations and surrounding gene directionality, these transcripts may serve to regulate expression of surrounding genes. Relative stability of locations, overall copy number and expression levels of the elements throughout the sequenced *P. gingivalis* genomes point to neutral or advantageous maintenance of BrickBuilt.

Further sequencing projects and phylogenomics will be necessary to determine which other species and strains contain the BrickBuilt element and at what evolutionary point these species and/or strains diverged. Additionally, strain-specific experimental evolution and plasmid-based recombination systems could be employed to determine mutation rates within BrickBuilt in comparison to the rest of the genome, whether and how 23 nt repeats expand and contract at a given locus, and how recombinogenic the elements are.

With respect to 'mobility' of a whole or partial BrickBuilt element, several experimental setups could be

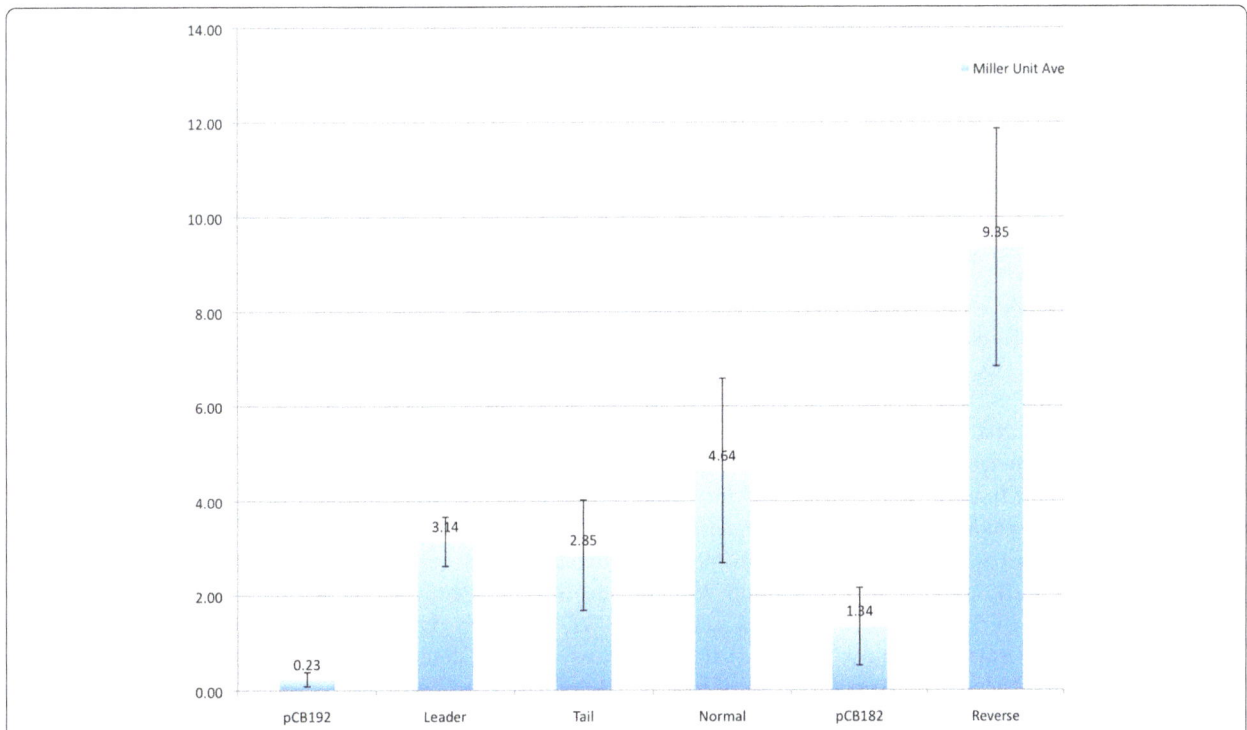

Fig. 10 ONPG assays for promoter capabilities of BrickBuilt_5 based on *lacZ* promoter probe constructs. Promoter-less *lacZ* backbone vectors pCB182 and pCB192 give low apparent β –Galactosidase activity. β –Galactosidase activity measured through ONPG cleavage after 3 h incubation at 28 °C. Error bars represent standard deviation between biological replicates in triplicate. Statistical significance determined by *t*-test ($p < 0.05$)

considered. First, inducing expression of endogenous transposases in order to mobilize BrickBuilt elements. One would need to initially determine under what conditions each transposase type in *P. gingivalis* is expressed, then induce expression and either PCR or sequence target BrickBuilt locations. Additionally, whole genome sequence could be employed to find the locations of element movement or duplication. Second, exogenous transposases could be expressed in *P. gingivalis*. Given the specificity of some transposases, a panel may need to be tested. Third, BrickBuilt elements could be introduced on plasmids into other bacterial species, specifically other *Bacteroidetes*, in order to try to obtain insertion into the heterologous host.

Adding BrickBuilt to the list of transposable and repetitive elements types in *P. gingivalis* brings the current total to 4 MITEs (or MITE-like elements), 11 ISs, 2 Ctn and 1 Tn. The ORFs and total base pairs encompassed by these elements constitute an impressive proportion of the genome. When compiled, the total percent of the *P. gingivalis* genome encoded by MITEs is 1 %; 0.44 % from BrickBuilt elements, 0.39 % from MITEPgRS elements, and the remaining 0.17 % from MITE700 and MITE239 family elements.

The ability of several of these elements being involved in genome evolution has been established [47, 48]. However, the full effects of these elements on genome stability and evolution as well as transcriptional, translational and post-translational response to stimuli remains to be experimentally determined.

Methods
Genomes/Strains
Genome sequence FASTA and GenBank files were downloaded from the NCBI database. At the time of this research, strains ATCC 33277, TDC60, W83, HG66 and JCVI SC001 were available as completed sequencing and assembly projects (ATCC 33277, TDC60 and W83 as 'gapless chromosome' status, HG66 as a single contig, and JCVI SC001 as a draft of many stitched contigs) [10, 13–16]. The five sequenced wild-type strains are disparate based on origin or lineage: W83 isolated in Germany (1950's) from an oral lesion; ATCC 33277 was isolated in the USA (1980's) from subgingival plaque; TDC60 was isolated in Japan (2011) from an oral lesion; HG66 isolated in the USA (1989) from a dental school patient; and JCVI SC001 was isolated in the USA (2013) from a hospital sink. The sequencing projects utilized different sequencing and assembly methods; each providing a *de novo* assembly. The JCVI SC001 genome sequence contained unidentified bases and residual gaps in the sequencing after the completed project.

Sequence analysis, clustering, alignment, phylogeneics/phylogenomics
NCBI BLAST suites were utilized to determine locations, structure and potential protein-coding capacity of the MITEs [43]. Query inputs were FASTA sequences taken directly from NCBI genome sequencing projects. For initial characterizations prior to determining species-specificity of the elements, the entire NCBI sequence database was queried. Following determination that the elements were only found (as of 11/2013) in the genomes of *P. gingivalis* strains, subsequence queries were focused to either the *P. gingivalis* species as a whole or specific *P. gingivalis* strains. Megablast, discontiguous megablast and BLASTn program selections for search optimization were all used in determining species-specificity as well as genome localizations.

MultAlin, Clustal Omega and the MEME suite were used to perform DNA-based and amino acid-based multiple alignments of the MITEs to determine conserved nucleotides and the start and stop points of the elements as well as proteins surrounding the MITEs. Amino acid-based alignments were used to determine whether the surrounding genes had structural domains at either the 5′ or 3′ ends that could potentially account for or facilitate MITE localization [75–77].

The BioCyc sequence pattern search tool 'PatMatch' was used to determine the number and genomic location of *P. gingivalis* MITEs in strain ATCC 33277 [78]. PatMatch indentifies potential sites given variations in the consensus sequences of the MITE direct repeats, TIRs, 'leader' and 'tail' regions because different mismatch numbers are allowed. Query inputs were nucleotide consensus sequences determined for each of the given parts of the MITE. Both DNA strands, as well as intergenic and coding sequences, were queried separately. Mismatches of '0' through '3' were allowed, with the constraint of the 'mismatch type' being a substitution.

The Tandem Repeats Database software was used for determining all types of tandem repeat elements in the *P. gingivalis* genomes (strains ATCC 33277 and W83 hosted on the server as of 12/2013), and to determine whether the tandem repeats or MITE as a whole was conserved in other sequenced species [79]. BLAST query of the entire bacterial and viral tandem repeat database was carried out using the FASTA sequence downloaded from NCBI for *P. gingivalis* strain ATCC 33277 MITE. The Tandem Repeats Finder software was used to determine the composition of the *P. gingivalis* tandem repeat element [80]. 'Basic' sequence analysis was selected for queries. Tandem Repeats Finder was also used to determine repeat conservation within and between loci as well as where a given element started and ended.

The Geneious software platform (version R8) was used to download, store, deposit, manipulate and query *P. gingivalis* genomes and BrickBuilt MITEs [81].

MITE and surrounding coding Sequences' nucleic acid and protein motif analysis

The Pfam and InterProScan databases and programs software were used to determine the presence and characteristics of nucleic acid and protein motifs [82, 83]. Query inputs were FASTA sequences from NCBI download files. For Pfam, an E-value of 1.0 and checking Pfam-B motifs were selected options prior to submission.

ExPASy Translate Tool software was used to determine whether the MITEs potentially encoded proteins, and thus are not strictly nucleic acid elements [84]. Genetic code option 'standard' was used for all queries. All six possible frames of translation were considered.

Modeling and structure prediction programs Mfold, RNAstructure and RegRNA2.0 were used to predict potential 2-D structures formed by MITE DNA and RNA [53–55]. Default options for programs in regard to structure prediction were chosen.

Cloning and reporter strains, media and growth conditions

Escherichia coli DH5α and TOP10 were used for cloning, plasmid maintenance and transcriptional assays. Ampicillin (100 μg/ml) was used when appropriate for prevention of contamination as well as isolation and maintenance of transformants containing plasmids. Strains were grown and maintained on LB [Lennox] agar or in LB [Lennox] broth (Invitrogen).

PCR primers containing BamHI and XmaI (NEB) restriction sites were designed immediately flanking the BrickBuilt_5 MITE associated with PGN_0361 (Additional file 2: Table S2). PCR products were generated using GoTaqLong Master Mix (Promega) and resultant bands were cloned into vector pCR TOPO-XL (Invitrogen), which was transformed into *E. coli* DH5α. Transformants were selected for kanamycin resistance and clones were confirmed by restriction digest and sequencing.

To generate constructs using the promoter probe vectors pCB182 and pCB192, pCR2.1 and pCR-TOPO-XL cloned BrickBuilt MITE constructs and pCB182/pCB192 were double-digested with BamHI (NEB) and XbaI (NEB) or BamHI and XmaI (NEB) and then transformed into *E. coli* TOP10. Transformants were selected for ampicillin resistance generated by an insertion event. Clones were confirmed by restriction digest and sequencing.

Transcriptional analyses

BROP, specifically the 'Genomics Tools for Oral Pathogens' and 'Microbial Transcriptome Database' sections of the resource, were used to determine genome location, characteristics of coding sequencings surrounding MITEs, differences between strains as well as transcriptome data [85]. Over the course of the research, two different variations of the RNAseq data for *P. gingivalis* strain W83 were supported, one directly on BROP and then a later form on JBrowse. The JBrowse form gives greater functionality in displaying data and visualization [86]. Under the 'Genomics Tools for Oral Pathogens' subset, the 'GenomeViewer' function was used to compare genome arrangements of *P. gingivalis* strains ATCC 33277 and W83 with relation to MITEs, as well as display the previously performed microarray data (under the strain W83 section) for MITE-associated genome areas under the three different nutrient conditions (the same conditions performed in the RNAseq) [9, 70].

β –galactosidase assays

Escherichia coli strains were grown and maintained in Luria-Bertani (LB) media supplemented with ampicillin (100 μg/l) as required. PCR primers and synthesized oligos used for strain constructions are listed in Additional file 2: Table S2. The pCB182 and pCB192 vectors lack promoters but contain translational start codons [87]. As such, gene expression of *lacZ*, and in turn protein expression of LacZ read out through β –galactosidase activity, should be the result of promoter activity from fragments cloned into the vector. β -galactosidase assays were performed under plate-based (X-Gal) and broth-based (ONPG) setups. For plate-based assays, frozen stock cultures of the BrickBuilt MITE derivatives transcriptionally fused to *lacZ* in their respective *E. coli* strains were plated onto LB agar containing X-gal and ampicillin. For broth-based assays, cultures of the BrickBuilt MITE derivatives transcriptionally fused to *lacZ* in their respective *E. coli* strains were grown in LB broth for 3 h with shaking at 37 °C. An aliquot of each culture (500 μl) was added to a lysis and assay solution mixture (500 μl), vortexed, and then incubated at 28 °C for 3 h. Color development was measured spectrophotometrically at OD_{420} nm and cell debris at OD_{550} nm. Respective Miller units were calculated as previously described [88].

All data, genome sequences as well as RNAseq and microarray, are currently available in public repositories and publications related to these data have been referenced. Locations of the MITE sequences in *P. gingivalis* and *P. gulae* strains will be deposited to NCBI such that identifiers and notes can be amended to the graphic outputs of sequence files.

Additional files

Additional file 1: Figure S1. Tandem Repeat Finder analysis of *P. gingivalis* strain ATCC 33277 BrickBuilt_5. Overall statistics of the repeats/repeat region found within BrickBuilt_5; 23 nt repeat indicies of the element relative to the entire element, period size, copy number, consensus size, percent matches, percent InDels, alignment score, percent composition for each nucleotide and entropy measure based on percent composition. The individual locations of mismatches and InDels within BrickBuilt_5 are shown as positions marked by stars (*). (PDF 92 kb)

Additional file 2: Table S1. Loci and nucleotide sites of BrickBuilt elements across four sequenced and annotated *P. gingivalis* strains. BrickBuilt elements are situated intergenically between the genes noted. Grayed-out boxes represent loci at which BrickBuilt is aberrant. **Table S2.** Primers for sequencing and cloning. Sequence of oligos ordered for cloning directly into promoter-probe vectors. (XLSX 66 kb)

Additional file 3: Figure S2. *P. gingivalis* strain SJD2 assembly showing an 'assembly gap' at the site of BrickBuilt elements from strains ATCC 33277, W83, TDC60 and HG66. The same genes flanking the assembly gap are found flanking BrickBuilt_11 in strains ATCC 33277, W83, TDC60 and HG66. The top dark green track depicts individual contigs (a total of 140 for this strain), the second dark green track depicts individual scaffolds (a total of 117 for this strain), the yellow track shows predicted coding sequences, the light green track shows predicted genes, and the bottom track shows the assembly gap as a rightward-pointing triangle. (PNG 19 kb)

Additional file 4: Figure S3. BrickBuilt_5 region MAFFT alignment and PHYML tree of *P. gulae* and *P. gingivalis* strains. All COT_052 *P. gulae* strains were sequenced/deposited during the preparation of the manuscript. Additionally, all *P. gulae* strains are currently scaffold or contig assemblies; none are completed chromosomes and thus are also not available for default BLASTn query on NCBI. The aligned bases between 2,500-3,800 contain the BrickBuilt_5 MITE; flanking regions contain the same 2 upstream and downstream genes in all strains. Of the 12 nodes in the tree, 8 have a bootstrap value of greater than 85 (100 bootstrap iterations). In the 'consensus identity' track, green indicates sites of complete conservation, yellow of partial conservation and red of little conservation. Within each of the 15 strain tracks the black lines or blocks indicate sites that deviate from the consensus at that given site. (PDF 108 kb)

Additional file 5: Figure S4. Microarray display of transcripts in BROP MTD database of BrickBuilt_5 and surrounding area in strain W83. Tracts represent positive and negative strand blood agar (top), tryptic soy broth (middle) and minimal media (bottom), respectively. BrickBuilt_5 noted with black bracket.

Additional file 6: Figure S5. X-gal and ONPG assays of promoter capabilities of BrickBuilt_5 based on *lacZ* promoter probe constructs. Promoter-less *lacZ* vectors pCB182 and pCB192 give no apparent β –Galactosidase activity. BrickBuilt_5 leader and tail oligos (Eurofins Operon) were cloned into vector pCB192. Full-length BrickBuilt_5 was cloned into pCB192, 'Normal', with the tail segment of the element upstream of *lacZ* facing in the same orientation (tail abutting *lacZ*). Full-length BrickBuilt_5 was cloned into pCB182, 'Reverse', with the leader segment of the element upstream of *lacZ* facing in the same orientation (flipped leader abutting *lacZ*). X-gal activity visualized after 24 h incubation. Top of two liquid assays of ONPG activity visualized after 3 h incubation; bottom after 24 h incubation.

Abbreviations

MITE: Miniature Inverted-repeat Transposable Element; RE: Repetitive Element; TE: Transposable Element; TIR: Terminal Inverted Repeat; TSD: Target Site Duplication; PCR: Polymerase Chain Reaction; BROP: Bioinformatics Resource Oral Pathogens; MTD: Microbial Transcriptome Database; BLAST: Basic Local Alignment Search Tool; OD: Optical Density; X-gal: 5-bromo-4-chloro-3-indolyl-beta-D-galacto-pyranoside; ONPG: O-Nitrophenyl-β-D-galactopyranoside.

Competing interests

The authors declare that they have no competing interests.

Authors' contributions

BAK conceived of the study, participated in its design and coordination, carried out molecular genetics, carried out bioinformatic analyses and drafted the manuscript. TC participated in study design and coordination, carried out bioinformatic analyses and drafted the manuscript. AK participated in study design and coordination and carried out molecular genetics. JCS participated in study design and coordination and carried out molecular genetics. MJD participated in study design and coordination and drafted the manuscript. LTH conceived of the study, participated in its design and coordination and drafted the manuscript. All authors read and approved the final manuscript.

Acknowledgements

Drs. Andrea Pauli and Eivind Valen working in the laboratory of Dr. Alexander Schier at Harvard University for their help and knowledge in non-coding RNA biology and RNA regulation. Dr. Michael Malamy at Tufts University for his expertise in mobile genetic elements and plasmid biology. Dr. Carla Cugini at Rutgers School of Dental Medicine for her expertise *P. gingivalis* biology.

Funding

This project was supported by grants from the National Institute of Dental & Craniofacial Research, F31 DE022491 (BAK), R01 DE024308 (LTH and MJD) and R01 DE015931 (MJD). The content is solely the responsibility of the authors and does not necessarily represent the official views of the National Institute of Dental & Craniofacial Research or the National Institutes of Health.

Author details

[1]Department of Molecular Biology and Microbiology, Tufts University Sackler School of Biomedical Sciences, Boston, MA 02111, USA. [2]Department of Microbiology, The Forsyth Institute, Cambridge, MA 02142, USA.

References

1. Hajishengallis G, Liang S, Payne MA, Hashim A, Jotwani R, Eskan MA, et al. Low-abundance biofilm species orchestrates inflammatory periodontal disease through the commensal microbiota and complement. Cell Host Microbe. 2011;10:497–506.
2. Curtis MA, Zenobia C, Darveau RP. The relationship of the oral microbiotia to periodontal health and disease. Cell Host Microbe. 2011;10:302–6.
3. Maley J, Roberts IS. Characterisation of IS1126 from *Porphyromonas gingivalis* W83: A new member of the IS4 family of insertion sequence elements. FEMS Microbiol Lett. 1994;123:219–24.
4. Wang C, Bond VC, Genco CA. Identification of a second endogenous *Porphyromonas gingivalis* insertion element. J Bacteriol. 1997;179:3808–12.
5. Lewis JP, Macrina FL. IS195, an insertion sequence-like element associated with protease genes in *Porphyromonas gingivalis*. Infect Immun. 1998;66:3035–42.
6. Sawada K, Kokeguchi S, Hongyo H, Sawada S, Miyamoto M, Maeda H, et al. Identification by subtractive hybridization of a novel insertion sequence specific for virulent strains of *Porphyromonas gingivalis*. Infect Immun. 1999;67:5621–5.
7. Califano JV, Kitten T, Lewis JP, Macrina FL, Fleischmann RD, Fraser CM, et al. Characterization of *Porphyromonas gingivalis* insertion sequence-like element ISPg5. Infect Immun. 2000;68:5247–53.
8. Califano JV, Arimoto T, Kitten T. The genetic relatedness of *Porphyromonas gingivalis* clinical and laboratory strains assessed by analysis of insertion sequence (IS) element distribution. J Periodontal Res. 2003;38:411–6.
9. Chen T, Hosogi Y, Nishikawa K, Abbey K, Fleischmann RD, Walling J, et al. Comparative whole-genome analysis of virulent and avirulent strains of *Porphyromonas gingivalis*. J Bacteriol. 2004;186:5473–9.
10. Naito M, Hirakawa H, Yamashita A, Ohara N, Shoji M, Yukitake H, et al. Determination of the genome sequence of *Porphyromonas gingivalis* strain ATCC 33277 and genomic comparison with strain W83 revealed extensive genome rearrangements in *P. gingivalis*. DNA Res. 2008;15:215–25.
11. Naito M, Sato K, Shoji M, Yukitake H, Ogura Y, Hayashi T, et al. Characterization of the *Porphyromonas gingivalis* conjugative transposon

CTnPg1: Determination of the integration site and the genes essential for conjugal transfer. Microbiology. 2011;157:2022–32.

12. Bainbridge BW, Hirano T, Grieshaber N, Davey ME. Deletion of a 77-base-pair inverted repeat element alters the synthesis of surface polysaccharides in *Porphyromonas gingivalis*. J Bacteriol. 2015;197:1208–20.

13. Nelson KE, Fleischmann RD, DeBoy RT, Paulsen IT, Fouts DE, Eisen JA, et al. Complete genome sequence of the oral pathogenic bacterium *Porphyromonas gingivalis* strain W83. J Bacteriol. 2003;185:5591–601.

14. Watanabe T, Maruyama F, Nozawa T, Aoki A, Okano S, Shibata Y, et al. Complete genome sequence of the bacterium *Porphyromonas gingivalis* TDC60, which causes periodontal disease. J Bacteriol. 2011;193:4259–60.

15. McLean JS, Lombardo M, Ziegler MG, Novotny M, Yee-Greenbaum J, Badger JH, et al. Genome of the pathogen *Porphyromonas gingivalis* recovered from a biofilm in a hospital sink using a high-throughput single-cell genomics platform. Genome Res. 2013;23:867–77.

16. Siddiqui H, Yoder-Himes DR, Mizgalska D, Nguyen KA, Potempa J, Olsen I. Genome sequence of *Porphyromonas gingivalis* strain HG66 (DSM 28984). Genome Announc. 2014;2:doi:10.1128/genomeA.00947-14.

17. Treangen TJ, Abraham AL, Touchon M, Rocha EP. Genesis, effects and fates of repeats in prokaryotic genomes. FEMS Microbiol Rev. 2009;33:539–71.

18. Zhou K, Aertsen A, Michiels CW. The role of variable DNA tandem repeats in bacterial adaptation. FEMS Microbiol Rev. 2014;38:119–41.

19. Padeken J, Zeller P, Gasser SM. Repeat DNA in genome organization and stability. Curr Opin Genet Dev. 2015;31:12–9.

20. Siguier P, Gourbeyre E, Chandler M. Bacterial insertion sequences: Their genomic impact and diversity. FEMS Microbiol Rev. 2014;38:865–91.

21. Piégu B, Bire S, Arensburger P, Bigot Y. A survey of transposable element classification systems - A call for a fundamental update to meet the challenge of their diversity and complexity. Mol Phylogenet Evol. 2015;86:90–109.

22. Gonzalez J, Petrov D. MITEs - the ultimate parasites. Science. 2009;325:1352–3.

23. Ilyina TS. Miniature repetitive mobile elements of bacteria: Structural organization and properties. Mol Genet Microbiol Virol. 2010;25:139–47.

24. Fattash I, Rooke R, Wong A, Hui C, Luu T, Bhardwaj P, et al. Miniature inverted-repeat transposable elements: Discovery, distribution, and activity. Genome. 2013;56:475–86.

25. Darmon E, Leach DR. Bacterial genome instability. Microbiol Mol Biol Rev. 2014;78:1–39.

26. Feschotte C, Jiang N, Wessler SR. Plant transposable elements: Where genetics meets genomics. Nat Rev Genet. 2002;3:329–41.

27. Naito K, Zhang F, Tsukiyama T, Saito H, Hancock CN, Richardson AO, et al. Unexpected consequences of a sudden and massive transposon amplification on rice gene expression. Nature. 2009;461:1130–4.

28. Lu C, Chen J, Zhang Y, Hu Q, Su W, Kuang H. Miniature inverted-repeat transposable elements (MITEs) have been accumulated through amplification bursts and play important roles in gene expression and species diversity in oryza sativa. Mol Biol Evol. 2012;29:1005–17.

29. Wang X, Tan J, Bai J, Deng X, Li Z, Zhou C, et al. Detection and characterization of miniature inverted-repeat transposable elements in "candidatus liberibacter asiaticus". J Bacteriol. 2013;195:3979–86.

30. Bureau TE, Wessler SR. Tourist: A large family of small inverted repeat elements frequently associated with maize genes. Plant Cell. 1992;4:1283–94.

31. Bureau TE, Wessler SR. Stowaway: A new family of inverted repeat elements associated with the genes of both monocotyledonous and dicotyledonous plants. Plant Cell. 1994;6:907–16.

32. Chen SL, Shapiro L. Identification of long intergenic repeat sequences associated with DNA methylation sites in *Caulobacter crescentus* and other alpha-proteobacteria. J Bacteriol. 2003;185:4997–5002.

33. Han Y, Korban SS. Spring: A novel family of miniature inverted-repeat transposable elements is associated with genes in apple. Genomics. 2007;90:195–200.

34. Nelson WC, Bhaya D, Heidelberg JF. Novel miniature transposable elements in thermophilic *Synechococcus* strains and their impact on an environmental population. J Bacteriol. 2012;194:3636–42.

35. Deng H, Shu D, Luo D, Gong T, Sun F, Tan H. Scatter: A novel family of miniature inverted-repeat transposable elements in the fungus *Botrytis cinerea*. J Basic Microbiol. 2013;53:815–22.

36. Sampath P, Murukarthick J, Izzah NK, Lee J, Choi HI, Shirasawa K, et al. Genome-wide comparative analysis of 20 miniature inverted-repeat transposable element families in *Brassica rapa* and *B. oleracea*. PLoS ONE. 2014;9.

37. Coates BS, Kroemer JA, Sumerford DV, Hellmich RL. A novel class of miniature inverted repeat transposable elements (MITEs) that contain hitchhiking (GTCY)n microsatellites. Insect Mol Biol. 2011;20:15–27.

38. Satovic E, Plohl M. Tandem repeat-containing MITEs in the clam *Donax trunculus*. Genome Biol Evol. 2013;5:2549–59.

39. Parkhill J, Achtman M, James KD, Bentley SD, Churcher C, Klee SR, et al. Complete DNA sequence of a serogroup A strain of *Neisseria meningitidis* Z2491. Nature. 2000;404:502–6.

40. Xia Q, Wang T, Taub F, Park Y, Capestany CA, Lamont RJ, et al. Quantitative proteomics of intracellular *Porphyromonas gingivalis*. Proteomics. 2007;7:4323–37.

41. Kuboniwa M, Hendrickson EL, Xia Q, Wang T, Xie H, Hackett M, et al. Proteomics of *Porphyromonas gingivalis* within a model oral microbial community. BMC Microbiol. 2009;9:98,2180-9-98.

42. Maeda K, Nagata H, Ojima M, Amano A. Proteomic and transcriptional analysis of interaction between oral microbiota *Porphyromonas gingivalis* and *Streptococcus oralis*. J Proteome Res. 2015;14:82–94.

43. Altschul SF, Gish W, Miller W, Myers EW, Lipman DJ. Basic local alignment search tool. J Mol Biol. 1990;215:403–10.

44. Tatusova T, Ciufo S, Fedorov B, O'Neill K, Tolstoy I. RefSeq microbial genomes database: New representation and annotation strategy. Nucleic Acids Res. 2015;43:3872.

45. Coil DA, Alexiev A, Wallis C, O'Flynn C, Deusch O, Davis I, et al. Draft genome sequences of 26 *Porphyromonas* strains isolated from the canine oral microbiome. Genome Announc. 2015;3:doi:10.1128/genomeA.00187-15.

46. Coenye T, Vandamme P. Characterization of mononucleotide repeats in sequenced prokaryotic genomes. DNA Res. 2005;12:221–33.

47. Duncan MJ. Genomics of oral bacteria. Crit Rev Oral Biol Med. 2003;14:175–87.

48. Tribble GD, Kerr JE, Wang B. Genetic diversity in the oral pathogen *Porphyromonas gingivalis*: Molecular mechanisms and biological consequences. Future Microbiol. 2013;8:607–20. Accessed 2 May 2015.

49. Hikosaka A, Kawahara A. Lineage-specific tandem repeats riding on a transposable element of MITE in *Xenopus* evolution: A new mechanism for creating simple sequence repeats. J Mol Evol. 2004;59:738–46.

50. Yang HP, Barbash DA. Abundant and species-specific DINE-1 transposable elements in 12 *Drosophila* genomes. Genome Biol. 2008;9:R39,2008-9-2-r39. Epub 2008 Feb 21.

51. Koressaar T, Remm M. Characterization of species-specific repeats in 613 prokaryotic species. DNA Res. 2012;19:219–30.

52. Halász J, Kodad O. Hegedus A. Plant Journal: Identification of a recently active Prunus-specific non-autonomous mutator element with considerable genome shaping force; 2014.

53. Zuker M. Mfold web server for nucleic acid folding and hybridization prediction. Nucleic Acids Res. 2003;31:3406–15.

54. Bellaousov S, Reuter JS, Seetin MG, Mathews DH. RNAstructure: Web servers for RNA secondary structure prediction and analysis. Nucleic Acids Res. 2013;41:W471–4.

55. Chang TH, Huang HY, Hsu JB, Weng SL, Horng JT, Huang HD. An enhanced computational platform for investigating the roles of regulatory RNA and for identifying functional RNA motifs. BMC Bioinformatics. 2013;14 Suppl 2:S4,2105-14-S2-S4. Epub 2013 Jan 21.

56. Nawrocki EP, Burge SW, Bateman A, Daub J, Eberhardt RY, Eddy SR, et al. Rfam 12.0: Updates to the RNA families database. Nucleic Acids Res. 2015;43:D130–7.

57. Zhou F, Tran T, Xu Y. Nezha, a novel active miniature inverted-repeat transposable element in *Cyanobacteria*. Biochem Biophys Res Commun. 2008;365:790–4.

58. Delihas N. Impact of small repeat sequences on bacterial genome evolution. Genome Biol Evol. 2011;3:959–73.

59. Zhang HH, Xu HE, Shen YH, Han MJ, Zhang Z. The origin and evolution of six miniature inverted-repeat transposable elements in *Bombyx mori* and *Rhodnius prolixus*. Genome Biol Evol. 2013;5:2020–31.

60. De Gregorio E, Silvestro G, Petrillo M, Carlomagno MS, Di Nocera PP. Enterobacterial repetitive intergenic consensus sequence repeats in *Yersiniae*: Genomic organization and functional properties. J Bacteriol. 2005;187:7945–54.

61. Petrillo M, Silvestro G, Di Nocera PP, Boccia A, Paolella G. Stem-loop structures in prokaryotic genomes. BMC Genomics. 2006;7:170.

62. Bertels F, Rainey PB. Within-genome evolution of REPINs: A new family of miniature mobile DNA in bacteria. PLoS Genet. 2011;7, e1002132.

63. Sato K. Por secretion system of *Porphyromonas gingivalis*. J Oral Biosci. 2011;53:187–96.

64. Seers CA, Slakeski N, Veith PD, Nikolof T, Chen YY, Dashper SG, et al. The RgpB C-terminal domain has a role in attachment of RgpB to the outer

membrane and belongs to a novel C-terminal-domain family found in
Porphyromonas gingivalis. J Bacteriol. 2006;188:6376–86.

65. Matus-Garcia M, Nijveen H, Van Passel MWJ. Promoter propagation in
prokaryotes. Nucleic Acids Res. 2012;40:10032–40.

66. Bennetzen JL, Wang H, eds. The Contributions of Transposable Elements to the
Structure, Function, and Evolution of Plant Genomes.; 2014; No. 65:505–30.

67. Slotkin RK, Martienssen R. Transposable elements and the epigenetic
regulation of the genome. Nat Rev Genet. 2007;8:272–85.

68. Feschotte C. Transposable elements and the evolution of regulatory
networks. Nat Rev Genet. 2008;9:397–405.

69. Lisch D. How important are transposons for plant evolution? Nat Rev Genet.
2013;14:49–61.

70. Høvik H, Wen-Han Y, Olsen I, Chen T. Comprehensive transcriptome analysis
of the periodontopathogenic bacterium *Porphyromonas gingivalis* W83.
J Bacteriol. 2012;194:100–14.

71. Phillips P, Progulske-Fox A, Grieshaber S. Grieshaber N. FEMS Microbiology
Letters: Expression of Porphyromonas gingivalis small RNA in response to
hemin availability identified using microarray and RNA-seq analysis; 2013.

72. Nishikawa K, Yoshimura F, Duncan MJ. A regulation cascade controls
expression of *Porphyromonas gingivalis* fimbriae via the FimR response
regulator. Mol Microbiol. 2004;54:546–60.

73. Hirano T, Beck DA, Demuth DR, Hackett M, Lamont RJ. Deep sequencing of
Porphyromonas gingivalis and comparative transcriptome analysis of a LuxS
mutant. Front Cell Infect Microbiol. 2012;2:79.

74. Chen S, Li X. Transposable elements are enriched within or in close
proximity to xenobiotic-metabolizing cytochrome P450 genes. BMC Evol
Biol. 2007;7:46.

75. Corpet F. Multiple sequence alignment with hierarchical clustering. Nucleic
Acids Res. 1988;16:10881–90.

76. Bailey TL, Boden M, Buske FA, Frith M, Grant CE, Clementi L, et al. MEME
SUITE: Tools for motif discovery and searching. Nucleic Acids Res.
2009;37:W202–8.

77. Sievers F, Wilm A, Dineen D, Gibson TJ, Karplus K, Li W, et al. Fast, scalable
generation of high-quality protein multiple sequence alignments using
clustal omega. Mol Syst Biol. 2011;7:539.

78. Karp PD, Paley SM, Krummenacker M, Latendresse M, Dale JM, Lee TJ, et al.
Pathway tools version 13.0: Integrated software for pathway/genome
informatics and systems biology. Brief Bioinform. 2010;11:40–79.

79. Gelfand Y, Rodriguez A, Benson G. TRDB–the tandem repeats database.
Nucleic Acids Res. 2007;35:D80–7.

80. Benson G. Tandem repeats finder: A program to analyze DNA sequences.
Nucleic Acids Res. 1999;27:573–80.

81. Kearse M, Moir R, Wilson A, Stones-Havas S, Cheung M, Sturrock S, et al.
Geneious basic: An integrated and extendable desktop software platform
for the organization and analysis of sequence data. Bioinformatics.
2012;28:1647–9.

82. Finn RD, Bateman A, Clements J, Coggill P, Eberhardt RY, Eddy SR, et al.
Pfam: The protein families database. Nucleic Acids Res. 2014;42:D222–30.

83. Jones P, Binns D, Chang HY, Fraser M, Li W, McAnulla C, et al. InterProScan
5: Genome-scale protein function classification. Bioinformatics.
2014;30:1236–40.

84. Artimo P, Jonnalagedda M, Arnold K, Baratin D, Csardi G, de Castro E, et al.
ExPASy: SIB bioinformatics resource portal. Nucleic Acids Res. 2012;40:W597–603.

85. Chen T, Abbey K, Deng W, Cheng M. The bioinformatics resource for oral
pathogens. Nucleic Acids Res. 2005;33:W734–40.

86. Westesson O, Skinner M, Holmes I. Visualizing next-generation sequencing
data with JBrowse. Brief Bioinform. 2013;14:172–7.

87. Schneider K, Beck CF. Promoter-probe vectors for the analysis of divergently
arranged promoters. Gene. 1986;42:37–48.

88. Wang XG, Lin B, Kidder JM, Telford S, Hu LT. Effects of environmental
changes on expression of the oligopeptide permease (opp) genes of
Borrelia burgdorferi. J Bacteriol. 2002;184:6198–206.

6

Ribosomal protein and biogenesis factors affect multiple steps during movement of the *Saccharomyces cerevisiae* Ty1 retrotransposon

Susmitha Suresh[1,3], Hyo Won Ahn[2], Kartikeya Joshi[1], Arun Dakshinamurthy[1,4], Arun Kannanganat[1], David J. Garfinkel[2] and Philip J. Farabaugh[1*] (iD)

Abstract

Background: A large number of *Saccharomyces cerevisiae* cellular factors modulate the movement of the retrovirus-like transposon Ty1. Surprisingly, a significant number of chromosomal genes required for Ty1 transposition encode components of the translational machinery, including ribosomal proteins, ribosomal biogenesis factors, protein trafficking proteins and protein or RNA modification enzymes.

Results: To assess the mechanistic connection between Ty1 mobility and the translation machinery, we have determined the effect of these mutations on ribosome biogenesis and Ty1 transcriptional and post-transcriptional regulation. Lack of genes encoding ribosomal proteins or ribosome assembly factors causes reduced accumulation of the ribosomal subunit with which they are associated. In addition, these mutations cause decreased Ty1 + 1 programmed translational frameshifting, and reduced Gag protein accumulation despite at least normal levels of Ty1 mRNA. Several ribosome subunit mutations increase the level of both an internally initiated Ty1 transcript and its encoded truncated Gag-p22 protein, which inhibits transposition.

Conclusions: Together, our results suggest that this large class of cellular genes modulate Ty1 transposition through multiple pathways. The effects are largely post-transcriptional acting at a variety of levels that may include translation initiation, protein stability and subcellular protein localization.

Keywords: Retrotransposition, Host factors, Programmed frameshifting, Ribosomal protein insufficiency, Ribosome biogenesis

Background

The *Saccharomyces cerevisiae* Ty (**T**ransposons of **y**east) retrotransposons are members of the LTR (long terminal repeat) group and are similar to retroviruses both structurally and functionally [1, 2]. Like retroviruses, Ty elements undergo reverse transcription that occurs within virus-like particles (VLPs) formed from structural and enzymatic proteins encoded by two genes, *GAG* and *POL*. Ty elements are valuable as models for human retroviruses; several groups have exploited yeast genetic tools to identify genes encoding Ty host factors that modulate transposition. Knowing how these factors affect Ty retro-transposition can provide clues as to what host processes affect retrovirus or retrotransposon replication and pathogenicity. Genome-wide forward genetic screens identified host factors that are required for (cofactor genes) or prevent (restriction genes) retrotransposition by Ty1 [3–7]. The most salient feature of the genes identified in these screens is the diversity of function of their encoded products, including roles in transcription, chromatin structure and modification, intracellular signaling, cytoplasmic protein synthesis, DNA repair, RNA processing and cell cycle regulation among others. Among the most statistically overrepresented host cofactor genes are those encoding

* Correspondence: farabaug@umbc.edu
[1]Department of Biological Sciences and Program in Molecular and Cell Biology, University of Maryland Baltimore County, Baltimore, MD 21250, USA
Full list of author information is available at the end of the article

cytoplasmic ribosomal proteins [7] suggesting that Ty transposition might depend on efficient biogenesis of ribosomes. Host factors for other plus stranded viruses in yeast have not been found to be as diverse. Prominent among these is the endogenous L-A virus of *S. cerevisiae*. It supports the replication of satellite dsRNA molecules, one of which encodes a peptide toxin lethal to uninfected cells [8]. Maintenance of L-A and the satellites depends on availability of the large (60S) ribosomal subunit [9], implying a more global role of protein synthesis for positive stranded viruses. Because, unlike Ty1, L-A has no integrated DNA form, it does not share a dependence on genes such as those involved in transcription, chromatin recombination and DNA repair. Its dependence on 60S abundance may relate to the L-A mRNA not being polyadenylated since polyA tails facilitate 60S joining during translation initiation [10]. Thus, reduced 60S availability could reduce L-A mRNA translation relative to bulk poly(A)$^+$ mRNA (reviewed in [11]). Ty1 expresses an abundant, poly(A)$^+$ mRNA and depends on both 40S and 60S availability so its dependence on the translation machinery may have a different origin. Also, only three Ty1 cofactor genes were also identified as L-A host factors—*SKI1/ KEM1/XRN1, SKI2* and *SKI8*—and their Ty1 phenotype is opposite to their effect of L-A virus; these factors are required for Ty1 mobility but restrict L-A propagation. Therefore, Ty1 and L-A occupy distinct genetic niches with respect to their dependence on host proteins.

Ty elements, as well as many viruses and virus-like elements including L-A, employ an unusual translational control mechanism—programmed translational frameshifting [12]. The Ty and L-A frameshift mechanisms are distinct. Ty elements employ +1 frameshifting, in which translation shifts one base in the downstream or 3′ direction, while L-A uses -1 frameshifting, shifting one base in the opposite direction. The Ty1-encoded enzymatic (Pol) protein is encoded as a fusion to the upstream-encoded Gag structural protein by +1 frameshifting at a 7 nt RNA signal [13]. A similar or identical signal is used in all but the Ty5 element. The frequency of Ty1 frameshifting is approximately 40 % measured in a reporter gene construct containing only the frameshift signal [13]. In the intact Ty1 element the Gag-Pol protein is expressed at 3 % the amount of the Gag protein, suggesting a further ~10-fold reduction in expression of Gag-Pol, which may result from either a translational effect during elongation through the *POL* gene or reduced stability of Gag-Pol relative to Gag protein; changes to this ratio blocked retrotransposition [14]. Altered Gag to Gag-Pol stoichiometry also reduces transposition of many other viruses [15–20]. Because retrotransposition frequency requires a specific level of programmed frameshifting, that process could explain the dependence of retrotransposition on efficient ribosome biogenesis.

In addition to cellular cofactor and restriction genes that affect Ty1 transposition, a protein expressed from subgenomic internally initiated Ty1i transcripts (Gag-p22) containing the C-terminal half of Gag is a self-encoded restriction factorthat inhibits transposition and controls Ty1 copy number [21]. Gag-p22 antagonizes VLP function by interfering with assembly of VLPs and assembly foci [22], called T-bodies [23] or retrosomes [24]. Well-known Ty1 cofactors such as *SPT3* and *XRN1*, which are implicated in full-length transcription [25], and RNA turnover and VLP function [26–28], respectively, influence the level of Ty1i RNA [21]. However, additional cellular genes that modulate Ty1i/Gag-p22 expression remain to be discovered, and in fact, may be present in Ty1 cofactor or restriction gene collections. A clue to what types of factors might influence this effect is the fact that formation of retrosomes requires co-translational insertion of the Ty1 Gag protein into the endoplasmic reticulum (ER) [22]. Interfering with ER insertion blocks formation of retrosomes and the Gag protein produced is more rapidly degraded. This suggests that some Ty1 cofactor genes might encode factors required for Gag ER insertion.

To gain a more thorough understanding of the relationship between ribosome biogenesis and Ty1 transposition, we analyzed the effect on Ty1 transposition of chromosomal deletions that remove structural proteins of the 40S and 60S ribosomal subunits as well as proteins involved in ribosomal processing or protein synthesis. We show that translation-associated cofactor deletion mutants affect Ty1 transposition through a combination of mechanisms. Most of the mutants tested show reduced accumulation of the corresponding ribosomal subunit, significantly decreased +1 programmed translational frameshifting at the Ty1 site, and reduced expression of Gag protein despite expressing at least normal amounts of Ty1 mRNA. Interestingly, several ribosome subunit mutants also express more Ty1i RNA relative to Ty1 mRNA and significant amounts of Gag-p22 and its C-terminally processed product, Gag-p18, consistent with the idea that producing more of the transpositional inhibitor Gag-p22 contributes to the Ty1 defects in these mutants [21]. Together, our results suggest that multiple post-transcriptional processes are required for optimal Ty1 transposition.

Methods
Media and yeast strains
Yeast genetic techniques and media were used as described previously [29, 30]. Strains from the haploid *MAT*α deletion collection [31] were obtained from Invitrogen (Carlsbad, CA). The mutant strains, constructed in BY4742 (*MAT*α *his3-Δ1 leu2-Δ0 lys2-Δ0 ura3-Δ0*) [32] were transformed with pJC573, a *URA3*-based integrating plasmid carrying an active Ty1 element tagged

with a modified indicator gene *his3-AI*, which cannot recombine with the *his3-Δ1* allele present in BY4742 to generate a functional *HIS3* gene [5]. The centromere-based Ty1 overexpression plasmid pGTy1*his3-AI* [21] was also introduced into BY4742 and an isogenic *rpl1BΔ* mutant.

Frequency of Ty1*his3-AI* mobility

Mobility of Ty elements in each mutant strain was determined essentially as described [5, 33]. Strains were streaked for single colonies on SC –Ura plates at 20 °C and a single colony suspended in SC –Ura liquid and ~10^3 cells inoculated into each of six tubes and incubated at 20 °C to saturation. Aliquots were plated on SC –Ura and SC –His –Ura and incubated at 30 °C. The frequency of Ty1*his3-AI* was calculated by dividing the average number of His$^+$ Ura$^+$ cells per milliliter by the average number of Ura$^+$ cells per milliliter. Mobility of cells expressing a *GAL1*-promoted Ty1*his3-AI* plasmid (pGTy1*his3-AI*) was determined as described by Saha et al. [21].

Ty1 frameshifting efficiency

Ty1 programmed +1 frameshifting efficiency was measured as described [13]. Briefly, the assay employs two reporter plasmids that include a translational fusion of the first 30 codons of the yeast *HIS4* gene to the *Escherichia coli lacZ* gene, which encodes β-galactosidase. In the plasmid pMB38-9merWT, a short linker connecting the two genes includes the Ty1 heptameric frameshifting site fused to *lacZ* in the +1 reading frame. In a second plasmid, pMB38-9merFF, a single nucleotide deletion in the heptamer places the *lacZ* gene in the 0 reading frame so its expression does not require frameshifting. The two plasmids are transformed separately into the recipient strain. Frameshifting efficiency is calculated as the ratio of expression from pMB38-9merWT to that of pMB38-9merFF.

Polysome analysis

Sucrose gradient analysis of yeast ribosomes was performed essentially as described [34]. Briefly, 200 ml of each strain were grown in YPED medium to mid-exponential phase and harvested after addition of 10 mg cycloheximide. After washing, cells were lysed with glass beads and 40 A$_{260}$ units of supernatant was layered on a 10 to 50 % sucrose gradient and centrifuged in an SW40 rotor for 4 h at 41,000 rpm. Fractions were collected and continuously analyzed for absorption at 260 nm using an ISCO Foxy Jr fraction collector.

Northern analysis

The steady-state level of Ty1 mRNA was determined essentially as described [35]. Total cell RNA was isolated by the acid-phenol method [36] and 5 μg was separated by electrophoresis in 1 % agarose-glyoxal-DMSO gels and blotted to Brightstar-Plus positively charged nylon membranes (Life Sciences). For poly(A)$^+$ RNA purification, total RNA was prepared using the MasturePure yeast RNA purification kit (Epicentre Biotechnologies, Madison, WI). Poly(A)$^+$ RNA was isolated from 250 μg total RNA using the NucleoTrap mRNA purification kit (Clontech, Mountain View, CA). A DNA probe obtained as a 1.6 kb PvuII-ClaI fragment of the Ty1 *POL* gene and as a 1.4 kb EcoRI-XbaI fragment of the *PYK1* gene were labeled by random priming using α-[^{32}P]dATP using the Deca Prime II kit (Life Sciences). In vitro transcription of Ty1 *GAG* (nt 1266-1601) was performed using a MAXIscript kit (Life Technologies, Carlsbad, CA) and α-[^{32}P]UTP (3,000 Ci/mmol; Perkin Elmer, Waltham, MA). Hybridization was visualized by autoradiography or by image analysis using a STORM 840 phosphor imager (GE Healthcare). The experimental results shown in the figure are representative of three experiments performed.

Western blot analyses

Three-milliliter SC-Ura liquid cultures were grown at 20 °C until saturated, which occurred between 24 and 48 h for different mutants. Strains were grown under similar conditions but split into different groups according to growth rate, and each group contained a wild type control. Total cell protein was prepared as previously described [37]. Protein isolated by trichloroacetic acid (TCA) extraction [38] from wild type and the *rpl1bΔ* mutant either expressing pGTy1 or not was also subjected to immunoblotting. Galactose-induction of cells containing pGTy1 was performed as previously described [6]. Protein concentration was determined using Coomassie Plus (Bradford) Assay Reagent (Thermo scientific, Waltham, MA). Protein samples were separated on a 10 % SDS-PAGE gel, and then transferred onto polyvinylidene difluoride (PVDF) membranes. Membranes were blocked in 5 % powdered milk–Tris buffered saline (100 mM Tris–HCl, 150 mM NaCl pH 7.5) with 0.1 % Tween 20 (TBST) and then incubated with primary antibody for 1 h at room temperature. Rabbit antisera directed against Ty1 VLPs (used to detect Gag; gift of Alan J. Kingsman), recombinant p18 (used to detect Gag and p22/p18) [21], and the control protein Hts1p (Gift of Thomas L. Mason) were used at 1:7,000, 1:5000 and 1:40,000 dilutions, respectively. Blots were washed three times for 10 min each in TBST. Primary antibody was detected with ECL anti-rabbit IgG, Horseradish peroxide linked whole antibody from donkey (GE healthcare, Pittsburgh, PA) at a 1:4,000 dilution in TBST for 1 h. Blots were washed three times for 10 min each in TBST, visualized by ECL Western Blotting Detection Reagents (GE healthcare, Pittsburgh, PA) and exposed to X-ray film. The experimental results shown are representative of two or three experiments performed.

Results

Identifying Ty1 host cofactor and restriction genes involved in protein synthesis

We have previously described screens to identify Ty1 cellular cofactor [6] and restriction genes [5] using a Ty1 mobility assay. The assay employs a plasmid (pJC573) bearing a modified Ty1 element, Ty1his3-AI [33]. HIS3 is inserted downstream of the POL gene opposite to the direction of Ty1 transcription and is transcribed from its own promoter; the gene is interrupted by the artificial intron (AI), which is oriented in the direction of Ty1 transcription. The Ty1 RNA expressed from this construct is spliced before undergoing reverse transcription, removing the disruption of the HIS3 gene and resulting after its reintegration into the genome in complementation of the chromosomal his3 deletion (His+). Most His+ cells result from transposition of the element. A minority of retromobility events can occur by homologous recombination with an endogenous Ty1 transposon [39].

Among 457 Ty1 cofactor genes isolated in various systematic screens of the viable deletion mutants [3–7], 71 encode ribosomal proteins, ribosome biogenesis factors and translation factors including 33 ribosomal proteins genes: RPL1B, RPL4A, RPL6A, RPL7A, RPL14A, RPL15B, RPL16B, RPL18A, RPL19A, RPL19B, RPL20B, RPL21A, RPL21B, RPL27A, RPL31A, RPL33B, RPL34A, RPL37A, RPL39, RPL40A, RPL41B, RPL43A, RPP1A, RPP2B, RPS0B, RPS9B, RPS10A, RPS11A, RPS19A, RPS19B, RPS25A, RPS27B and RPS30A. On the other hand, of the 91 identified Ty1 restriction genes only three are translation related [3, 5]. ASC1 is an integral ribosomal protein of the 40S ribosomal subunit and is the yeast homolog of the mammalian Receptor of Activated C Kinase 1 (RACK1) protein [40]. ARC1 is a cofactor for aminoacyl-tRNA synthetases [41] and TRM7 encodes a tRNA 2′-O-ribose methyltransferase [42].

Quantitative assays of Ty1 mobility were performed as described [5] to assess the severity of the transposition defects caused by deleting 16 identified ribosome-associated Ty1 cofactor genes, eight encoding large subunit (60S) subunit proteins, five encoding small subunit (40S) proteins, two encoding biogenesis proteins of the 60S (rrp6 [43]) or 40S (rrp8 [44, 45]) subunit and a karyopherin gene (kap123 [46]) functioning in nuclear export of 60S subunits. These 19 strains were chosen for study based on their showing a strong mobility defect in the initial screen. Two control genes that reduce transposition by a mechanism not known to be associated with translation were also tested: BEM4, involved in budding, cell polarity and in maintenance of telomere length [47], and SPE3, encoding

spermidine synthase [48]. The assay employs a Ty1-his3AI transposon integrated upstream of the HIS4 gene. As shown in Table 1, the frequency of transposition was significantly reduced for all of the twenty deletion mutant strains. For 16 ribosome-associated cofactor mutant strains the frequency of transposition averaged 4.0×10^{-7} or 10-fold lower than the wild type frequency of 4.0×10^{-6}. The frequencies varied from a minimum of 1.9×10^{-8} (for rpl39Δ) to a maximum of 8.1×10^{-7} (for rpl41BΔ). The mobility of the two control strains was also much less than wild type. For two mutant strains, kap123Δ and bem4Δ, we were unable to observe any mobility events and so can only estimate a upper bound for the mobility frequency that is at least 73 and 340-fold below wild type, respectively. This secondary screen validates the identification of these genes as Ty1 cofactor genes.

Most translation associated Ty1 cofactor mutants impair ribosome biogenesis or function

Deficits in ribosomal proteins generally results in impaired ribosome biogenesis (reviewed in [49]). These defects include blocks to ribosome biogenesis events

Table 1 Quantitative Ty1 mobility is reduced in ribosomal protein gene Ty1 cofactor mutants

Strain	Function[a]	Ty1 mobility ± SEM ($\times 10^{-7}$)[b]	Fold reduced from WT
WT	–	40 ± 2.0	–
rpl1BΔ	LSU protein	4.7 ± 1.1	8.5
rpl4AΔ	"	1.9 ± 0.56	21
rpl15BΔ	"	2.6 ± 0.78	16
rpl21AΔ	"	4.9 ± 1.3	8.1
rpl27AΔ	"	6.5 ± 1.6	6.2
rpl39Δ	"	0.19 ± 0.14	210
rpl41BΔ	"	8.1 ± 1.9	4.9
rpp2BΔ	"	3.0 ± 1.0	13
rps0BΔ	SSU protein	5.6 ± 0.81	7.2
rps9BΔ	"	4.7 ± 0.79	8.5
rps10AΔ	"	4.8 ± 0.66	8.3
rps19BΔ	"	1.4 ± 0.27	29
rps25AΔ	"	7.8 ± 1.4	5.1
rrp6Δ	LSU processing	2.8 ± 0.56	14
rrp8Δ	SSU processing	4.0 ± 0.79	10
kap123Δ	60S nuclear export	<0.54[c]	>74
bem4Δ	Budding	<0.11[c]	>340
spe3Δ	Polyamine synthesis	0.67 ± 0.67	59

[a]LSU = large (60S) ribosomal subunit; SSU = small (40S) ribosomal subunit
[b]Mobility was calculated from a minimum of five repeated experiments as number of His+ cells per number of total viable cells
[c]No observed mobility; the maximum mobility is less than assuming mobility calculated if one event had occurred in the number of assays performed

including rRNA processing, binding of other ribosomal proteins to the pre-ribosome and transport to the cytoplasm. We therefore expected that the translation-associated Ty1 cofactor mutants would show effects on ribosome biogenesis. Many of the mutants showed slowed growth rates, consistent with reduced ribosome availability, however since most ribosomal protein gene deletion mutants grow at a normal rate [50] this is a poor test of their effect on biogenesis. Therefore, we directly assessed the effect of a subset of the cofactor mutants by analyzing polysomes from the wild type control and 11 of the translation-associated cofactor gene deletions using sucrose density centrifugation [51].

Most of the mutants tested had obvious defects in subunit abundance. Mutants of 40S ribosomal proteins genes (*RPS0B* or *RPS10A*) or a 40S subunit processing factor gene (*RRP8*) were severely impaired in 40S assembly (Fig. 1). Each had 40S/60S ratios less than one-tenth of the wild type reflecting the near absence of free 40S subunits and increased amounts of 60S. The lack of Rrp8 was previously shown to cause reduced accumulation of mature 18S rRNA of the 40S subunit [44].

Similarly, lack of most large subunit protein genes tested showed evidence of reduced 60S accumulation or activity. Deletions of 60S ribosomal protein genes (*RPL1B* and *RPL27A*) or a 60S subunit assembly factor gene involved in 3′-end processing of 5.8S rRNA (*RRP6*) all resulted in reduced amounts of 60S subunits and all three showed evidence of "halfmers", which are secondary peaks indicating complexes with masses slightly greater than a 70S or polysome peak. Halfmers are caused by the presence of mRNA-bound 43S pre-initiation complexes to which 60S subunits have failed to assemble in addition to one or more 80S ribosomes on an mRNA [52]. These peaks are direct evidence of slowed 60S subunit recruitment.

Four mutants displayed profiles resembling the wild type; the deletion of these genes does not appear to grossly alter the rate of assembly of either subunit. These genes encode a 40S ribosomal protein (*RPS25A*) two 60S proteins (*RPL15B, RPL41B*) and a ribosome nuclear export factor (*KAP123*). Previous work showed that lack of the 60S Ty1 cofactor gene *RPP2B* also does not alter subunit abundance [53]. Of the five encoded proteins, only Rpl15 is essential for viability; the proteins encoded by the other four can be eliminated without affecting viability although only a strain lacking Rpl41 grows at a normal rate [50]. These five proteins must affect Ty1 mobility without altering ribosome biogenesis.

Most ribosomal protein gene deletions cause a significant reduction in Ty1 frameshift activity

An obvious reason for the connection between translation and Ty1 retrotransposition could be the Ty1 + 1 programmed frameshifting event responsible for expression of the Gag-Pol fusion protein. The stoichiometry of Gag to Gag-Pol sensitively controls transposition efficiency and even slight changes in the ratio of Gag to Gag-Pol proteins can block retrotransposition [12]. For Ty1, increasing frameshifting blocks transposition by causing incomplete proteolytic processing of the Gag-Pol polyprotein leading to formation of defective VLPs [14]. Reducing Ty1 frameshifting also blocks transposition [17] although the mechanism of this blockage is not known.

Frameshift activity in mutant strains was determined using a well characterized β-galactosidase reporter construct [13]. The construct has the first 33 codons of the *HIS4* gene fused to the β-galactosidase gene through a minimal Ty1 frameshift site with expression of the enzyme requiring frameshifting. The percent frameshift activity is expressed as the ratio of the frameshift activity to that of a frame fusion control in which the genes are in one open reading frame so expression does not require frameshifting. The use of the frame fusion control eliminates other transcriptional, post-transcriptional and translational effects on the activity of the enzyme.

Figure 2 shows the frameshift efficiency of the wild type (white column) and the same 18 Ty1 cofactor deletion strains tested in Table 1. We found that 12 of the mutant strains showed significantly lower frameshifting efficiency than the wild type ($P < 0.05$ or 0.005). These included five small subunit genes, six large subunit genes and two ribosome biogenesis genes. These deletions each had decreased frameshift efficiencies that averaged 2.2-fold lower than wild type and varied from 1.5 to 4.5-fold. The deletion mutants not known to be involved in ribosome function or assembly were included as controls: *SPE3* and *BEM4* had no significant effect on frameshift efficiency. Four translation-associated cofactor deletion mutants (*rps25A*Δ, *rpl15B*Δ, *rpl41B*Δ and *kap123*Δ) also had frameshift efficiencies that were not significantly different from wild type; these mutants were the four that also showed near normal polysome profiles. The complete correspondence between these two phenotypes suggests a mechanistic connection between reduced ribosome biogenesis and reduced programmed frameshifting. The reduction in frameshift efficiency in about half of the tested mutants is 2-fold or more. Similar changes in stoichiometry blocked replication in the L-A virus [18] suggesting that this change in stoichiometry could significantly reduce transposition. Several of the mutants showed a less than 2-fold decrease (*rpl4A*Δ, *rpl21A*Δ and *rrp6*Δ) and the four mentioned above showed no decrease at all. We conclude that reduced frameshift efficiency probably contributes to the Ty1 mobility defect in most of the mutants but the full decrease in mobility may result from other abnormalities.

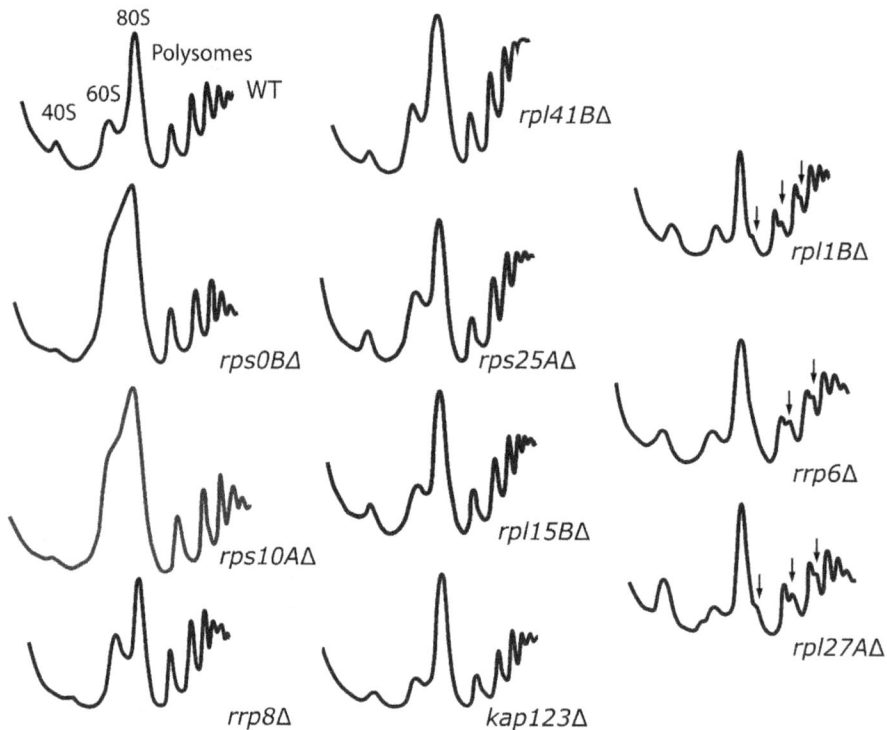

Fig. 1 Most translation-associated cofactor mutants show reductions in the relevant ribosomal subunit. Sucrose gradient analyses for wild type (WT) and mutant strains. The Y-axis represents the absorbance at 260 nm (A_{260}), proportional to RNA concentration, and the X-axis denotes increasing sucrose concentration with the lightest species eluting first (7–50 % sucrose). The 40S, 60S subunits, 80S monosome and polysomes in each of the profile are labeled and the presence of halfmers is denoted (black arrowhead)

Comparison of the frameshift and transposition phenotypes of each of the mutants showed no significant correlation (Pearson's correlation coefficient, $r = 0.07$). If we exclude the mutants that had no effect on frameshifting or the polysome profile the correlation is better ($r = 0.27$) but still weak. This statistical analysis suggests that the magnitude of the effect on transposition does not correlate well with the magnitude of the reduction of frameshifting, suggesting that effects beyond frameshift efficiency explain the reduction in Ty1 mobility. Clearly, for translation-associated cofactor mutants that do not alter either frameshifting or the polysome profile the effect on transposition must be from another cause, possibly extraribosomal [54].

Post-transcriptional regulation reduces Ty1 Gag protein accumulation in most Ty1 translation-associated cofactor mutants

Because most translation-associated Ty1 cofactor genes reduce 80S availability and programmed translational frameshifting, we suspected that they might also affect translation efficiency, in particular of the Ty1 Gag protein. The steady state level of Ty1 Gag, therefore, was determined for the wild type and 15 of the mutants tested in Table 1 by Western blotting using an anti-VLP

antibody that strongly reacts with Gag [55]. Based on expression studies with cells expressing a Ty1 on a *GAL1*-promoted pGTy1 plasmid, we expected to see both the primary translational product, Gag-p49 and the mature, C-terminally processed form, Gag-p45 [56]. However, recent work suggests that alternate forms of endogenous Gag in addition to p45 are detected in normal cells [22]. As shown in Fig. 3, the wild type consistently showed approximately a 2-fold excess of endogenous p45 over the slower migrating bands that contain altered forms of Gag and perhaps p49 (denoted as Gag+). Surprisingly, we were unable to detect either endogenous Gag protein in the *rpl1BΔ* and *rpl39Δ* cofactor mutants. The lack of Gag might predict an extremely severe Ty1 mobility deficit and the *rpl39Δ* mutant does have the lowest frequency of mobility of the deletion mutants tested, 210-fold lower than wild type and 25-fold lower than the average of the other mutants (Table 1). By contrast, the mobility frequency of the *rpl1BΔ* mutant was near the average of all mutants tested. To determine if Gag might be insoluble in the *rpl1BΔ* mutant, total protein from wild type and the mutant was prepared by TCA extraction and immunoblotted. Even under these harsher extraction conditions, which were developed to monitor transport of proteins into mitochondria [57], endogenous Gag was not detected

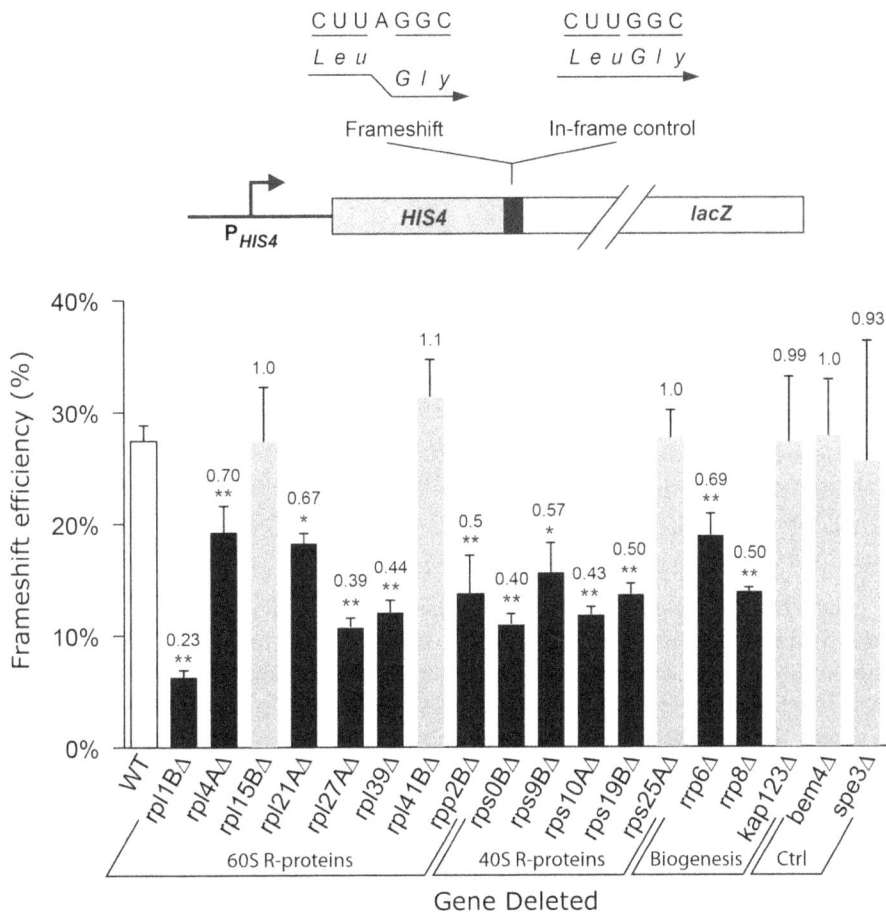

Fig. 2 Analysis of translation-associated cofactor mutants for Ty1 + 1 programmed frameshift efficiency. The frameshift activity was measured using a β-galactosidase construct. The percent frameshift efficiency is reported as the activity of the frameshift construct relative to the frame fusion control construct. The unshaded column represents the frameshift activity for the wild type (WT) strain BY4741. The assays were repeated at least three times, each assay performed in triplicate. Asterisks indicate the frameshift activities of deletion strains that were significantly different from wild type activity as measured using ANOVA followed by the Tukey's test (one asterisk, $P \leq 0.05$; two asterisks $P \leq 0.005$). Error bars represent the SEM. The ratio of the frameshift efficiency to the wild type appears above each column

in $rpl1B\Delta$ (Additional file 1: Figure S1). However, Gag was detected if Ty1 was overexpressed from pGTy1 and, even though Gag was mostly present in the altered form in the mutant, Ty1 mobility was restored to almost wild type levels. Similar results were also obtained with an antipeptide Gag antibody (data not shown). It is unclear why $rpl1B\Delta$ does not show a more severe phenotype given the severity of the putative Gag accumulation deficit. Perhaps $rpl1B\Delta$ Gag is present in a modified form that fails to enter the gel, is poorly electroblotted, or no longer reacts with Gag antisera.

Previously, we reported that the 40S assembly factor mutant $bud22\Delta$ accumulated lower amounts of Gag and higher amounts of Gag-p49 or Gag^+ compared to Gag-p45 [6]. We confirmed the $bud22\Delta$ phenotype (Fig. 3) and found that three of the mutants tested here showed a similar phenotype ($rps10A\Delta$, $rps19B\Delta$ and $rps25A$). Three other mutants, $rps0B\Delta$, $rps9B\Delta$ and $rrp6$,

accumulated similarly reduced amounts of Gag but do not show reduced processing of Gag-p49 to Gag-p45. All of these five genes encode 40S ribosomal proteins or 40S biogenesis proteins implying a link between 40S availability and the Bud22 phenotype and suggesting that the phenotype results specifically from a reduction in 40S availability.

The phenotypes of the remaining 60S ribosomal protein mutants are quite different than those of the 40S mutants. First, none of the 60S ribosomal protein mutants show an obvious deficit in p45 processing; the amount of the processed Gag-p45 is consistently greater than that of unprocessed Gag-p49/Gag^+. Second, the 60S mutants, other than $rpl1B\Delta$ and $rpl39$, clearly accumulate higher amounts of Gag than the 40S mutants although most express less than wild type (>50 % of wild type level). The co-occurrence of these two effects suggests that Gag processing or other posttranslational

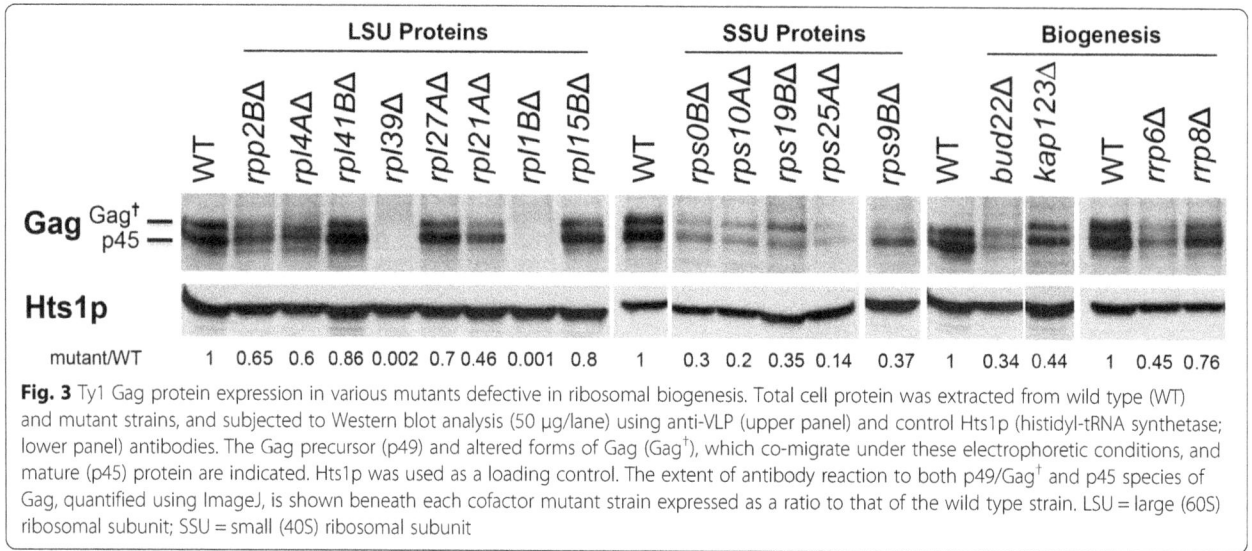

Fig. 3 Ty1 Gag protein expression in various mutants defective in ribosomal biogenesis. Total cell protein was extracted from wild type (WT) and mutant strains, and subjected to Western blot analysis (50 µg/lane) using anti-VLP (upper panel) and control Hts1p (histidyl-tRNA synthetase; lower panel) antibodies. The Gag precursor (p49) and altered forms of Gag (Gag†), which co-migrate under these electrophoretic conditions, and mature (p45) protein are indicated. Hts1p was used as a loading control. The extent of antibody reaction to both p49/Gag† and p45 species of Gag, quantified using ImageJ, is shown beneath each cofactor mutant strain expressed as a ratio to that of the wild type strain. LSU = large (60S) ribosomal subunit; SSU = small (40S) ribosomal subunit

modifications [22] may explicitly depend on a sufficient supply of Gag protein. Reduced numbers of Gag proteins relative to Ty1 genomic RNA in the 40S mutants could result in increased production of defective VLPs. Combined with a reduction in the frequency of programmed translational frameshifting in these mutants, which reduces the ratio of Gag to Gag-Pol to far less than 50:1, many of these VLPs might then actually lack the protease activity of Ty1 Pol protein, blocking processing of the Gag and presumably Pol proteins as well. For the 60S mutants, the increased amounts of Gag and Gag-Pol per genomic RNA might ensure the presence of protease in more of the VLPs.

The steady state level of full-length Ty1 mRNA cannot explain observed differences in Gag accumulation

The Ty1 promoter is complex and extends over the first approximately 1000 bp of the element, including sites upstream of the start site of transcription and downstream within the *GAG* gene [58]. The region includes binding sites for six transcription factors: Gcn4, Gcr1, Mot3, Ste12/Tec1, Mcm1, Tea1 and Rap1. Since Gcn4 and Rap1/Gcr1 regulate ribosomal protein genes [59, 60], it was reasonable to suspect that deficits in ribosomal protein accumulation might indirectly affect transcriptional regulation of chromosomal Ty1 elements. To test this idea, total RNA was isolated from ribosomal protein deletion strains grown at 20°C, a temperature conducive to Ty1 transposition [61] and was hybridized with radiolabeled probes specific to Ty1 (RT domain) and *PYK1*, which served as a normalization control (Fig. 4). The Spt3 protein, a component of SAGA complex [25], is required for chromosomal Ty1 element transcription [62]. Although these strains contained a plasmid-borne copy of Ty1*his3-AI* integrated near *HIS4* [5], the level of Ty1*his3-AI* RNA was not easily detected due to the much higher

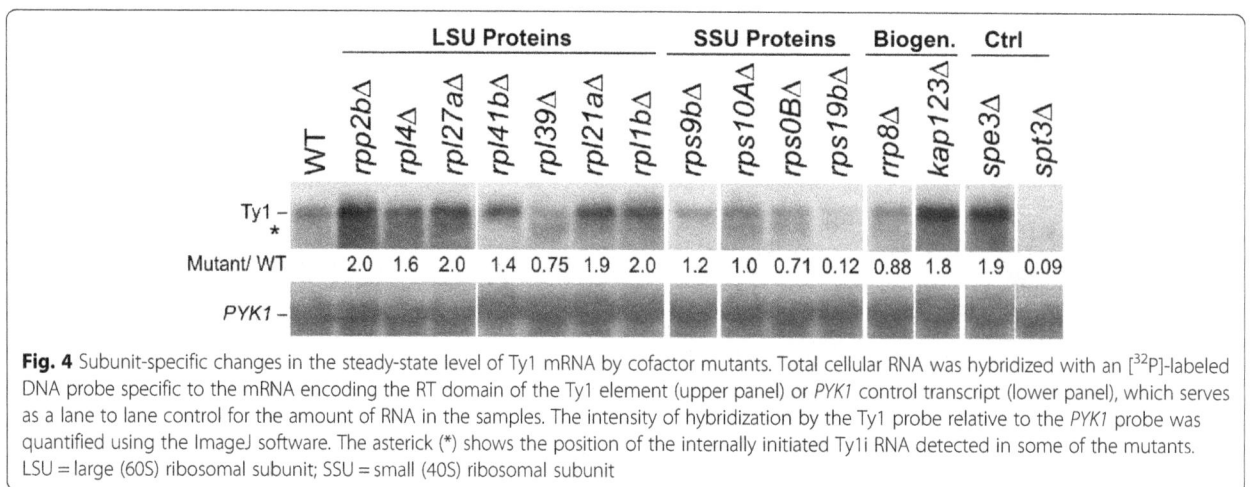

Fig. 4 Subunit-specific changes in the steady-state level of Ty1 mRNA by cofactor mutants. Total cellular RNA was hybridized with an [^{32}P]-labeled DNA probe specific to the mRNA encoding the RT domain of the Ty1 element (upper panel) or *PYK1* control transcript (lower panel), which serves as a lane to lane control for the amount of RNA in the samples. The intensity of hybridization by the Ty1 probe relative to the *PYK1* probe was quantified using the ImageJ software. The asterick (*) shows the position of the internally initiated Ty1i RNA detected in some of the mutants. LSU = large (60S) ribosomal subunit; SSU = small (40S) ribosomal subunit

amount of total Ty1 RNA, which is similar to earlier results obtained with chromosomal Ty1*his3-AI* elements [33]. As expected, endogenous Ty1 transcript production is severely reduced in an *spt3Δ* strain, providing the negative control in this experiment. The steady state amount of Ty1 transcript in the 60S subunit mutants was about 2-fold greater than wild type. An exception is *rpl39Δ*, which accumulates about 75 % of wild type. Mutants affecting 40S subunit proteins, by contrast, accumulated slightly reduced amounts of Ty1 mRNA corresponding to 75–100 % of wild type. In general, the Ty1 mRNA levels of these mutants correlated poorly with the amount of Ty1 Gag produced from each. As shown in Fig. 3, none of the 60S subunit protein mutants expressed increased Gag (especially true for the two mutants with severely reduced Gag accumulation, *rpl1BΔ* and *rpl39Δ*) and the 40S mutants accumulated far less than predicted by mRNA accumulation. The lack of correlation indicates that the reduction in Gag accumulation is post-transcriptional, however, the mechanism of this effect remains to be determined. The simplest model is that the proportion of Ty1 mRNA in the translated pool may be reduced in the mutants, possibly because of reduced availability of 40S or 60S subunits.

Ty1i RNA increases relative to Ty1 mRNA in several ribosomal subunit mutants

A subgenomic RNA (Fig. 4, denoted by the asterisk) was detected in several mutants, including *rpl27AΔ*, *rpl21AΔ*, *rps0BΔ*, and *rpl39Δ*, that may correspond to the newly discovered Ty1i transcript involved in controlling Ty1 copy number [21]. Therefore, we repeated the Northern analyses (Fig. 5) for wild type, *spt3Δ*, and selected ribosomal subunit mutants using total (Fig. 5a) and poly(A)$^+$ (Fig. 5b) RNA and a ^{32}P-labeled riboprobe derived from Ty1 *GAG* (nt 1266-1601). As reported previously [21], wild type cells contain a low level of the 4.9-kb Ty1i transcript, which increases in abundance in an *spt3Δ* mutant. Interestingly, the level of Ty1i RNA increased in the four ribosomal subunit mutants tested. Hybridization signals were more evident in poly(A)$^+$ RNA when compared with total RNA, perhaps because the relative amounts of polyadenylated Ty1 mRNA and Ty1i RNA differs [23]. Since recent results suggest that the relative levels of Ty1i and Ty1 mRNA are effective readouts for inhibition of Ty1 transposition by the Ty1i encoded product p22 [21], we compared the relative levels of Ty1i RNA and Ty1 mRNA. The *spt3Δ* mutation greatly increased the Ty1i/Ty1 RNA ratio since Spt3p is required for Ty1 mRNA but not Ty1i RNA transcription. The Ty1i/Ty1 transcript ratio also increased to a lesser extent in the ribosomal subunit mutants.

Ty1i mRNA is translated to produce an N-terminally truncated Gag-p22, which is likely the primary

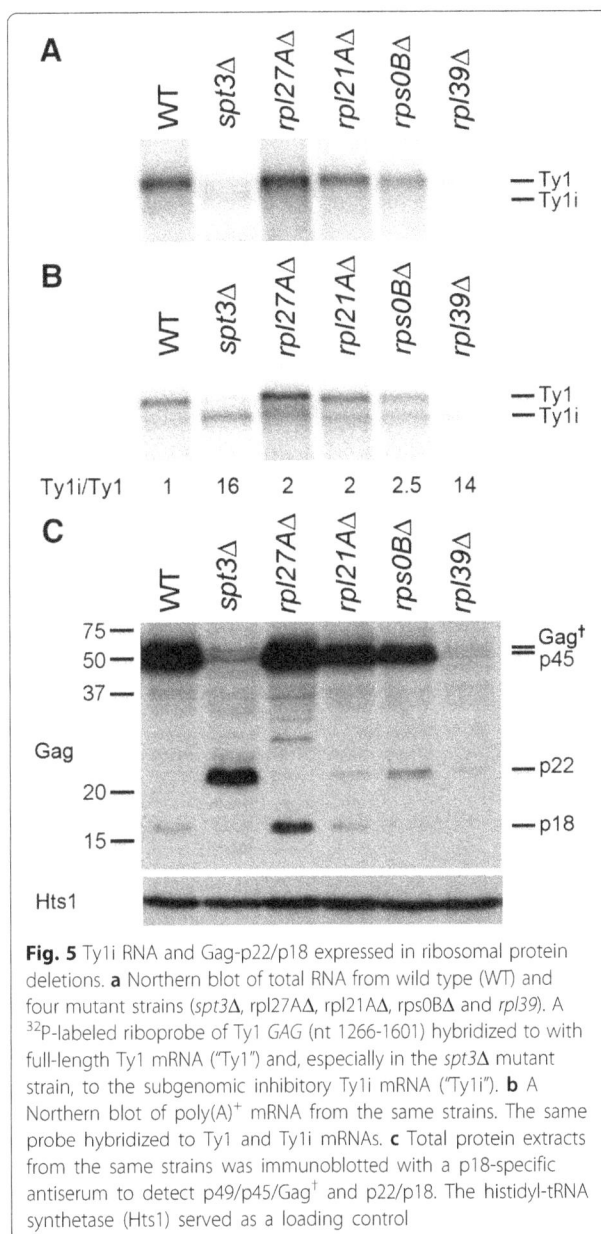

Fig. 5 Ty1i RNA and Gag-p22/p18 expressed in ribosomal protein deletions. **a** Northern blot of total RNA from wild type (WT) and four mutant strains (*spt3Δ*, rpl27AΔ, rpl21AΔ, rps0BΔ and *rpl39*). A ^{32}P-labeled riboprobe of Ty1 *GAG* (nt 1266-1601) hybridized to with full-length Ty1 mRNA ("Ty1") and, especially in the *spt3Δ* mutant strain, to the subgenomic inhibitory Ty1i mRNA ("Ty1i"). **b** A Northern blot of poly(A)$^+$ mRNA from the same strains. The same probe hybridized to Ty1 and Ty1i mRNAs. **c** Total protein extracts from the same strains was immunoblotted with a p18-specific antiserum to detect p49/p45/Gag† and p22/p18. The histidyl-tRNA synthetase (Hts1) served as a loading control

translational product, and a C-terminally processed Gag-p18 protein; the presence of these proteins leads to defective VLP assembly and function, and reduced Ty1 mobility [21]. We performed immunoblotting with an antibody raised against recombinant p18 on the same set of wild type and mutant strains and found evidence for p22/p18 in all of the mutants (Fig. 5c). As expected, the *spt3Δ* mutant accumulated large amounts of p22; because little Ty1-encoded proteins are expressed in this strain, no p18 was detected. In the four Ty1 ribosome-associated mutants there was evidence of p22 and/or p18. The amount of p18 was correlated with the amount of full-length Gag protein, which indicated the extent of translation of full-length Ty1 proteins. The *rpl27AΔ* mutant had both the highest amount of both Gag-p49/p45/Gag⁺ and Gag-p22. The increasingly lower amounts of Gag-p49/45 in the *rpl21A∆*, *rps0B∆* and *rps39∆* mutants correlated with increasingly lower amounts of Gag-p18 and an increased ratio of Gag-p22 to Gag-p18. These results confirm that the mutants tested express substantial amounts of Ty1i mRNA and the inhibitory Gag-p22/p18 proteins and suggest that the reduced Ty1 mobility in these strains in part results from this inhibitory mechanism.

Discussion

The number and diversity of genes identified as host factors for Ty1 retrotransposition reflects the complexity of the Ty1 lifecycle [2]. Many host restriction and cofactor genes encode proteins involved in basic processes of cellular information transfer with those involved in protein synthesis being significantly overrepresented among the Ty1 cofactor genes, which are required for optimal level of transposition [7]. Similarly, mutations targeting 60S ribosomal proteins are also required for propagation of the yeast L-A double-stranded RNA virus [9]; L-A shares several features with retroviruses and retrotransposons (reviewed in [11]) so this shared mode of control may reflect similar mechanisms of propagation. Quantitative assays of Ty1 mobility (Table 1) validate the requirement for 13 ribosomal proteins genes and three ribosome biogenesis genes (*RRP6, RRP8* and *KAP123*). As expected, most but not all mutants lacking 40S or 60S structural proteins or biogenesis factors are deficient in the corresponding subunit. It has long been recognized that biogenesis of the two subunits diverges early in biogenesis with a 90S pre-ribosome containing immature forms of both subunits dividing into a pre-40S and pre-60S complexes (reviewed in [63]). No comprehensive test of the effect of ribosomal protein depletion on ribosome biogenesis has been performed but most of the proteins have been tested and in all cases the lack of a ribosomal protein blocks maturation and accumulation of the corresponding subunit [49, 64].

The role of individual ribosomal proteins in ribosome biogenesis appears to be regional with proteins that bind in similar locations on the ribosome having roles in early, middle or late subunit biogenesis [49, 64]. Paradoxically, only a subset of ribosomal protein genes has been identified as Ty1 cofactors, totaling 33 of the 138 ribosomal protein genes. If this subset were a discrete group based on their function in ribosome function or biogenesis their protein products would be expected to cluster in a similar fashion in the ribosome. We have tested the Ty1 mobility phenotype of other ribosomal protein genes not previously characterized as Ty1 cofactor genes, 19 using the qualitative test and 12 of those using the quantitative test and found that each had

reduced Ty1 mobility (Additional file 2: Table S1), expanding the number of ribosomal protein cofactor genes to 52 and suggesting strongly that most ribosomal proteins may in fact be encoded by cofactor genes that may have escaped detection because of differences in mutant strain backgrounds, transposition assays, or strength of the Ty1 mobility phenotype. These 52 genes encode 37 proteins, representing 47 % of the 80S proteins, are distributed throughout the structure of the 80S ribosome with no obvious evidence of clustering (Fig. 6a-d). This distribution strongly implies that their function in Ty1 mobility has little or nothing to do with their location on the ribosome, or any specific role in biogenesis or during translation. The Ty1 cofactor phenotype may be a generic effect of mutations that cause a significant reduction in subunit availability. A comprehensive analysis of the effect of ribosomal protein depletion on Ty1 mobility would confirm this conjecture but is outside the scope of this study.

Based on Gag protein expression, we can divide these genes into three groups: the 40S ribosomal protein genes (strongly reduced Gag accumulation and processing), the majority of the 60S genes (slightly reduced Gag accumulation but normal processing) and *rpl1B∆* and *rpl39* (complete loss of detectable Gag; see Table 2 for a summary of all phenotypes). The reduction in accumulation of Gag is likely not transcriptional because of the lack of correlation between the accumulation of Ty1 mRNA and Gag protein. The effect could result from decreased protein stability or aberrant ER translocation [22] as we suspect in one case noted below, but given the primary defect is in availability of ribosome subunits, the most likely model is that translational insufficiency reduces Gag accumulation. The distinct effect on accumulation in the 40S and most 60S mutants could reflect a difference in the way that mRNAs compete for the two subunits. Binding of 40S subunits to individual mRNAs can differ widely among cellular transcripts based on sequence and structure with some mRNAs competing much more efficiently than others (reviewed in [65]). Recruitment of the 60S subunit should be less context dependent since the 60S mainly recognizes the 40S subunit once the initiation factors making up the 43S preinitiation complex have dissociated; there is no reason to suppose that some mRNAs compete more effectively at that stage of initiation. The greater reduction caused by reduced availability of 40S subunits, then, suggests that the Ty1 mRNA competes much less effectively for 43S preinitiation complex than does the average yeast mRNA. One reason for this could be that the Ty1 mRNA has an unusual structure and recent work has demonstrated that the 5′ end of the Ty1 mRNA forms a phylogenetically conserved RNA pseudoknot [66]. Mutational destabilization of the pseudoknot causes a modest increase in Gag accumulation, suggesting

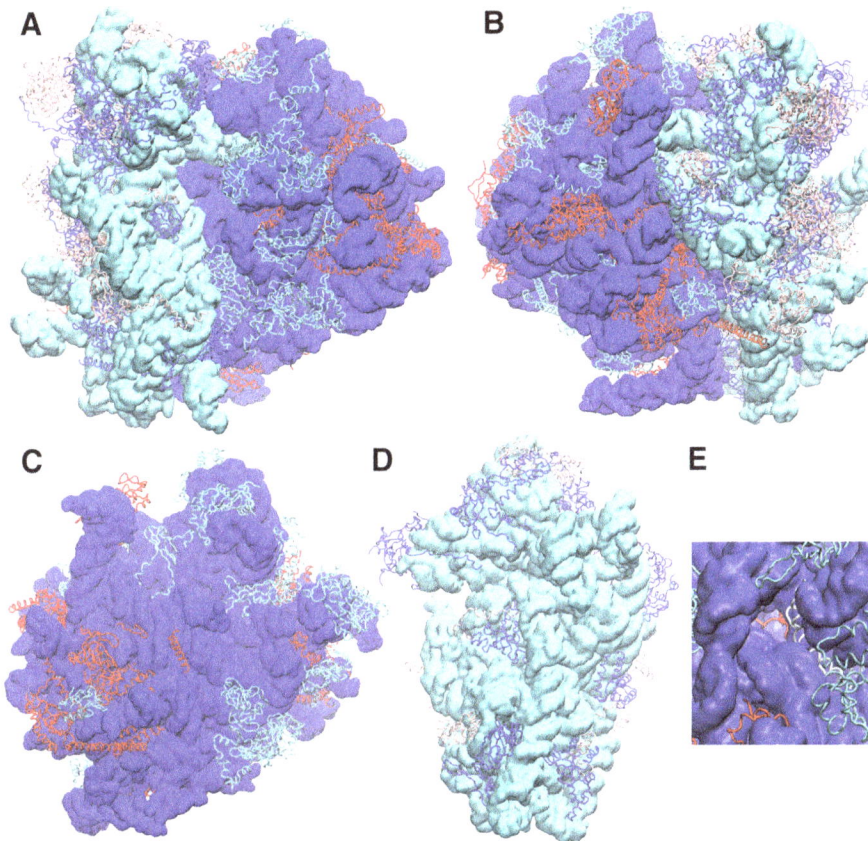

Fig. 6 Distribution of Ty1 cofactor ribosomal proteins on the structure of the *S. cerevisiae* ribosome. The structure of the yeast ribosome [69] is derived from structure files 3J78 deposited in the Protein Data Bank (PDB; http://www.rcsb.org/pdb/) [70]. The structure was modeled using the VMD Molecular Graphics Viewer (http://www.ks.uiuc.edu/Research/vmd/) [71]. The structure of the rRNAs are shown as a surface with the large subunit rRNAs colored blue and the small subunit rRNA colored cyan. The ribosomal proteins are shown in cartoon mode with those encoded by Ty1 cofactor genes in red (large subunit) or pink (small subunit) and all others in cyan (large subunit) or blue (small subunit). **a** The 80S subunit seen from A site side. **b** The 80S subunit rotated 180° to view from the E site side. **c** The 60S subunit showing the surface that is in contact with the 40S in the 80S complex. **d** The 40S subunit showing the 60S interface surface. **e** The end of the nascent peptide channel from the peptide exits showing Rps39 (*in white*) located immediately inside the end on the right side of the exit channel; beyond Rps39, deep within the exit channel, the tip of a loop on Rps4 can be seen (*in red*)

that the pseudoknot may inhibit Ty1 mRNA translation [66]. The same destabilizing mutations have the opposite effect—strongly decreasing Ty1 transposition—implying that the 5′ pseudoknot may play a structural role during the retrotransposition process [66]. These observations, however, do not contradict our finding that reducing 40S availability strongly reduces Gag accumulation. The highly structured nature of the 5′ end of the mRNA should reduce the efficiency of 43S complex binding, thus reducing the amount of Gag available to form virus-like particles. We do not imagine any direct effect of 40S availability on the role played by the 5′ pseudo-knot during retrotransposition.

The phenotype of the *rpl1BΔ* and *rpl39Δ* mutants is quite distinctive. The lack of accumulation of Gag in these mutants despite the presence of Ty1 mRNA strongly suggests a post-transcriptional block. The striking inability

to detect Gag protein in the *rpl39Δ* mutant is consistent with it having the lowest Ty1 mobility frequency of any of the mutants, 210-fold less than wild type. The *rpl1BΔ* mutant with a similar Gag accumulation phenotype, however, supports Ty1 mobility slightly more than the average of the mutants tested. Because the *rpl1BΔ* mutation causes a relatively small decrease in mobility we suspected that the low level of Gag detected in the *rpl1BΔ* mutant results from its being sequestered and not easily extracted. A harsher method of extraction detected no more Gag protein but overexpressing the Ty1 mRNA in this mutant background using a Gal-driven element resulted in significantly more Gag detected but also restored near normal Ty1 mobility. We cannot exclude that Gag is sequestered in the *rpl1BΔ* mutant and we are unable to explain why, despite their similar Gag phenotype, the *rpl1BΔ* and *rpl39* mobility phenotypes are so different. The location of the

Table 2 Summary of experimental results

Gene	Function[a]	Ty1 moblility relative to WT	Frameshifting relative to WT	Polysome effects	Gag relative to WT	Ty1 mRNA relative to WT
RPL1B	LSU protein	0.12	0.23	↓60S/halfmers	0.001	2.0
RPL4A	"	0.48	0.70	↓60S/halfmers[b]	0.60	1.6
RPL15B	"	0.07	1.0	~WT	0.80	n.d.
RPL21A	"	0.12	0.67	n.d.	0.46	1.9
RPL27A	"	0.16	0.39	↓60S/halfmers	0.70	2.0
RPL39	"	0.005	0.44	↓60S[c]	0.002	0.75
RPL41B	"	0.20	1.1	~WT	0.86	1.4
RPP2B	"	0.08	0.50	WT[d]	0.65	2.0
RPS0B	SSU protein	0.14	0.40	↓40S	0.30	0.71
RPS9B	"	0.12	0.57	n.d.	0.37	1.2
RPS10A	"	0.12	0.43	↓40S	0.20	1.0
RPS19B	"	0.04	0.50	↓40S	0.35	0.12
RPS25A	"	0.20	1.0	~WT[e]	0.14	n.d.
RRP6	LSU processing	0.07	0.69	↓60S/halfmers	0.45	n.d.
RRP8	SSU processing	0.10	0.50	↓40S	0.76	0.88
KAP123	60S nuclear export	<0.01	0.99	~WT	0.44	1.8

[a]LSU = large (60S) ribosomal subunit; SSU = small (40S) ribosomal subunit
[b]Ohtake et al. [72] [c]Sachs & Davis [73] [d]Cardenas et al. [53] [e]Léger-Silvestre et al. [74]

Rpl39 protein in the ribosomal subunit provides a clue to the origin of the difference. *RPL39* being a single copy gene, the deletion mutant accumulates 60S subunits lacking the protein. Rpl39 is located at the opening of the peptide exit tunnel (see Fig. 6e) and interacts with the hydrophobic signal anchor sequence of a nascent protein during co-translational insertion into the endoplasmic reticulum (ER) [67]; this interaction appears to be important for targeting proteins to the ER [68]. Doh et al. [22] demonstrated that VLP assembly sites are nucleated by targeting of ribosomes translating Ty1 mRNAs to the ER by contranslational insertion of Ty1 proteins into the ER. The formation of cytoplasmic foci [22], called T-bodies [23] or retrosomes [24] may be necessary for efficient formation of VLPs and therefore for maximal Ty1 mobility. Ty1 Gag is synthesized but is less stable when targeting to the ER is blocked. This suggests that a block to ER targeting caused by the absence of Rpl39 inhibits VLP assembly and Gag accumulation, resulting in a severe transposition defect.

Conclusions

The overall conclusion of this work is that failure in ribosome biogenesis results in reduced Ty1 mobility with distinct phenotypes for mutants deficient in 40S and 60S subunit proteins. The effect shows no clear connection to a particular step in biogenesis or the position of the protein within either subunit. The effect is largely translational involving both decreased programmed translational

frameshifting, reduced efficiency of translation and possibly increased instability of newly synthesized Ty1 Gag protein. A connection has been made to co-translational insertion of Ty1 Gag protein into the endoplasmic reticulum both by the severe phenotype of a mutant lacking the Rpl39 protein, which plays a role in targeting co-translational ER insertion and our demonstration of the accumulation of the Ty1i protein and Gag-p22/p18 in several of the translation-associated Ty1 cofactor mutant strains. Experiments are continuing to determine whether the connection between ribosome subunit sufficiency and Ty1 mobility is through the disruption of this newly discovered step in the Ty1 transposition process. Future studies will address how the individual pathways identified here modulate Ty1 gene expression and function, and whether similar processes also affect other retroelements.

Additional files

Additional file 1: Figure S1. Ty1 Gag level and Ty1*his3-AI* mobility in an *rpl1B*Δ mutant. Total cell protein was prepared by trichloroacetic acid extraction form wild type and *rpl1B*Δ mutant cells that were induced for expression of pGTy1*his3-AI* or not. Ty1 Gag-p45 and slower migrating forms of Gag (†) were detected by immunoblotting with VLP antiserum. Histidyl tRNA synthetase (Hts1) served as a loading control. Relative Ty1*his3-AI* mobility from galactose-induced cells was determined by dividing the frequency of Ty1*his3-AI* mobility obtained in the *rpl1B*Δ mutant [6.4×10^{-4} (0.6)] by the wild type [9×10^{-4} (0.9)] as described previously [21]. (TIFF 813 kb)

Additional file 2: Table S1. Identification of novel Ty1 mobility genes. Given the over representation of ribosomal protein gene deletions among the identified Ty1 mobility genes, we extended screen to include ribosomal protein genes that had not previously been identified in any large scale screen. Each of 19 additional genes were screened first qualitatively with all showing defects in transposition as shown. The mobility phenotypes of most of these mutants were then quantitated as shown. (DOC 3310 kb)

Abbreviations
ER: endoplasmic reticulum; Gag: group specific antigen; L-A: a Large double-stranded RNA virus of *Saccharomyces cerevisiae*; LSU: large ribosomal subunit; LTR: long terminal repeat; POL: POLymerase; the retroviral polyprotein processed to produce the protease, reverse transcriptase, RNase H and integrase enzymes; RT: reverse transcriptase; SSU: small ribosomal subunit; Ty: transposon of yeast; VLP: virus-like particle.

Competing interests
The authors declare that they have no competing interests.

Authors' contributions
SS carried out the experiments on ribosome biogenesis, Ty1 RNA blotting and, with KJ, the translational frameshifting. AK performed the quantitative mobility experiments. HA performed the Western analyses and Ty1i RNA blotting. DJG and PJF designed and supervised the study, with input from AD. PJF wrote the manuscript with input on revision from SS, AH and DJG. All authors read and approved the final manuscript.

Acknowledgements
We thank Stephen Hajduk for sharing equipment, and Jessica M. Tucker, Agniva Saha and Yuri Nishida for helpful discussions. Thanks to Thomas L. Mason for the gift of anti-Hts1 antibody. This work was supported by NIH grants GM095622 (D.J.G) and GM029480 (P.J.F.).

Author details
[1]Department of Biological Sciences and Program in Molecular and Cell Biology, University of Maryland Baltimore County, Baltimore, MD 21250, USA. [2]Department of Biochemistry & Molecular Biology, University of Georgia, Athens, GA 30602, USA. [3]Present address: Division of Infectious Diseases, Department of Internal Medicine, Stanford University School of Medicine, Stanford, California 94305, USA. [4]Present address: Department of Nanosciences and Technology, Karunya University, Karunya Nagar, Coimbatore 641 114, Tamil Nadu, India.

References
1. Voytas DF, Boeke JD. Yeast retrotransposon revealed. Nature. 1992;358:717.
2. Curcio MJ, Lutz S, Lesage P. The Ty1 LTR-retrotransposon of budding yeast, *Saccharomyces cerevisiae*. Microbiol Spectr. 2015;3:1–35.
3. Scholes DT, Banerjee M, Bowen B, Curcio MJ. Multiple regulators of Ty1 transposition in Saccharomyces cerevisiae have conserved roles in genome maintenance. Genetics. 2001;159:1449–65.
4. Griffith JL, Coleman LE, Raymond AS, Goodson SG, Pittard WS, Tsui C, et al. Functional genomics reveals relationships between the retrovirus-like Ty1 element and its host Saccharomyces cerevisiae. Genetics. 2003;164:867–79.
5. Nyswaner KM, Checkley MA, Yi M, Stephens RM, Garfinkel DJ. Chromatin-associated genes protect the yeast genome from Ty1 insertional mutagenesis. Genetics. 2008;178:197–214.
6. Dakshinamurthy A, Nyswaner KM, Farabaugh PJ, Garfinkel DJ. BUD22 affects Ty1 retrotransposition and ribosome biogenesis in Saccharomyces cerevisiae. Genetics. 2010;185:1193–205.
7. Risler JK, Kenny AE, Palumbo RJ, Gamache ER, Curcio MJ. Host co-factors of the retrovirus-like transposon Ty1. Mob DNA. 2012;3:12.
8. Wickner RB. Double-stranded RNA viruses of Saccharomyces cerevisiae. Microbiol Rev. 1996;60:250–65.
9. Ohtake Y, Wickner RB. Yeast virus propagation depends critically on free 60S ribosomal subunit concentration. Mol Cell Biol. 1995;15:2772–81.
10. Searfoss A, Dever TE, Wickner R. Linking the 3′ poly(A) tail to the subunit joining step of translation initiation: relations of Pab1p, eukaryotic translation initiation factor 5b (Fun12p), and Ski2p-Slh1p. Mol Cell Biol. 2001;21:4900–8.
11. Wickner RB, Fujimura T, Esteban R. Viruses and prions of Saccharomyces cerevisiae. Adv Virus Res. 2013;86:1–36.
12. Farabaugh P. Post-transcriptional regulation of transposition by Ty retrotransposons of Saccharomyces cerevisiae. J Biol Chem. 1995;270:10361–4.
13. Belcourt MF, Farabaugh PJ. Ribosomal frameshifting in the yeast retrotransposon Ty: tRNAs induce slippage on a 7 nucleotide minimal site. Cell. 1990;62:339–52.
14. Kawakami K, Pande S, Faiola B, Moore D, Boeke J, Farabaugh P, et al. A rare tRNA-Arg(CCU) that regulates Ty1 element ribosomal frameshifting is essential for Ty1 retrotransposition in Saccharomyces cerevisiae. Genetics. 1993;135:309–20.
15. Felsenstein K, Goff S. Expression of the *gag-pol* fusion protein of Moloney murine leukemia virus without *gag* protein does not induce virion formation or proteolytic processing. J Virol. 1988;62:2179–82.
16. Hung M, Patel P, Davis S, Green SR. Importance of ribosomal frameshifting for human immunodeficiency virus type 1 particle assembly and replication. J Virol. 1998;72:4819–24.
17. Xu H, Boeke JD. Host genes that influence transposition in yeast: the abundance of a rare tRNA regulates Ty1 transposition frequency. Proc Natl Acad Sci U S A. 1990;87:8360–4.
18. Dinman JD, Wickner RB. Ribosomal frameshifting efficiency and *gag/gag-pol* ratio are critical for yeast M_1 double-stranded RNA virus propagation. J Virol. 1992;66:3669–76.
19. Park J, Morrow CD. Overexpression of the *gag-pol* precursor from human immunodeficiency virus type 1 proviral genomes results in efficient poteolytic processing in the absence of virion production. J Virol. 1992;3:12.
20. Karacostas V, Wolffe EJ, Nagashima K, Gonda MA, Moss B. Overexpression of the HIV-1 gag-pol polyprotein results in intracellular activation of HIV-1 protease and inhibition of assembly and budding of virus-like particles. Virology. 1993;193:661–71.
21. Saha A, Mitchell JA, Nishida Y, Hildreth JE, Ariberre JA, Gilbert WV, et al. A trans-dominant form of Gag restricts Ty1 retrotransposition and mediates copy number control. J Virol. 2015;89:3922–38.
22. Doh JH, Lutz S, Curcio MJ. Co-translational localization of an LTR-retrotransposon RNA to the endoplasmic reticulum nucleates virus-like particle assembly sites. PLoS Genet. 2014;10:e1004219.
23. Malagon F, Jensen TH. The T body, a new cytoplasmic RNA granule in Saccharomyces cerevisiae. Mol Cell Biol. 2008;28:6022–32.
24. Sandmeyer SB, Clemens KA. Function of a retrotransposon nucleocapsid protein. RNA Biol. 2010;7:642–54.
25. Grant PA, Duggan L, Cote J, Roberts SM, Brownell JE, Candau R, et al. Yeast Gcn5 functions in two multisubunit complexes to acetylate nucleosomal histones: characterization of an Ada complex and the SAGA (Spt/Ada) complex. Genes Dev. 1997;11:1640–50.
26. Berretta J, Pinskaya M, Morillon A. A cryptic unstable transcript mediates transcriptional trans-silencing of the Ty1 retrotransposon in S. cerevisiae. Genes Dev. 2008;22:615–26.
27. Checkley MA, Nagashima K, Lockett SJ, Nyswaner KM, Garfinkel DJ. P-body components are required for Ty1 retrotransposition during assembly of retrotransposition-competent virus-like particles. Mol Cell Biol. 2010;30:382–98.
28. Dutko JA, Kenny AE, Gamache ER, Curcio MJ. 5′ to 3′ mRNA decay factors colocalize with Ty1 gag and human APOBEC3G and promote Ty1 retrotransposition. J Virol. 2010;84:5052–66.
29. Sherman F, Fink GR, Hicks JB. Methods in yeast genetics. Cold Spring Harbor: Cold Spring Harbor Laboratoy; 1986.
30. Guthrie C, Fink GR. Guide to yeast genetics and molecular biology. In: Abelson JN, Simon MI, editors. Methods in Enzymology. San Diego, California: Academic; 1991.
31. Giaever G, Chu AM, Ni L, Connelly C, Riles L, Veronneau S, et al. Functional profiling of the Saccharomyces cerevisiae genome. Nature. 2002;418:387–91.
32. Brachmann CB, Davies A, Cost GJ, Caputo E, Li J, Hieter P, et al. Designer deletion strains derived from Saccharomyces cerevisiae S288C: a useful set of strains and plasmids for PCR-mediated gene disruption and other applications. Yeast. 1998;14:115–32.
33. Curcio MJ, Garfinkel DJ. Single-step selection for Ty1 element retrotransposition. Proc Natl Acad Sci U S A. 1991;88:936–40.
34. Deshmukh M, Tsay YF, Paulovich AG, Woolford Jr JL. Yeast ribosomal protein L1 is required for the stability of newly synthesized 5S rRNA and the assembly of 60S ribosomal subunits. Mol Cell Biol. 1993;13:2835–45.

35.   Lee BS, Lichtenstein CP, Faiola B, Rinckel LA, Wysock W, Curcio MJ, et al. Posttranslational inhibition of Ty1 retrotransposition by nucleotide excision repair/transcription factor TFIIH subunits Ssl2p and Rad3p. Genetics. 1998;148:1743–61.

36.   Schmitt ME, Brown TA, Trumpower BL. A rapid and simple method for preparation of RNA from Saccharomyces cerevisiae. Nucleic Acids Res. 1990;18:3091–2.

37.   Braiterman LT, Monokian GM, Eichinger DJ, Merbs SL, Gabriel A, Boeke JD. In-frame linker insertion mutagenesis of yeast transposon Ty1: phenotypic analysis. Gene. 1994;139:19–26.

38.   Lawler Jr JF, Merkulov GV, Boeke JD. A nucleocapsid functionality contained within the amino terminus of the Ty1 protease that is distinct and separable from proteolytic activity. J Virol. 2002;76:346–54.

39.   Sharon G, Burkett TJ, Garfinkel DJ. Efficient homologous recombination of Ty1 element cDNA when integration is blocked. Mol Cell Biol. 1994;14:6540–51.

40.   Gerbasi VR, Weaver CM, Hill S, Friedman DB, Link AJ. Yeast Asc1p and mammalian RACK1 are functionally orthologous core 40S ribosomal proteins that repress gene expression. Mol Cell Biol. 2004;24:8276–87.

41.   Simos G, Segref A, Fasiolo F, Hellmuth K, Shevchenko A, Mann M, et al. The yeast protein Arc1p binds to tRNA and functions as a cofactor for the methionyl- and glutamyl-tRNA synthetases. EMBO J. 1996;15:5437–48.

42.   Pintard L, Lecointe F, Bujnicki JM, Bonnerot C, Grosjean H, Lapeyre B. Trm7p catalyses the formation of two 2'-O-methylriboses in yeast tRNA anticodon loop. EMBO J. 2002;21:1811–20.

43.   Briggs MW, Burkard KT, Butler JS. Rrp6p, the yeast homologue of the human PM-Scl 100-kDa autoantigen, is essential for efficient 5.8 S rRNA 3' end formation. J Biol Chem. 1998;273:13255–63.

44.   Bousquet-Antonelli C, Vanrobays E, Gelugne JP, Caizergues-Ferrer M, Henry Y. Rrp8p is a yeast nucleolar protein functionally linked to Gar1p and involved in pre-rRNA cleavage at site A2. RNA. 2000;6:826–43.

45.   Peifer C, Sharma S, Watzinger P, Lamberth S, Kotter P, Entian KD. Yeast Rrp8p, a novel methyltransferase responsible for m1A 645 base modification of 25S rRNA. Nucleic Acids Res. 2013;41:1151–63.

46.   Sydorskyy Y, Dilworth DJ, Yi EC, Goodlett DR, Wozniak RW, Aitchison JD. Intersection of the Kap123p-mediated nuclear import and ribosome export pathways. Mol Cell Biol. 2003;23:2042–54.

47.   Askree SH, Yehuda T, Smolikov S, Gurevich R, Hawk J, Coker C, et al. A genome-wide screen for Saccharomyces cerevisiae deletion mutants that affect telomere length. Proc Natl Acad Sci U S A. 2004;101:8658–63.

48.   Hamasaki-Katagiri N, Tabor CW, Tabor H. Spermidine biosynthesis in Saccharomyces cerevisae: polyamine requirement of a null mutant of the SPE3 gene (spermidine synthase). Gene. 1997;187:35–43.

49.   de la Cruz J, Karbstein K, Woolford Jr JL. Functions of ribosomal proteins in assembly of eukaryotic ribosomes in vivo. Annu Rev Biochem. 2015;84:93–129.

50.   Steffen KK, McCormick MA, Pham KM, MacKay VL, Delaney JR, Murakami CJ, et al. Ribosome deficiency protects against ER stress in Saccharomyces cerevisiae. Genetics. 2012;191:107–18.

51.   Sagliocco FA, Moore PA, Brown AJ. Polysome analysis. Methods Mol Biol. 1996;53:297–311.

52.   Helser TL, Baan RA, Dahlberg AE. Characterization of a 40S ribosomal subunit complex in polyribosomes of Saccharomyces cerevisiae treated with cycloheximide. Mol Cell Biol. 1981;1:51–7.

53.   Cardenas D, Revuelta-Cervantes J, Jimenez-Diaz A, Camargo H, Remacha M, Ballesta JP. P1 and P2 protein heterodimer binding to the P0 protein of Saccharomyces cerevisiae is relatively non-specific and a source of ribosomal heterogeneity. Nucleic Acids Res. 2012;40:4520–9.

54.   Warner JR, McIntosh KB. How common are extraribosomal functions of ribosomal proteins? Mol Cell. 2009;34:3–11.

55.   Adams SE, Mellor J, Gull K, Sim RB, Tuite MF, Kingsman SM, et al. The functions and relationships of TyVLP proteins in yeast reflect those of mammalian retroviral proteins. Cell. 1987;49:111–9.

56.   Merkulov GV, Swiderek KM, Brachmann CB, Boeke JD. A critical proteolytic cleavage site near the C terminus of the yeast retrotransposon Ty1 Gag protein. J Virol. 1996;70:5548–56.

57.   Ohashi A, Gibson J, Gregor I, Schatz G. Import of proteins into mitochondria. The precursor of cytochrome c1 is processed in two steps, one of them heme-dependent. J Biol Chem. 1982;257:13042–7.

58.   Servant G, Pennetier C, Lesage P. Remodeling yeast gene transcription by activating the Ty1 long terminal repeat retrotransposon under severe adenine deficiency. Mol Cell Biol. 2008;28:5543–54.

59.   Joo YJ, Kim JH, Kang UB, Yu MH, Kim J. Gcn4p-mediated transcriptional repression of ribosomal protein genes under amino-acid starvation. EMBO J. 2011;30:859–72.

60.   Deminoff SJ, Santangelo GM. Rap1p requires Gcr1p and Gcr2p homodimers to activate ribosomal protein and glycolytic genes, respectively. Genetics. 2001;158:133–43.

61.   Paquin CE, Williamson VM. Temperature effects on the rate of ty transposition. Science. 1984;226:53–5.

62.   Winston F, Durbin KJ, Fink GR. The SPT3 gene is required for normal transcription of Ty elements in S. cerevisiae. Cell. 1984;39:675–82.

63.   Tschochner H, Hurt E. Pre-ribosomes on the road from the nucleolus to the cytoplasm. Trends Cell Biol. 2003;13:255–63.

64.   Woolford Jr JL, Baserga SJ. Ribosome biogenesis in the yeast Saccharomyces cerevisiae. Genetics. 2013;195:643–81.

65.   Gale Jr M, Tan SL, Katze MG. Translational control of viral gene expression in eukaryotes. Microbiol Mol Biol Rev. 2000;64:239–80.

66.   Huang Q, Purzycka KJ, Lusvarghi S, Li D, Legrice SF, Boeke JD. Retrotransposon Ty1 RNA contains a 5'-terminal long-range pseudoknot required for efficient reverse transcription. RNA. 2013;19:320–32.

67.   Zhang Y, Wolfle T, Rospert S. Interaction of nascent chains with the ribosomal tunnel proteins Rpl4, Rpl17, and Rpl39 of Saccharomyces cerevisiae. J Biol Chem. 2013;288:33697–707.

68.   Lin PJ, Jongsma CG, Liao S, Johnson AE. Transmembrane segments of nascent polytopic membrane proteins control cytosol/ER targeting during membrane integration. J Cell Biol. 2011;195:41–54.

69.   Svidritskiy E, Brilot AF, Koh CS, Grigorieff N, Korostelev AA. Structures of yeast 80S ribosome-tRNA complexes in the rotated and nonrotated conformations. Structure. 2014;22:1210–8.

70.   Berman HM, Westbrook J, Feng Z, Gilliland G, Bhat TN, Weissig H, et al. The protein data bank. Nucleic Acids Res. 2000;28:235–42.

71.   Humphrey W, Dalke A, Schulten K. VMD: visual molecular dynamics. J Mol Graph. 1996;14:33–8.

72.   Ohtake Y, Wickner RB. KRB1, a suppressor of mak7-1 (a mutant RPL4A), is RPL4B, a second ribosomal protein L4 gene, on a fragment of Saccharomyces chromosome XII. Genetics. 1995;140:129–37.

73.   Sachs AB, Davis RW. The poly(A)–binding protein is required for poly(A) shortening and 60S ribosomal subunit dependent translation intiation. Cell. 1989;58:857–67.

74.   Leger-Silvestre I, Caffrey JM, Dawaliby R, Alvarez-Arias DA, Gas N, Bertolone SJ, et al. Specific role for yeast homologs of the Diamond Blackfan Anemia-associated Rps19 Protein in Ribosome Synthesis. J Biol Chem. 2005;280:38177–85.

7

The making of a genomic parasite - the *Mothra* family sheds light on the evolution of *Helitrons* in plants

Stefan Roffler, Fabrizio Menardo and Thomas Wicker[*]

Abstract

Background: Helitrons are Class II transposons which are highly abundant in almost all eukaryotes. However, most Helitrons lack protein coding sequence. These non-autonomous elements are thought to hijack recombinase/helicase (RepHel) and possibly further enzymes from related, autonomous elements. Interestingly, many plant Helitrons contain an additional gene encoding a single-strand binding protein homologous to Replication Factor A (RPA), a highly conserved, single-copy gene found in all eukaryotes.

Results: Here, we describe the analysis of *DHH_Mothra*, a high-copy non-autonomous Helitron in the genome of rice (*Oryza sativa*). *Mothra* has a low GC-content and consists of two distinct blocs of tandem repeats. Based on homology between their termini, we identified a putative mother element which encodes an RPA-like protein but has no *RepHel* gene. Additionally, we found a putative autonomous sister-family with strong homology to the *Mothra* mother element in the RPA protein and terminal sequences, which we propose provides the RepHel domain for the *Mothra* family. Furthermore, we phylogenetically analyzed the evolutionary history of RPA-like proteins. Interestingly, plant Helitron RPAs (PHRPAs) are only found in monocotyledonous and dicotyledonous plants and they form a monophyletic group which branched off before the eukaryotic "core" RPAs.

Conclusions: Our data show how erosion of autonomous Helitrons can lead to different "levels" of autonomy within Helitron families and can create highly successful subfamilies of non-autonomous elements. Most importantly, our phylogenetic analysis showed that the PHRPA gene was most likely acquired via horizontal gene transfer from an unknown eukaryotic donor at least 145–300 million years ago in the common ancestor of monocotyledonous and dicotyledonous plants. This might have led to the evolution of a separate branch of the Helitron superfamily in plants.

Keywords: Transposon, Helitron, RPA, Rice, Horizontal transfer

Background

Helitrons are a superfamily of transposable elements (TEs) in eukaryotes which was discovered only relatively recently in *Arabidopsis thaliana*, *Caenorhabditis elegans* and *Oryza sativa* [1]. They have since been found in many genomes of flowering plants [1, 2], mosses [3], fungi [4–6] but also many animals such as sea urchin [7], fish [8, 9] and bats [10]. A recent *in silico* analysis using the program *Helsearch* [2] estimates the number of Helitrons in rice and sorghum to approximately 7000 and 5000, respectively, covering several megabases of their hosts' genomes. The most extensively studied genome regarding Helitrons is the one of maize, where approximately 2000 intact Helitrons and more than 20,000 Helitron fragments were found. Based on high homology between individual elements they are thought to still be very active [11]. As for most DNA transposons, the majority of Helitron elements are non-autonomous and do not encode any proteins. These non-autonomous elements presumably depend for their transposition on enzymes encoded by "mother" or "master" elements elsewhere in the genome.

One reason why Helitrons remained undiscovered for a long time is their limited diagnostic features. They lack

* Correspondence: wicker@botinst.uzh.ch
Institute of Plant Biology, University of Zürich, Zollikerstrasse 107, Zürich CH-8008, Switzerland

terminal inverted repeats (TIRs) and the only motifs common to all Helitrons are the dinucleotide TC at the 5' end as well as a CTRR motif at the 3' end. Additionally, almost all Helitrons have a G/C rich 15–20 bp hairpin motif approximately 10–12 bp upstream of the 3' end, which is thought to serve as a stop signal in the transposition process [1]. Finally, Helitrons have a strong preference to insert between the bases A and T or sometimes between two Ts [1].

The transposition mechanism of Helitrons and the involved proteins differ from those of the well described DDE transposases. Autonomous Helitrons encode a RepHel protein of 1000–3000 amino acids (aa) length, which is thought to initiate the replication. The RepHel constitutes a replication initiation domain (RCR/Rep) followed by a helicase enzyme (Hel) of approximately 400 aa [12]. Because of structural homology with the catalytic core of HUH endonucleases of a bacterial rolling-circle transposons [13], it was suggested that Helitrons use a rolling-circle mechanism involving a single-stranded DNA intermediate for transposition and replication [1, 12]. Li and Dooner [14], however, clearly showed excisions of Helitrons from 0.4 to 6 kb size in somatic Maize tissue. This challenges the current model and suggests an alternative mode of transposition involving excision and repair similar to TIR transposons. Indeed, it is possible that single stranded DNA transposition can result in the elimination of that copy from that locus when occurring during S phase of meiosis 1 [15].

Even though Helitrons are ascribed to the Class II (DNA) transposons, they remain unique due to their exclusive structural features and transposition mechanism and belong to a separate subclass within the DNA transposons [16]. However, rolling-circle transposition mechanisms have been described for gemini viruses [17], plasmids and some bacterial transposons [18]. Structural homology between their transposases suggests very ancient origin of Helitrons [1].

In plants, some Helitrons have been reported to also encode a distant homolog of the Replication Protein A (RPA), a protein ubiquitous in eukaryotes [19, 20]. RPA has several single-strand DNA binding sites and is involved in processes such as DNA replication and repair. RPA homologs have also been identified in Helitrons from zebrafish and sea anemone [12] and in Helentrons (a sub-type of Helitrons) in *Drosophila melanogaster* [21].

At least in maize, Helitrons seem to acquire close by gene fragments very frequently. Several studies showed an ongoing gene movement, gene shuffling and transcriptional read-troughts, which is attributed to Helitron activity [22, 23]. In the maize line B73, approximately 11,000 such chimeric transcripts have been found to be expressed which represents almost one quarter of all genes [24]. Therefore, it is thought that Helitrons contributed substantially to the recent diversification observed in the maize genus. Moreover, frequent gene capturing mediated by Helitrons was also reported in the silk worm *Bombyx mori* [25] and in the bat *Myotis lucifugus* [26].

In this study we describe the analysis and origin of a high-copy Helitron family in rice, which we named *DHH_Mothra*. Non-autonomous *Mothra* elements are present in hundreds or even thousands of copies in multiple rice species, which merited an in-depth analysis of this TE family. We identified a putative mother element for the *Mothra* family that encodes an RPA homolog but no RepHel protein. We moreover identified a closely related Helitron family, which we propose to be the donor for the lacking RepHel enzyme of *Mothras*. According to our model, this introduces an additional level of autonomy. We furthermore investigated the evolutionary background of Helitron RPA acquisition in plants and suggest horizontal transfer most likely from a unicellular eukaryote into the common ancestor of mono- and dicotyledonous plants.

Results

Mothra is a high-copy non-autonomous Helitron

In a previous study [27] we compared the two closely related rice species *O. sativa*, the Asian rice, with its relative *O. glaberrima*, the African rice, and investigated presence/absence polymorphisms of Class II transposons of the TIR subclass. While scanning polymorphic TE sites, we repeatedly encountered a sequence which was obviously of repetitive nature but we were unable to classify it at that time. Now, we found that it was in fact a non-autonomous TE of the Helitron order which we called *Mothra*.

We identified a total of 1,682 *Mothra* elements from which we manually deduced consensus sequences of 22 sub-types. The 22 *Mothra* sub-types share the same terminal and internal sequence motifs but vary in size between 1252 and 2741 bp (see Methods). The differences in size between the sub-types are due to differences in the order, length and/or orientation of blocs of tandem repeats (see below). From these 22 sub-types, we created a single consensus sequence of 1993 bp in length which we refer to as consensus of the non-autonomous *Mothra* elements (Fig. 1a). As described for other Helitrons, *Mothra* elements show the characteristic dinucleotide TC at its 5' end and the four bases CTAG at the 3' end. Additionally, we found the characteristic hairpin motif of 16 bp length located 13 bp upstream of the 3' end of the elements. From this, we concluded that *Mothra* is in fact is a non-autonomous TE of the Helitron order.

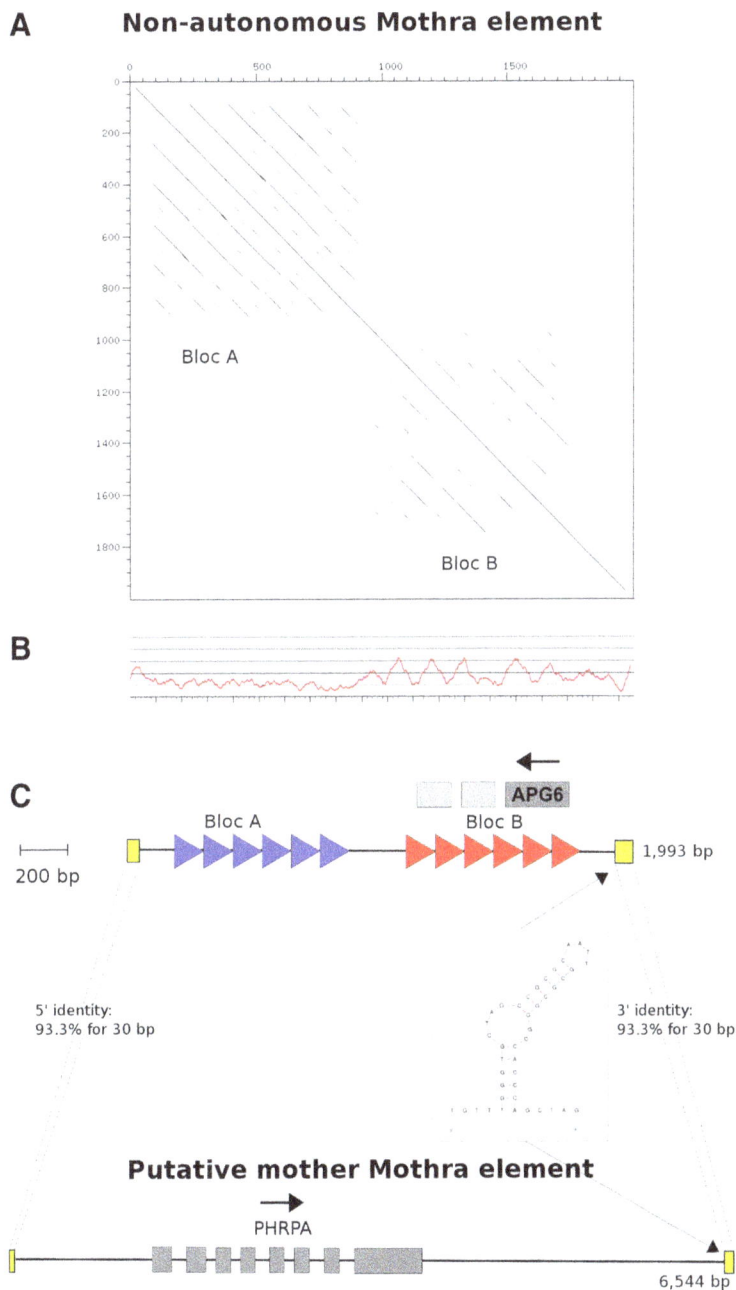

Fig. 1 Overview of the non-autonomous *Mothra* consensus sequence and its putative mother element. **a** Dot-plot of the non-autonomous *Mothra* consensus sequence against itself shows the two repetitive Blocs A and B. **b** GC-plot of the non-autonomous element. Note that Bloc A shows a unusual low GC-content of approximately 20 %. **c** Schematic overview of the non-autonomous *Mothra* and its putative mother element below. Both elements share the characteristic hairpin structure at the 3' end. The termini of the putative mother element and the non-autonomous consensus are conserved (in yellow). Furthermore, the non-autonomous elements shows the putative ORF of 96 amino acids. Note here, that the putative mother element of *Mothras* encodes for a RPA homolog, which we named PHRPA, but no RepHel protein

Mothra contains tandem repeats and gene fragments

Mothra contains two distinct sequence blocs (Bloc A and B, Fig. 1a). Bloc A, which ranges approximately from position 80 to position 900 in the consensus sequence, consists of six direct repeats and shows a very low GC content of 20 %. Bloc B ranges from position 950 – 1860 and consists of six different, less conserved direct repeats and exhibits an average GC content of about 40 % (Fig. 1b). There is great variety in the number of the repeat units within the Blocs A and B among the individual copies. In some cases, the order of the blocs is even reversed. In other cases, additional sequence is present between or sometimes even within one of the two blocs.

By definition, non-autonomous elements do not encode any proteins. But interestingly, the *Mothra* consensus sequence contains a putative open reading frame (ORF) of 96 amino acids in reverse orientation in Bloc B. The predicted protein shows sequence homology to the APG6 domain (Pfam ID: pfam04111, e-value: 2,2e-03) which has been described to be involved in autophagy and vascular sorting pathways in yeast [28]. Because of the repeat structure of Bloc B, this homology is partially repeated two more times downstream of this ORF. These additional copies, however, lack start codons and therefore do not constitute intact ORFs. We assume that this ORF is the result of gene fragment capture but probably has no function. The fact that this gene fragment is part of the *Mothra* consensus sequence indicates that the gene capture event occurred before the radiation of the *Mothra* family.

The putative *Mothra* mother element lacks a *RepHel* gene

Usually, non-autonomous TEs share their terminal sequences with their autonomous "mother" elements. That is why we scanned the genome of *O. sativa* using the first 50 and the last 80 bp of the non-autonomous element, respectively, as queries. We extracted 323 sequences where we identified both ends in the same orientation located within 25 kb from each other. We scanned the 323 fragments for the presence of transposases and helicases but could not identify a single one. However, we identified one sequence of 6544 bp in length that encodes an RPA homolog (Fig. 1c). This RPA sequence was annotated in the rice genome as hypothetical protein (LOC_Os11g47400). The predicted protein contains several generic single-stranded DNA-binding sites. After manual re-annotation of the protein we were able to extend the putative protein length from 296 aa to 472 aa

and the number of exons from four to eight. Interestingly, this sequence was the only one among all 323 analyzed fragments containing a putative complete gene between the two *Mothra* ends. The sequence homology between the termini of this putative mother element and the non-autonomous *Mothra* consensus is very high (93,3 % of the terminal 30 bp, and 81,2 % and 80,2 %, respectively for the terminal 100 bp). According to Yang et al. [2], this makes them not only members of the same family but also of the same sub-family. Moreover, we identified a deletion derivative of the putative mother element that shows homology to almost the entire element but lacks the RPA domain (Fig. 2). This indicates that we indeed identified a distinct element rather than an RPA homolog that is flanked by chance by two fragments of termini from non-autonomous *Mothra* elements. Therefore, we propose this element, even if we did not find an ORF encoding an RepHel protein, to be the mother element of the numerous non-autonomous *Mothras*. Thus, in the strict sense, the putative *Mothra* mother element might itself not be autonomous (see before).

Polymorphisms between *O. sativa* and *O. glaberrima* demonstrate recent activity of *Mothra* elements

In a previous study we produced an alignment of approximately 60 % of the *O. sativa* and *O. glaberrima* genomes for identification of presence/absence polymorphisms of TIR transposons [27]. Now, we searched this alignment for polymorphisms related to *Mothra* elements. Out of a total of 856 *Mothra*-related polymorphisms, we investigated 148 manually. Most of them turned out not to be actual presence/absence polymorphisms, but rather variations in the number of repeat units between orthologous *Mothras* of the two species. Most of these differences probably arose from mechanisms such as unequal crossing-over or repeat slippage rather than from transposition activity (Fig. 3). Thus, the vast majority of *Mothra* copies are found in the same position in both rice species,

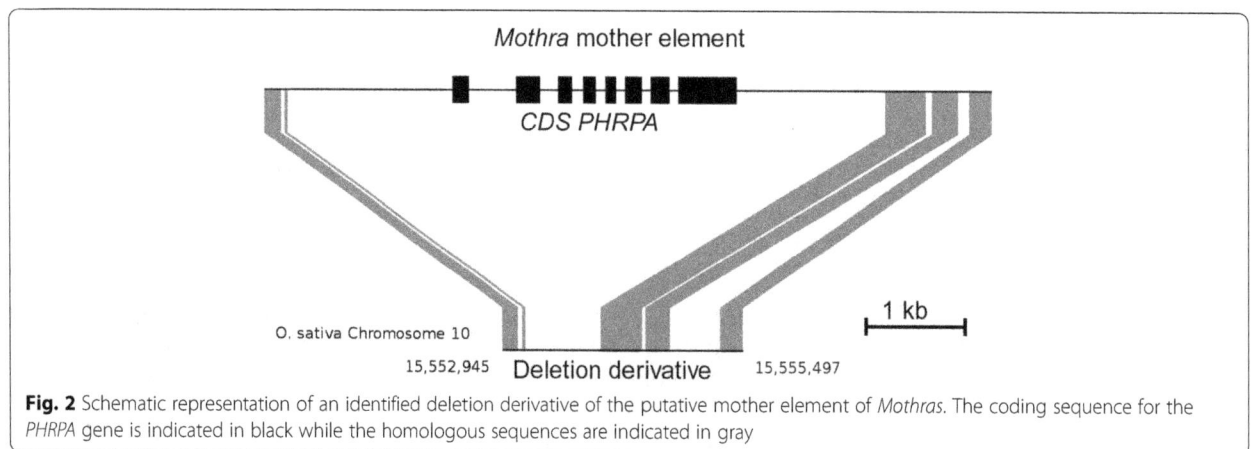

Fig. 2 Schematic representation of an identified deletion derivative of the putative mother element of *Mothras*. The coding sequence for the *PHRPA* gene is indicated in black while the homologous sequences are indicated in gray

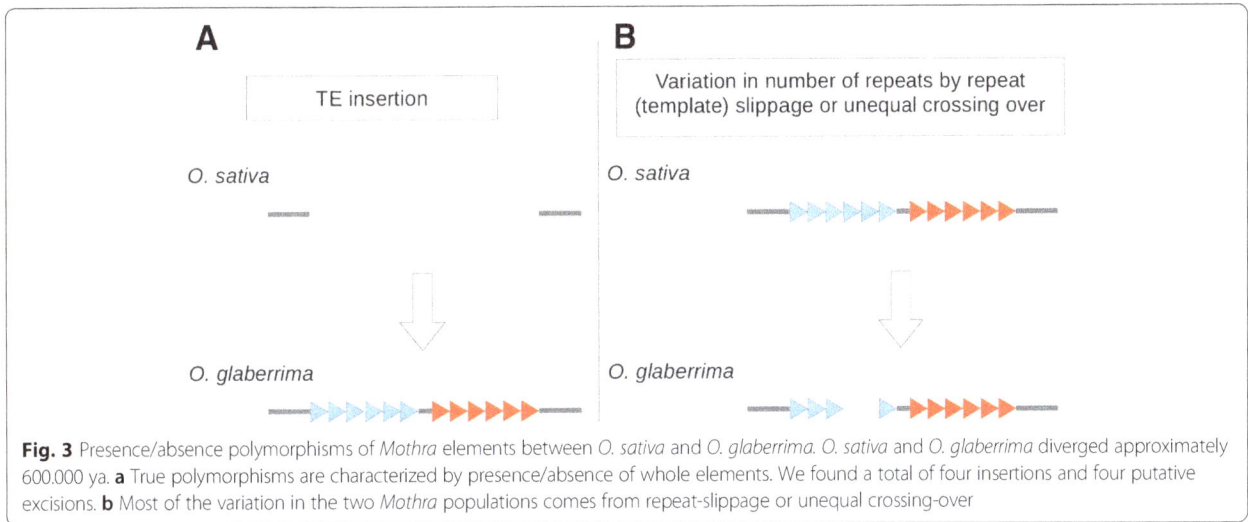

Fig. 3 Presence/absence polymorphisms of *Mothra* elements between *O. sativa* and *O. glaberrima*. *O. sativa* and *O. glaberrima* diverged approximately 600.000 ya. **a** True polymorphisms are characterized by presence/absence of whole elements. We found a total of four insertions and four putative excisions. **b** Most of the variation in the two *Mothra* populations comes from repeat-slippage or unequal crossing-over

meaning that they inserted before the two species diverged approximately 600,000 years ago [29]. Therefore, we can say that most of the copies are older than 600,000 years.

However, we also identified eight sites where we found putative insertion/excision polymorphisms of non-autonomous *Mothras* between the two rice species (Fig. 4a). In four cases, we found the *Mothra* element located between the characteristic nucleotides A and T present in *O. sativa* but not in *O. glaberrima*. Because Helitrons do not generate target site duplications, these events probably represent typical insertions in *O. sativa*. Interestingly, we found four sites where we suspect putative *Mothra* excisions. We conclude this based on the DNA repair patterns which are similar to those described for TIR DNA transposon excisions [30] (Fig. 4b). In two cases, we observed incomplete excision events whereas the other two cases went along with a deletion and the introduction of filler DNA, respectively.

The eight polymorphic elements correspond to 5,4 % of subset of 148 manually investigated polymorphisms. Considering that we identified a total of 856 insertion/deletion polymorphisms between the two species, we extrapolate that a total of approximately 46 *Mothra* elements have moved since the two species diverged about 600,000 years ago [29]. However, this number is based on approximately 60 % of the genome which was aligned. Thus, the actual number of transposed elements might be even higher. Compared to the previously investigated TIR transposons [27], we conclude that *Mothra* has a level of activity similar to that of highly active DTT-Mariner elements.

Phylogenetic analysis of the *Mothra* RPA homolog family

RPA proteins are involved in crucial processes such as DNA-replication and -repair. Furthermore, this "core"

RPA is a single copy gene and highly conserved among eukaryotes. This makes RPA useful for phylogenetic analysis and, thus, to study the origin of the plant Helitron RPA homolog (PHRPA). We used the the original "core" RPA as well as identified *Mothra* PHRPA of *O. sativa* as queries for NCBI blast searches against representatives from all major eukaryotic branches. We also included species from the largely under-sampled unicellular eukaryotic clades, such as Alveolata, Amoebae, Oomycetes and Rhizaria. Furthermore, we include two RPA homologs from Helentrons that were identified in *Drosophila melanogaster* [21] to investigate their relationship to PHRPAs. As an outgroup, we used some distant homologs from archaea (Fig. 5). Except in monocotyledonous and dicotyledonous plants, we usually found exactly one *RPA* gene (see below). The final dataset comprised 72 proteins from 62 species.

Our results show that most major eukaryotic clades cluster in monophyletic groups. We observe a clear grouping into plants, animals, fungi and Oomycetes. The phylogeny within these clades is consistent with the established taxonomy of eukaryotes [31]. For example plant RPAs first split into algae, mosses and later into monocots and dicots (Fig. 5). Because of the robustness of the tree and the great concordance with the taxonomy, these proteins most probably represent the intrinsic, eukaryotic "core" RPAs.

Most clades have exactly one RPA gene but there are exceptions. Interestingly, one of the two copies obtained from the Alveolata, *Cryptosporidium*, also clusters at the root of the plant branch. However, the other copy we find, as expected, in the clade of Alveolates, which are even more distant to the core RPA clade than the PHRPA family. Furthermore, we found two RPA paralogs in the genomes of *Physcomitrella*, a Moss, and the

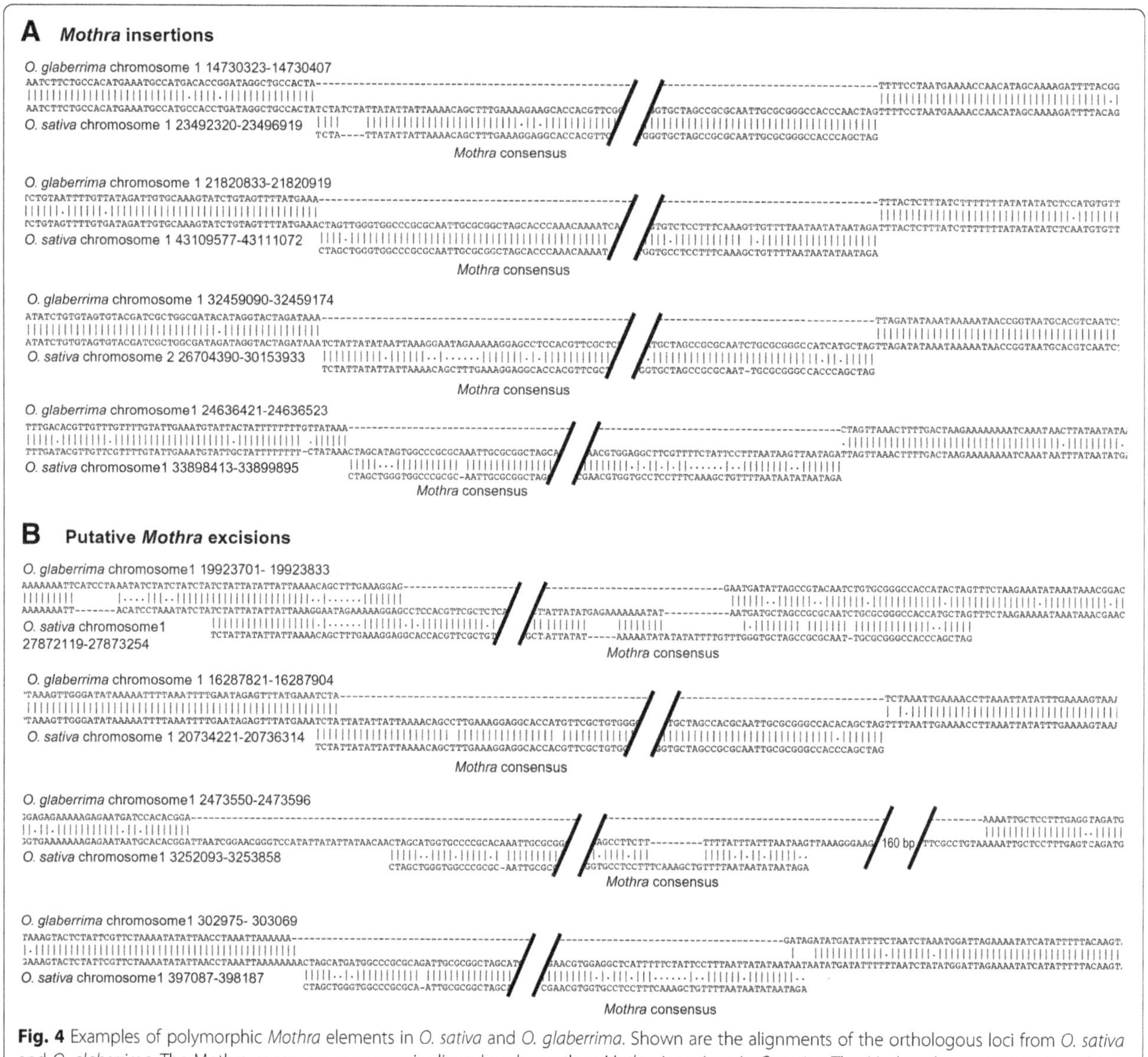

A *Mothra* insertions

O. glaberrima chromosome 1 14730323-14730407

O. sativa chromosome 1 23492320-23496919

Mothra consensus

O. glaberrima chromosome 1 21820833-21820919

O. sativa chromosome 1 43109577-43111072

Mothra consensus

O. glaberrima chromosome 1 32459090-32459174

O. sativa chromosome 2 26704390-30153933

Mothra consensus

O. glaberrima chromosome1 24636421-24636523

O. sativa chromosome1 33898413-33899895

Mothra consensus

B Putative *Mothra* excisions

O. glaberrima chromosome1 19923701- 19923833

O. sativa chromosome1 27872119-27873254

Mothra consensus

O. glaberrima chromosome 1 16287821-16287904

O. sativa chromosome 1 20734221-20736314

Mothra consensus

O. glaberrima chromosome1 2473550-2473596

O. sativa chromosome1 3252093-3253858

Mothra consensus

O. glaberrima chromosome1 302975- 303069

O. sativa chromosome 1 397087-398187

Mothra consensus

Fig. 4 Examples of polymorphic *Mothra* elements in *O. sativa* and *O. glaberrima*. Shown are the alignments of the orthologous loci from *O. sativa* and *O. glaberrima*. The Mothra consensus sequence is aligned underneath. **a** *Mothra* insertions in *O. sativa*. The *Mothra* elements insert into the genome without producing a target site duplication. **b** Putative excision events in *O. glaberrima*. DNA repair patterns are similar to those found for DNA transposons. They include incomplete excision of the element (top two alignments), deletions in the flanking regions (third alignment) or insertion of filler sequences (bottom alignment)

green alga *Chlorella*, to form a monophyletic group on the same level as the PHRPAs. It is possible that there are contaminations since these organisms are difficult to isolate and cultivate.

Most importantly, we find the PHRPAs to form a separate, monophyletic group outside the core RPA clade. Thus, we conclude that the PHRPA ancestor protein has evolved very early in the transition from prokaryotes to eukaryotes. Interestingly, we only find representatives of mono- and dicotyledonous plants in the PHRPA clade. Moreover, PHRPAs are more diverse than core RPAs. Indeed, PHRPA proteins are on average 21 % identical to each other, while core RPAs show an average of 39 % sequence identity (Additional file 1: Figure S1). Also the branch lengths of the PHRPA clade are noticeable long. This suggests diversification of new, independent gene subfamilies. The possible reasons why these proteins are only found in monocots and dicots which diverged approximately 145–300 million years ago (mya) [32, 33] are discussed below (see discussion). Moreover, our analysis reveals that the RPA proteins acquired by Helentrons seem to be of another, even more distant origin. These proteins form a separate group which branches off before the radiation of eukaryotes (Fig. 5).

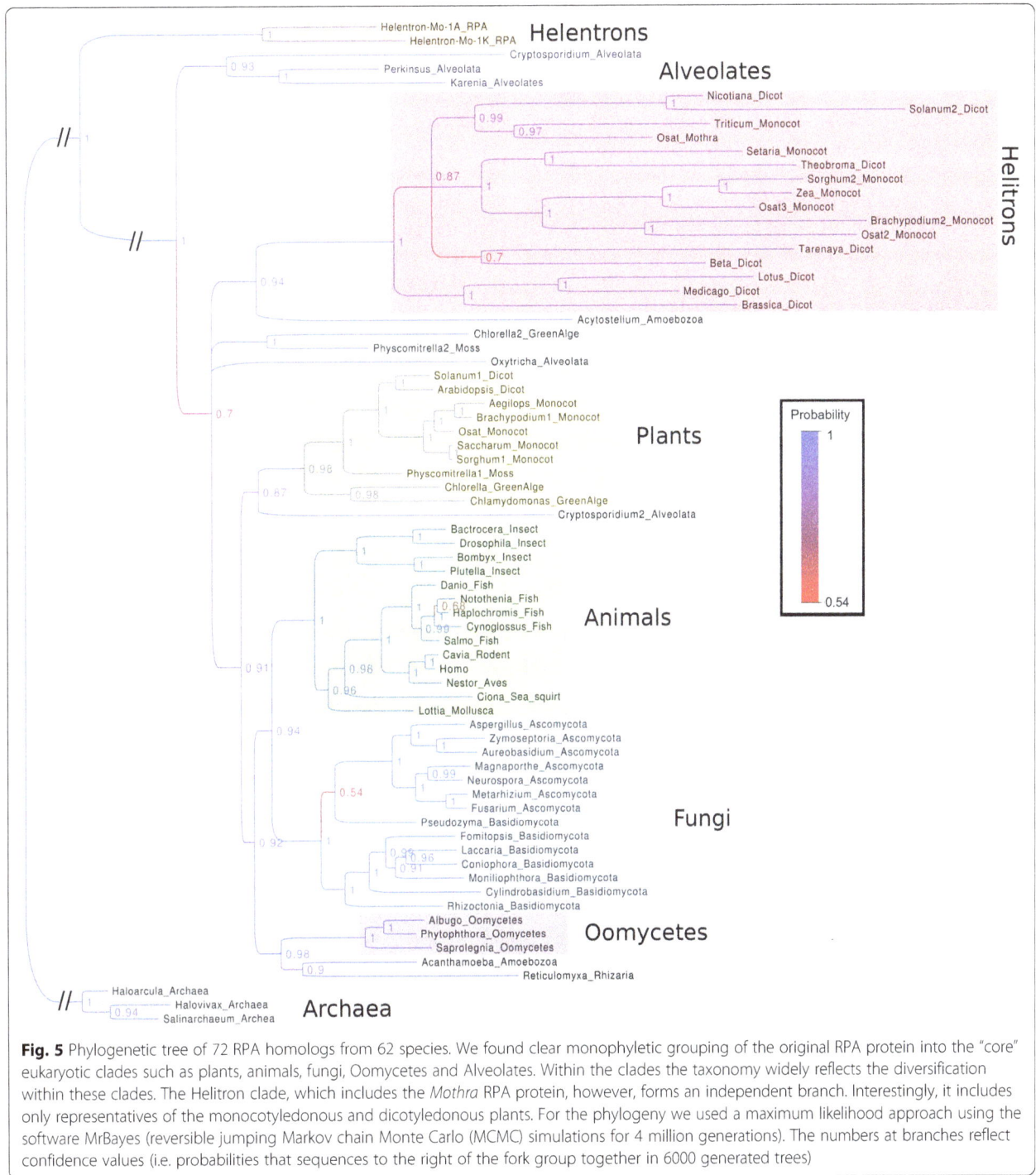

Fig. 5 Phylogenetic tree of 72 RPA homologs from 62 species. We found clear monophyletic grouping of the original RPA protein into the "core" eukaryotic clades such as plants, animals, fungi, Oomycetes and Alveolates. Within the clades the taxonomy widely reflects the diversification within these clades. The Helitron clade, which includes the *Mothra* RPA protein, however, forms an independent branch. Interestingly, it includes only representatives of the monocotyledonous and dicotyledonous plants. For the phylogeny we used a maximum likelihood approach using the software MrBayes (reversible jumping Markov chain Monte Carlo (MCMC) simulations for 4 million generations). The numbers at branches reflect confidence values (i.e. probabilities that sequences to the right of the fork group together in 6000 generated trees)

Mothras might use the RepHel protein of closely related Helitrons

Above, we describe that the putative mother element of the non-autonomous *Mothras* encoded an PHRPA protein but not for a RepHel protein. This raises the question of how these elements would actually transpose. As it has been described for non-autonomous elements, that they recruit closely related transposases, we suspect that RepHel from a closely related Helitron family would be used by *Mothra* elements. Therefore, we scanned the *O. sativa* genome for homologs of the PHRPA protein and extracted 21 fragments including 20 kb up- and downstream of the protein. Out of these we identified nine sequences with sizes from 8064 to 15,513 bp that all contain a *PHRPA* homolog and an adjacent *RepHel* gene.

Based on sequence homology we could clearly differentiate them into three groups. While we found five copies of group 1 elements, there were two copies each for groups 2 and 3, respectively. The PHRPA of group 1 is most similar to that of the Mothra mother element (46.1 % similarity compared to 21.6 % and 22.6 % for groups 2 and 3, respectively). Moreover, the elements of group 1 and *Mothras* nearly fulfill the criteria of Yang et al. [2] to belong to the same family (73 % identity over 30 bp at the 5' end and 77 % identity at the 3' end). Because of this and the strong homology of their RPA proteins, we henceforth refer to these Helitrons of as the sister-family of *Mothra* (Fig. 6). Interestingly, when we compared the five copies of the sister-family with those in *O. glaberrima*, we found all of them to be polymorphic (Table 1), indicating recent activity of the *Mothra* sister-family. Thus, we propose that *Mothra* elements recruit the RepHel protein of their sister-family to transpose. For both, the *PHRPA* gene of the *Mothra* mother element and *PHRPA* and *RepHel* of the sister-family, we found transcripts in NCBI, suggesting that both might still be active (Additional file 2: Table S1).

Discussion

The goal of our study was to characterize the origin and evolution of the high-copy Helitron family *Mothra* in rice. Although Helitrons are found in nearly all eukaryotic genomes they are much less well understood than other TE superfamilies. Despite their considerable role in exon shuffling and gene movement in plants [22–24], only few studies are available that shed light on their transposition mechanism. Initially, it was proposed that Helitrons replicate via a rolling-circle mechanism [1]. However, this was challenged by the discovery of Helitron excisions in somatic maize tissue [20]. Our data also suggest that some of the presence/absence polymorphism in rice might represent Helitron excisions. While Li and Dooner [14] mainly found repair patterns introducing TA micro-satellites as "filler" DNA, our putative excision events were also associated with deletions of the flanking sequences. These footprints strongly resemble those of TIR transposon excisions [27, 29, 34, 35]. Thus, these combined findings suggest the existence of at least one alternative transposition pathway to the proposed rolling-circle mechanism.

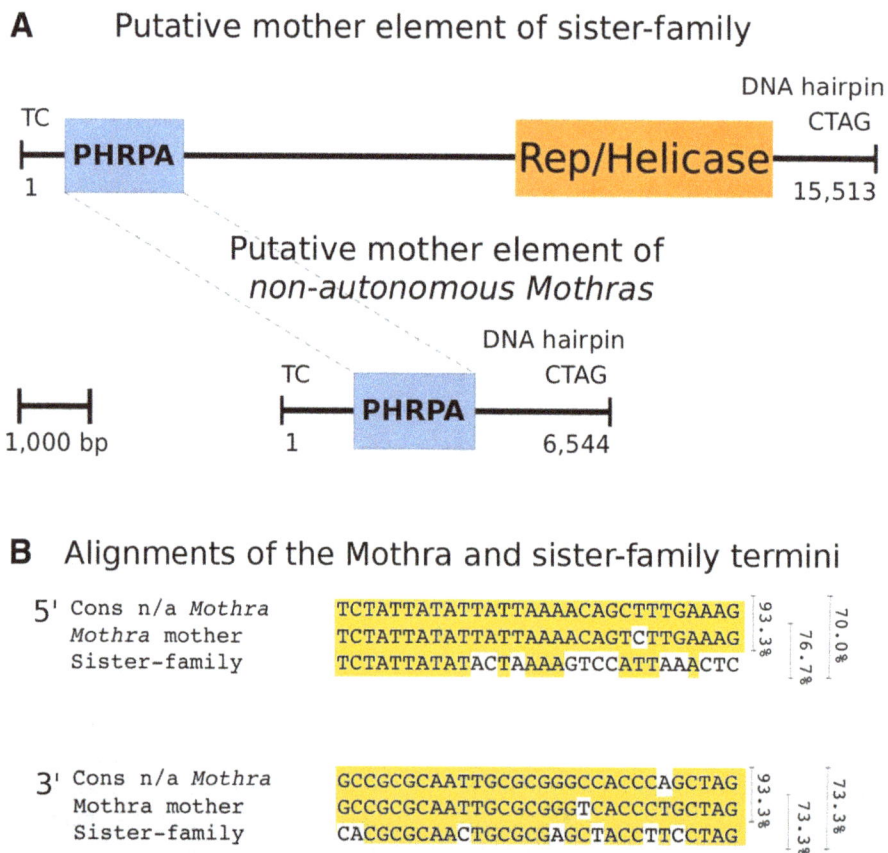

Fig. 6 Schematic overview of a putative autonomous sister element of *Mothra*. **a** The element encodes a RPA protein closely related with the *Mothra* RPA (PHRPA) and, most importantly, also a RepHel protein. **b** The alignment shows the terminal sequences of the non-autonomous *Mothra* consensus, the putative mother element and the *Mothra* sister-family

Table 1 Overview of all identified copies of the putatively autonomous elements of the *Mothra* sister-family in *O. sativa* and *O. glaberrima*

Mothra sister-family copies

O. sativa

Chromosome	Start pos.	End pos.	Comment
11	26,634,911	26,619,399	Reverse
11	22,184,151	22,168,642	Reverse
11	24,183,680	24,199,188	Forward
5	592,132	606,965	Forward
5	25,964,570	25,949,203	Reverse

O. glaberrima

Chromosome	Start pos.	End pos.	Comment
11	19,460,159	19,467,758	No RPA

Despite these open questions, the main findings of our study provided insight into the evolution of different levels of non-autonomous elements and, more importantly, of the Helitron superfamily in plants in general. Our main conclusions are discussed in the following.

Sequence composition of non-autonomous *Mothras* elements might play a role in transposition efficiency

Non-autonomous transposons can create hundreds or even thousands of copies in only few generations [36]. Loss of protein coding sequences and thereby autonomy has happened in all major Class II TE superfamilies. It can be explained by the fact that hosts regulate TEs via epigenetic silencing. Thus, constant reshaping, shortening and the accumulation of "nonsense" sequences might be mechanisms to avoid RNA silencing [37]. Alternatively, the presence of an active functional copy might release selection pressure on other copies, allowing for non-autonomous derivatives to emerge. Still, non-autonomous elements retain the ability to cross-mobilize related transposases. This type of trans-acting system has best been described in detail for the TIR transposons of the *DTT-Mariner* superfamily [36]. Transient expression experiments in yeast showed that the affinity for the autonomous element was determined by the TIR region. The efficiency of transposition, however, was influenced dramatically, positively or negatively, by different compositions of internal sequences.

We suspect that the great success of *Mothra* elements might have to do with their unusual sequence composition (see Fig. 1). The Blocs A and B of the non-autonomous element are unique to *Mothra* elements and their high conservation within the *Mothra* family suggests functional importance. When we screened the genomes of *O. sativa* and *O. glaberrima* for *Mothra*

related polymorphisms (see above), we found that the majority of the differences were variations in the number of repeat units. Most likely these were caused by repeat slippage or unequal crossing-over for which the repeat arrays of Blocs A and B served as templates. Thus, these repeat arrays may be a sources of plasticity and permanent turnover within non-autonomous *Mothra* elements.

The *Mothra* RPA homolog likely originated from horizontal transfer

In our phylogenetic analysis of RPA proteins we found clear monophyletic clustering of the "core" RPAs in all major eukaryotic groups which broadly reflects the separation of early eukaryotes into distinct lineages (see Fig. 5). Interestingly, the clade representing the RPA homologs from plant Helitrons (PHRPAs) branches off even before the separation of plants, animals, fungi and Oomycetes, indicating a very ancient origin of these proteins. It is the more surprising that this clade only includes proteins from monocotyledonous and dicotyledonous plants which only separated approximately 145–300 mya [32, 33]. Previous studies proposed that plant Helitrons hijacked and modified the eukaryotic core *RPA* gene which later became the plant Helitron RPA [1, 38]. However, the clear monophyletic origin of PHRPAs outside the core RPA clade challenges this model.

There are two possible explanations for the phylogenetic position of PHRPAs: First, PHRPA proteins were originally present in all other eukaryotes and were lost in all lineages except the monocots and dicots. We consider this highly unlikely. The second explanation (which we clearly favor) is horizontal gene transfer. Typical characteristics of horizontal gene transfer are phylogenetic incongruence and/or unusually high sequence identity of proteins from otherwise distantly related species. In our case, we found very well supported phylogenetic incongruence. However, we could not identify a putative donor of PHRPA. This donor was obviously not sampled in our collection. We propose that PHRPA was transferred from this unknown and distantly related eukaryote into the progenitor of monocots and dicots. This horizontal transfer must have occurred before monocots and dicots diverged 145–300 mya [32, 33], since we have not found PHRPAs in any other plant group that diverged earlier. Our data indicate that the progenitor of all eukaryotic RPA genes was already present during eukaryogenesis, but it remains unclear if the last eukaryotic common ancestor had one or several RPA homologs (Fig. 7), because in several organisms such as *Physcomitrella*, *Chlorella*, *Acanthamoeba* and *Cryptosporidium* we find both, a core RPA and a homolog that is equally distant from the core RPA as the PHRPA clade. We therefore suspect that the donor of plant Helitron RPA

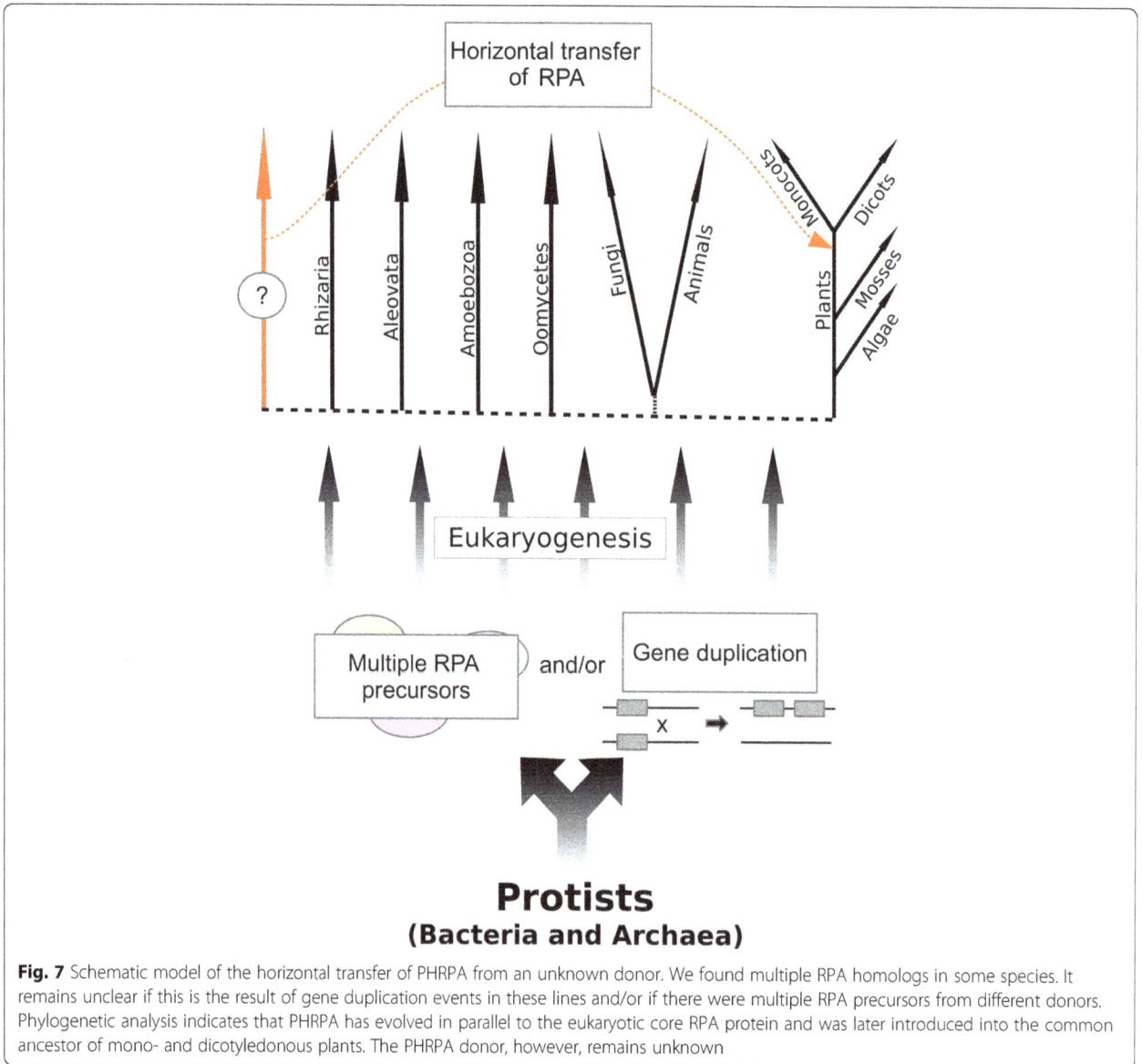

Fig. 7 Schematic model of the horizontal transfer of PHRPA from an unknown donor. We found multiple RPA homologs in some species. It remains unclear if this is the result of gene duplication events in these lines and/or if there were multiple RPA precursors from different donors. Phylogenetic analysis indicates that PHRPA has evolved in parallel to the eukaryotic core RPA protein and was later introduced into the common ancestor of mono- and dicotyledonous plants. The PHRPA donor, however, remains unknown

homologs was probably a basal eukaryote similar to those mentioned above.

In Prokaryotes (bacteria and archaea), horizontal gene transfer is common and and it is believed to be a major mechanism for adaptation [39]. It becomes more and more evident that horizontal transfer is also a common process in eukaryotes. For example the extremophilic red alga *Galdieria sulphuraria* exhibits a enormous metabolic flexibility which it acquired by various genes from different bacteria and archaea [40]. Like genes, also TEs (if they are not the vector for gene transfer themselves) can be transfered between hosts. Often this involves intermediate vectors such as blood feeding insects or pathogens carrying bacteria or viruses to their new hosts. For example in 24 species of the insect order Lepidoptera two non-autonomous *Helitrons* were identified

which were also found in the genomes of several double-stranded DNA polydnaviruses [41]. In plants, up to two million horizontal TE transfers only of LTR-retrotransposons were suggested by a comparative analysis among flowering plants [42].

However, what makes the case of PHRPA special is that the proposed horizontal transfer resulted in a successful new type of TE whose widespread distribution in monocots and dicots suggests advantages over normal Helitrons lacking this gene. Indeed, Dong et al. [43] described how stepwise acquisition of gene fragments can produce elements of increasing complexity.

Interestingly, our analysis also suggests that RPA homologs in *Drosophila*, called Helentrons, might also have been acquired though horizontal transfer. But the phylogenetic analysis indicates that they are of an even more distant origin. Furthermore, highly divergent RPA homologs were also found in Helitrons of zebrafish and starlet sea anemone [12]. However, here we were not able to identify any homology to PHRPAs, which is why they were not included in our phylogenetic analysis. Thus, it appears that Helitrons acquired single-strand binding proteins at least three times independently during evolution, suggesting convergent evolution.

A model for the evolution of semi-autonomous and non-autonomous plant Helitrons

Our data suggest that the numerous non-autonomous *Mothra* elements are mobilized by a single mother element. Surprisingly, this putative mother element encodes for PHRPA but not for a RepHel protein. We speculate that the mother element might itself be depending on a related and fully autonomous element. Indeed, we found one candidate Helitron family that shows strong homology

with the RPA protein and the termini of the *Mothra* mother element. We referred to that Helitron family as the *Mothra* sister-family.

Based on these observations, we propose a model which introduces the putative mother element as an additional level of "semi-autonomy" (Fig. 8). We assume that the ancient Helitron consisted of a *RepHel* gene and probably the structural features like the 3' hairpin that we find to be common in all Helitrons. According to our model, the PHRPA protein was then introduced in the common ancestor of mono- and dicots via horizontal transfer 145–300 mya [32, 33] where it got acquired by the progenitor of all RPA containing plant Helitrons (discussed above). We propose that at a later point, one Helitron lineage lost its *RepHel* gene, resulting in the putative *Mothra* mother element that only contains the *PHRPA* gene. This semi-autonomous element would still fulfill some functions in the transposition process but would rely

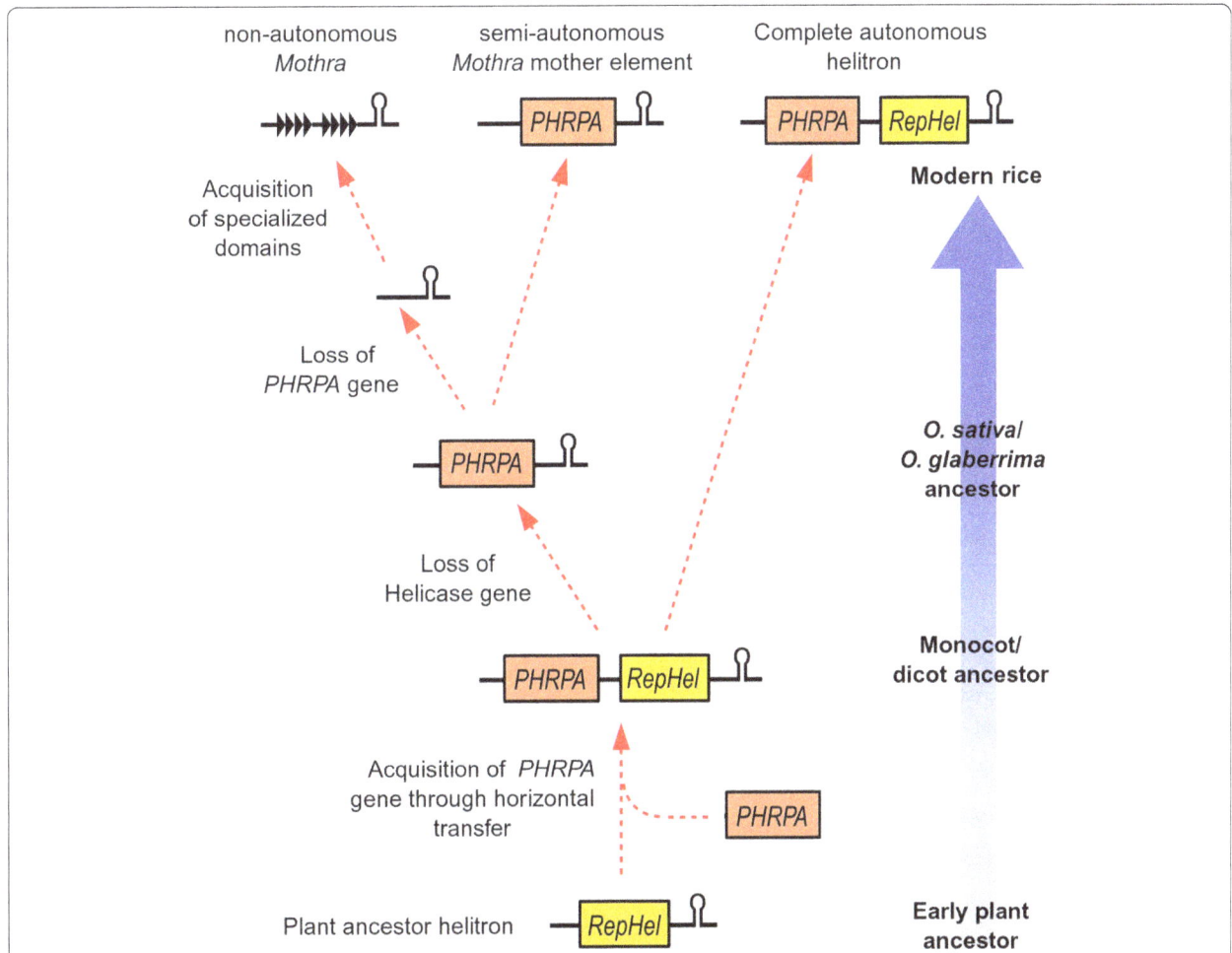

Fig. 8 Schematic model of the evolution of plant Helitrons. We propose that the progenitor of plant Helitrons contained only a RepHel domain. Later, the PHRPA protein was introduced into the progenitor of mono- and dicotyledonous plants approximately 145 – 300 mya via horizontal transfer to form the first plant RPA-like Helitron. In the case of *Mothra* elements, the RepHel domain was later lost, thus introducing an additional level of "semi-autonomy" between the non-autonomous elements and the fully autonomous Helitrons

on the RepHel protein provided by the *Mothra* sister-family. Loss of internal sequences is common during transposition of Helitrons [43]. Furthermore, the evolution of non-autonomous transposable elements has been described in virtually all TE superfamilies [16].

According to our model, the next step in *Mothra* evolution was the loss of the *PHRPA* gene, resulting in a completely non-autonomous element that relies both on the *Mothra* mother element and functional copies of the *Mothra* sister-family (Fig. 8). Finally, the non-autonomous Mothra element acquired the complex tandem repeat blocs which, we propose, improved its transposition efficiency. This proposed stepwise evolution ultimately led to the situation we find in modern rice species where all three types of elements (fully autonomous, semi-autonomous and non-autonomous) exist side-by-side. However, biochemical assays will be needed to confirm the functional relationship between the described elements.

Conclusion

Analysis of the *Mothra* family of Helitrons has provided unexpected insight in to the early evolution of plant Helitrons through the identification of a putative horizontal gene transfer that resulted in a successful sub-group of the Helitron superfamily. Furthermore, the great success of the non-autonomous *Mothra* elements suggests that combinations of different levels of transposition autonomy might be particularly efficient in Helitrons.

Methods

Mothra annotation

To generate the *Mothra* consensus sequence, we extracted and aligned 100 putative copies including 5 kb of flanking sequence which we used to manually determine the boundaries of the element. The identified termini matched the previously described canonical Helitron termini [16]. To deduce the consensus sequences for the sub-types and finally the consensus sequence of the non-autonomous *Mothra* element, we used the multiple alignment software Clustal X [44], the graphical dot-matrix program Dotter from the SeqTools package (https://www.sanger.ac.uk/resources/software/seqtools/) and in-house Perl scripts which are available upon request. To annotate *Mothra* elements we used the *Mothra* consensus sequence in Blastn searches against the *O. sativa Nipponbare* cultivar genome (Version 5) provided by the International Rice Genome Sequencing Project (IRGSP) (plantbiology.msu.edu/pub/data/) [45]. We included hits with a minimum length of 80 basepairs and at least 80 % identity. Because we found many fragments, we merged all hits that were found within 200 bps of flanking sequence to single hits.

To identify the *Mothra* mother element we used Blastn searches of the first 50 and the last 80 bps of the *Mothra* consensus sequence. We considered fragments where we found both ends in the same orientation and that were located within 25 kb from each other. We used the online NCBI platform (http://blast.ncbi.nlm.nih.gov/Blast.cgi) to perform Blastn and Blastx searches against the 323 putative sequences to identify the *RPA* gene. To identify the polymorphisms between *O. sativa* and *O. glaberrima* we used the whole genome alignment produced in a previous study [27].

Phylogenetic tree

The sequences for the phylogenetic tree were retrieved from the NCBI database (http://www.ncbi.nlm.nih.gov/). We used the sequences of the identified Mothra RPA and the core RPA of *O. sativa* as queries and searched each of the main eukaryotic groups, animal, fungi, plants, Alveolata, Amoebae, Rhizaria, Oomycetes and archaea separately. We aligned them using Clustal X [44] with the following parameters for multiple alignments: Gap opening penalty of 10 and Gap extension penalty of 0.1. The phylogenetic tree was generated using MrBayes 3.2.2 [46]. We conducted two runs with 4 chains, each for 4 million generations, sampling every 500 generations. We used all the protein models available in MrBayes and used a reversible jump Monte Carlo Markov Chain (MCMC) [47]. Heterogeneity of substitution rates among different sites was modeled with a gamma distribution. The first quartile of generations was discarded (burn-in) and convergence was evaluated with the average standard deviation of split frequencies (0.002). To illustrate and re-root the tree we used the program Figtree (http://tree.bio.ed.ac.uk/software/figtree/).

Data access

Sequences of *Mothra* elements were deposited in the TREP database (http://www.botinst.uzh.ch/research/genetics/thomasWicker/TREP.html). Sequence alignments that were used for phylogenetic analyses as well as in-house Perl scripts are available upon request.

Abbreviations
TE: Transposable element; TIR: Terminal inverted repeat; aa: Amino acid; RPA: Replication protein A; ORF: Open reading frame; PHRPA: Plant helitron replication protein A; mya: Million years ago.

Competing interests
The authors declare that they have no competing interests.

Authors' contributions
SR performed the analysis and wrote the paper. TW designed the study and wrote the paper. FM helped with the phylogenetic analysis. All authors have read and approved the final version of the paper.

Acknowledgements
This study was supported by the Swiss National Foundation grant # 31003A_138505/1.

References
1. Kapitonov VV, Jurka J. Rolling-circle transposons in eukaryotes. PNAS. 2001;98(15):8714–9.
2. Yang L, Bennetzen JL. Structure-based discovery and description of plant and animal Helitrons. PNAS. 2009;106(31):12832–7.
3. Rensing SA, Lang D, Zimmer AD, Terry A, Salamov A, Shapiro H, et al. The *Physcomitrella* Genome Reveals Evolutionary Insights into the Conquest of Land by Plants. Science. 2008;319:64–9.
4. Hood ME. Repetitive DNA in the automictic fungus *Microbotryum violaceum*. Genetica. 2005;124(1):1–10.
5. Poulter RTM, Goodwin TJD, Butler MI. Vertebrate helentrons and other novel Helitrons. Gene. 2003;313:201–12.
6. Nierman WC, Pain A, Anderson MJ, Wortman JR, Kim HS, Arroyo J, et al. Genomic sequence of the pathogenic and allergenic filamentous fungus *Aspergillus fumigatus*. Nature. 2005;438:1151–6.
7. Kapitonov VV, Jurka J. Helitron-1_SP, a family of autonomous Helitrons in the sea urchin genome. Repbase Rep. 2005;5:393.
8. Zhou Q , Froschauer A, Schultheis C, Schmidt C, Bienert GP, Wenning M, etal. Helitron transposons on the sex chromosomes of the platyfish *Xiphophorus maculatus* and their evolution in animal *genomes*. Zebrafish. 2006;3:39–52.
9. Ennio C, De Iorio S, Capriglione T. Identification of a novel helitron transposon in the genome of Antarctic fish. Mol Phylogenet Evol. 2011;58(3):439–46.
10. Pritham EJ, Feschotte C. Massive amplification of rolling-circle transposons in the lineage of the bat *Myotis lucifugus*. PNAS. 2007;104(6):1895–900.
11. Yang L, Bennetzten JL. Distribution, diversity, evolution, and survival of *Helitrons* in the maize genome. PNAS. 2009;106(47):19922–7.
12. Kapitonov VV, Jurka J. *Helitrons* on a roll: eukaryotic rolling-circle transposons. Science. 2007;23(10):521–9.
13. Chandler M, de la Cruz F, Dyda F, Hickman AB, Moncalian G, Ton-Hoang B. Breaking and joining single-stranded DNA: the HUH endonuclease superfamily. Nat Rev Micro. 2013;11(8):525–38.
14. Li Y, Dooner HK. Excision of Helitron transposons in maize. Genetics. 2009;182(1):399–402.
15. Thomas J, Pritham EJ. Helitrons, the eukaryotic rolling-circle transposable elements. Microbiol Spectrum. 2015;3(4):MDNA3-0049-2014.
16. Wicker T, Sabot F, Hua-Van A, Bennetzen JL, Capy P, Chalhoub B, et al. A unified classification system for eukaryotic transposable elements. Nat Rev Genet. 2003;8:973–82.
17. Stenger DC, Revington GN, Stevenson MC, Bisaro DM. Replicational release of geminivirus genomes from tandemly repeated copies: evidence for rolling-circle replication of a plant viral DNA. PNAS. 1991;88(18):8029–33.
18. Mendiola MV, Bernales I, De La Cruz F. Differential roles of the transposon termini in IS91 transposition. PNAS. 1994;91(5):1922–6.
19. Oakley GG, Patrick SM. Replication protein A: directing traffic at the intersection of replication and repair. Front Biosci. 2010;15:883.
20. Wold MS. Replication protein A: a heterotrimeric, single-stranded DNA-binding protein required for eukaryotic DNA metabolism. Annu Rev Biochem. 1997;66(1):61–92.
21. Thomas J, Vadnagara K, Pritham EJ. DINE-1, the highest copy number repeats in Drosophila melanogaster are non-autonomous endo-nuclease-encoding rolling-circle transposable elements (Helentrons). Mob DNA. 2014;5:18.
22. Lai J, Li Y, Messing J, Dooner HK. Gene movement by Helitron transposons contributes to the haplotype variability of maize. PNAS. 2005;102(25):9068–73.
23. Morgante M, Brunner S, Pea G, Fengler K, Zuccolo A, Rafalski A. Gene duplication and exon shuffling by helitron-like transposons generate intraspecies diversity in maize. Nat Genet. 2005;37(9):997–1002.
24. Barbaglia AM, Klusman KM, Higgins J, Shaw JR, Hannah LC, Lal SK. Gene capture by Helitron transposons reshuffles the transcriptome of maize. Genetics. 2012;190(3):965–75.
25. Han MJ, Shen YH, Xu MS, Liang HY, Zhang HH, Zhang Z. Identification and Evolution of the Silkworm Helitrons and their Contribution to Transcripts. DNA Res. 2013;20:471–84.
26. Thomas J, Phillips CD, Baker RJ, Pritham EJ. Rolling-Circle Transposons Catalyze Genomic Innovation in a Mammalian Lineage. Genome Biol Evol. 2014;6(10):2595–610.
27. Roffler S, Wicker T. Genome-wide comparison of Asian and African rice reveals high recent activity of DNA transposons. Mob DNA. 2015;6(1):8.
28. Kemetaka S, Okano T, Ohsumi M, Ohsumi Y. Apg14p and Apg6/Vps30p Form a Protein Complex Essential for Autophagy in the Yeast, *Saccheromyces cerevisiae*. J Biol Chem. 1998;273(35):22284–91.
29. Wang M, Yu Y, Haberer G, Marri PR, Fan C, Goicoechea JL, et al. The genome sequence of African rice (Oryza glaberrima) and evidence for independent domestication. Nat Genet. 2014;46:982–8.
30. Kikuchi K, Terauchi K, Wada M, Hirano HY. The plant MITE mPing is mobilized in anther culture. Nature. 2003;421:167–70.
31. Koonin EV. The origin and early evolution of eukaryotes in the light of phylogenomics. Genome Biol. 2010;11(5):209.
32. Kawai Y, Otsuka J. The deep phylogeny of land plants inferred from a full analysis of nucleotide base changes in terms of mutation and selection. J Mol Evol. 2004;58:479–89.
33. Zimmer A, Lang D, Richardt S, Frank W, Reski R, Rensing SA. Dating the early evolution of plants: detection and molecular clock analyses of orthologs. Mol Genet Genomics. 2007;278:393–402.
34. Buchmann JP, Matsumoto T, Stein N, Keller B, Wicker T. Interspecies sequence comparison of Brachypodium reveals how transposon activity corrodes genome colinearity. Plant J. 2012;488:213–7.
35. Yang G, Weil CF, Wessler SR. A rice Tc1/mariner-like element transposes in yeast. Plant Cell. 2006;18:2469–78.
36. Yang G, Nagel DH, Feschotte C, Hancock CN, Wessler SR. Tuned for transposition: Molecular determinants underlying the hyperactivity of a *Stowaway* MITE. Science. 2009;325(5946):1391–4.
37. Lisch D. Epigenetic regulation of transposable elements in plants. Plant Biol. 2009;60:43–66.
38. Feschotte C, Wessler SR. Treasures in the attic: rolling circle transposition discovered in eukaryotic genomes. PNAS. 2001;98(16):8923–4.
39. Rocha EPC. With a little help from prokaryotes. Science. 2013;339(6124):1154–5.
40. Schönknecht G, Chen WH, Ternes CM, Barbier GG, Shrestha RP, Stanke M, et al. Gene transfer from bacteria and archaea facilitated evolution of an extremophilic eukaryote. Science. 2013;339(6124):1207–10.
41. Coates BS. Horizontal transfer of a non-autonomous Helitron among insect and viral genomes. BMC Genomics. 2015;16(1):137.
42. El Baidouri M, Carpentier MC, Cooke R, Gao D, Lasserre E, Llauro C, et al. Widespread and frequent horizontal transfers of transposable elements in plants. Genome Res. 2014;24(5):831–8.
43. Dong Y, Lu X, Song W, Shi L, Zhang M, Zhao H, et al. Structural characterization of helitrons and their stepwise capturing of gene fragments in the maize genome. BMC Genomics. 2011;12:609.
44. Larkin MA, Blackshields G, Brown NP, Chenna R, McGettigan PA, McWilliam H, et al. Clustal W and Clustal X version 2.0. Bioinformatics. 2007;23:2947–8.
45. International Rice Genome Sequencing Project. The map-based sequence of the rice genome. Nature. 2005;436:793–800.
46. Ronquist F, Teslenko M, van der Mark P, Ayers DL, Darling A, Höhna S, et al. MrBayes 3.2: Efficient Bayesian Phylogenetic Inference and Model Choice Across a Large Model Space. Syst Biol. 2012;61(3):539–42.
47. Huelsenbeck JP, Larget B, Alfaro ME. Bayesian phylogenetic model selection using reversible jump Markov chain Monte Carlo. Mol Biol Evol. 2004;21(6):1123–33.

8

Phylogenomic analysis reveals genome-wide purifying selection on TBE transposons in the ciliate *Oxytricha*

Xiao Chen[1] and Laura F. Landweber[2]*

Abstract

Background: Transposable elements are a major player contributing to genetic variation and shaping genome evolution. Multiple independent transposon domestication events have occurred in ciliates, recruiting transposases to key roles in cellular processes. In the ciliate *Oxytricha trifallax*, the telomere-bearing elements (TBE), a Tc1/*mariner* transposon, occupy a significant portion of the germline genome and are involved in programmed genome rearrangements that produce a transcriptionally active somatic nucleus from a copy of the germline nucleus during development.

Results: Here we provide a thorough characterization of the distribution and sequences of TBE transposons in the *Oxytricha* germline genome. We annotate more than 10,000 complete and 24,000 partial TBE sequences. TBEs cluster into four major families and display a preference for either insertion into DNA segments that are retained in the somatic genome or their maintenance at such sites. The three TBE-encoded genes in all four families display dN/dS ratios much lower than 1, suggesting genome-wide purifying selection. We also identify TBE homologs in other ciliate species for phylogenomic analysis.

Conclusions: This paper provides genome-wide characterization of a major class of ciliate transposons. Phylogenomic analysis reveals selective constraints on transposon-encoded genes, shedding light on the evolution and domesticated functions of these transposons.

Keywords: Transposable element, Transposon domestication, Ciliates, Genome rearrangement, Purifying selection

Background

Transposable elements (TEs) are genomic parasites present in all eukaryotic genomes. There exist multiple different classes of TEs, which occupy distinct fractions of the genome and show a wide variety of genomic activity. Despite the drastic differences, TEs play important roles in shaping the genome and facilitating genome evolution by processes that can promote genome rearrangements, contribute to the origin of new genes and alter gene expression [1–4].

Ciliates are unicellular eukaryotes that possess two types of nuclei, a transcriptionally active somatic nucleus and an archival germline nucleus [5]. The somatic nucleus develops from a copy of the germline through extensive

genome rearrangements. In *Oxytricha*, the somatic macronucleus (MAC) is extremely gene dense, with ~16,000 short "nanochromosomes" that average 3.2 kb, and most encode a single gene [6]. The germline micronucleus (MIC), on the other hand, exhibits a highly fragmented and complex genome architecture, with short gene segments (Macronuclear Destined Sequences, MDSs) interrupted by brief noncoding sequences (Internal Eliminated Sequences, IESs). These DNA segments are the information that is retained in the soma after development; intriguingly, the DNA segments are often present in a permuted order or inverse orientation in the germline. Therefore, correct assembly of functional genes in the soma requires precise deletion of noncoding sequences and extensive reordering and inversion of gene segments that are "scrambled" in the germline. The somatic genome is free of transposons, although it contains some transposase-like genes [6]. Nearly 20 % of the germline

* Correspondence: lfl@princeton.edu
[2]Department of Ecology and Evolutionary Biology, Princeton University, Princeton, NJ 08544, USA
Full list of author information is available at the end of the article

genome is occupied by TEs [7], which are all eliminated during somatic development.

Ciliates provide novel model systems to study transposable elements because multiple TEs, especially the transposases they encode, have been recruited to provide important cellular functions for somatic development [8, 9]. The macronuclear genomes of *Tetrahymena* and *Paramecium* encode a homolog of the PiggyBac transposase that is expressed during development. Knockdown of the PiggyBac transposase results in a developmental defect, implicating its role in nuclear development [10, 11]. Tc1/*mariner* transposons are the most prevalent transposons in ciliate germline genomes, including the Tec elements in *Euplotes* [12] and *Tennessee, Sardine* and *Thon* elements in *Paramecium* [13, 14]. The terminal sequences of *Paramecium* IESs resemble the terminal inverted repeats of Tec elements in *Euplotes* [12, 15] and the ends of Tc1/*mariner* transposons [16], leading to the hypothesis that many IESs are remnants of TE insertions [17].

In *Oxytricha*, the telomere-bearing elements (TBEs) are another group of Tc1/*mariner* DNA transposons that have long been studied in ciliate germline genomes [18]. There is also phylogenetic evidence for recent insertion of TBEs [19]. TBEs encode three open reading frames (ORFs), a 42kD transposase, a 22kD ORF with unknown function and a 57kD ORF with zinc finger and kinase domains but unknown function (Fig. 1a). The 42kD transposase, together with the transposase encoded by *Euplotes* Tec elements and other Tc1/*mariner* transposases, belong to a superfamily of transposase genes with a common DDE catalytic motif [20]. Similar to the PiggyBac transposase, knockdown of the TBE transposase also leads to developmental defects, such as accumulation of unprocessed DNA and incorrectly rearranged nanochromosomes [21], suggesting that the TBE transposase has acquired an essential function in genome rearrangement. Because the transposase gene is present in many thousands of copies in the germline, this experiment was unique in knocking down such a high copy target. Nowacki *et al.* concluded that the 42kD transposase has likely been recruited for its DNA cleaving activity or another role in eliminating noncoding sequences, including their own elimination [21, 22].

A few studies have suggested that purifying selection is acting on the 42kD transposase encoded by TBEs [21, 23, 24]. However, these studies were limited by the small number (up to 100) of TBE sequences that were previously available. The levels of selection acting on the 22kD and 57kD ORFs have not been reported before and here we investigate their properties genome-wide. With the recent sequencing and assembly of the *Oxytricha* micronuclear genome [7], we are able to provide a thorough characterization of TBE sequences in the germline, including their genomic distribution and sequence features. We also infer the levels of selective

constraints acting on the three transposon-encoded ORFs, and we discovered homologs of TBE transposons in other ciliate genomes. Together, these results provide insights into the origin and evolution of TBE transposons in *Oxytricha*.

Results

TBE sequences in the micronuclear genome cluster into four major families

We annotated TBE sequences in the micronuclear genome using the translated protein sequences of the three ORFs as query. In total we annotated 10,109 complete TBEs and 24,702 partial TBEs (Table 1, Additional file 1). The complete TBE sequences (those that encode all three ORFs) cluster into four major families, which correspond to the previously published TBE1 and TBE3 families [21], as well as two subfamilies within the TBE2 family. The two TBE2 subfamilies encode 42kD transposases and 22kD ORFs that are indistinguishable from each other, with comparable pairwise similarity either within or between TBE2.1 and TBE2.2 (Table 2); however, they encode distinct 57kD ORFs (% pairwise similarity 53.5 %, Table 2). Phylogenetic analysis confirms that the TBE2.1 and TBE2.2 42kD and 22kD genes do not form separate monophyletic clades (though there is some resolution of TBE2.1 and TBE2.2, especially for the 22kD gene, which may imply recent diversification) (Fig. 1b and c), whereas the 57kD genes are clearly distinguishable between TBE2.1 and TBE2.2 (Fig. 1d). The orientation of the three ORFs is consistent among the four TBE families, with the 22kD ORF in the reverse orientation relative to the other two ORFs (Fig. 1a). All TBEs contain a ~200 bp region with short tandem repeats between the 22kD ORF and the 57kD ORF.

Most annotated, complete TBEs are flanked by two terminal inverted repeats (TIRs) (Table 1). Apart from differences in the sequences of the three ORFs, the four TBE families also have distinct TIRs, with variation in both sequence and length (Table 3). All TIRs contain the *Oxytricha* telomeric repeat, $CA_4C_4A_4C_4$, with the exception of TBE2.1 which contains $CA_4C_4A_4C_3$. TBE2.2 transposons have two distinct types of TIRs, one of which is a 21 bp shorter version of the TBE2.1 TIR. The two TBE2.2 TIRs (117 bp and 112 bp) are 92.5 % similar to each other. The protein sequences of TBE2.2 transposons with these two TIRs are indistinguishable from each other (percent pairwise sequence similarity between 57kD genes of the two types: 87 ± 4.3, vs. 86.6 ± 5.9 and 89.2 ± 6.3 % within each group). Each family also exhibits unique distances between the TIR and the start of the first ORF (42kD) and between the TIR and the end of the last ORF (57kD) (Table 3). Curiously, the TIR of TBE2.1 ends right before the start codon of the 42kD ORF. For the two types of TIRs within TBE2.2, although they are shorter than the TIR of TBE2.1, the distance

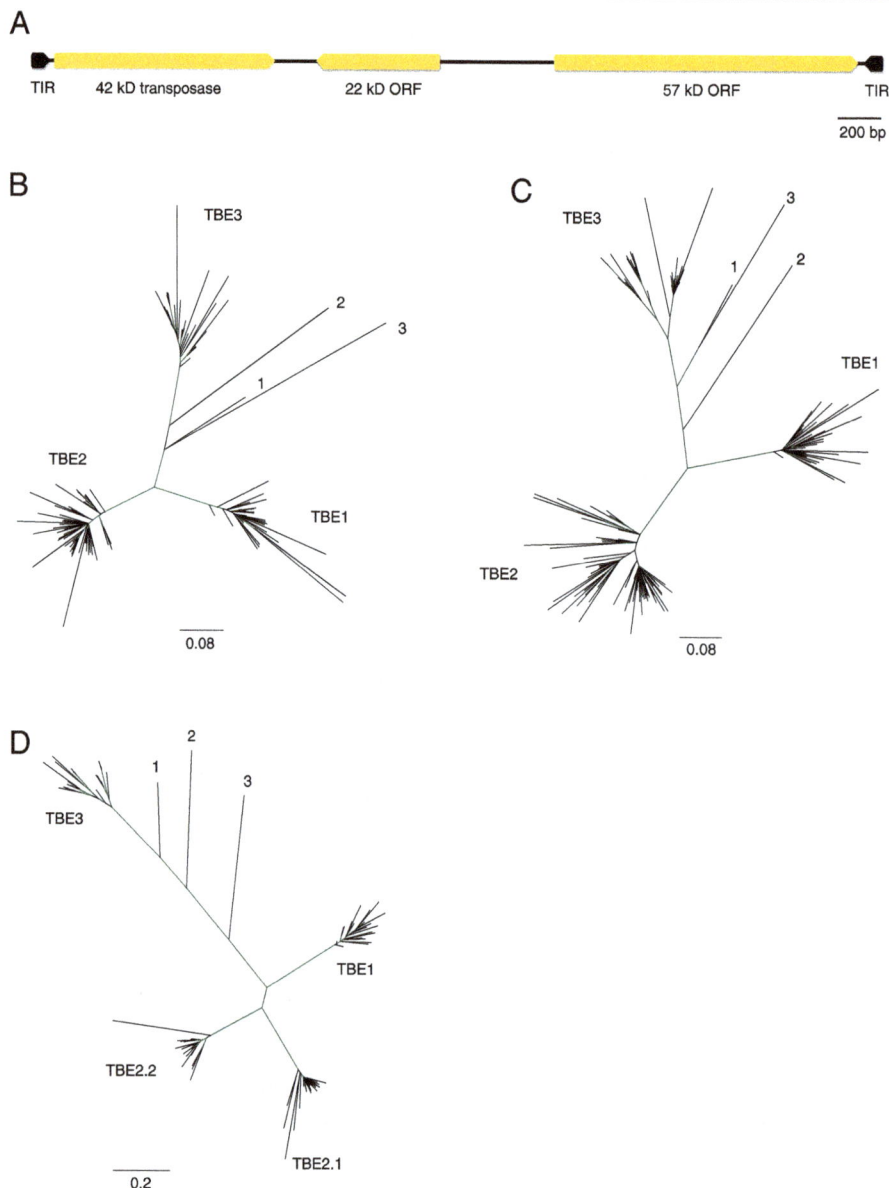

Fig. 1 Phylogeny of sampled *Oxytricha* TBE genes and orthologs identified in three other stichotrich ciliates. **a** Schematic map of TBE transposons. Gray arrows represent terminal inverted repeats (TIR). Orange arrows represent ORFs encoded by TBEs. **b** Phylogeny constructed with TBE 42kD transposases (29 TBE1, 27 TBE2.1, 26 TBE2.2 and 25 TBE3 42kD protein sequences). Clades formed by TBE1, TBE2 and TBE3 are labeled accordingly. TBE2.1 representatives are indicated in red and TBE2.2 in blue. Internal branches supported by posterior probability higher than 0.9 are colored in green. **c** Phylogeny constructed with TBE 22kD ORFs (32 TBE1, 39 TBE2.1, 30 TBE2.2 and 28 TBE3 22kD protein sequences). Colors are as above. **d** Phylogeny constructed with TBE 57kD ORFs (27 TBE1, 26 TBE2.1, 23 TBE2.2 and 21 TBE3 57kD protein sequences). Clades formed by TBE1, TBE2.1, TBE2.2 and TBE3 are labeled accordingly; colors as above. The multiple sequence alignment was produced with MAFFT v6.956b and trimmed with trimAl v1.2 to remove excess gaps and poorly aligned regions. The unrooted Bayesian trees were produced with MrBayes v3.2.2 [35]. The three TBE orthologs are 1: *Sterkiella histriomuscorum*; 2: *Tetmemena sp.*; 3: *Laurentiella sp.*. All posterior probability values are above 0.5. The scale below the phylogeny illustrates branch substitutions per site

between the end of TIR and the 42kD ORF is longer, such that the total distance between the 5′ terminus of a TBE2 and the start of the 42kD ORF is precisely the same among most TBE2 sequences. It is possible that in the TBE2.2 subfamily, the selective constraints on TIRs are weaker so that the TIR becomes shorter, leaving the

sequence between the TIR and the start of the 42kD ORF more flexible to accumulate substitutions.

Distribution of TBEs in the micronuclear genome

Annotated TBEs occupy ~13.3 % of the micronuclear genome. This is slightly smaller than the previously reported

Table 1 Genomic distribution of TBEs in the *Oxytricha* germline genome

	Complete								Partial			Total	
	#	Length (Mb)	% of complete TBEs by length	TIR			Near MDSs		#	Length (Mb)	% of partial TBEs by length	Length (Mb)	% of all TBEs by length
				2	1	0	#	% total copies					
TBE1	2502	9.9	24.75 %	2005	228	269	521	28.2 %	6216	6.5	24.8 %	16.4	24.8 %
TBE2.1	2484	10.0	25 %	2166	58	260	354	19.2 %	3129	3.9			
TBE2.2	1087	4.3	10.75 %	916	30	141	197	10.7 %	1146	1.3			
TBE2	3571	14.3	35.75 %	3082	88	401	551	29.9 %	9898 (TBE2.1 + TBE2.2 + unclassified partial TBE2)	10.5	40.1 %	24.8	37.5 %
TBE3	3946	15.8	39.5 %	3148	358	440	773	41.9 %	8588	9.2	35.1 %	25.0	37.7 %
Total	10,019	40.0	100 %	8235	674	1110	1845	100 %	24,702	26.2	100 %	66.2	100 %

estimate of 15 % [7] because the previous annotation is based on RepeatMasker (http://www.repeatmasker.org/), which uses sequence similarity at the nucleotide level, often including short nucleotide matches. Here, our annotation approach is based on sequence similarity at the protein level, and therefore sequences other than the three ORFs, such as terminal inverted repeats and spacer regions between the three ORFs, may have been missed, especially for partial TBEs. Among all annotated TBEs, 24.8 % are TBE1, 37.5 % are TBE2, with the ratio between TBE2.1 and TBE2.2 approximately 2.3:1, and 37.7 % are TBE3 (Table 1).

Annotated partial TBEs are more likely to be located within 500 bp of contig ends (57.7 %) than complete TBEs (19.4 %), suggesting that the original PacBio and Illumina-based genome assembly algorithm [7] had difficulty spanning repetitive sequences. Therefore, improvements in the genome assembly would be expected to lead to completion of these terminal, partial TBEs. On the other hand, partial TBEs located internal to a contig have lower sequence similarity to the protein sequence consensus of each family than those at contig termini (for example, % protein sequence similarity for internal partial TBE1 42kD genes: 71.6 ± 18.5; vs. terminal partial TBE1 42kD genes: 90.1 ± 9.3), suggesting that a significant portion of internal, partial TBEs are degenerate copies that are truly partial TBEs due to loss of one or two ORFs.

TBE sequences display a preference for insertion into MDSs (precursor DNA segments that are incorporated into the somatic genome), with more frequent distribution near MDSs (18.3 % within 500 bp) than the 11.1 % estimate of the genome space occupied by MDSs (Table 1, Chi-squared test, *p*-value = 6.304e-05). The short noncoding elements (IESs) that interrupt MDSs have long been proposed to be remnants of ancient transposon insertions [17]. Since the TBE transposase has been implicated in IES removal and genome rearrangement [21], this enrichment near MDSs may facilitate the removal of both the transposons, themselves, and IESs. Among the TBEs that are near MDSs, there is a slight enrichment for members of TBE1 and TBE3, accompanied by a slight depletion of both TBE2 representatives (Table 1). Satellite repeats are another major class of repetitive sequences in the germline genome. There is no significant preference for TBE insertions near satellite repeats. Only 80 (0.79 %) complete TBEs reside within 500 bp of 380 bp repeats (which occupy 1.4 % of the genome) and 97 (0.96 %) complete TBEs reside near 170 bp repeats (which represent 1.2 % of the genome). Therefore, TBEs are more often associated with MDSs. Either their preferential insertion or maintenance near MDS-rich regions is consistent with the inferred participation of TBEs in genome rearrangement events that reassemble MDSs [21].

Sequence analysis of TBE sequences

Complete TBE sequences are highly similar to each other within each family (Table 2), with ~90 % pairwise similarity for the 42kD and 22kD ORFs and a slightly lower similarity,

Table 2 Pairwise percent protein sequence similarity of TBE genes

	42kD				22kD				57kD			
	TBE1	TBE2.1	TBE2.2	TBE3	TBE1	TBE2.1	TBE2.2	TBE3	TBE1	TBE2.1	TBE2.2	TBE3
TBE1	89.3 ± 4.4	74.2 ± 3.6	74.9 ± 3.3	67.1 ± 3.1	90.4 ± 4.2	67.1 ± 3.6	68.5 ± 3.6	65.3 ± 3.4	86.4 ± 5.2	46.3 ± 2.7	49.5 ± 2.8	36.4 ± 1.7
TBE2.1		89.3 ± 4.3	89.3 ± 4.2	67.9 ± 3.2		89.5 ± 5.1	87.6 ± 5.2	64.4 ± 3.4		87.2 ± 6.0	53.5 ± 3.1	34.5 ± 1.6
TBE2.2			90.8 ± 4.5	68.7 ± 2.9			89.8 ± 4.8	65.5 ± 3.1			86.1 ± 5.9	38.2 ± 1.8
TBE3				90.9 ± 4.8				89.9 ± 5.4				86.1 ± 7.5

Table 3 Features of *Oxytricha* TBEs and complete TBEs identified in other stichotrich genomes

		Terminal inverted repeat (TIR) (Underline for telomeric repeat)	Target site duplication	Distance (bp) between (mode, % of mode)	
				TIR/42kD	57kD/TIR
Oxytricha trifallax	TBE1	<u>CAAAACCCCAAAACCCC</u>TTAATGAGGTTTA TAAGTGCTTTGATTTGTAGGGAATTTGTTA GGGGTTGGGGTTATTAAT (78 bp)	ANT	3 97.7 %	41 86.5 %
	TBE2.1	<u>CAAAACCCCAAAACCC</u>TTTCAGTAGTTTGA TTGAGTTTTTGATTGATAAAAGTAGACTAT TAGTGCATACTTTATTAGGGTTTTAATAGG GTTTATGTAGGGGTTTAATGTTTAAATATT AGTAATTTAAGTGAGTAT (138 bp)	ANT	0 99.1 %	23 82.3 %
	TBE2.2	<u>CAAAACCCCAAAACCC</u>TTTCAGTAGTTTGA TTGAGTTTTTGATTGATAAAAGTAGACTAT TAGTGCATACTTTATTAGGGTTTTAATAGG GTTTATGTAGGGGTTTAATGTTTAAAT (117 bp)	ANT	21 96.4 %	44 86 %
		<u>CAAAACCCCAAAACCCC</u>TGAAGTTGTTGA TTGAGTTTTTGATTGATGAAAGTAGACTAT TAACGCATGCTTTATTAGGGTTTTAATAGG GTTTATGTAGGGGTTTAGGGTT (112 bp)	ANT	26 97.6 %	49 92.1 %
	TBE3	<u>CAAAACCCCAAAACCCC</u>TTAGTGAGGTTTA TAAGTGCTTTGATTTGTAGGGTATAGTTGG GGTCTTATTGGGGTTAGTAGAGAAA (85 bp)	ANT	17 93.3 %	54 84.6 %
Sterkiella histriomuscorum		<u>CAAAACCCCAAAACCCC</u>TTCATGAGTTGTT TATGAGTTTTTGATTGTGTTGGGATTATTA GTGTTTATTAGGGTTTATTAATAATTGGGG TTAGTACACAAA (102 bp)	ANT	0	28
Tetmemena sp.		<u>CAAAACCCCAAAACCCC</u>ATAATATGATAAG AAAGTGAAAATAAGTTGTGTATAATTAATT TCTTTATTAATACTTATAATCATGC (85 bp)	ANT	−4	−6
Laurentiella sp.		<u>CCCAAAACCCC</u>AACTACTTATAAAATGTGA TTAATAATAAGAATTGATATATATTAATTT CATAATTATCAACGTTTTTAGAGTAATTAA ACTGCGATGAGTTATATAAA (110 bp)	ANT	1	11

~86 %, for the 57kD ORF. This high sequence similarity suggests that either their expansion and insertion occurred relatively recently or that each family is subject to strong selective constraints. TBE1 and TBE2 members are more similar to each other than to TBE3 (Table 2). Among different families, the 42kD transposase gene is more conserved than the 22kD ORF. The 57kD ORF is the least conserved compared to the other two ORFs (Table 2), with just 36.4 % similarity between TBE1 and TBE3, for example.

The terminal inverted repeat sequences are highly similar between both ends of a TBE (% sequence similarity: TBE1: 95.4 ± 3.6; TBE2.1: 93 ± 4.4; TBE2.2: 93.2 ± 4.4; TBE3: 95.9 ± 3.7), also consistent with either recent insertion or selective constraint.

We observe a prevalence of premature stop codons and frameshifts in TBE open reading frames (Table 4). 1360 TBE1, 1025 TBE2.1, 503 TBE2.2 and 1842 TBE3 elements encode three full-length proteins, but 96–98 % of these transposons contain premature stop codons and/or frameshifts in at least one of the three genes. The prevalence of stop codons is particularly prominent in the TBE3 42kD and 22kD ORFs, with an excess of stop codons occurring at a few specific sites. Among all TBE3 42kD genes (352 residues), 83.5 % contain a stop codon at residue 70, and 13.8 % contain a stop codon at residue 127. Among all TBE3 22kD genes (192 residues), 35.2 % contain a stop codon at residue 38, 22.6 % contain a stop codon at residue 39, and 83.2 % contain a stop codon at

Table 4 Prevalence of premature stop codons and frameshifts

	42kD		22kD		57kD	
	Stop codon	Frameshift	Stop codon	Frameshift	Stop codon	Frameshift
TBE1	30.4 %	76.8 %	14.9 %	59.0 %	38.1 %	87.1 %
TBE2.1	29.6 %	79.1 %	23.1 %	62.7 %	45.5 %	86.2 %
TBE2.2	29.4 %	74.8 %	16.8 %	60.8 %	36.7 %	87.3 %
TBE3	84.7 %	75.7 %	90.7 %	56.3 %	35.8 %	81.3 %

residue 186. While the prevalence of stop codons and frameshifts could be an artifact of less accurate genome assembly in repetitive regions, the enrichment of stop codons in TBE3s cannot be explained by such an assembly artifact alone, since assembly errors would result in stop codons that are randomly distributed across the coding sequence rather than enriched at specific sites.

Substitution rate analysis suggests that both groups of TBE sequences that do or do not contain stop codons or frameshifts have dN/dS ratios significantly lower than 1 (Fig. 2, Additional file 2). The overall dN/dS ratios are in the range of 0.1–0.3, suggesting genome-wide purifying selection acting on TBEs, which is consistent with earlier small-scale studies on the 42kD transposase [21, 23, 24] and unpublished studies from our lab of the other two ORFS [25]. Our study demonstrates that purifying selection is also acting on the other two TBE-encoded ORFs, indicating potential functional roles of these two genes. TBEs without a premature stop codon or frameshift display lower dN/dS values than those that contain

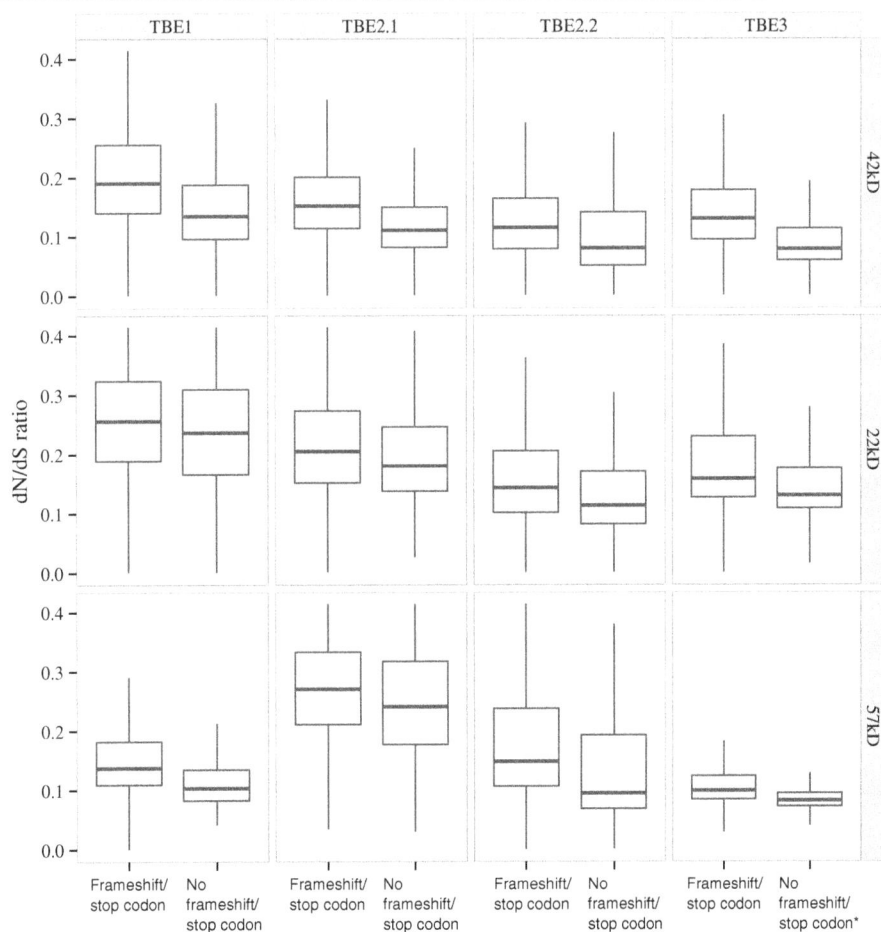

Fig. 2 TBE substitution rate variation. Box plots represent dN/dS values for the three TBE-encoded genes (with or without premature stop codons or frameshifts) among the four TBE families. The numbers of ORFs analyzed are summarized in Additional file 2. *For TBE3 42kD and 22kD genes, since very few sequences lack frameshifts or premature stop codons, we permitted the presence of the most frequent stop codons at residue 70 (42kD protein) and residue 186 (22kD protein) for the "No frameshift/stop codon" group

premature stop codons or frameshifts and thus are more likely to be functional transposon copies. For TBE3 42kD and 22kD genes, since very few copies lack stop codons (Table 4), we included in the "No frameshift/stop codon" group those sequences that contain just the most abundant stop codons listed above. This category also displays lower dN/dS ratios than those with other stop codons or frameshifts. In addition to pairwise dN/dS analysis, we also compared likelihoods of evolutionary models with estimated dN/dS ratios <1 and with dN/dS fixed at 1 (no selection) using a chi-squared test (Additional file 3). The former model fits significantly better in every case. The observed levels of purifying selection acting on TBE proteins that contain stop codons or frameshifts, especially the TBE3 42kD and 22kD genes, may suggest the presence of a biological mechanism to correct the stop codons and frameshifts so that functional proteins can be expressed.

TBEs in newly sequenced stichotrich genomes

We searched six newly sequenced stichotrich macronuclear genomes [26] for orthologous sequences of TBE transposons. Since TBEs are repetitive sequences that occupy a large portion of the micronuclear genome, their copy number in whole cell DNA is often comparable to nanochromosomes at high copy number in the macronuclear genome. DNA prepared from whole cell extracts therefore often contains some TBE sequences. We took advantage of this to extract TBE orthologs from macronuclear genome assemblies prepared from whole cell DNA.

We were able to identify complete or partial TBE sequences in the macronuclear genome assemblies of *Urostyla sp.*, *Paraurostyla sp.*, *Laurentiella sp.*, *Stylonychia lemnae*, *Tetmemena sp.* and *Sterkiella histriomuscorum* (the phylogeny of these species is discussed in [26]). Complete TBEs were found in *Laurentiella*, *Tetmemena* and *Sterkiella* (Table 3), with conserved orientation of the three ORFs but distinct terminal inverted repeats. Complete but degenerate TBEs were found in *Paraurostyla*, with no terminal inverted repeat and an inverted 57kD ORF. In *Stylonychia*, we could only identify incompletely assembled contigs containing TBE sequences. In *Urostyla*, we found only one sequence that exhibits weak protein sequence similarity to the *Oxytricha* 42kD transposase and we identified no homolog for the 22kD and 57kD ORFs. Similar to the DDE transposases in *Euplotes* Tec elements and the *Tetrahymena* and *Paramecium* genomes [9], the *Urostyla* DDE transposase homolog is very divergent from the *Oxytricha* 42kD transposase, exhibiting ~26 % sequence similarity in only a ~100 amino acid region containing the DDE motif towards the C-terminus.

Phylogenetic analysis supports the grouping of the assembled TBE orthologs in *Sterkiella*, *Tetmemena* and *Laurentiella* with *Oxytricha* TBE3 (Fig. 1b, c and d) (the incompletely assembled TBE sequences in the *Urostyla*, *Paraurostyla* and *Stylonychia* genomes also group with TBE3, data not shown). We found no premature stop codon in *Sterkiella* TBE orthologs of the 42kD and 22kD ORFs, the *Tetmemena* ortholog of the 22kD TBE ORF and the *Laurentiella* ortholog of the 22kD ORF. The *Tetmemena* and *Laurentiella* orthologs of the 42kD transposase both contain a premature stop codon, neither of which is present at a homologous position with each other nor with common sites of premature stop codons in the 42kD ORFs of *Oxytricha* TBE3. This suggests that the most common premature stop codons in the *Oxytricha* TBE3 genes may be specific to the *Oxytricha* lineage.

Since TBE orthologs group with TBE3, and TBE1 and TBE2 are more similar to each other than either is to TBE3 (Table 2), we infer that the TBE1 and TBE2 divergence and expansion most likely occurred recently in the *Oxytricha* lineage. Alternatively, the divergence may have occurred earlier but orthologous TBE1 and TBE2 sequences could be rare, or otherwise absent from the whole cell genome data for all other stichotrich genomes surveyed, or TBE 1 and 2 could have been lost from those lineages during evolution; however, these are all less parsimonious explanations than the conclusion that TBE 1 and 2 arose after TBE3 and underwent an expansion in the *Oxytricha* lineage. Furthermore, preliminary micronuclear genome sequence data from one of the outgroup species confirm the absence of TBE1/TBE2 orthologs in its micronuclear genome (Beh, Lindblad, Chen, Sebra, and Landweber, unpublished). Since the micronuclear genome sequences of most stichotrichs are not available and the DDE transposases in *Euplotes*, *Tetrahymena* and *Paramecium* are too divergent to provide outgroups, it is difficult to infer features of the ancestral TBE transposon that first invaded stichotrich germline genomes.

Discussion

We report a genome-wide characterization of the distribution and sequence features of TBE transposons in *Oxytricha* and provide phylogenomic evidence that the root among them may be in the TBE3 clade. The four major TBE families each have distinct terminal inverted repeats and spacer regions between TIRs and ORFs.

Of the three TBE-encoded genes, the 57kD ORF is much less conserved among different families than the 42kD and 22kD ORFs. It is possible that the structure and function of the 57kD protein allows it to be tolerant to more substitutions. Notably, the two subfamilies of TBE2 have similar 42kD and 22kD genes but very different 57kD genes, consistent with both increased variation in and the possible expanding roles of the 57kD protein. One type of the TBE2.2 terminal inverted repeat is precisely 21 bp shorter than the TBE2.1 TIR. It is possible that TBE2.1 is the ancestral form of TBE2 and that TBE2.2 diverged later from TBE2.1 with the acquisition of substitutions in the 57kD ORF, and that this was accompanied by shortening or altering the TIRs. While all three ORFs currently appear to be under purifying selection, the branch lengths in Fig. 1d suggest that the 57kD gene appears to have evolved rapidly under relaxed selective constraints after the divergence of the TBE families. This may have been a period when the diversification of the 57kD genes contributed to the functional differences among TBE families. Functional studies of the 57kD protein would provide insight into its biological roles in transposon elimination or genome rearrangement.

No TBE1 or TBE2 orthologs are found in related stichotrich ciliates, but future sequencing of their germline genomes would provide a better view of their germline transposons and help delineate the origin and evolution of TBE1 and TBE2, as well as TBE2.1 and TBE2.2 elements. Comparative germline genome sequences will also shed light on the evolutionary relationship between TBE3 and TBE1/2, and possibly permit inference of the ancestral TBE type that first invaded ciliate genomes.

Our analysis of transposons relies on the accuracy of genome assembly. The *Oxytricha* micronuclear genome was assembled using a hybrid approach, taking advantage of long PacBio reads that average ~7 kb [7]. A complete TBE sequence is ~4 kb and can be easily spanned by a PacBio read. Therefore, the accuracy of the assembly should be high for characterization of the genomic location and distribution of TBEs. However, PacBio reads were first error-corrected with high confidence unitigs assembled from Illumina reads before genome assembly [7], and Illumina reads, limited by their short length, can be ambiguous in repetitive regions. While Illumina unitigs are longer and more informative than Illumina reads, it is still possible that unitigs were ambiguous in resolving individual repeats, and hence that some PacBio reads deriving from repetitive regions may have not been corrected 100 % accurately. Therefore, TBE sequences will assemble less accurately than non-repetitive regions. The observed prevalence of premature stop codons and frameshifts may partially derive from this assembly artifact. However, such assembly artifacts could not explain the enrichment of stop codons at specific sites in the 42kD and 22kD ORFs of TBE3, since they would result in stop codons that are randomly distributed across the coding sequence. Assembly artifacts may have also contributed to the slightly higher dN/dS ratios that we identified, compared to ref. [21]. Another

factor contributing to the higher dN/dS ratios could be that we included all annotated TBEs (both active and inactive copies) in the analysis, whereas the previous study was based on a small set of known TBE sequences that are more likely to contain active copies.

Conclusions

This study provides the first genome-wide evolutionary analysis of ciliate transposons, suggesting the importance of all three TBE-encoded gene products, either in genome arrangement or other aspects of late nuclear differentiation, when the transposon genes are expressed. Sequencing and comparative analysis of more ciliate germline genomes will provide insights into the evolution and recruitment of domesticated transposons in genomes with complex genetic architecture.

Methods

Annotation and extraction of TBE sequences from the micronuclear genome

The protein sequences for the three ORFs (GenBank accession: AAB42034.1, AAB42016.1 and AAB42018.1) were used to query the *Oxytricha* micronuclear genome (GenBank accession: ARYC00000000) as well as the ciliate macronuclear genome assemblies (*Urostyla sp.*: LASQ02000000, *Paraurostyla sp.*: LASR02000000, *Laurentiella sp.*: LASS02000000, *Sterkiella histriomuscorum*: LAST02000000, *Tetmemena sp.*: LASU02000000 *and Stylonychia lemnae*: ADNZ03000000) with TBLASTN (BLAST+ [27], parameters: -db_gencode 6 -evalue 1e-7). TBE regions were annotated according to the TBLASTN output. Regions containing three ORFs in proximity (within 1 kb from each other) and in the correct orientation were annotated as complete TBEs, while those that do not contain all three ORFs were annotated as partial TBEs.

Clustering and alignment of TBE sequences

Complete TBE sequences were aligned to each other using an all-by-all BLASTN (BLAST+ [27], parameters: -word_size 50). Pairwise sequence similarity values were converted into input for MCL (parameter: -I 1.2) [28], which clustered TBE sequences into large clusters. Coding sequences were extracted using Exonerate [29] (parameters: −model protein2dna −geneticcode 6 −ryo " > %ti_%tab_%tae\n%tcs" −verbose 0 −showalignment no −showvulgar no). All-by-all BLASTP searches were performed on translated protein sequences (BLAST+ [27], E-value cutoff 10^{-7}) and pairwise protein sequence similarities were extracted from the BLASTP output. Terminal inverted repeats were determined by aligning the two ends of a TBE sequence using BLASTN, and clustering and consensus sequence generation were performed using UCLUST [30]. For *Oxytricha* TBEs, target site duplications were determined by comparing the MIC genome sequences immediately flanking TBEs. For TBEs identified in other ciliate macronuclear genomes, target site duplications were determined by mapping genomic reads to TBEs using BWA [31] (default parameters) and comparing the sequences flanking the terminal inverted repeats.

Construction of phylogenetic trees
Randomly sampled protein sequences of the three TBE ORFs were aligned with MAFFT [32] and excess gaps and poorly aligned regions were removed with trimAl (version 1.2, with the "-automated1" parameter) [33]. We used ProtTest [34] to determine the most suitable protein model (JTT + I + G). Phylogenetic trees were generated from the alignments using MrBayes v3.2.2 [35] (parameters: prset aamodelpr = fixed(jones); lset rates = invgamma). Trees were drawn using FigTree 1.4.2 (http://tree.bio.ed.ac.uk/software/figtree/).

Estimation of substitution rates
Pairwise protein alignments (MAFFT version 6.956b, [32]) were performed for each of the three genes (stop codons masked and frameshifts corrected) encoded by TBEs. Protein alignments were converted to coding sequence alignments using PAL2NAL [36]. The lengths of trimmed alignments are 344 codons (42kD protein), 187 codons (22kD) and 471 codons (57kD). Nonsynonymous to synonymous rate (dN/dS) ratios were calculated using the codeml program in PAML [37] (version 4.5) with parameters "icode = 5, runmode = −2, CodonFreq = 2". Synonymous substitution rates below 0.01 or above 5 were excluded from the analysis. In addition to pairwise dN/dS estimation, we randomly sampled 50 to 80 42kD, 22kD and 57kD ORFs and used codeml to compare likelihoods of models with estimated dN/dS (runmode = 0, fixed_omega = 0) and that with dN/dS = 1 (runmode = 0, fixed_omega = 1, omega = 1) (Additional file 3).

Abbreviations
TE: Transposable elements; TBE: Telomere bearing elements; MDS: Macronuclear destined sequence; IES: Internal eliminated sequence; ORF: Open reading frame; TSD: Target site duplication; TIR: Terminal inverted repeat.

Competing interests
The authors declare that they have no competing interests.

Authors' contributions
XC and LFL designed the study. XC performed the analyses. XC and LFL wrote the paper. Both authors read and approved the final manuscript.

Acknowledgements
We thank Clayton Schwarz for preliminary analysis, Tom Doak, and all members of the Landweber lab, and Seolkyoung Jung and Sean Eddy for discussion. This study was supported by NIH grants GM59708, GM109459 and the Human Frontier Science Program RGP004/2014 to L.F.L.

Author details
[1]Department of Molecular Biology, Princeton University, Princeton, NJ 08544, USA. [2]Department of Ecology and Evolutionary Biology, Princeton University, Princeton, NJ 08544, USA.

References
1. Kidwell MG, Lisch DR. Transposable elements and host genome evolution. Trends Ecol Evol. 2000;15:95–9.
2. Feschotte C, Pritham EJ. DNA transposons and the evolution of eukaryotic genomes. Annu Rev Genet. 2007;41:331–68.
3. Slotkin RK, Martienssen R. Transposable elements and the epigenetic regulation of the genome. Nat Rev Genet. 2007;8:272–85.
4. Fedoroff NV. Transposable elements, epigenetics, and genome evolution. Science. 2012;338:758–67.
5. Prescott DM. The DNA of ciliated protozoa. Microbiol Rev. 1994;58:233–67.
6. Swart EC, Bracht JR, Magrini V, Minx P, Chen X, Zhou Y, et al. The oxytricha trifallax macronuclear genome: a complex eukaryotic genome with 16,000 tiny chromosomes. PLoS Biol. 2013;11:e1001473.
7. Chen X, Bracht JR, Goldman AD, Dolzhenko E, Clay DM, Swart EC, et al. The architecture of a scrambled genome reveals massive levels of genomic rearrangement during development. Cell. 2014;158:1187–98.
8. Dubois E, Bischerour J, Marmignon A, Mathy N, Regnier V, Betermier M. Transposon invasion of the paramecium germline genome countered by a domesticated PiggyBac transposase and the NHEJ pathway. Int J Evol Biol. 2012;2012:e436196.
9. Vogt A, Goldman AD, Mochizuki K, Landweber LF. Transposon domestication versus mutualism in ciliate genome rearrangements. PLoS Genet. 2013;9:e1003659.
10. Baudry C, Malinsky S, Restituito M, Kapusta A, Rosa S, Meyer E, et al. PiggyMac, a domesticated piggyBac transposase involved in programmed genome rearrangements in the ciliate Paramecium tetraurelia. Genes Dev. 2009;23:2478–83.
11. Cheng C-Y, Vogt A, Mochizuki K, Yao M-C. A domesticated piggyBac transposase plays key roles in heterochromatin dynamics and DNA cleavage during programmed genome deletion in Tetrahymena thermophila. Mol Biol Cell. 2010;21:1753–62.
12. Jahn CL, Doktor SZ, Frels JS, Jaraczewski JW, Krikau MF. Structures of the Euplotes crassus Tec1 and Tec2 elements: identification of putative transposase coding regions. Gene. 1993;133:71–8.
13. Mouël AL, Butler A, Caron F, Meyer E. Developmentally regulated chromosome fragmentation linked to imprecise elimination of repeated sequences in paramecia. Eukaryot Cell. 2003;2:1076–90.
14. Arnaiz O, Mathy N, Baudry C, Malinsky S, Aury J-M, Denby Wilkes C, et al. The paramecium germline genome provides a niche for intragenic parasitic DNA: evolutionary dynamics of internal eliminated sequences. PLoS Genet. 2012;8:e1002984.
15. Doak TG, Witherspoon DJ, Jahn CL, Herrick G. Selection on the genes of Euplotes crassus Tec1 and Tec2 transposons: evolutionary appearance of a programmed frameshift in a Tec2 gene encoding a tyrosine family site-specific recombinase. Eukaryot Cell. 2003;2:95–102.
16. Klobutcher LA, Herrick G. Consensus inverted terminal repeat sequence of Paramecium IESs: resemblance to termini of Tc1-related and Euplotes Tec transposons. Nucleic Acids Res. 1995;23:2006–13.
17. Klobutcher LA, Herrick G. Developmental genome reorganization in ciliated protozoa: the transposon link. Prog Nucleic Acid Res Mol Biol. 1997;56:1–62.
18. Herrick G, Cartinhour S, Dawson D, Ang D, Sheets R, Lee A, et al. Mobile elements bounded by C4A4 telomeric repeats in Oxytricha fallax. Cell. 1985; 43(3 Pt 2):759–68.
19. Seegmiller A, Williams KR, Hammersmith RL, Doak TG, Witherspoon D, Messick T, et al. Internal eliminated sequences interrupting the Oxytricha 81 locus: allelic divergence, conservation, conversions, and possible transposon origins. Mol Biol Evol. 1996;13:1351–62.
20. Doak TG, Doerder FP, Jahn CL, Herrick G. A proposed superfamily of transposase genes: transposon-like elements in ciliated protozoa and a common "D35E" motif. Proc Natl Acad Sci U S A. 1994;91:942–6.
21. Nowacki M, Higgins BP, Maquilan GM, Swart EC, Doak TG, Landweber LF. A functional role for transposases in a large eukaryotic genome. Science. 2009; 324:935–8.
22. Williams K, Doak TG, Herrick G. Developmental precise excision of Oxytricha trifallax telomere-bearing elements and formation of circles closed by a copy of the flanking target duplication. EMBO J. 1993;12:4593–601.

23. Doak TG, Witherspoon DJ, Doerder FP, Williams K, Herrick G. Conserved features of TBE1 transposons in ciliated protozoa. Genetica. 1997;101:75–86.
24. Witherspoon DJ, Doak TG, Williams KR, Seegmiller A, Seger J, Herrick G. Selection on the protein-coding genes of the TBE1 family of transposable elements in the ciliates *Oxytricha fallax* and *O. trifallax*. Mol Biol Evol. 1997;14:696–706.
25. Schwarz C. The roles of transposon-encoded genes in a genome rearrangement process. Princeton University Senior Thesis. 2011.
26. Chen X, Jung S, Beh LY, Eddy SR, Landweber LF. Combinatorial DNA rearrangement facilitates the origin of new genes in ciliates. Genome Biol Evol. 2015;7(10):2859–70. doi:10.1093/gbe/evv172.
27. Camacho C, Coulouris G, Avagyan V, Ma N, Papadopoulos J, Bealer K, et al. BLAST+: architecture and applications. BMC Bioinformatics. 2009;10:421.
28. van Dongen S: A cluster algorithm for graphs. Tech. Report, National Research Institute for Mathematics and Computer Science in the Netherlands, Amsterdam 2000.
29. Slater GS, Birney E. Automated generation of heuristics for biological sequence comparison. BMC Bioinformatics. 2005;6:31.
30. Edgar RC. Search and clustering orders of magnitude faster than BLAST. Bioinformatics. 2010;26:2460–1.
31. Li H, Durbin R. Fast and accurate short read alignment with Burrows-Wheeler transform. Bioinformatics Oxf Engl. 2009;25:1754–60.
32. Katoh K, Misawa K, Kuma K, Miyata T. MAFFT: a novel method for rapid multiple sequence alignment based on fast Fourier transform. Nucleic Acids Res. 2002;30:3059–66.
33. Capella-Gutiérrez S, Silla-Martínez JM, Gabaldón T. trimAl: a tool for automated alignment trimming in large-scale phylogenetic analyses. Bioinformatics. 2009;25(15):1972–3. doi:10.1093/bioinformatics/btp348.
34. Darriba D, Taboada GL, Doallo R, Posada D. ProtTest 3: fast selection of best-fit models of protein evolution. Bioinformatics. 2011;27(8):1164–5. doi:10.1093/bioinformatics/btr088.
35. Huelsenbeck JP, Ronquist F. MRBAYES: Bayesian inference of phylogenetic trees. Bioinformatics Oxf Engl. 2001;17:754–5.
36. Suyama M, Torrents D, Bork P. PAL2NAL: robust conversion of protein sequence alignments into the corresponding codon alignments. Nucleic Acids Res. 2006;34(Web Server issue):W609–12.
37. Yang Z. PAML 4: phylogenetic analysis by maximum likelihood. Mol Biol Evol. 2007;24:1586–91.

9

Isolation and characterization of putative functional long terminal repeat retrotransposons in the *Pyrus* genome

Shuang Jiang[1,2], Danying Cai[3], Yongwang Sun[1,4,5] and Yuanwen Teng[1,4,5*]

Abstract

Background: Long terminal repeat (LTR)-retrotransposons constitute 42.4 % of the genome of the 'Suli' pear (*Pyrus pyrifolia* white pear group), implying that retrotransposons have played important roles in *Pyrus* evolution. Therefore, further analysis of retrotransposons will enhance our understanding of the evolutionary history of *Pyrus*.

Results: We identified 1836 LTR-retrotransposons in the 'Suli' pear genome, of which 440 LTR-retrotransposons were predicted to contain at least two of three gene models (*gag*, integrase and reverse transcriptase). Because these were most likely to be functional transposons, we focused our analyses on this set of 440. Most of the LTR-retrotransposons were estimated to have inserted into the genome less than 2.5 million years ago. Sequence analysis showed that the reverse transcriptase component of the identified LTR-retrotransposons was highly heterogeneous. Analyses of transcripts assembled from RNA-Seq databases of two cultivars of *Pyrus* species showed that LTR-retrotransposons were expressed in the buds and fruit of *Pyrus*. A total of 734 coding sequences in the 'Suli' genome were disrupted by the identified LTR-retrotransposons. Five high-copy-number LTR-retrotransposon families were identified in *Pyrus*. These families were rarely found in the genomes of *Malus* and *Prunus*, but were distributed extensively in *Pyrus* and abundance varied between species.

Conclusions: We identified potentially functional, full-length LTR-retrotransposons with three gene models in the 'Suli' genome. The analysis of RNA-seq data demonstrated that these retrotransposons are expressed in the organs of pears. The differential copy number of LTR-retrotransposon families between *Pyrus* species suggests that the transposition of retrotransposons is an important evolutionary force driving the genetic divergence of species within the genus.

Keywords: Retrotransposons, Insertion time, Distribution, Genetic diversity, *Pyrus*

Background

Repetitive sequences make up a large proportion of plant genomes. Among repetitive sequences are transposable elements [1, 2], which are broken into two main classes according to their transposition intermediate: Class I retrotransposons transpose via an RNA intermediate by a "copy and paste" mechanism; and Class II transposons transpose via a DNA intermediate by a "cut and paste" mechanism [2]. LTR-retrotransposons are Class I retrotransposons that have been found in all plant species investigated to date [2–4]. These retrotransposons are flanked by LTRs and undergo replicative transposition; thus, their copy numbers increase and occupy a large portion of the genome, especially in higher plants [5–7]. For example, retrotransposons make up more than 50 % of the maize and wheat genomes [8, 9]. Active LTR-retrotransposons increase the size of plant genomes. In *Oryza australiensis*, a wild relative of rice, transposition of retrotransposons led to a rapid two-fold increase in genome size during the last 3 million years [10], suggesting that rapid amplification of LTR-retrotransposons has played a major evolutionary role in genome expansion. Environmental stress and demethylation have been

* Correspondence: ywteng@zju.edu.cn
[1]Department of Horticulture, Zhejiang University, Hangzhou, Zhejiang 310058, China
[4]The Key Laboratory of Horticultural Plant Growth, Development and Quality Improvement, The Ministry of Agriculture of China, Hangzhou, Zhejiang 310058, China
Full list of author information is available at the end of the article

hypothesized to activate retrotransposons and induce duplication events in the genome [11–13]. The retrotransposons isolated from plants appear to be young—less than 5 million years old [14]. Therefore, pathways must exist for the removal of retrotransposons. The rice genome has lost a large number of retrotransposons, corresponding to a rapid reduction in genome size [15].

Retrotransposons can insert within or near transcriptionally active regions and can cause mutations by disrupting genes, altering gene expression levels, or by driving genomic rearrangements [16, 17]. Recent evidence indicated that a retrotransposon inserted into a *myb*-related gene was associated with pigmentation loss in grape [18]. In blood orange, insertion of a retrotransposon upstream of an anthocyanin biosynthesis-related gene caused color formation in its fruit to become cold-dependent [19]. Retrotransposons display extreme sequence diversity, and there are thousands or even tens of thousands of different retrotransposon families in plants [2, 5]. An autonomous retrotransposon is composed of two nearly sister LTR sequences flanked by target site duplications of usually 4–6 bp [1]. The internal region is usually composed of two open reading frames required for replication (in some cases, LTR retrotransposons possess one unique open reading frame, such as *Tnt1*, *Tto1*, or *Tos17*): the *pol* gene encodes products with the enzymatic functions of a protease (PR), reverse transcriptase (RT) and integrase (INT); and the *gag* gene encodes structural proteins involved in the maturation and packaging of retrotransposon RNA. Conserved sequence motifs, for example, the primer-binding site and the polypurine tract are also essential for retrotransposon replication. LTR-retrotransposons can be subdivided into the Ty1-*copia* and the Ty3-*gypsy* groups based on the order of the domains encoded within *pol* genes. The order in the Ty3-*gypsy* group is PR-RT-INT, and that in the Ty1-*copia* group is PR-INT-RT [2].

The *Pyrus* L. (pear) is believed to have originated in the Tertiary period in the mountainous regions of western and southwestern China [20]. According to its original distribution area, *Pyrus* can be divided geographically into two groups: the occidental pear group and the oriental pear group [21]. The major species of oriental pear are native to China [22]. The oriental pear group contains wild pea pears and cultivated species with large fruit. Their evolutionary history is still controversial [23]. Recently, the whole genome of *P. pyrifolia* Chinese white pear 'Suli' was sequenced. The assembled *P. pyrifolia* genome consists of 2103 scaffolds with an N50 of 540.8 kb, totaling 512.0 Mb with 194× coverage. Sequencing and assembly revealed that much of the *P. pyrifolia* genome is retrotransposon-derived [24]; 16.9 and 25.5 % of the genome was reported to be *copia* and *gypsy* retrotransposons, respectively. A large number of retrotransposons were also

found in other species in the Rosaceae family. For example, retrotransposons accounted for 37.6 and 18.6 % of the genomes of *Malus* and *Prunus* species, respectively [25, 26]. Jiang et al. (2015) reported that the retrotransposon *Ppcr1* was inserted in many loci in the genomes of cultivated *Pyrus* species, but only in a few loci in the genomes of wild *Pyrus* species [27]. This suggested that retrotransposons might play a major role in species evolution. Therefore, research on retrotransposons in *Pyrus* species will be helpful to understand the evolutionary history of *Pyrus*. Yin et al. (2014) reported that LTR retrotransposons in the *Pyrus* genome have complex structures [28], and that frequent recombination events followed by transposition of retrotransposons may have played a critical role in the evolution of *Pyrus* genomes. However, their study did not focus on the various retrotransposon families in *Pyrus* and their inner structural domains, nor did it involve the copy number of retrotransposon families in different *Pyrus* species.

In this study, we predicted the LTR-retrotransposons present in the 'Suli' genome, and annotated all LTR-retrotransposons with three inner functional domains (RT, INT and GAG) to identify putative functional LTR-retrotransposons. LTR-retrotransposons in the 'Suli' genome [24] were extremely divergent [27, 28], which made it difficult to analyze every predicted LTR-retrotransposon. Therefore, we focused on conserved LTR-retrotransposon families with a high copy number in 'Suli' genome, and investigated the distribution of these families in different *Pyrus* species and other closely related species to evaluate the roles of LTR-retrotransposon replication and mutation in the evolution of the *Pyrus* genome.

Results
Annotation and structure of LTR-retrotransposons in the 'Suli' genome
In previous study, a total of 1836 putative full-length LTR-retrotransposons were identified in the 'Suli' genome by LTRharvest. To determine which of these were most likely to be functional, we searched all identified LTR-retrotransposons for the conserved protein domains GAG, INT, and RT. A total of 440 putative LTR-retrotransposons (24.0 %) contained at least two domains and were analyzed further. Their positions in the 'Suli' genome and annotation information are listed in Additional file 1: Table S1 and Additional file 2, respectively. According to the order of the RT and INT domains, 373 and 67 retrotransposons belonged to the *copia* and *gypsy* groups, respectively (Table 1). *Copia*-type retrotransposons (average length, 5448 bp) were significantly shorter than *gypsy*-type retrotransposons (average length, 10,742 bp) ($p < 0.01$ by t-Test). The average LTR length of *copia* and *gypsy* retrotransposons was 374 and 542 bp, respectively.

Table 1 Characteristics of *copia* and *gypsy* putative full-length retrotransposons with more than two gene models identified in *Pyrus* genome

Type	Number	Length (nt) ± SE	5′ LTR length (nt) ± SE	3′ LTR length (nt) ± SE
Copia	373	5448.4 ± 1526.5	374.1 ± 138.9	374.9 ± 139.5
Gypsy	67	10742.0 ± 2823.7	542.4 ± 259.6	539.5 ± 259.6
t-Test		**	**	**

**means significant difference at the $p < 0.01$ level (*t*-Test)

Transposable elements can affect gene expression by disrupting functional genes or by inserting into the upstream or downstream regulatory regions of genes. We used BLAST to align our 440 conserved domain-containing LTR-retrotransposons to annotated introns in the 'Suli' genome, and used the Blast2GO annotation tool to assign probable gene ontology (GO) terms. A total of 734 genes aligned to LTR-retrotransposons, suggesting that they were disrupted. Of these, 531 unigenes could be annotated using GO. The unigenes were categorized into three main GO categories: biological process, cellular component, and molecular function (Fig. 1). These putatively disrupted genes were annotated using the NCBI nr database and listed in Additional file 3: Table S2. To further analyze putative retrotransposon-associated gene sequences, we searched 10,000-bp genome regions flanked by the predicted retrotransposons. A total of 2536 sequences were found, of which 1922

unigenes could be annotated using GO (data not shown).

To group the identified retrotransposons into families, we used each identified retrotransposon to conduct BLASTN searches against the whole dataset of 440 LTR-retrotransposons (coverage: 80 % and e-value: 10^{-5}). In this initial effort, we identified five LTR-retrotransposon families with high-copy numbers, which we investigated further (Table 2). BLASTN searches against the Repbase database were conducted to identify conserved repetitive elements in these five families. Similar sequences identified in Repbase and reference sequences in the *Pyrus* genome are listed in Table 2. The PFAM database has many gene models related to LTR-retrotransposons. In this study, three genes (*gag*, reverse transcriptase, and integrase) were predicted to be present in high copy numbers, while the other two genes (aspartic protease and RNase H) were infrequently identified in *Pyrus*

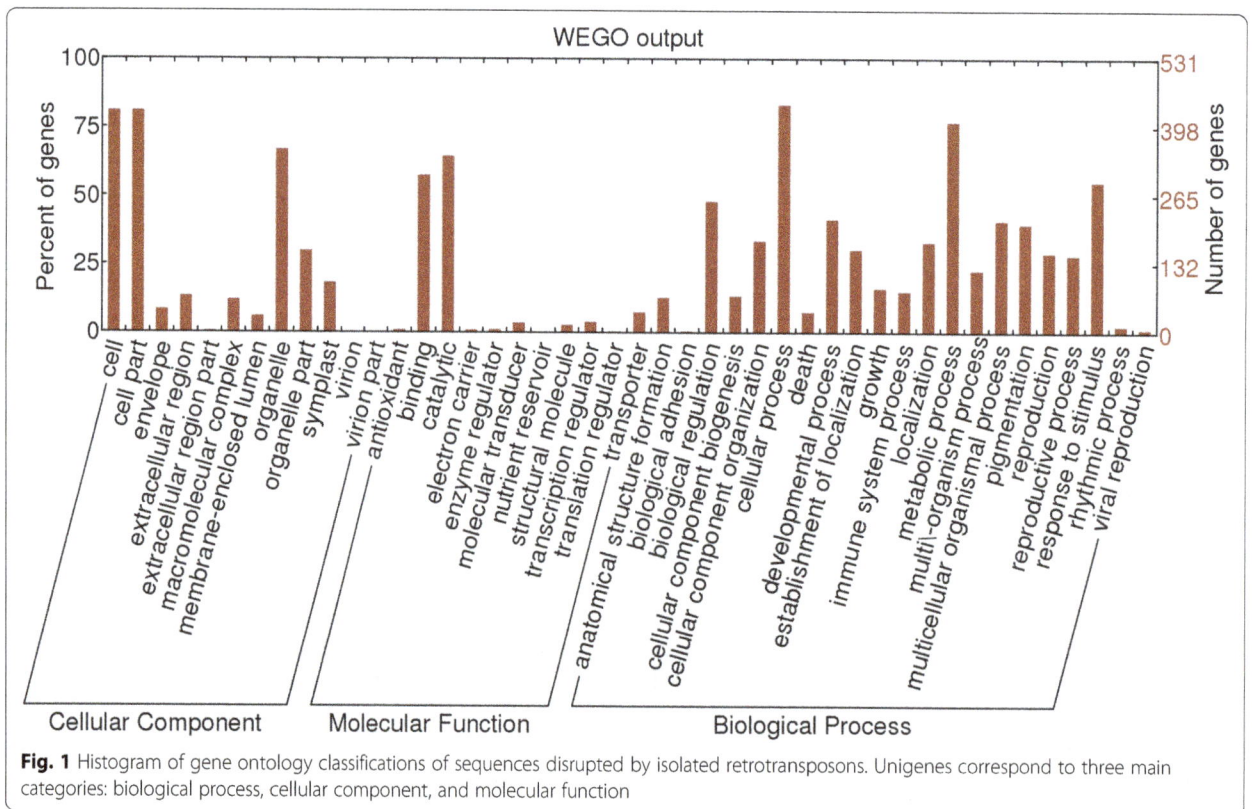

Fig. 1 Histogram of gene ontology classifications of sequences disrupted by isolated retrotransposons. Unigenes correspond to three main categories: biological process, cellular component, and molecular function

Table 2 LTR retrotransposon families investigated in this study

Family	Size (kb)	Copy number/total retrotransposons	Type	Ref Seq	ID of similar sequence in Repbase
Family I	5141	29/373	*copia*	AJSU01007348.1(8605–13,745 bp)	Copia-24_PX
Family II	5355	15/373	*copia*	AJSU01000113.1(27,402–32,756 bp)	Copia-106_Mad
Family III	6482	20/373	*copia*	AJSU01017137.1(16,748–23,229 bp)	Copia-90_Mad
Family IV	5123	14/373	*copia*	AJSU01025615.1(15,180–10,058 bp)	Copia-53_Mad
Family V	5670	5/67	*gypsy*	AJSU01016963.1(42,874–37,205 bp)	Gypsy-5_Mad

using the present gene models. Based on the predictions of three gene models, we described the structure of the five LTR-retrotransposon families isolated from *Pyrus* (Table 2, Additional file 4: Figure S1).

Putative insertion time of LTR-retrotransposons

The insertion time of LTR-retrotransposons was estimated by analyzing the divergence of sister LTRs. We used the molecular clock rate of 1.3×10^{-8} substitutions per site per year [29]. The insertion time can only be considered as a rough estimate, and only large differences should be considered significant. The divergence between sister LTRs ranged from 0 to 0.076, representing a maximum insertion time of 2.93 MYA. The predicted mean insertion time of the 440 LTR-retrotransposons analyzed in this study was 0.42 MYA. The predicted mean insertion time of *copia*-type LTR-retrotransposons was 0.35 MYA, significantly shorter than the predicted insertion time of 0.81 MYA ($p < 0.01$ by *t*-Test) for *gypsy*-type LTR-retrotransposons. Most of the retrotransposons were estimated to have inserted into the genome during the last 2.5 million years (Fig. 2). The peak of retrotransposon mobilization was observed at 0–0.5 MYA, indicating that our predicted retrotransposons were inserted relatively recently.

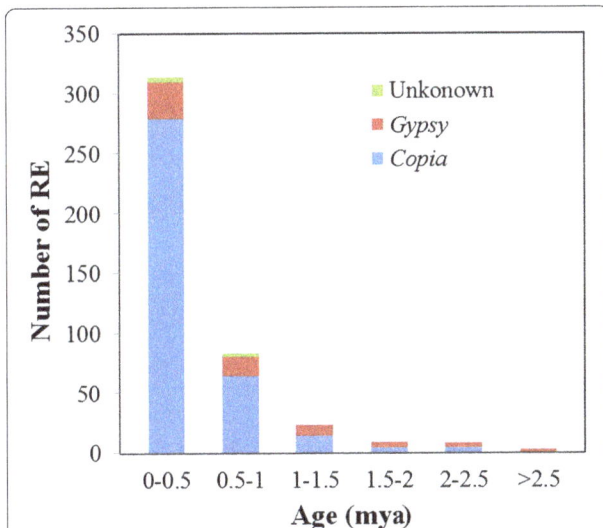

Fig. 2 Insertion time of 440 retrotransposons identified in 'Suli' genome

The mean insertion time of each member of the five LTR-retrotransposon families was estimated to be within the last 1 million years (Additional file 5: Figure S2). One member from Family I was inserted 1.75 MYA. In Families II, III, IV, and V, some members did not show LTR variations, indicating that they were inserted into the genome recently.

Phylogenetic relationships among isolated LTR-retrotransposons

The LTR-retrotransposons showed wide variations in their full-length sequences and could not be clustered. To evaluate the relationship among predicted LTR-retrotransposons, we used the neighbor-joining method to cluster the translated nucleotide sequences of reverse transcriptase (*rt*) in our identified LTR-retrotransposons with known TE families (Fig. 3). Both translated *copia*- and *gypsy*-type RT sequences clustered into many groups (Fig. 3). Although there was wide divergence among RT sequences, five and three conserved clades of RT sequences were identified among *copia* and *gypsy* retrotransposons, respectively. The average divergence of untranslated *copia*- and *gypsy*-type *rt* sequences was 0.64 and 0.55, respectively, indicating high heterogeneities among *rt* sequences (data not shown). Five *rt* sequences from each conserved clade of *copia* retrotransposons were aligned (Additional file 6: Figure S3), and the sequence divergence ranged from 0.068 to 0.691. *rt4* and *rt5* were similar. For the *gypsy* retrotransposons, the sequences of *rt6*, *rt7*, and *rt8* were aligned (Additional file 6: Figure S3), and their sequence divergences were 0.775, 0.898, and 0.98, respectively.

Transcriptional analysis of LTR-retrotransposons in various organs in *Pyrus*

Two transcriptomes assembled from RNA-Seq datasets were used in this study. A total of 116,182 sequences (62.6 Mb) assembled from 19,878,957 reads collected from buds of 'Suli' (SRX147917) and 36,495 sequences (15.8 Mb) assembled from 452,428,795 reads collected from fruit of *P. pyrifolia* 'Meirensu' (SAMN03857509-SAMN03857515) were aligned using BLAST to the 440 LTR-retrotransposons that we identified. LTR-retrotransposons were transcriptionally active in both

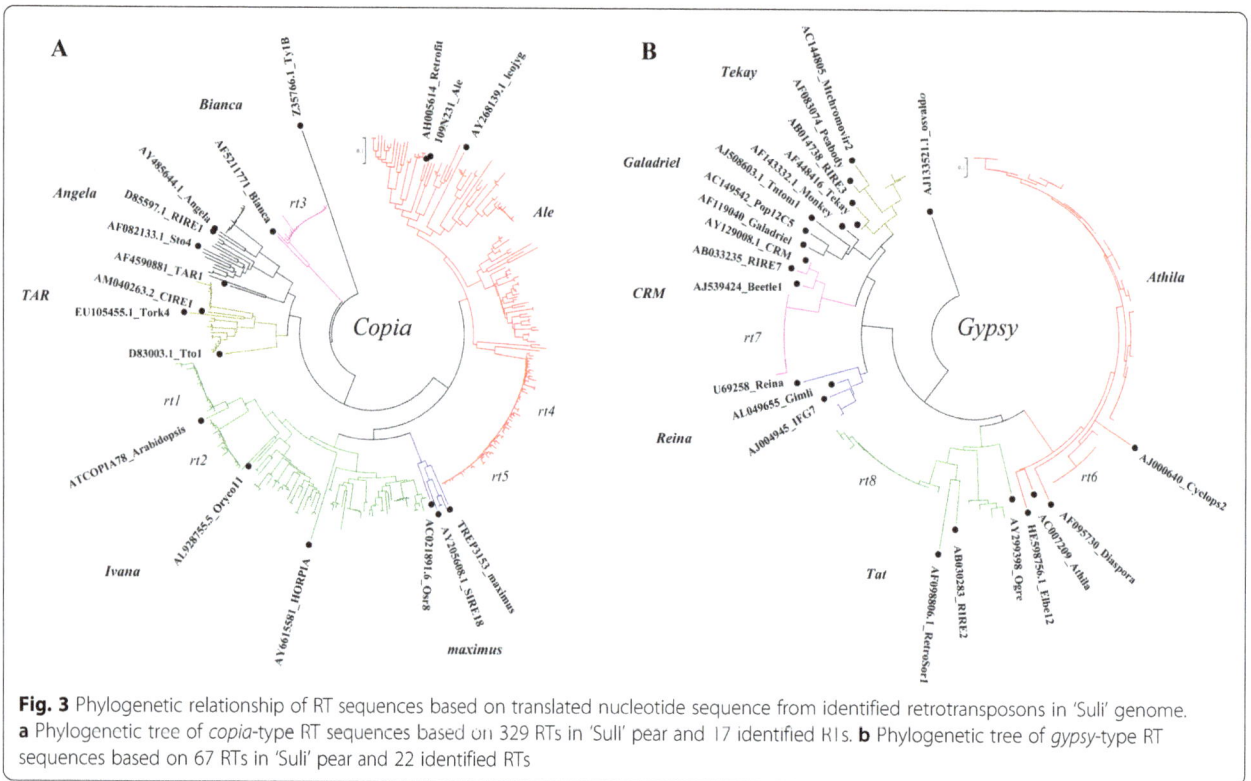

Fig. 3 Phylogenetic relationship of RT sequences based on translated nucleotide sequence from identified retrotransposons in 'Suli' genome. **a** Phylogenetic tree of *copia*-type RT sequences based on 329 RTs in 'Suli' pear and 17 identified RTs. **b** Phylogenetic tree of *gypsy*-type RT sequences based on 67 RTs in 'Suli' pear and 22 identified RTs

the fruit and bud (Fig. 4). A total of 266 *copia*-type and 66 *gypsy*-type LTR-retrotransposons aligned with transcripts from the bud of 'Suli' and 146 *copia*-type and 55 *gypsy*-type LTR-retrotransposons aligned with transcripts from the fruit of 'Meirensu', indicating that these retrotransposons were expressed (Fig. 4). Because the normalized expression values of individual retrotransposons were very low (data not shown), we only showed the reads per kilobase of gene model per million reads values of eight RT families (*rt1*–*rt8*). In

fruit of 'Meirensu', the high transcription level of *rt3* were represented.

Distribution of LTR-retrotransposon families among *Pyrus* species

To determine the exact copy number of LTR-retrotransposons, we used the reverse transcriptase gene model to search the database of protein sequences translated from 'Suli' genome data with Hmmer3.0. A total of 8144 *copia*-type RTs and 3748 *gypsy*-type RTs were

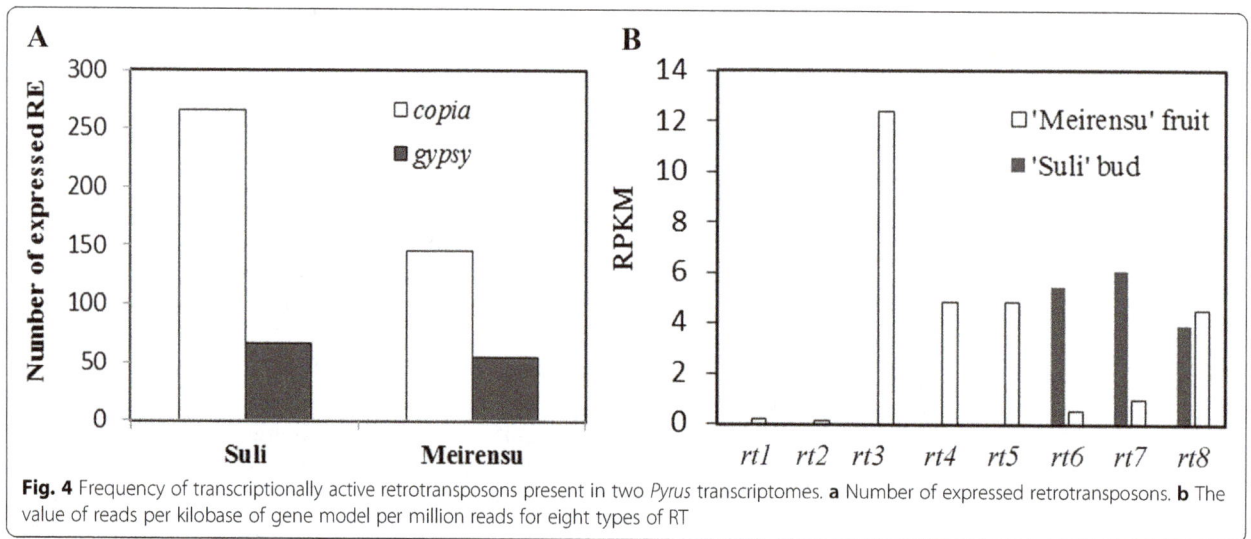

Fig. 4 Frequency of transcriptionally active retrotransposons present in two *Pyrus* transcriptomes. **a** Number of expressed retrotransposons. **b** The value of reads per kilobase of gene model per million reads for eight types of RT

identified. According to the average length of *copia* and *gypsy* retrotransposons (Table 1), *copia* and *gypsy* retrotransposons accounted for 8.8 % (42.3 Mb) and 8.0 % (38.4 Mb) of the genome, respectively.

The distribution of LTR-retrotransposon families was estimated in different *Pyrus* species and related species. *Pyrus* species exhibited little variation in genome size (Additional file 7: Table S3). We could not calculate the exact copy number of retrotransposons in *Pyrus*, but the relative copy number could be measured by real-time quantitative PCR (Q-PCR). Analyses of the LTR and inner sequences of five LTR-retrotransposon families showed that all LTR-retrotransposon families were present in all *Pyrus* species and *Malus × domestica*, but not in *Prunus persica* (Fig. 5). Families I and II were found infrequently in *Malus* genomes and two cultivated pear species (*P. pyrifolia* and *P. ussuriensis*), but they

were abundant in the genomes of three wild pear species (*P. pashia*, *P. betulaefolia*, and *P. nivalis*). Interestingly, families II, III, and IV in *P. elaeagrifolia* and *P. nivalis*, exhibited increased copy number of the inner sequence relative to LTRs of retrotransposons. The copy numbers of family III and V retrotransposons were higher in oriental pears than in occidental pears.

Discussion
Distribution and duplication of *copia* and *gypsy* retrotransposons in *Pyrus*
Recent evidence showed that a large proportion of retrotransposons were non-functional because of mutations in their protein-coding domains [30]. In this study, we identified predicted LTR-retrotransposons in the 'Suli' genome, and focused on LTR-retrotransposons that had the highest likelihood of being functional based on the

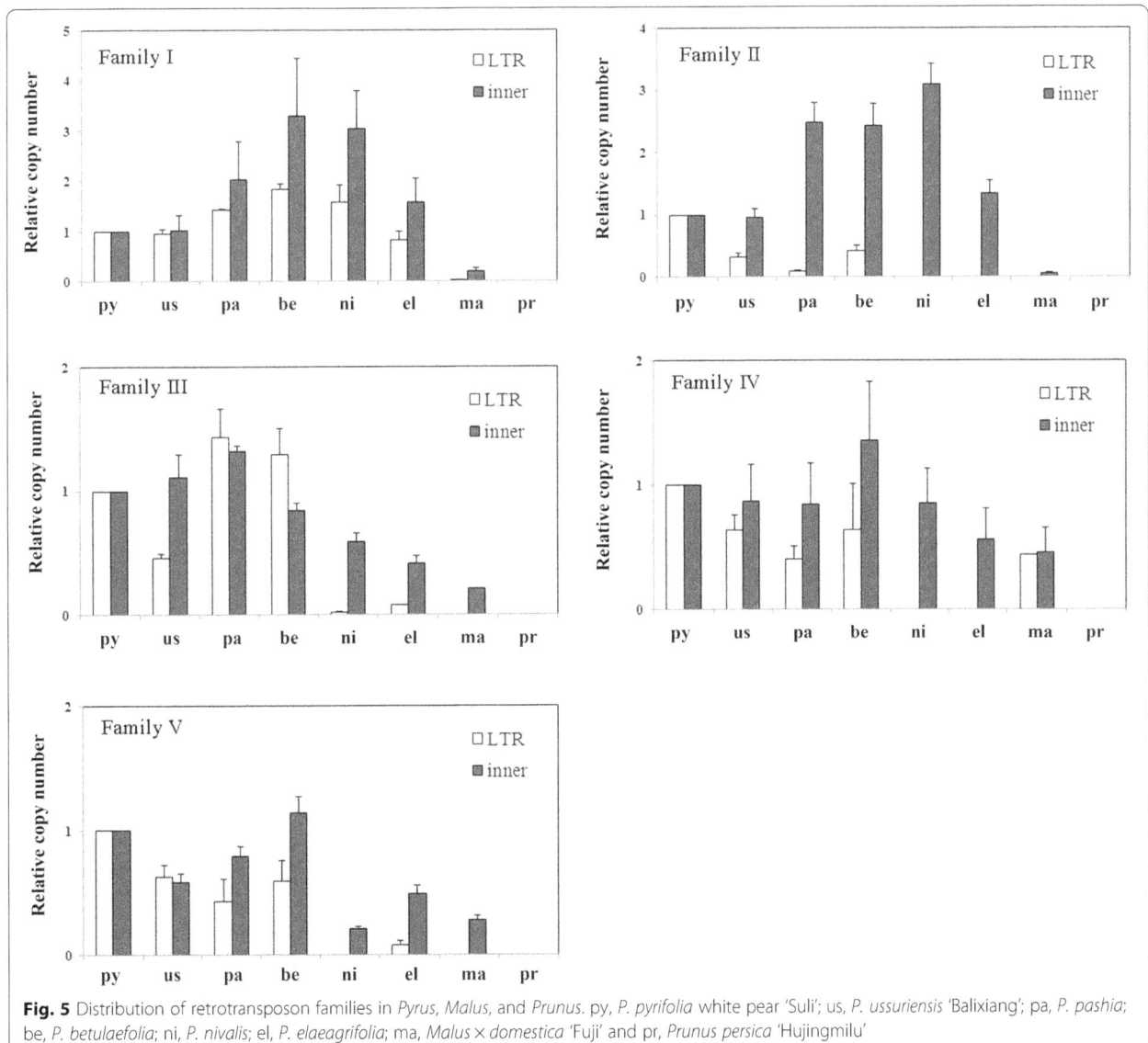

Fig. 5 Distribution of retrotransposon families in *Pyrus*, *Malus*, and *Prunus*. py, *P. pyrifolia* white pear 'Suli'; us, *P. ussuriensis* 'Balixiang'; pa, *P. pashia*; be, *P. betulaefolia*; ni, *P. nivalis*; el, *P. elaeagrifolia*; ma, *Malus × domestica* 'Fuji' and pr, *Prunus persica* 'Hujingmilu'

presence of annotated inner protein domains. Previously, we identified 1836 retrotransposons by running LTR-harvest based on two nearly sister LTR flanking sequences and some conserved sequence motifs [27]. However, the current study showed that only 440 retrotransposons had at least two inner protein domains. This finding suggests that there are very few full-length retrotransposons, and even fewer potentially functional LTR-retrotransposons in the *Pyrus* genome.

In a previous study, *copia* and *gypsy* retrotransposons were reported to account for 16.9 and 25.5 % (ratio, 0.66) of the genome of the 'Suli' pear, respectively [24]. However, in the present study, *copia* and *gypsy* retrotransposons were estimated to account for 8.8 and 8.0 % (ratio, 1.1) of the genome of the 'Suli' pear, respectively, based on RT gene models. Our predictions focused on the existence of *rt* gene in LTR retrotransposons, which is essential for retrotransposon transposition. Therefore, the retrotransposons predicted in this study may be functional, suggesting that at least 60 % of retrotransposons in the 'Suli' pear genome lack *rt* genes, and are therefore unable to replicate. Previous studies have established that lacking *rt* genes causes many LTR retrotransposons to be non-functional entities within host genomes [30].

High heterogeneity of LTR-retrotransposons in 'Suli' genome

The sequences and sequence length differed significantly among the full-length LTR-retrotransposons from the 'Suli' genome. We analyzed *rt* sequences to evaluate the diversity of retrotransposons. Our data showed that the average divergence of *rt* sequences in *copia*- and *gypsy*-family retrotransposons was 0.64 and 0.55, respectively. These findings indicate that the *rt* sequences from pear are highly heterogeneous (Fig. 3), like those in rice [31], strawberry [32] and masson pine [33]. There could be several reasons for the observed high sequence heterogeneity. First, gene mutation is the major cause of heterogeneity. In recent reports, many retrotransposons were existed in the genome for a long time [31, 34]. In this study, some retrotransposons were predicted to exist before the speciation of *Pyrus* and *Malus* based on sequence divergence (Fig. 5). The long period since the first retrotransposon insertion events is one potential source of variation. Both active and non-functional retrotransposons would have accumulated mutations over time, giving rise to a highly heterogeneous population [1]. Second, all transposons are integrated into chromosomal DNA. Therefore, mutated retrotransposon sequences, carrying mainly nonsense mutations are heritable, permitting a high degree of heterogeneity of retrotransposons between generations. Third, the high divergence between *rt* sequences of the LTR-retrotransposons we identified suggests a complex origin. For example, the divergence between *rt6* and *rt7* and between *rt6* and *rt8* was 0.898 and 0.98, respectively, suggesting that the origin of these related retrotransposons was complex, rather than from a single source. High sequence heterogeneity is the main obstacle that makes it difficult to classify retrotransposons as *copia*- or *gypsy*-types. In this study, we identified five related families of LTR-retrotransposons (Table 2). The members of each family showed high similarity and were strongly conserved, suggesting that these families have duplicated many times in recent years.

The insertion time of LTR-retrotransposon in 'Suli' genome

The divergence of sister LTR sequences was used to estimate the insertion time of retrotransposons. When an LTR-retrotransposon is inserted into the genome, the similarity of LTR sequences is 100 %. As time passes, mutations occur within the two LTRs, resulting in a larger genetic distance between them. In this study, only putative full-length LTR-retrotransposons were analyzed, and annotation of LTRs was performed by LTRharvest, which is known to be biased toward recent insertions of LTR-retrotransposons. Therefore, only recently inserted LTR-retrotransposons might be identified in our study. Our data showed that the majority of the retrotransposons we identified in the 'Suli' genome were inserted into the genome over the last 2.5 million years (Fig. 2). It was estimated that *Pyrus* and *Malus* diverged from each other between 5.4 and 21.5 MYA [24], suggesting that mobilization of these retrotransposons occurred frequently in the evolution of *Pyrus* species after the divergence of *Malus* and *Pyrus*. Within the retrotransposon families, the majority of members of families I–IV were estimated to have inserted into the genome over the last 1 million years (Additional file 5: Figure S2), confirming that these retrotransposons in *Pyrus* were inserted into the genome only recently.

Transcription of LTR-retrotransposons in pear organs

The expression of LTR-retrotransposons is likely to be silent in plant tissue during normal development. Many retrotransposons are expressed and transposed in protoplasts [35], and some are activated by abiotic stresses [11, 36]. In our study, the isolated retrotransposon sequences were aligned against the assembled transcriptomes of 'Suli' pear buds (SRX147917) and 'Meirensu' pear fruit (SAMN03857509-SAMN03857515) using BLAST. The expression of retrotransposons was detected in the fruit and buds of *Pyrus* cultivars (Fig. 4), which suggested that retrotransposons are expressed in *Pyrus* organs under normal conditions of growth and development. The expression of retrotransposons is advantageous for replication of retrotransposons, and retrotransposon transposition commonly results in mutation [18, 19]. In pear fruit and buds, retrotransposons showed transcriptional activity, which could increase their copy number in the genome. The mutations in buds and seeds could be transmitted to the next generation. The high rates of retrotransposon expression and transposition may contribute to the large proportion of retrotransposons in the *Pyrus* genome (as high as 42.4 %) [24].

Genetic diversity of LTR-retrotransposons in *Pyrus* and other close-related genera

Multiple studies support the hypothesis that retrotransposons might be associated with the evolution of plant genomes [7, 15]. In *Pyrus*, we identified 440 full-length LTR-retrotransposons that differed significantly from each other (Fig. 3). Five high copy-number retrotransposon families (four from the *copia* group and one from the *gypsy* group) were identified to further analyze the diversity of retrotransposons in *Pyrus* and other closely related genera. All five LTR-retrotransposon families were detected in six *Pyrus* species (Fig. 5), among which *P. betulaefolia* and *P. pashia* are believed to be the ancestral species in the genus *Pyrus* [23, 37]. The detection of a large number of retrotransposons indicates that these retrotransposons have widely existed in pear species for a long time. However, these five LTR-retrotransposon families were rare in *Malus*, and absent from *Prunus* (Fig. 5), indicating that they were duplicated and increased their copy number in *Pyrus* genomes after the differentiation of *Pyrus* and *Malus*. Both *Malus* and *Prunus* genomes contain a large number of retrotransposons [25], which are likely descended from different families than those found in *Pyrus*. These results suggest that the evolution of retrotransposons has varied among the different genera in the Rosaceae family.

Retrotransposons have played a major role in changing the size of genomes by either increasing genome size [10] or promoting rapid genomic DNA loss [15]. In *Pyrus*, the genome size does not vary greatly among species (Additional file 7: Table S3). Therefore, we can estimate the relative copy number of retrotransposon families in different *Pyrus* species. Our result shows that the copy number of retrotransposon families differs in *Pyrus* species. For example, *P. nivalis*, *P. pashia* and *P. betulaefolia* have a higher copy number of family I and II LTR-retrotransposons than *P. pyrifolia*, *P. ussuriensis*, and *P. elaeagrifolia*. In addition, *P. nivalis* has a low copy number of family III and V, implying these families were lost in *P. nivalis* evolution. The changes in the number of retrotransposon families might cause genetic divergence in *Pyrus* species. In *P. betulaefolia*, all five LTR-retrotransposon families showed high copy numbers in

the genome, indicating that this species has a larger proportion of retrotransposons in the genome than other *Pyrus* species. *Pyrus nivalis* and *P. elaeagrifolia* have a low copy number of the LTR regions of retrotransposons in families II, III and IV. The LTR region of these families might be lost and formed solo LTRs, or this region might have mutated. We inferred that the retrotransposon families have mutated and duplicated highly during the evolution of *Pyrus*.

Conclusions

We predicted 440 full-length LTR-retrotransposons from the 'Suli' pear genome, and annotated three inner protein domain sequences (GAG, INT, and RT) in retrotransposons, suggesting that the isolated retrotransposons might be functional. The analysis of three RNA-Seq databases of buds and fruit in different *Pyrus* cultivars showed retrotransposons were still active in pear organs. The isolated retrotransposons were highly heterogeneous. They had existed in *Pyrus* species for a long time, but have rapidly expanded during the last 2.5 million years after the divergence of *Malus* and *Pyrus*. Our results showed that the copy number of retrotransposon families varied among *Pyrus* species. To our knowledge, this is the first investigation of genetic variation of retrotransposons within the genus *Pyrus*. These findings support that retrotransposon transposition is an important evolutionary force driving the genetic divergence of species within the genus *Pyrus*.

Methods
Plant materials and DNA extraction
The plant materials used in this study consisted of six *Pyrus* accessions (two oriental cultivars: *P. pyrifolia* Chinese white pear 'Suli' and *P. ussuriensis* 'Balixiang', two oriental wild species: *P. pashia* and *P. betulaefolia*, and two occidental wild species: *P. nivalis* and *P. elaeagrifolia*), *Malus × domestica* 'Fuji', and *Prunus persica* 'Hujingmilu'. Genomic DNA was extracted from the young leaves of each specimen using the modified CTAB protocol described by JJ Doyle and JL Doyle [38] The precise concentration of DNA was detected using DNAQF-1KT (Sigma, St Louis, MO, USA). The DNA concentration of each sample was diluted to 1 ng·μl^{-1}, and 1 μl was used as a template for real-time quantitative PCR analysis.

Identification and annotation of LTR-retrotransposons
In a previous study, 1836 full-length LTR-retrotransposons were mined from the whole-genome data of *Pyrus* (AJSU00000000) [27]. The details of each retrotransposon were obtained from the output of LTRharvest. All retrotransposons were translated into proteins in all six possible reading frames using an in-house Perl script. All of the *copia* and *gypsy* gene models were downloaded from

the PFAM database (*gag*, PF03732; integrase, PF00665; reverse transcriptase, PF00078 and PF07727). Each gene model was used to search all of the proteins translated from retrotransposons with Hmmer3.0 software. To describe the genes around retrotransposons, 10,000 bp upstream and downstream of each LTR-retrotransposon were annotated with the BLAST algorithm using Blast2GO, and the results were visualized using the WEGO tool [39]. In the 'Suli' genome, a total of 42,812 coding genes were identified [24], and we searched gene introns isolated from the *Pyrus* genome to detect genes that were disrupted by retrotransposons.

Phylogenetic analyzes

According to the position of *rt* in the Hmmer3.0 results, we calculated the start and end of the *rt* sequences in the assembled 'Suli' genome. An in-house Perl script was used to extract nucleotide sequences from the whole-genome data, and translated them to amino acid sequences. The amino acid sequences of RT in *copia* and *gypsy* retrotransposons were aligned with known TE families, including *Maximus, Ivana, Ale, Angela, TAR, Bianca* in *copia* elements and *Athila, Tat, Tekay, CRM, Reina, Galadriel* in *gypsy* elements separately using ClustalW, and a neighbor-joining tree was constructed based on their genetic distance using Mega 5.2 software [40].

Estimation of insertion time of full-length LTR-retrotransposons

Bioperl scripts were used to automate the process of estimating the time of retrotransposon insertion. The two LTRs of each isolated retrotransposon were first aligned using ClustalW 2.0 [41], and genetic divergence between the two LTRs was estimated using the baseml module of PAML4 [42]. The insertion time (T) was estimated for each LTR-retrotransposon using the formula $T = k / 2r$, where k is the divergence between two LTRs and r is the substitution rate of 1.3×10^{-8} substitutions/site/year [29].

Estimation of LTR-retrotransposon copy number by Q-PCR

Q-PCR was used to estimate the copy number of retrotransposons in the genome [43]. We aligned five retrotransposon families with the *Malus* and *Prunus* genomes using BLAST, and designed Q-PCR primers (Additional file 8: Table S4) in the conserved region of LTR and inner domain using Primer 3 software (http://primer3.ut.ee/). The reaction solution (total volume, 20 μl) consisted of 10.0 μl SYBR Premix Ex Taq (Takara, Shiga, Japan), 0.4 μl each primer (10 μM), 1 μl DNA (1 ng·μl^{-1}), and 7.2 μl double distilled water. The reaction, performed on a LightCycler 1.5 instrument (Roche, Mannheim, Germany), started with a preliminary step of 95 °C for 30 s followed by 40 cycles of 95 °C for 5 s and 60 °C

for 20 s. A template-free control for each primer pair was set for each run. Three biological replicates were used and three measurements were performed on each replicate. The relative copy number of each sample was calculated using the Ct value [43].

Transcriptional analysis of retrotransposons in various organs/tissues of *Pyrus*

The Illumina RNA-Seq data from two samples were downloaded from NCBI. Data from buds (*P. pyrifolia* CWP 'Suli', SRX147917) and fruits (*P. pyrifolia* 'Meirensu', SAMN03857509-SAMN03857515) were analyzed to identify the transcriptional patterns of isolated retrotransposons. Raw sequence data in fastq format were filtered to remove reads containing adaptors, reads with more than 5 % unknown nucleotides, and low-quality reads with more than 20 % bases with a quality value of ≤10. Only clean reads were used in the following analyzes. Transcriptome *de novo* assembly was carried out using the short-read assembly program Trinity [44]. Two transcript databases were obtained for BLAST searches, and the isolated LTR-retrotransposons were used to identify the activity of each retrotransposon.

Additional files

Additional file 1: Table S1. Annotation of 440 isolated LTR retrotransposons. (XLSX 88 kb)

Additional file 2: The nucleotide sequences of 440 isolated LTR retrotransposons analyzed in the study. (FASTA 2783 kb)

Additional file 3: Table S2. List of disrupted genes. (XLSX 40 kb)

Additional file 4: Figure S1. Structure of five retrotransposon families in *Pyrus*. (TIF 729 kb)

Additional file 5: Figure S2. Insertion times of members of retrotransposon families I–V. (TIF 43 kb)

Additional file 6: Figure S3. Alignment of five *rt* sequences from each conserved clade of *copia* retrotransposons and three *rt* sequences from each conserved clade of *gypsy* retrotransposons. (TIF 264 kb)

Additional file 7: Table S3. Genome size of *Pyrus* species and related species. (DOCX 19 kb)

Additional file 8: Table S4. Primers used in this study. (XLSX 10 kb)

Abbreviations
LTR: Long terminal repeat; *rt*: Reverse transcriptase.

Competing interests
The authors declare that they have no competing interests.

Authors' contributions
SJ performed the experiments and wrote the manuscript. DC and YS helped with the data analysis and revised the manuscript. YT designed the research and wrote the manuscript. All authors read and approved the manuscript.

Acknowledgments
This work was financed by a Grant from the National Natural Science Foundation of China (No. 31201592), and a Grant for Innovative Research Team of Zhejiang Province of China (2013TD05).

Author details

[1]Department of Horticulture, Zhejiang University, Hangzhou, Zhejiang 310058, China. [2]Forest & Fruit Tree Institute, Shanghai Academy of Agricultural Sciences, Shanghai 201403, China. [3]Institute of Horticulture, Zhejiang Academy of Agricultural Sciences, Hangzhou, Zhejiang 310021, China. [4]The Key Laboratory of Horticultural Plant Growth, Development and Quality Improvement, The Ministry of Agriculture of China, Hangzhou, Zhejiang 310058, China. [5]Zhejiang Provincial Key Laboratory of Horticultural Plant Integrative Biology, Hangzhou, Zhejiang 310058, China.

References

1. Kumar A, Bennetzen JL. Plant retrotransposons. Annu Rev Genet. 1999;33(1):479–532.

2. Wicker T, Sabot F, Hua-Van A, Bennetzen JL, Capy P, Chalhoub B, et al. A unified classification system for eukaryotic transposable elements. Nat Rev Genet. 2007;8(12):973–82.

3. Sabot F, Schulman AH. Parasitism and the retrotransposon life cycle in plants: a hitchhiker's guide to the genome. Heredity (Edinb). 2006;97(6):381–8.

4. SanMiguel P, Tikhonov A, Jin YK, Motchoulskaia N, Zakharov D, Melake-Berhan A, et al. Nested retrotransposons in the intergenic regions of the maize genome. Science. 1996;274(5288):765–8.

5. Havecker ER, Gao X, Voytas DF. The diversity of LTR retrotransposons. Genome Bio. 2004;5(6):225.

6. Peterson DG, Schulze SR, Sciara EB, Lee SA, Bowers JE, Nagel A, et al. Integration of Cot analysis, DNA cloning, and high-throughput sequencing facilitates genome characterization and gene discovery. Genome Res. 2002;12(5):795–807.

7. SanMiguel P, Gaut BS, Tikhonov A, Nakajima Y, Bennetzen JL. The paleontology of intergene retrotransposons of maize. Nat Genet. 1998;20(1):43–5.

8. Daron J, Glover N, Pingault L, Theil S, Jamilloux V, Paux E, et al. Organization and evolution of transposable elements along the bread wheat chromosome 3B. Genome Biol. 2014;15(12):546.

9. Meyers BC, Tingley SV, Morgante M. Abundance, distribution, and transcriptional activity of repetitive elements in the maize genome. Genome Res. 2001;11(10):1660–76.

10. Piegu B, Guyot R, Picault N, Roulin A, Saniyal A, Kim H, et al. Doubling genome size without polyploidization: dynamics of retrotransposition-driven genomic expansions in Oryza australiensis, a wild relative of rice. Genome Res. 2006;16(10):1262–9.

11. De Felice B, Wilson RR, Argenziano C, Kafantaris I, Conicella C. A transcriptionally active copia-like retroelement in Citrus limon. Cell Mol Biol Lett. 2009;14(2):289–304.

12. Hirochika H, Okamoto H, Kakutani T. Silencing of retrotransposons in Arabidopsis and reactivation by the ddm1 mutation. Plant Cell. 2000;12(3):357–69.

13. Tsukahara S, Kobayashi A, Kawabe A, Mathieu O, Miura A, Kakutani T. Bursts of retrotransposition reproduced in Arabidopsis. Nature. 2009;461(7262):423–6.

14. El Baidouri M, Panaud O. Comparative genomic paleontology across plant kingdom reveals the dynamics of TE-driven genome evolution. Genome Biol Evol. 2013;5(5):954–65.

15. Ma J, Devos KM, Bennetzen JL. Analyses of LTR-retrotransposon structures reveal recent and rapid genomic DNA loss in rice. Genome Res. 2004;14(5):860–9.

16. Feschotte C, Jiang N, Wessler SR. Plant transposable elements: where genetics meets genomics. Nat Rev Genet. 2002;3(5):329–41.

17. Shapiro JA. Retrotransposons and regulatory suites. BioEssays. 2005;27(2):122–5.

18. Kobayashi S, Goto-Yamamoto N, Hirochika H. Retrotransposon-induced mutations in grape skin color. Science. 2004;304(5673):982.

19. Butelli E, Licciardello C, Zhang Y, Liu J, Mackay S, Bailey P, et al. Retrotransposons control fruit-specific, cold-dependent accumulation of anthocyanins in blood oranges. Plant Cell. 2012;24(3):1242–55.

20. Rubstov GA. Geographical distribution of the genus Pyrus and trends and factors in its evolution. Am Nat. 1944;78:358–66.

21. Bailey L. Standard cyclopedia of horticulture. Vol. 5. New York, USA: Macmillan Press; 1917. p. 2865–78.

22. Teng Y, Tanabe K. Reconsideration on the origin of cultivated pears native to East Asia. Acta Horticult. 2004;634:175–82.

23. Zheng X, Cai D, Potter D, Postmand J, Liu J, Teng Y. Phylogeny and evolutionary histories of Pyrus L. revealed by phylogenetic trees and networks based on data from multiple DNA sequences. Mol Phylogenet Evol. 2014;80:54–65.

24. Wu J, Wang Z, Shi Z, Zhang S, Ming R, Zhu S, et al. The genome of the pear (Pyrus bretschneideri Rehd.). Genome Res. 2013;23(2):396–408.

25. Velasco R, Zharkikh A, Affourtit J, Dhingra A, Cestaro A, Kalyanaraman A, et al. The genome of the domesticated apple (Malus x domestica Borkh.). Nat Genet. 2010;42(10):833–9.

26. Verde I, Abbott AG, Scalabrin S, Jung S, Shu SQ, Marroni F, et al. The high-quality draft genome of peach (Prunus persica) identifies unique patterns of genetic diversity, domestication and genome evolution. Nat Genet. 2013;45(5):487–94.

27. Jiang S, Zong Y, Yue X, Postman J, Teng Y, Cai D. Prediction of retrotransposons and assessment of genetic variability based on developed retrotransposon-based insertion polymorphism (RBIP) markers in Pyrus L. Mol Genet Genomics. 2015;290(1):225–37.

28. Yin H, Du JC, Li LT, Jin C, Fan L, Li M, et al. Comparative genomic analysis reveals multiple long terminal repeats, lineage-specific amplification, and frequent interelement recombination for Cassandra retrotransposon in pear (Pyrus bretschneideri Rehd.). Genome Biol Evol. 2014;6(6):1423–36.

29. Ma J, Bennetzen JL. Rapid recent growth and divergence of rice nuclear genomes. Proc Natl Acad Sci U S A. 2004;101(34):12404–10.

30. Navarro-Quezada A, Schoen DJ. Sequence evolution and copy number of Ty1-copia retrotransposons in diverse plant genomes. Proc Natl Acad Sci U S A. 2002;99(1):268–73.

31. Baucom RS, Estill JC, Leebens-Mack J, Bennetzen JL. Natural selection on gene function drives the evolution of LTR retrotransposon families in the rice genome. Genome Res. 2009;19(2):243–54.

32. Ma Y, Sun H, Zhao G, Dai H, Gao X, Li H, et al. Isolation and characterization of genomic retrotransposon sequences from octoploid strawberry (Fragaria x ananassa Duch.). Plant Cell Rep. 2008;27(3):499–507.

33. Fan FH, Wen XP, Ding GJ, Cui BW. Isolation, identification, and characterization of genomic LTR retrotransposon sequences from masson pine (Pinus massoniana). Tree Genet Genomes. 2013;9(5):1237–46.

34. Cossu RM, Buti M, Giordani T, Natali L, Cavallini A. A computational study of the dynamics of LTR retrotransposons in the Populus trichocarpa genome. Tree Genet Genomes. 2012;8(1):61–75.

35. Pearce SR, Kumar A, Flavell AJ. Activation of the Ty1-copia group retrotransposons of potato (Solanum tuberosum) during protoplast isolation. Plant Cell Rep. 1996;15(12):949–53.

36. Tapia G, Verdugo I, Yanez M, Ahumada I, Theoduloz C, Cordero C, et al. Involvement of ethylene in stress-induced expression of the TLC1.1 retrotransposon from Lycopersicon chilense Dun. Plant Physiol. 2005;138(4):2075–86.

37. Zheng X, Hu CY, Spooner D, Liu J, Cao JS, Teng Y. Molecular evolution of Adh and LEAFY and the phylogenetic utility of their introns in Pyrus (Rosaceae). BMC Evol Bio. 2011;11:255.

38. Doyle JJ, Doyle JL. A rapid DNA isolation procedure for small quantities of fresh leaf tissue. Phytochem Bull. 1987;19:11–5.

39. Ye J, Fang L, Zheng H, Zhang Y, Chen J, Zhang Z, et al. WEGO: a web tool for plotting GO annotations. Nucleic Acids Res. 2006;34(Web Server issue):W293–7.

40. Tamura K, Peterson D, Peterson N, Stecher G, Nei M, Kumar S. MEGA5: molecular evolutionary genetics analysis using maximum likelihood, evolutionary distance, and maximum parsimony methods. Mol Biol Evol. 2011;28(10):2731–9.

41. Thompson JD, Gibson TJ, Higgins DG. Multiple sequence alignment using ClustalW and ClustalX. Curr Protoc Bioinformatics. 2002;Chapter 2:Unit 2.3.

42. Yang Z. PAML 4: phylogenetic analysis by maximum likelihood. Mol Biol Evol. 2007;24(8):1586–91.

43. Wilhelm J, Pingoud A, Hahn M. Real-time PCR-based method for the estimation of genome sizes. Nucleic Acids Res. 2003;31(10), e56.

44. Grabherr MG, Haas BJ, Yassour M, Levin JZ, Thompson DA, Amit I, et al. Full-length transcriptome assembly from RNA-Seq data without a reference genome. Nat Biotechnol. 2011;29(7):644–52.

10

Active recombinant *Tol2* transposase for gene transfer and gene discovery applications

Jun Ni[1,2], Kirk J. Wangensteen[3,4], David Nelsen[3], Darius Balciunas[3,5], Kimberly J. Skuster[1], Mark D. Urban[1] and Stephen C. Ekker[1*]

Abstract

Background: The revolutionary concept of "jumping genes" was conceived by McClintock in the late 1940s while studying the *Activator/Dissociation* (*Ac/Ds*) system in maize. Transposable elements (TEs) represent the most abundant component of many eukaryotic genomes. Mobile elements are a driving force of eukaryotic genome evolution. McClintock's *Ac*, the autonomous element of the *Ac/Ds* system, together with *hobo* from *Drosophila* and *Tam3* from snapdragon define an ancient and diverse DNA transposon superfamily named *hAT*. Other members of the *hAT* superfamily include the insect element *Hermes* and *Tol2* from medaka. In recent years, genetic tools derived from the 'cut' and 'paste' *Tol2* DNA transposon have been widely used for genomic manipulation in zebrafish, mammals and in cells in vitro.

Results: We report the purification of a functional recombinant *Tol2* protein from *E.coli*. We demonstrate here that following microinjection using a zebrafish embryo test system, purified *Tol2* transposase protein readily catalyzes gene transfer in both somatic and germline tissues in vivo. We show that purified *Tol2* transposase can promote both in vitro cutting and pasting in a defined system lacking other cellular factors. Notably, our analysis of *Tol2* transposition in vitro reveals that the target site preference observed for *Tol2* in complex host genomes is maintained using a simpler target plasmid test system, indicating that the primary sequence might encode intrinsic cues for transposon integration.

Conclusions: This active *Tol2* protein is an important new tool for diverse applications including gene discovery and molecular medicine, as well as for the biochemical analysis of transposition and regulation of *hAT* transposon/genome interactions. The measurable but comparatively modest insertion site selection bias noted for *Tol2* is largely determined by the primary sequence encoded in the target sequence as assessed through studying *Tol2* protein-mediated transposition in a cell-free system.

Keywords: *Tol2* transposase, *hAT* superfamily, Recombinant transposase protein, Zebrafish, Transposition site preference

Background

Our understanding of transposable elements (TEs) begins with McClintock's revolutionary work with the *Activator/Dissociation* (*Ac/Ds*) system in maize [1]. *Ac*, the autonomous element of the *Ac/Ds* system, together with *hobo* from *Drosophila* and *Tam3* from snapdragon define an ancient and diverse DNA transposon super-family named *hAT* [2–4]. Widespread in plants and animals, *hAT* transposons are the most abundant DNA transposons in humans [5]. However, none of the human *hAT* elements have been active during the past 50 million years [5].

The first active DNA transposon discovered in vertebrates was the medaka fish (*Oryzias latipes*)-derived *hAT* element *Tol2* [6]. *Tol2* shares a number of features with other *hAT* members including transposases with a DDE (aspartate-aspartate-glutamate) catalytic motif, short terminal inverted repeats (TIRs) and formation of 8-bp host duplications upon transposition [2, 7]. Because

* Correspondence: ekker.stephen@mayo.edu
[1]Department of Biochemistry and Molecular Biology, Mayo Clinic, 200 1st St SW, 1342C Guggenheim, Rochester, MN 55905, USA
Full list of author information is available at the end of the article

derivatives from *Tol2* have high cargo-capacity and low susceptibility to over-production inhibition, they have become popular gene transfer agents in a variety of animal systems, including zebrafish, African frog, chicken, mouse and human cell cultures including primary T cells [8], and for various genome biology applications (for reviews, see [9, 10].

TEs generally display very diverse patterns of target site selectivity. Studying the mechanisms for such selection is useful to gain insights on the biology of transposition, shedding light on the genome structure and designing better transposon tools for specific applications. Previous research has suggested that the mechanisms of target site selection are very complex and varied from mobile element to mobile element. In many cases, it involves the direct interaction between the transposase/recombinase and the target DNA or their indirect communication through accessory proteins [11]. However, specific factors that contribute to *hAT* element integration preference are largely unknown.

Our knowledge of transposition is largely based on bacterial TEs. Less is known about eukaryotic transposase proteins since they have been historically more difficult to express and reconstitute in vitro. In the present work, we establish a recombinant *Tol2* protein-based system to serve as a tool for genome engineering and to probe the transposition mechanism of this vertebrate *hAT* transposon, focusing on the integration steps. We directly compare known *Tol2* isoforms for activity in both human cells and zebrafish in vivo. We demonstrate that the highest activity variant can be epitope-tagged and retain full activity, and we purify epitope-tagged *Tol2* protein (*His-Tol2*) from E. coli. The functionality of this recombinant *Tol2* transposase is demonstrated in vivo in zebrafish using both somatic and germline transposition assays. Thus *His-Tol2* protein is a viable new source of transposase for molecular medicine and genome engineering applications.

We further show that purified *His-Tol2* can carry out both the excision and integration steps of transposition in vitro in the absence of any cellular co-factors. *Tol2* displays a modest preference for AT-rich DNA in vivo [12]. Such insertion bias is also noted in a cell-free and defined assay when the insertion distribution of *miniTol2* into a target plasmid was measured. *miniTol2* contains the transposon end sequences necessary and sufficient in vivo for excision and integration [13, 14]. *miniTol2* insertion is accompanied by 8-bp target site duplications as occurs in vivo and displays an insertion site preference in vitro, with a higher likelihood of insertion into AT-rich sequences similar to that noted for in vivo integrations [12]. These results suggest the target selection mechanism is at least in part maintained in this much simpler system.

Results

The 649 amino acid *Tol2* isoform is the most active transposase in vivo and in vitro

Different coding sequences and activities have been described in previous works for the *hAT* transposase *Tol2*. Two distinct endogenous *Tol2* mRNAs were identified in the original medaka fish isolate [15]. The shorter mRNA (*Tol2-S*, GenBank accession number AB031080) is encoded by exons 2–4 and results in a predicted protein of 576 amino acids [15] (Fig. 1a). The longer *Tol2* mRNA (*Tol2-L*, GenBank accession number AB031079) is encoded by exons 1–4, and a protein of 685 amino acids is predicted using the first in-frame start codon [15] (Fig. 1a). *Tol2-S* inhibits excision by the *Tol2-L* protein, this inhibition was probably not through competition of DNA-binding but some other unknown mechanism, as *Tol2-S* protein lacks the majority of the BED zinc finger motif [16] (Fig. 1a). This also suggests that the sequences encoded by exon 1 are critical to *Tol2* transposase function. When a copy of medaka *Tol2* genomic sequence was introduced into zebrafish cells that do not harbor any endogenous *Tol2* elements, a third *Tol2* isoform of different length was identified (*Tol2-M*) [17] (Fig. 1a). This mRNA encodes a predicted protein of 649 amino acids that is shorter at the N-terminus than *Tol2-L* (Fig. 1a). This latter, heterologous mRNA (*Tol2-M*) is used in nearly all currently available *Tol2*-based genetic tools [13, 18–20]. The activity of the naturally occurring *Tol2-L* has not been explored in detail or compared to *Tol2-M* under similar conditions.

To determine whether the isoforms of *Tol2* might encode different functional outcomes, we compared *Tol2* isoforms in human HeLa cell transposition assays upon co-transfection of transposase helper and transposon donor plasmids [13]. For *Tol2-M*, the gene-transfer activity is positively related to the amount of *Tol2* transposase provided as plasmids (Fig. 1c). In addition, through codon optimization [21], we increased gene transfer in HeLa cells by 'humanizing' the *Tol2-M* codon usage (*hTol2-M*). The same strategy was applied to *Tol2-L*. However, even this 'humanized' *Tol2-L* (*hTol2-L*) showed reduced activity compared to *Tol2-M* in human cells (Fig. 1c). *Tol2-M* also showed higher activity than *Tol2-L* in zebrafish transposition assays [13] (data not shown). Thus the 649 amino acid *Tol2-M* isoform shows the highest activity of the three described protein forms (Fig. 1a), and will be referred to hereafter as *Tol2* unless otherwise stated.

Tol2 can be modified on either the N- or C- terminus by 6XHis and retain full activity

The ability to identify and readily purify a protein is critical for the development of a system for biochemical analysis. Targeted modification of vertebrate transposase enzymes has been a challenge in the field. The addition

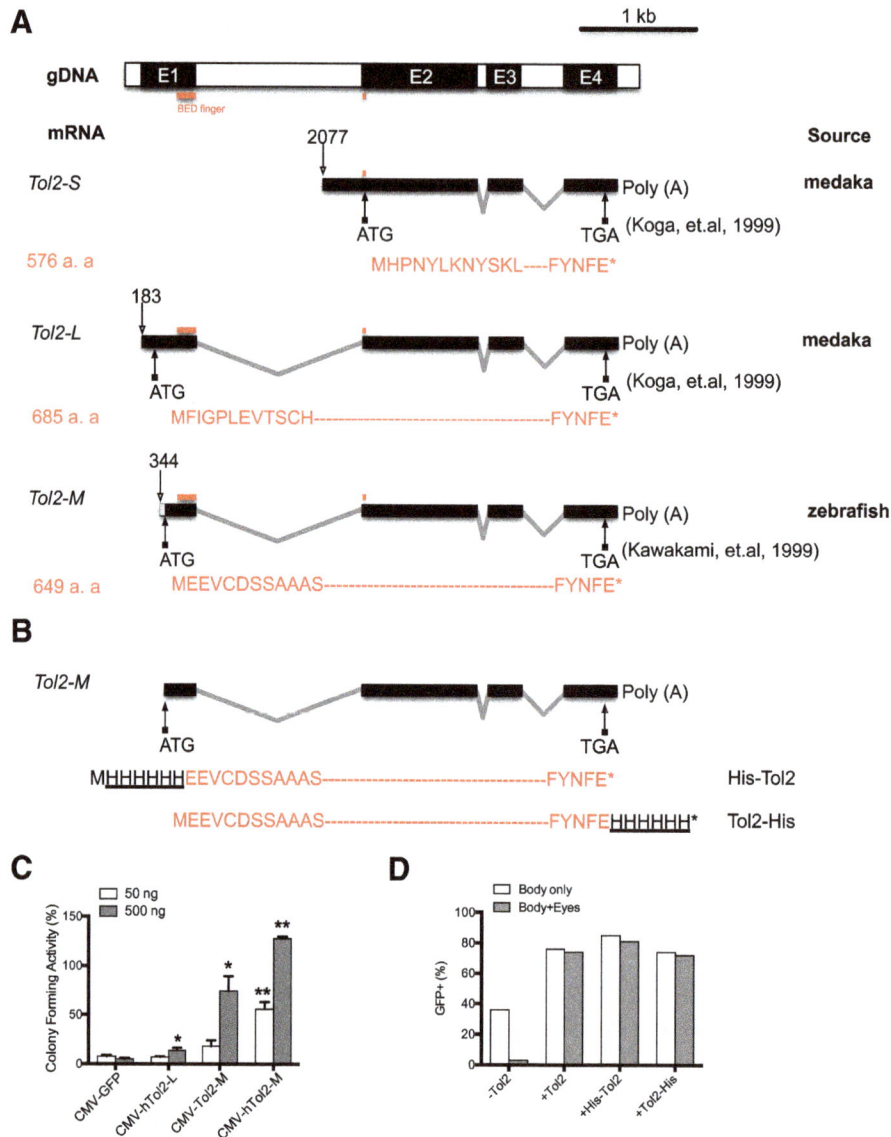

Fig. 1 The highest activity *Tol2* isoform and its modification by 6X His tag. **a** *Tol2* genomic DNA (GenBank: D84375.2) structure and expressed mRNA from various sources (*Tol2-S* GenBank: AB031080; *Tol2-L* GenBank: AB031079; *Tol2-M* GenBank: AB032244). The first nucleotide location is numbered in relation to genomic DNA sequence in each mRNA and indicated by an open-ended arrow. The start and stop codons of the longest open reading frame in each mRNA are also indicated. *Tol2-L* spans all 4 exons (black bar, E1-4) while *Tol2-S* covers only the last three (E2-4). *Tol2-M* starts further into exon 1 than *L*. The predicted *Tol2-M* protein is slightly shorter at the amino-terminus than *Tol2-L*. Corresponding sequences encoding the BED zinc finger DNA binding motif are indicated on genomic DNA and mRNA. **b** In this paper we used a modified *Tol2-M* mRNA [13] encoding the same protein as *Tol2-M*, but all the nucleotides 5' to the first start codon in *Tol2-M* (Fig. 1a, *Tol2-M* gray box) were deliberately removed to eliminate any possible effects from an out-of-frame ATG within those sequences. 6X His tags were added to either 5'- or 3' of the modified mRNA, and the deduced N-terminus or C-terminus modified *Tol2* protein sequences were outlined. **c** Human codon-optimized *Tol2-M* was most effective for gene transfer in HeLa cells. *Tol2-M* was tested for activity in human cell culture by colony forming assay. A *Tol2* transposon vector carrying a Zeocin resistance gene cassette was co-transfected with different version of *Tol2* or GFP control driven by CMV promoter at 50 ng or 500 ng. The numbers of Zeocin-resistant colonies formed in each treatment were compared to the treatment with the highest colony numbers (500 ng CMV-hTo2-M) as percentages. Bars represent mean values ± SEM of three independent experiments. * or ** indicate data is significantly different from respective CMV-GFP controls (* t-test $P < 0.05$; ** t-test $P < 0.01$). **d** *Tol2* transposase could be modified by 6XHis epitope without loss of activity. A GFP-reporter [13] was co-injected with synthesized mRNA corresponding to untagged *Tol2* or *Tol2* tagged with 6X-His at either N-terminus (*His-Tol2*) or C-terminus (*Tol2-His*). Somatic GFP expression was scored at 3 dpf and the percentages of injected fish demonstrating GFP only in the body (body only) or demonstrating strong GFP in eyes and anywhere in the body (body + eyes) were recorded

of new sequences normally results in either reduced or limited enzymatic activity, as was the case for *Sleeping Beauty*, the most studied vertebrate transposable element [22, 23] or with *Tol2* in prior studies [24, 25]. Here we chose a small 6X-His protein tag for use with *Tol2* (Fig. 1b).

To test if such modification will compromise transposase activity, zebrafish embryos were injected with synthetic *Tol2* transposase mRNA encoding *Tol2*, 5′-His (6X) tagged *Tol2*, or 3′-His (6X) tagged *Tol2* as well as with a transposon reporter vector [13]. Embryos were scored for GFP-positive cells or eyes with green fluorescence at 3 days post-fertilization (dpf) as described [13]. Visualization of net gene transfer into the eyes of injected zebrafish embryos is a convenient and robust assay system for determining *Tol2*-mediated transposition in somatic tissues [13]. Importantly, the 6X-His tagged versions of *Tol2* showed the same activity as the untagged *Tol2* proteins in this in vivo transposition assay (Fig. 1d).

Purified *His-Tol2* from *E.coli*

We examined the expression of both N- and C-terminally tagged *Tol2* variants in *E. coli*. The C-terminally tagged version gave predominantly the full-length form with multiple, smaller protein products. In contrast, the amino-terminally tagged *Tol2* (Fig. 2a) yielded a single protein product at the predicted size (~74 kD) when purified from *E. coli* (Fig. 2b, c). This *His-Tol2* protein was used in all subsequent studies unless noted otherwise.

Recombinant *Tol2* protein is a fully functional *hAT* transposase in vivo

Transposon tools are often used as a two-component system in gene discovery and gene delivery applications (Fig. 3a). One component is donor DNA containing a genetic cargo of interest flanked by transposon terminal sequences, and the other component is a transposase source. In previously described vertebrate applications, transposase was provided by either a DNA-encoding plasmid or in vitro transcribed transposase-encoding

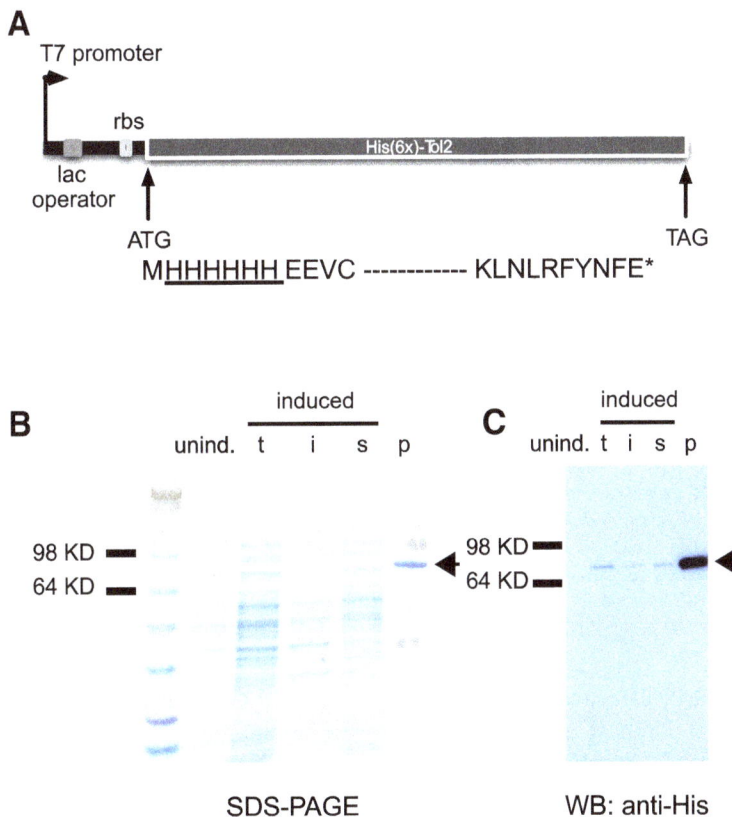

Fig. 2 Expression and detection of recombinant *Tol2* transposase protein. **a** *Tol2* transpoase sequence was cloned into an N-terminal 6XHis expression vector pET-21a (Novagen). The expression cassette driven by T7 promoter is shown, and a fusion protein of ~74 kD was expected. rbs: ribosome binding site. **b** Expression of *His-Tol2* in *E.coli* BL21-AI strain. unind: cell lysate from uninduced cells harboring *Tol2*-expression vector; t: total induced crude cell lysate; i: insoluble protein fraction from induced cell lysate; s: soluble protein fraction from induced cell lysate. Equal volume of cell culture was loaded in each lane. p: ~ 350 ng purified *His-Tol2* protein (arrow head). **c** Immunoblot analysis of *His-Tol2* expression with anti-His antibody. The same protein samples as in (**b**) were loaded except that only ~ 6 ng purified *His-Tol2* was used for detection by western blot

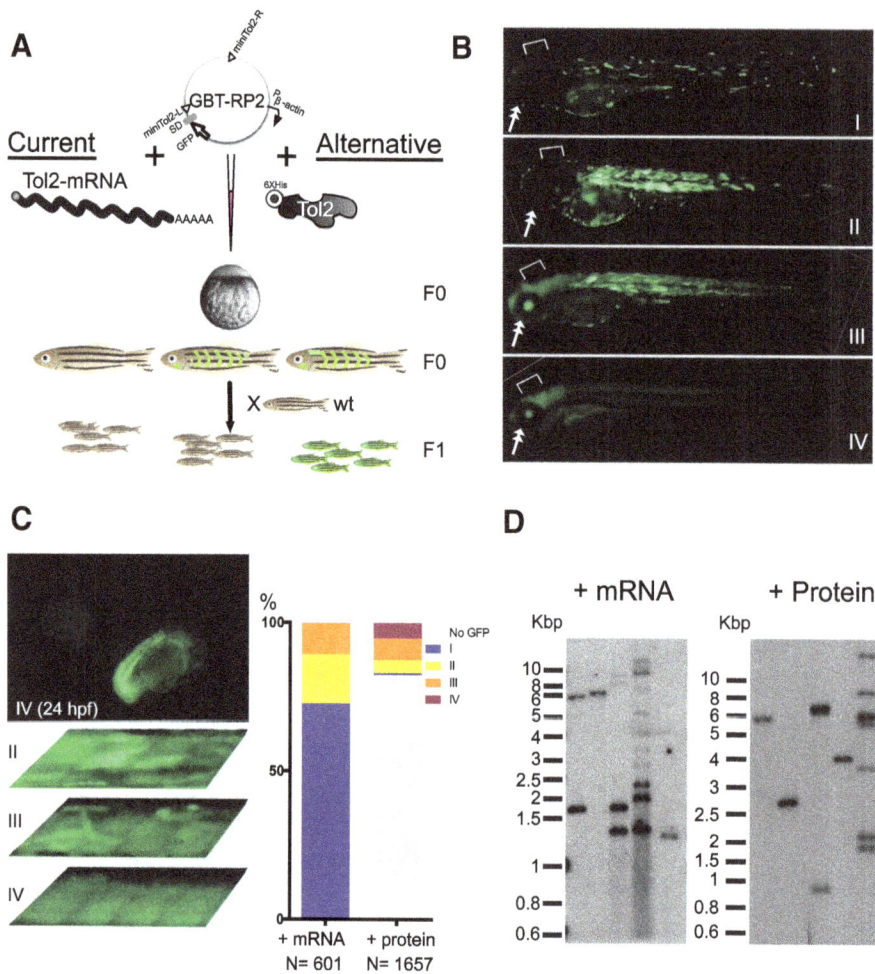

Fig. 3 *His-Tol2* is a fully functional transposase in vivo. **a** Diagram showing microinjection of zebrafish embryo at one-cell stage to generate transgenic animals. GBT-RP2 plasmid containing GFP-reporter gene trap was co-injected with either *Tol2* mRNA or *His-Tol2* protein. Mosaic somatic GFP signals were seen in F0 injected fish. F0 embryos were raised to adult and outcrossed to obtain F1 generation. Ubiquitous GFP would be detected in F1 fish if the integration was due to a gene trap event and was passed through the germline. Molecular analysis of transposon insertion numbers and genomic locations was carried out on tissue from F1 generation animals. P$_{\beta\text{-actin}}$: β-actin promoter; SD: splice donor. **b** Representative images of F0 embryos demonstrating a catalog of GFP-positive somatic patterns that could be observed at 3 dpf from microinjections of different methods. Notice the signal difference in eyes (double arrow) and brain regions (open end arrow) for each category. With vector-only injections, notice only Cat. I and II were observed. **c** A small number of *His-Tol2* protein injected embryos showing ubiquitous whole body green fluorescence as early as 24 hpf. Those fish generally displayed uniform GFP expression throughout the body, with few uneven myotome GFP stripes at 3 dpf (Cat. IV). Also shown are 3 dpf injected fish somites at higher magnification, demonstrating the difference in uniformity of GFP signals in tissues from various injection categories. Injection GFP pattern distribution from typical RNA or protein mediated injection was shown. **d** GFP-positive F1 fish from either founder family generated by mRNA injection or *His-Tol2* injection were subjected to fin-clip and Southern blot analysis with a GFP probe. Each lane represents one individual F1 adult fish. Southern blots from 5 adult fish selected from random founder family are shown for each method. See also Additional file 1

mRNA. To determine if purified recombinant *His-Tol2* protein is fully functional and robust enough to be an alternative transposase source for practical genetic applications, we tested *His-Tol2* function in vivo using this two-component system in both somatic and germline gene transfer assays.

For gene transfer testing in somatic cells, we used a high stringency transposition test system in which insertions of a specialized *miniTol2* element, GBT-RP2 [26], in transcriptionally competent genomic loci result in

enhanced Green Fluorescent Protein (GFP) expression in the resulting animals. The GBT-RP2 element was based on the 3′ gene trap derived from previously described gene-breaking transposon (GBT) vectors [27, 28]. Transposon DNA was co-injected with either *Tol2* mRNA or *His-Tol2* protein into one-cell zebrafish embryos (Fig. 3a). The F0 somatic GFP patterns of injected embryos generally can be characterized into different categories based on signal uniformity at 3 days post-fertilization (dpf): a category I pattern is characterized by embryos showing minimal GFP signals, usually only a couple of bright stripes in the myotomes; a category II pattern is characterized by large areas of the fish body with GFP signals, but with little or no GFP-positive signals in the eyes or brain region; and a category III pattern is characterized by whole body green fluorescence, including the brain and eyes (Fig. 3b).

Background, non-transposase mediated gene transfer was measured by the injection of the RP2 vector without transposase, resulting in sporadic GFP signals seen in myotomal tissues of some F0 fish by 3 dpf. Usually, the majority of the background GFP-positive fish displayed category I patterns, and very few displayed category II patterns (Fig. 3b).

When transposase was provided by *Tol2* mRNA, the overall somatic GFP signals were notably increased, especially with a higher number of embryos exhibiting the category III pattern, which was a very rare observation without the co-addition of transposase (Fig. 3b, c).

With *His-Tol2* protein, the majority of the injected fish did not display any GFP signal. Among the GFP-positive F0 embryos, GFP patterns representing all three categories were noted. However, the number of category I animals, which were less likely to carry transposon-mediated germ-line integration, was much lower than the GFP-positive category II or category III (Fig. 3b, c). We also conducted multi-generational testing for germline gene transfer rate testing. F0 fish displaying category III somatic patterning from either mRNA or *His-Tol2* transposase resulted in a similar gene trapping rate and low mosaicism (Table 1).

Interestingly, an additional pattern found only in protein-injected embryos was observed. A small portion of these injected animals (~5–10 %) exhibited ubiquitous GFP expression as early as 24 h after injection (Fig. 3c), and the signals were very uniform throughout the body at 3 dpf (Fig. 3c). Such F0 embryos were designated category IV (Table 1). Zebrafish embryos undergo rapid cell

divisions every 20 min during the early development and reach ~ 1000 cells by 3 h post fertilization [29]. Thus the timing of transposition potentially plays a critical role in determining the mosaicism of resulting animals. We hypothesized that the uniform GFP signals of the category IV fish could be the result of very early integrations of the GFP reporter gene into the genome. Indeed, a higher percentage of germline transmission was obtained from such F0 fish (Table 1). More importantly, the average F0 germline mosaicism was much reduced for category IV fish (Table 1). Our observations of *His-Tol2* mediated injection suggest that even though transposition events were rarer under current condition than when compared to mRNA as a source of transposase, protein injection can "jump start" integration inside the early zebrafish embryo.

An important additional measurement of the germline gene transfer rate is copy number. Transposon copy number was estimated by Southern blot analysis on individual fish (Fig. 3d). An average of ~ 4 unique transposon insertions per haploid F0 founder germline were observed (Additional file 1). The fact that category IV (Additional file 1, family 1,2,3,5,6 and 8) did not contain more transposon insertions than in other groups suggests that the uniform fluorescence is most likely due to earlier incorporation of the reporter gene into the embryo genome.

Transposase-mediated insertion can be distinguished from other insertion mechanisms by the target site duplications flanking the transposon sequences. We sequenced the insertion sites of the *His-Tol2* protein-mediated RP2 transposon insertions, and the *hAT* signature 8-bp genomic duplications were present in all checked cases (Table 2). We conclude that *His-Tol2* protein is a fully functional *hAT* transposase in both somatic and germline lineages in vivo.

Recombinant *Tol2* protein can mediate concerted target joining in vitro

We continued to explore whether this recombinant *Tol2* protein is functional in vitro, focusing on the target joining step by directly assaying the ability of *His-Tol2* to join both ends of a DNA segment flanked by *Tol2* end sequences to a target plasmid (Fig. 4a). *Tol2* terminal sequences consisting of 261 bp from the left arm and 192 bp from the right arm (*miniTol2*) are sufficient transposon end sequences for efficient *Tol2* transposition in vivo [13, 14]. We used a PCR-generated Kan^R gene flanked by these *Tol2* ends (Additional file 2), *mini-Tol2-Kan^R* together with *His-Tol2* protein to assess target joining in vitro. After incubation of blunt end *miniTol2-Kan^R* with an ampicillin resistant target plasmid (pGL) in the presence of *His-Tol2* protein and Mn^{2+} ions, DNA was precipitated, transformed into *E. coli*,

Table 1 *His-Tol2* and *Tol2* mRNA as sources of transposase in vivo

Somatic F0 pattern	*Tol2* form	Germline trapping and expression frequency	Average F0 germline mosaicism
Cat II.	mRNA	65 % (13/20)	27 %
Cat III.	mRNA	88 % (35/40)	39 %
Cat III.	Protein	76 % (13/17)	39 %
Cat IV.	Protein	90 % (9/10)	51 %

Analyses of somatic and germline transmission of reporter genes are shown. Fish injected with either mRNA or *His-Tol2* were characterized by different F0 somatic patterns and assessed for corresponding germline trapping frequency, estimated as the percentage of F0 fish producing GFP-positive offspring. The number of fish screened is shown in parentheses. The F0 germline mosaicism rate was determined by the percentage of GFP-positive F1 offspring from a founder outcross. The average mosaicism rate for founder fish is listed from each different category

Table 2 Transposase-mediated germline integration

Site	Sequence	Chr.	Refseq
1	aaatatttac**caagcaac**-5'-Tol2-3'-**caagcaac**acgttcagtg	8	Intron of gfra2
2	ataatttcct**cttatttg**-5'-Tol2-3'-**cttatttg**catgtcagat	13	Intergenic
3	cgcatgctaa**cttataga**-5'-Tol2-3'-**cttataga**ggaggtgccc	8	Exon of wu: fb79a07
4	aaacgttcct**cctaacac**-5'-Tol2-3'-**cctaacac**agttagatgg	3	Intergenic
5	caacacatga**ctcgttgg**-5'-Tol2-3'-**ctcgttgg**ccatatgcta	15	Intergenic
6	gggaatatgt**gttattaa**-5'-Tol2-3'-**gttattaa**ctgcgtccca	4	Repetitive sequences
7	agctgtctct**tctgtgtc**-5'-Tol2-3'-**tctgtgtc**attcagtctc	3	Intron of LOC557901
8	tgtcagagat**ctaggtca**-5'-Tol2-3'-**ctaggtc**agatggaggaa	25	Intergenic

Tol2 insertion sites generated by *His-Tol2* protein co-injection were cloned from F1 transgenic fish. Examples of 8-bp genomic sequence duplication (in bold) were shown with insertion loci information

and plated on LB-Amp$^+$ plates to determine the total number of target plasmids and on LB-Kan$^+$ plates to identify integration products (Fig. 4a, b). The number of *Tol2* integrants increased as the number of transposons in the reaction increased (Fig. 4c, Additional file 3), and

we detected up to 0.8 % of integration ratio among the conditions tested. We sequenced large numbers of integrants obtained in multiple experiments and found that the 8-bp host duplications at the insertion site characteristic of *hAT* transposition occurred in 86 % of insertions

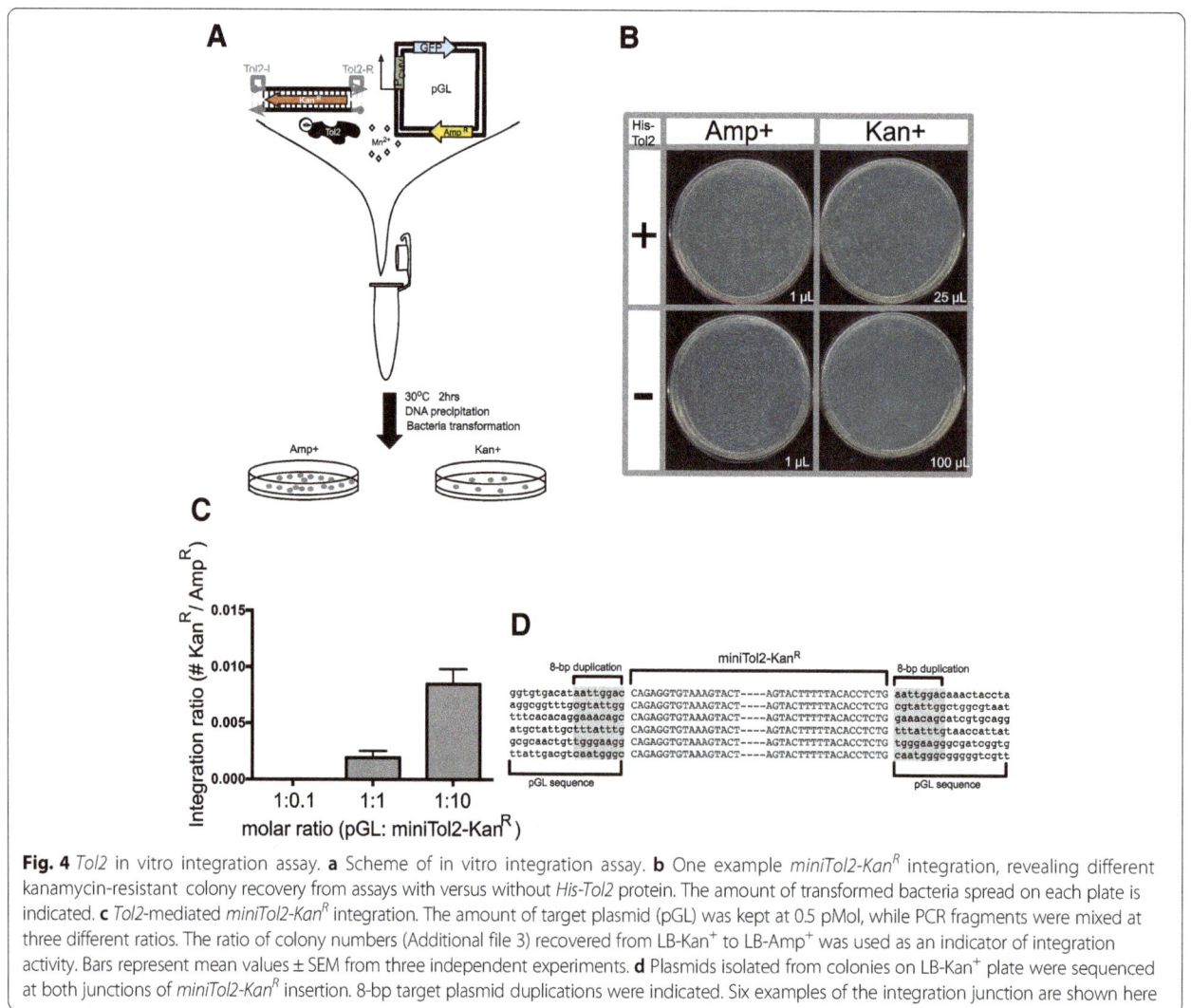

Fig. 4 *Tol2* in vitro integration assay. **a** Scheme of in vitro integration assay. **b** One example *miniTol2-Kan^R* integration, revealing different kanamycin-resistant colony recovery from assays with versus without *His-Tol2* protein. The amount of transformed bacteria spread on each plate is indicated. **c** *Tol2*-mediated *miniTol2-Kan^R* integration. The amount of target plasmid (pGL) was kept at 0.5 pMol, while PCR fragments were mixed at three different ratios. The ratio of colony numbers (Additional file 3) recovered from LB-Kan$^+$ to LB-Amp$^+$ was used as an indicator of integration activity. Bars represent mean values ± SEM from three independent experiments. **d** Plasmids isolated from colonies on LB-Kan$^+$ plate were sequenced at both junctions of *miniTol2-Kan^R* insertion. 8-bp target plasmid duplications were indicated. Six examples of the integration junction are shown here

(Total $N = 333$) (Fig. 4d). Thus *His-Tol2* integration in vitro recapitulates *Tol2* integration in vivo. The distribution of recovered insertions in the target plasmid was non-random with *miniTol2* element insertion biased towards the AT-rich (AT content 64 %) SV40 polyA segment of the target vector (Fig. 5a; see also below). We also noticed that *Tol2* protein was able to excise a *miniTol2*-flanking fragment out of a plasmid donor (data not shown), indicating *His-Tol2* is fully competent gene transfer vector in vitro as well as in vivo.

Target site preference of *Tol2* is conserved using a simplified target plasmid system

As described above, the *miniTol2-Kan^R*-insertions generated in vitro were not randomly distributed on the target plasmid, and they clustered at the SV40 poly (A) rich region. Analysis of these patterns by Monte Carlo simulation demonstrated that the bias was highly significant (p-value by simulation < 0.0001, see Methods) in each of four independent experiments (Fig. 5a; Additional file 4 and Additional file 5). This observed preference in vitro is

in agreement with prior work showing that *Tol2* has a preference for AT-rich regions in vivo. More importantly, the in vivo observed integration signature, TNA**(C/G)**TTA-TAA**(G/C)**TNA (bold: insertion site) [12], was similar to the findings reported here using the plasmid insertion system (Fig. 5b). This observation suggests a primary sequence of the target site is likely to play a critical role in *Tol2/hAT* integration site selection.

Discussion

In the past decade, the *Tol2*-based transposition system has become a versatile tool for gene delivery in commonly used model organisms, especially in zebrafish, which has seen many cases of successful genetic screening with various highly efficient Tol2 systems [30, 31]. Recently, *Tol2* transposon tools have been applied to pioneer gene transfer in novel biological systems, such as haplochromine cichlid [32, 33] and African killifish [33]; or as an alternative to viral vectors showing promise in engineered T-cell therapy [34]. Here we report the purification of recombinant *Tol2 hAT* transposase from

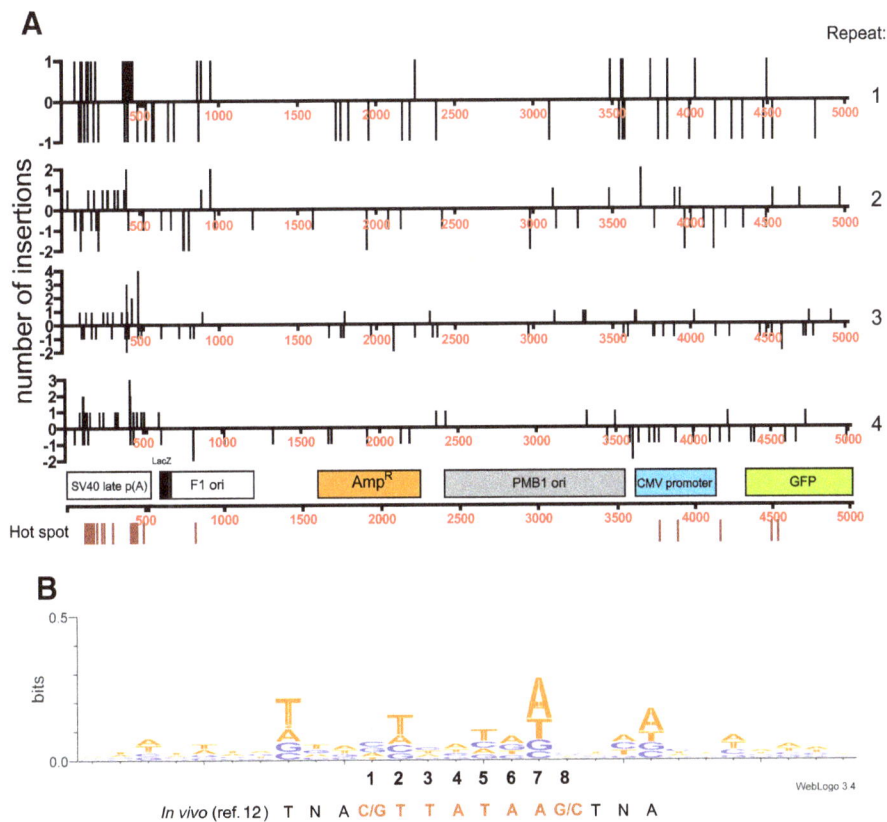

Fig. 5 *miniTol2* insertion sites and consensus sequence. a *Tol2*-mediated insertions from four independent experiments were mapped to target plasmid (pGL) for *miniTol2-Kan^R* (n = 266). Y-axis indicates the number of insertion events and orientation (positive: sense orientation; negative: anti-sense orientation) Major features of the pGL vector were also indicated. Red lines below the sequences indicate insertion "hot spots", defined as the same locations that are discovered in more than one experiment, regardless of the insertion orientation or locations in one experiment that have transposons inserted in both directions (also see Additional file 5). b *Tol2* integration site motif analyzed by WebLog (version 3.0) (n = 266) and aligned to a previous indentified weak AT-rich consensus in vivo [12]

E. coli and demonstrate that it is fully functional in vivo and in vitro. A functional recombinant *Tol2* protein will be a useful addition to the *Tol2*-based genome engineering kit. Researchers will have more choices over what form of transposase will work best for the organisms/cells of their interest. Transposase provided as readily made protein, in some systems, may be advantageous over an mRNA-based delivery method. For example, when the translation of the *Tol2* mRNA is not efficient or optimized for the intended heterologous system. And even in zebrafish where mRNA is an effective method for Tol2 delivery, we noted advantages of using the functional Tol2 transposase protein.

We observed robust somatic and germline transposition in zebrafish using *E.coli* -purified *Tol2* transposase. Our experiments suggest using *His-Tol2* may be advantageous in generating low mosaic transgenic animals with higher germline transmission rate. Transposition using a protein source of transposase may "jumpstart" transposition as it avoids possible delays from transcription, translation and assembly of a functional transpososome containing transposase multimers when transposase is supplied in a ready form. Transposase multimers are the active form of transposase in other studied systems [7, 35, 36]. Static light scattering analysis has shown that purified *Tol2* is a dimer (A. Voth and F. Dyda, personal communication). This recombinant *Tol2* protein is an excellent tool for future experimental methods for gene delivery. As documented using this initial system, the percentage of GFP-positive F0 embryos was lower using protein as a transposase source when compared to the well-established mRNA co-injection methods,. However, the current protein co-injection method has yet to be fully optimized, such as experimentally establishing the optimal protein to donor plasmid ratio.

Functional *Tol2* transposase protein is also of use for studying transposon biology science, enabling experimental testing of this *hAT* element both in vivo and in vitro. Such work will shed more light on our understanding of how this and related transposons work in the future. In this paper, we demonstrated that Tol2 protein could be used in an assay to probe the integration site preference for the transposon and identified a weak consensus that are largely conserved between in vivo chromosomes and a cell-free target plasmid. Two other commonly used transposases for genome engineering, *Sleeping Beauty* and *piggyBac* use obligate TA and TTAA core consensus targeting sequence respectively [37, 38]. In contrast, *Tol2* target site selection is still considered promiscuous on the level of primary DNA sequence [39]. The preference noticed towards AT-rich DNA context could also reflect a preference towards integration into regions of higher DNA flexiblity as suggested previously [40–42].

Conclusions

We presented here a fully functional recombinant transposase protein, both in vitro and in vivo, for the popular vertebrate gene transfer transposon *Tol2*. Our initial work in zebrafish indicated that *His-Tol2* has a high potential for genome engineering applications. *His-Tol2* can also serve as a tool for probing transposon biology in vitro and help confirm an integration site consensus shared between in vivo and cell-free systems.

Methods
Ethics statement
Zebrafish larvae were raised within the Mayo Clinic's Zebrafish Facility with adherence to the NIH Guide for the Care and Use of Laboratory Animals and approval by Mayo Clinic's Institutional Animal Care and Use Committee.

Important constructs
Details of how important constructs were made are included in Additional file 6: Supplemental Experimental Procedures.

HeLa cell transfection assays
Transfection of HeLa cells with a Zeocin resistant vector (pTol2miniZeo) with either control vector pCMV-GFP or vectors providing *Tol2* transposase (pCMV-Tol2-M, pCMV-hTol2-M or pCMV-hTol2-L) was following the same protocol as described in [21]. Zeocin-resistant colony forming assay was conducted as in [21] as well.

Zebrafish somatic transposition assays
To compare transposition of a GFP-reporter mediated by tagged vs. untagged *Tol2* mRNA (Fig. 1d), zebrafish embryos were injected at one-cell stage with a mixture of 25 pg pTol2-S2EF1α-GM2 (pDB591, [13]) plasmid DNA with 25 pg or 100 pg T3TS-Tol2-M (*Tol2*), T3TS-His-Tol2, or T3TS-His-Tol2 RNA (Fig. 1d) respectively. Surviving normal-looking embryos were scored for GFP fluorescence at 3 dpf.

Protein expression and immunoblotting
Small scale detection of protein expression was conducted according to Qiagen recommended protocol. Detail procedures are included in Additional file 6: Supplemental Experimental Procedures. *His-Tol2* protein expression was detected by penta-His antibody (Qiagen) following the recommended western protocol. An adapted protocol [43] was used to generate large-scale nuclease-free protein fractions.

Zebrafish one-cell GBT microinjection
One-cell stage zebrafish microinjection was performed as described [44]. The injected amount of reagents were 25 pg GBT-RP2 plasmid co-injected with 25 pg *T3TS-Tol2* mRNA [13]; or co-injected with ~ 0.2 ng *His-Tol2* protein.

Southern blot

Genomic DNA isolation and Southern blot hybridization was conducted as previously described [28]. Tail fins from individual GFP-positive F1 fish generated by co-injection of *Tol2* mRNA or *His-Tol2* protein at F0 were clipped and genomic DNA was isolated followed by digestion with BamHI and BglII for Southern blotting.

Molecular analysis of genomic integration sites

Cloning of insertion site in F1 individual fish was conducted through adapted inverse and linker-mediated PCR (LM-PCR) [44]. Information of primer sequences and PCR conditions are included in Additional file 6: Supplemental Experimental Procedures.

In vitro integration assay

Blunt end *miniTol2-Kan^R* fragments were generated by PCR with platinum Pfx DNA polymerase (Invitrogen) with forward primer Tol2-mini5′ (5′- CAGAGGTGTAAAGT ACTTGA-3′), reverse primer Tol2-mini3′ (5′- CAG AGGTGTAAAAAGTACTC-3′) and a template (pTol2mi-niKan(-)amp). PCR product was treated with DpnI to get rid of the template before purified by QIAquick PCR purification kit (Qiagen). Freshly prepared plasmids and PCR products were used in the assay. In a 20 μL reaction, 0.5 pMol plasmid pGL, corresponding PCR fragments and 27 pMol *His-Tol2* (~2 μg) were mixed in MOPS buffer (25 mM MOPS, pH 7.0; 1 mM $MnCl_2$; 50 mM NaCl; 5 % glycerol; 2 mM DTT; 100 ng/ul BSA). The reaction was incubated for 2 h at 30 °C. DNA was phenol-chloroform extracted using Maxtract High Density columns (Qiagen) and resuspended in 5 μL TE buffer. 1 μL or 5 μL DNA was used to transform Top10 competent cells (Invitrogen). The number of colonies grew on Amp-resistant plates was used to calculate the total number recovered plasmids, and the number of colonies that grew on Kan-resistant plates was used to calculate the number of plasmid with Kan^R integration. MOPS buffer with $MgCl_2$ or no added cations were tested as well, and the integration activity was modestly higher with $MnCl_2$.

Insertion site distribution statistics

The simulation distributions of the number of insertions over the target features was done in R. We assume that every feature gets a number of distributions proportional to its size in bp. We simulated 75 counts with the probability of a count falling into a feature being the length of that feature divided by the length of all features types. We then ran this simulation 10,000 times. To check where the experimental data lie with respect to the simulated data, we determined how often the simulated counts are bigger or smaller than the experimental data.

Additional files

Additional file 1: Estimation of transposon insertion numbers by Southern analysis. Individual GFP-positive F1 fish from random selected F0 founder families were fin-clipped for gDNA isolation. Southern blotting with a GFP probe was conducted for each fish gDNA and the number of clear-labeled Southern bands was recorded as estimation of transposon insertions. Total number of unique Southern bands was used to estimate total transposon insertions for each F0 founder family. An average insertion number over all the families analyzed was calculated for both injection methods. (DOCX 91 kb)

Additional file 2: *miniTol2* sequences of the in vitro assays. DNA sequences of the left arm and right arm of *miniTol2* flanking the Kan^R used in the in vitro integration assays. (DOCX 55 kb)

Additional file 3: Colony counting numbers for in vitro integration assays. Colony numbers counted for the three independent experiments in Fig. 4c of different target plasmid and PCR insert ratios. (DOCX 49 kb)

Additional file 4: *P*-value by simulation. For each independent repeat of *miniTol2* mediated Kan^R insertion distribution, p-value was generated by simulation (see Methods) for different features on the target plasmid. (DOCX 61 kb)

Additional file 5: Insertion "hot spots" for *miniTol2*. "Hot spots" were arranged according to the insertion loci along the target plasmid. 8-bp target duplication and flanking sequnces were shown. Forward (F) or reverse (R) insertion copies were counted. Repeats coverage indicate the "hot spot" indentified by how many independent experiments. (XLSX 45 kb)

Additional file 6: Supplemental Experimental Procedures. (DOCX 131 kb)

Abbreviations

Ac/Ds: Activator/Dissociation; TEs: transposable elements; TIRs: terminal inverted repeats; GBT: gene-breaking transposon; dpf: days post-fertilization; Kan^R: kanamycin resistance.

Competing interests

The authors declare that they have no competing interests.

Authors' contributions

JN, KJW, DN, DB, KJS and MU carried out the experiments, and JN, KJW, DN, DB and SCE analyzed the data. KJW, DN and DB tested Tol2 isoforms and modifications, and KJW and DN conducted the Tol2 protein preparation. KJS and MU participated in the in vitro insertion site sequencing work and tests to determine transposition efficiency in vivo. JN and SCE drafted the manuscript. All authors read and approved the final manuscript.

Acknowledgments

We thank Gary Moulder and Dr. Karl Clark for help with zebrafish injections and Southern blotting. We thank Drs. Liqin Zhou˙ Xianghong Li and Nancy L. Craig for active scientific discussions on this project. We thank Ekker lab members, especially Tammy Greenwood and Dr. Henning Schneider for critical evaluation of the manuscript. The work was supported by the Mayo Foundation and the U.S. National Institutes of Health grants DA14546, HG006431 and GM63904 to Stephen C. Ekker.

Author details

[1]Department of Biochemistry and Molecular Biology, Mayo Clinic, 200 1st St SW, 1342C Guggenheim, Rochester, MN 55905, USA. [2]Department of Chemical and Systems Biology, Stanford University School of Medicine, Stanford, CA 94305, USA. [3]Department of Biochemistry, Molecular Biology, and Biophysics, University of Minnesota, Minneapolis, MN 55455, USA. [4]University of Pennsylvania, 9 Penn Tower, 3400 Spruce ST, Philadelphia, PA 19104, USA. [5]Department of Biology, Temple University, Philadelphia, PA 19122, USA.

References

1. McClintock B. The origin and behavior of mutable loci in maize. Proc Natl Acad Sci U S A. 1950;36(6):344–55.

2. Calvi BR, Hong TJ, Findley SD, Gelbart WM. Evidence for a common evolutionary origin of inverted repeat transposons in Drosophila and plants: hobo, Activator, and Tam3. Cell. 1991;66(3):465–71.

3. Rubin E, Lithwick G, Levy AA. Structure and evolution of the hAT transposon superfamily. Genetics. 2001;158(3):949–57.

4. Warren WD, Atkinson PW, O'Brochta DA. The Hermes transposable element from the house fly, Musca domestica, is a short inverted repeat-type element of the hobo, Ac, and Tam3 (hAT) element family. Genet Res. 1994;64(2):87–97.

5. Lander ES et al. Initial sequencing and analysis of the human genome. Nature. 2001;409(6822):860–921.

6. Koga A, Suzuki M, Inagaki H, Bessho Y, Hori H. Transposable element in fish. Nature. 1996;383(6595):30.

7. Hickman AB et al. Molecular architecture of a eukaryotic DNA transposase. Nat Struct Mol Biol. 2005;12(8):715–21.

8. Huang X, et al. Gene transfer efficiency and genome-wide integration profiling of Sleeping Beauty, Tol2, and piggyBac transposons in human primary T cells. Mol Ther. 2010;18(10):1803–13.

9. Kawakami K. Transposon tools and methods in zebrafish. Dev Dyn. 2005;234(2):244–54.

10. Kawakami K. Tol2: a versatile gene transfer vector in vertebrates. Genome Biol. 2007;8 Suppl 1:S7.

11. Craig NL. Target site selection in transposition. Annu Rev Biochem. 1997;66:437–74.

12. Kondrychyn I, Garcia-Lecea M, Emelyanov A, Parinov S, Korzh V. Genome-wide analysis of Tol2 transposon reintegration in zebrafish. BMC Genomics. 2009;10:418.

13. Balciunas D et al. Harnessing a high cargo-capacity transposon for genetic applications in vertebrates. PLoS Genet. 2006;2(11):e169.

14. Urasaki A, Morvan G, Kawakami K. Functional dissection of the Tol2 transposable element identified the minimal cis-sequence and a highly repetitive sequence in the subterminal region essential for transposition. Genetics. 2006;174(2):639–49.

15. Koga A, Suzuki M, Maruyama Y, Tsutsumi M, Hori H. Amino acid sequence of a putative transposase protein of the medaka fish transposable element Tol2 deduced from mRNA nucleotide sequences. FEBS Lett. 1999;461(3):295–8.

16. Tsutsumi M, Koga A, Hori H. Long and short mRnas transcribed from the medaka fish transposon Tol2 respectively exert positive and negative effects on excision. Genet Res. 2003;82(1):33–40.

17. Kawakami K, Shima A. Identification of the Tol2 transposase of the medaka fish Oryzias latipes that catalyzes excision of a nonautonomous Tol2 element in zebrafish Danio rerio. Gene. 1999;240(1):239–44.

18. Parinov S, Kondrichin I, Korzh V, Emelyanov A. Tol2 transposon-mediated enhancer trap to identify developmentally regulated zebrafish genes in vivo. Dev Dyn. 2004;231(2):449–59.

19. Kawakami K, Shima A, Kawakami N. Identification of a functional transposase of the Tol2 element, an Ac-like element from the Japanese medaka fish, and its transposition in the zebrafish germ lineage. Proc Natl Acad Sci U S A. 2000;97(21):11403–8.

20. Kawakami K, et al. A transposon-mediated gene trap approach identifies developmentally regulated genes in zebrafish. Dev Cell. 2004;7(1):133–44.

21. Keng VW et al. Efficient transposition of Tol2 in the mouse germline. Genetics. 2009;183(4):1565–73.

22. Yant SR, Huang Y, Akache B, Kay MA. Site-directed transposon integration in human cells. Nucleic Acids Res. 2007;35(7):e50.

23. Wu SC, et al. piggyBac is a flexible and highly active transposon as compared to sleeping beauty, Tol2, and Mos1 in mammalian cells. Proc Natl Acad Sci U S A. 2006;103(41):15008–13.

24. Shibano T, et al. Recombinant Tol2 transposase with activity in Xenopus embryos. FEBS Lett. 2007;581(22):4333–6.

25. Meir YJ, et al. Genome-wide target profiling of piggyBac and Tol2 in HEK 293: pros and cons for gene discovery and gene therapy. BMC Biotechnol. 2011;11:28.

26. Clark KJ, et al. In vivo protein trapping produces a functional expression codex of the vertebrate proteome. Nat Methods. 2011. In press.

27. Sivasubbu S, et al. Gene-breaking transposon mutagenesis reveals an essential role for histone H2afza in zebrafish larval development. Mech Dev. 2006;123(7):513–29.

28. Petzold AM, et al. Nicotine response genetics in the zebrafish. Proc Natl Acad Sci U S A. 2009;106(44):18662–7.

29. Kimmel CB, Ballard WW, Kimmel SR, Ullmann B, Schilling TF. Stages of embryonic development of the zebrafish. Dev Dyn. 1995;203(3):253–310.

30. Asakawa K, Kawakami K. The Tol2-mediated Gal4-UAS method for gene and enhancer trapping in zebrafish. Methods. 2009;49(3):275–81.

31. Clark KJ, et al. In vivo protein trapping produces a functional expression codex of the vertebrate proteome. Nat Methods. 2011;8(6):506–15.

32. Juntti SA, Hu CK, Fernald RD. Tol2-mediated generation of a transgenic haplochromine cichlid, Astatotilapia burtoni. PLoS One. 2013;8(10):e77647.

33. Valenzano DR, Sharp S, Brunet A. Transposon-Mediated Transgenesis in the Short-Lived African Killifish Nothobranchius furzeri, a Vertebrate Model for Aging. G3 (Bethesda). 2011;1(7):531–8.

34. Tsukahara T, et al. The Tol2 transposon system mediates the genetic engineering of T-cells with CD19-specific chimeric antigen receptors for B-cell malignancies. Gene Ther. 2015;22(2):209–15.

35. Mizuuchi M, Baker TA, Mizuuchi K. Assembly of the active form of the transposase-Mu DNA complex: a critical control point in Mu transposition. Cell. 1992;70(2):303–11.

36. Sakai J, Chalmers RM, Kleckner N. Identification and characterization of a pre-cleavage synaptic complex that is an early intermediate in Tn10 transposition. EMBO J. 1995;14(17):4374–83.

37. Wilson MH, Coates CJ, George Jr AL. PiggyBac transposon-mediated gene transfer in human cells. Mol Ther. 2007;15(1):139–45.

38. Vigdal TJ, Kaufman CD, Izsvak Z, Voytas DF, Ivics Z. Common physical properties of DNA affecting target site selection of sleeping beauty and other Tc1/mariner transposable elements. J Mol Biol. 2002;323(3):441–52.

39. Grabundzija I, et al. Comparative analysis of transposable element vector systems in human cells. Mol Ther. 2010;18(6):1200–9.

40. Vrljicak P, et al. Genome-Wide Analysis of Transposon and Retroviral Insertions Reveals Preferential Integrations in Regions of DNA Flexibility. G3. 115.026849; Early Online: 2016.

41. Yant SR, et al. High-resolution genome-wide mapping of transposon integration in mammals. Mol Cell Biol. 2005;25(6):2085–94.

42. Liu G, et al. Target-site preferences of Sleeping Beauty transposons. J Mol Biol. 2005;346(1):161–73.

43. Zhou L, et al. Transposition of hAT elements links transposable elements and V(D)J recombination. Nature. 2004;432(7020):995–1001.

44. Davidson AE, et al. Efficient gene delivery and gene expression in zebrafish using the Sleeping Beauty transposon. Dev Biol. 2003;263(2):191–202.

11

Convergent evolution of tRNA gene targeting preferences in compact genomes

Thomas Spaller[1], Eva Kling[1], Gernot Glöckner[2,3], Falk Hillmann[4] and Thomas Winckler[1*]

Abstract

Background: In gene-dense genomes, mobile elements are confronted with highly selective pressure to amplify without causing excessive damage to the host. The targeting of tRNA genes as potentially safe integration sites has been developed by retrotransposons in various organisms such as the social amoeba *Dictyostelium discoideum* and the yeast *Saccharomyces cerevisiae*. In *D. discoideum*, tRNA gene-targeting retrotransposons have expanded to approximately 3 % of the genome. Recently obtained genome sequences of species representing the evolutionary history of social amoebae enabled us to determine whether the targeting of tRNA genes is a generally successful strategy for mobile elements to colonize compact genomes.

Results: During the evolution of dictyostelids, different retrotransposon types independently developed the targeting of tRNA genes at least six times. DGLT-A elements are long terminal repeat (LTR) retrotransposons that display integration preferences ~15 bp upstream of tRNA gene-coding regions reminiscent of the yeast Ty3 element. Skipper elements are chromoviruses that have developed two subgroups: one has canonical chromo domains that may favor integration in centromeric regions, whereas the other has diverged chromo domains and is found ~100 bp downstream of tRNA genes. The integration of *D. discoideum* non-LTR retrotransposons ~50 bp upstream (TRE5 elements) and ~100 bp downstream (TRE3 elements) of tRNA genes, respectively, likely emerged at the root of dictyostelid evolution. We identified two novel non-LTR retrotransposons unrelated to TREs: one with a TRE5-like integration behavior and the other with preference ~4 bp upstream of tRNA genes.

Conclusions: Dictyostelid retrotransposons demonstrate convergent evolution of tRNA gene targeting as a probable means to colonize the compact genomes of their hosts without being excessively mutagenic. However, high copy numbers of tRNA gene-associated retrotransposons, such as those observed in *D. discoideum*, are an exception, suggesting that the targeting of tRNA genes does not necessarily favor the amplification of position-specific integrating elements to high copy numbers under the repressive conditions that prevail in most host cells.

Keywords: *Dictyostelium*, Chromo domain, Chromovirus, Ty3, RNA polymerase III

Abbreviations: APE, Apurinic/apyrimidinic endonuclease; CHD, Chromatin organization modifier domain (chromo domain); CSD, Chromo shadow domain; DGLT-A, *Dictyostelium* gypsy-like transposable element; DIRS, *Dictyostelium* intermediate repeat sequence; GAG, Group-specific antigen; IN, Integrase; ITR, Inverted terminal repeat; LTR, Long terminal repeat; mya, Million years ago; ORF, Open reading frame; PBS, Primer binding site; PPy, Polypyrimidine stretch; RNH, Ribonuclease H; RT, Reverse transcriptase; TRE, tRNA gene-associated retroelement

* Correspondence: t.winckler@uni-jena.de
[1]Institute of Pharmacy, Department of Pharmaceutical Biology, Friedrich Schiller University Jena, Semmelweisstraße 10, Jena 07743, Germany
Full list of author information is available at the end of the article

Background

Mobile elements are obligate genomic parasites that amplify as selfish DNA and play important roles in driving the evolution of their hosts [1–5]. Retrotransposons mobilize by reverse transcription of RNA intermediates and integration of the resulting DNA copies at new locations of their host's genomes. Retrotransposons encode proteins that mediate their mobility and they can be distinguished by their overall structures and retrotransposition mechanisms [6]. The supergroup of retrotransposons bearing long terminal repeats (LTRs) is classified into vertebrate retroviruses (Retroviridae), hepadnaviruses, caulimoviruses, Ty1/copia (Pseudoviridae), Ty3/gypsy (Metaviridae), BEL, and DIRS (Dictyostelium intermediate repeat sequence) [7–9]. Non-LTR retrotransposons are a diverse group of mobile elements that lack LTRs and can be further distingushied by structural features such as the presence of an encoded apurinic or apyrimidinic site DNA repair endonuclease or a type IIS restriction endonuclease instead of a retroviral integrase and the presence or absence of a ribonuclease H (RNH) domain as part of the reverse transcriptase (RT) [10, 11].

Dictyostelids are soil-dwelling protists that belong to the supergroup of Amoebozoa [12, 13]. Unfavorable environmental conditions, such as a lack of food, triggers social behaviors in single cells that aggregate and form fruiting bodies to spread some of the population as dormant spores into the environment [14, 15]. Dictyostelium discoideum, the model organism in studying the biology of social amoebae, has a 34-Mb haploid genome in which two thirds of the chromosomal DNA code for proteins and intergenic regions are mostly below 1 kb in length [16]. The gene density of this genome limits the available space for transposable elements to expand without causing damage to the host. Therefore, it is remarkable that the genome of D. discoideum is interspersed with a variety of mobile elements that add up to nearly 10 % of nuclear DNA [17].

The D. discoideum DIRS-1 element has inverted terminal repeats instead of LTRs and a complex arrangement of open reading frames (ORFs) that include an RT/RNH and a tyrosine recombinase (YR) instead of a canonical integrase (IN) [18, 19] (Fig. 1). DIRS-1 has a strong preference to integrate into existing DIRS-1 copies by a mechanism that probably involves YR-mediated homologous recombination [20]. Therefore, DIRS-1 forms complex clusters located near chromosome ends and contributes ~50 % of centromeric DNA of D. discoideum chromosomes [21].

DGLT-A and Skipper are related Ty3/gypsy-type LTR retrotransposons with strikingly different integration preferences. Skipper contains two ORFs coding for enzymatic activities required for retrotransposition arranged in the order RT-RNH-IN [22] (Fig. 1). Skipper

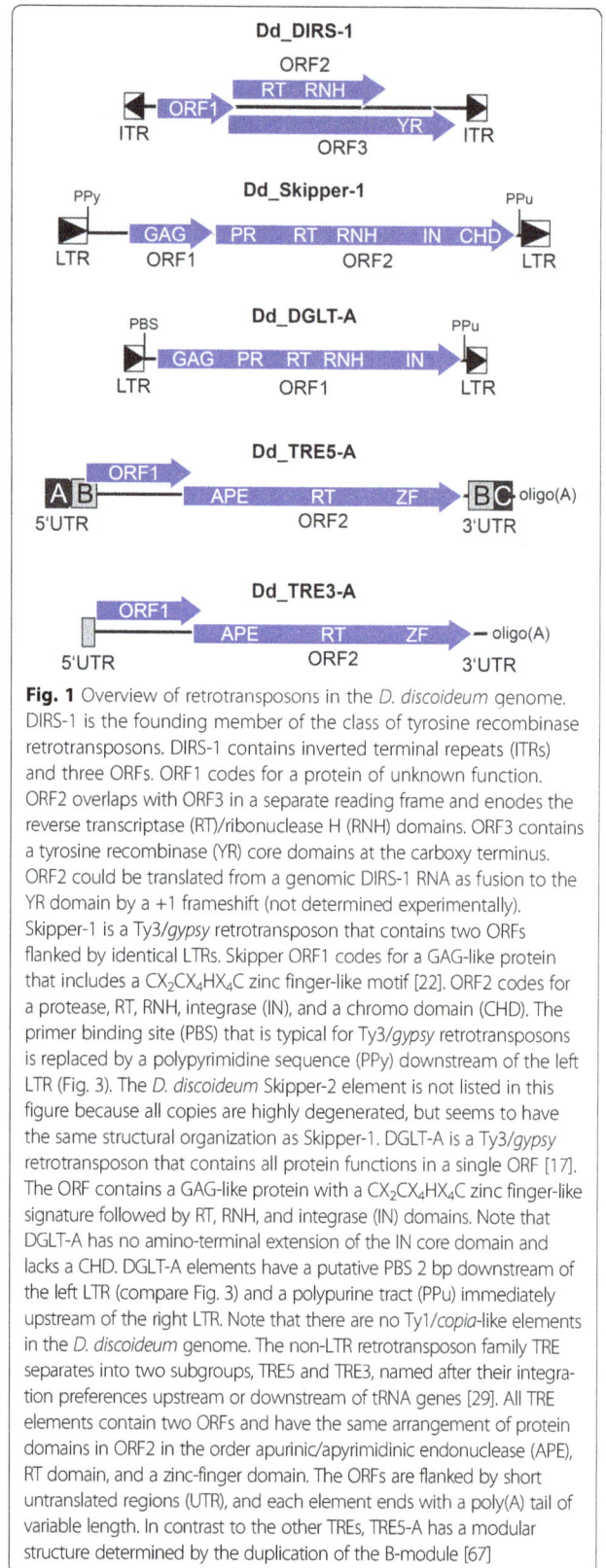

Fig. 1 Overview of retrotransposons in the *D. discoideum* genome. DIRS-1 is the founding member of the class of tyrosine recombinase retrotransposons. DIRS-1 contains inverted terminal repeats (ITRs) and three ORFs. ORF1 codes for a protein of unknown function. ORF2 overlaps with ORF3 in a separate reading frame and enodes the reverse transcriptase (RT)/ribonuclease H (RNH) domains. ORF3 contains a tyrosine recombinase (YR) core domains at the carboxy terminus. ORF2 could be translated from a genomic DIRS-1 RNA as fusion to the YR domain by a +1 frameshift (not determined experimentally). Skipper-1 is a Ty3/gypsy retrotransposon that contains two ORFs flanked by identical LTRs. Skipper ORF1 codes for a GAG-like protein that includes a CX₂CX₄C zinc finger-like motif [22]. ORF2 codes for a protease, RT, RNH, integrase (IN), and a chromo domain (CHD). The primer binding site (PBS) that is typical for Ty3/gypsy retrotransposons is replaced by a polypyrimidine sequence (PPy) downstream of the left LTR (Fig. 3). The *D. discoideum* Skipper-2 element is not listed in this figure because all copies are highly degenerated, but seems to have the same structural organization as Skipper-1. DGLT-A is a Ty3/gypsy retrotransposon that contains all protein functions in a single ORF [17]. The ORF contains a GAG-like protein with a CX₂CX₄HX₄C zinc finger-like signature followed by RT, RNH, and integrase (IN) domains. Note that DGLT-A has no amino-terminal extension of the IN core domain and lacks a CHD. DGLT-A elements have a putative PBS 2 bp downstream of the left LTR (compare Fig. 3) and a polypurine tract (PPu) immediately upstream of the right LTR. Note that there are no Ty1/copia-like elements in the *D. discoideum* genome. The non-LTR retrotransposon family TRE separates into two subgroups, TRE5 and TRE3, named after their integration preferences upstream or downstream of tRNA genes [29]. All TRE elements contain two ORFs and have the same arrangement of protein domains in ORF2 in the order apurinic/apyrimidinic endonuclease (APE), RT domain, and a zinc-finger domain. The ORFs are flanked by short untranslated regions (UTR), and each element ends with a poly(A) tail of variable length. In contrast to the other TREs, TRE5-A has a modular structure determined by the duplication of the B-module [67]

is the prototype chromovirus in the *D. discoideum* genome as it contains a chromo domain (CHD) in the carboxy-terminal extension of the IN protein. The CHD may be responsible for targeting the element to centromeric regions where it contributes to ~10 % of centromer length [21]. It is known that centromeric DNA in *D. discoideum* has properties of heterochromatin including the presence of H3K9 methylation [23]. Retrotransposon CHDs may bind to methylated H3K9 and mediate their accumulation in heterochromatin [24], but it has not yet been determined experimentally whether Skipper is tethered to centromers via binding of its CHD to H3K9 methylation marks.

D. discoideum DGLT-A contains a single ORF and lacks a carboxy-terminal extension of the IN including a CHD as found in Skipper (Fig. 1). DGLT-A is related to Skipper but shows a completely different genomic distribution [17]; it does not accumulate in centromeric DNA but displays a strong preference to integrate within a window of 13–33 bp upstream of the mature coding sequences of tRNA genes [17]. The average distance of DGLT-A to the first nucleotide of a tRNA gene is 15 bp. This is remarkably similar to the integration preference of the yeast Ty3 element, considering that Ty3 inserts 1–4 bp upstream of the transcription start sites of tRNA genes [25], which is ~12 bp upstream of the first nucleotide of mature tRNAs [26]. It is not known whether the molecular mechanism of tRNA gene recognition of DGLT-A resembles that of Ty3, which identifies integration sites by binding of the IN to tDNA-bound transcription factor TFIIIB [27, 28].

The "tRNA gene targeted retroelements" (TREs) form two subfamilies of non-LTR retrotransposons (Fig. 1) that can be distinguished by phylogenetic analysis of their ORF2 proteins [17] and their integration preferences near tRNA genes [29]. TRE5 elements are strictly associated with regions ~50 bp upstream of tRNA genes, whereas TRE3 elements are always found ~100 bp downstream of tRNA genes. All TREs contain two ORFs. ORF1 proteins of TREs have no similarity among each other or with proteins of non-LTR retrotransposons such as the mammalian L1, in which the ORF1 protein is involved in binding the retroelement's RNA as part of the pre-integration complex and contributes to the integration process [30, 31]. In *D. discoideum*, the ORF1 protein may be involved in the recognition of tRNA genes as integration sites by binding to subunits of RNA polymerase III transcription factor TFIIIB [32]. The TRE-encoded ORF2 proteins contain related apurinic/apyrimidinic endonuclease (APE) and RT domains (Fig. 1) that mediate retrotransposition.

It was of interest to trace the evolution of tRNA gene-associated mobile elements in social amoebae to understand how different tRNA gene-directed integration preferences emerged. In this study, we analyzed the annotated genomes of *D. discoideum*, *D. purpureum*, *D. lacteum*, *D. fasciculatum*, and *P. pallidum*, which represent the entire evolutionary history of social amoebae [16, 33, 34]. We found that the targeting of tRNA genes has independently developed at least six times through different mobile elements in the evolution of dictyostelids.

Results

Retrotransposons have excessively expanded in the *D. discoideum* genome

Hallmarks of the *D. discoideum* genome are the high gene density and the presence of retrotransposons that closely associate with tRNA genes, likely as a means to avoid insertional mutagenesis of host genes upon retrotransposition. This characteristic of the *D. discoideum* genome is similar to the yeast *Saccharomyces cerevisiae*, which has an even higher gene density than *D. discoideum* [35] and accommodates only retrotransposons that feature position-specific integration either near tRNA genes or in heterochromatin [36]. It has been of interest to compare integration preferences in yeast and dictyostelid genomes to evaluate whether tRNA gene-targeted integration presents an example of convergent evolution that enables mobile elements to settle in intergenic regions of compact genomes.

We evaluated retrotransposon families in the annotated genomes of *D. purpureum*, *D. lacteum*, *P. pallidum*, and *D. fasciculatum* in comparison with the model organism *D. discoideum*. The last common ancestor of all dictyostelids is estimated to date back approximately 600 million years and all examined species featured a long period of separate evolution [33] (Fig. 2), which must be considered when interpreting the relationships among transposable elements both within and outside the dictyostelids. We determined the retrotransposon contents of dictyostelid genomes by performing TBLASTX searches based on *D. discoideum* retrotransposon sequences of the tyrosine recombinase retrotransposon DIRS-1, the LTR retrotransposons Skipper and DGLT-A, and the non-LTR retrotransposons TRE5-A and TRE3-A (the structures of these elements are summarized in Fig. 1). The identified elements were reconstructed as consensus sequences. We also determined whether any of the identified retrotransposons may have a preference for integrating near tRNA genes by searching for tRNA genes within a distance of up to 3000 bp upstream and downstream of identified retroelements. A retrotransposon was considered to display active targeting to tRNA genes if several copies were found in a similar distance to tRNA genes. To ensure that we did not miss tRNA gene-targeting retrotransposons in this analysis, we performed a parallel search in which we first listed all tRNA genes of a given genome and then

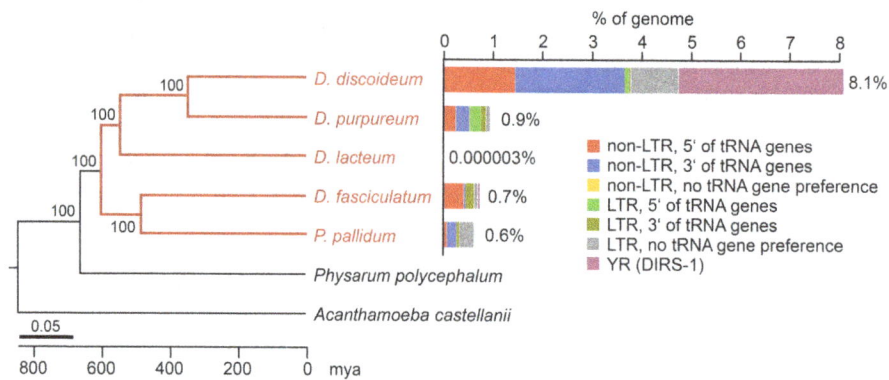

Fig. 2 Phylogenetic relationships between dictyostelids. A genome-based phylogenetic tree was constructed on concatenated sequences of 32 orthologous proteins (redrawn from [13]). The retrotransposon content in each dictyostelid genome is plotted separated by the class of retrotransposon and integration preference near tRNA genes. YR: tyrosine recombinase retrotransposon (DIRS-1)

inspected 3000 bp upstream and downstream sequences for the presence of repetitive elements.

With the exception of *D. lacteum*, which has a particularly small and compact genome, all analyzed dictyostelids have comparable genome sizes of ~30 Mb and gene densities of close to 400 genes/Mb of genomic DNA (Additional file 1: Table S1). A notable difference between the genome of *D. discoideum* and any other examined species is the total retrotransposon content (Fig. 2, Additional file 1: Table S1). Whereas retrotransposons have expanded to 8 % of the *D. discoideum* genome, they have been kept below 1 % in other species.

DIRS-1 has strongly amplified in *D. discoideum* and constitutes 3.3 % of the genome in this organism [17]. The expansion of Skipper to 1.0 % of the *D. discoideum* genome may be linked to the amplification of DIRS-1, because both elements reside in centromeric DNA and may have adopted centromer function in this species [21]. Centromeric accumulation of DIRS-1 or Skipper is not observed in any other dictyostelid species except *D. fasciculatum*, which may form small centromeric DIRS clusters that contribute to only 0.1 % of genome size [33]. DIRS-1 is even missing in the assembled sequences of *P. pallidum* and *D. purpureum*. The data suggest that a putative centromere function of DIRS-1 (and Skipper) as observed in *D. discoideum* is deeply rooted in the social amoebae, even though the majority of species may have evolved deviant strategies to organize their centromeres without allowing the accumulation of selfish mobile elements in these regions.

A notable trend to increase the number of tRNA genes is observed in *D. discoideum* and *D. purpureum* relative to other dictyostelids (Additional file 1: Table S1). This observation is of interest considering that it may be easier for tRNA gene-targeting retrotransposons to expand if more potential safe integration sites are available. Whereas the tRNA gene-targeting DGLT-A-like retrotransposons

are present in low copy numbers in all dictyostelds, a particularly strong amplification in *D. discoideum* relative to other species is observed in the TRE family (Fig. 2, Additional file 1: Table S1). Such expansion is not observed in the genome of *D. purpureum*, which has a comparable amount of tRNA genes. Thus, targeting preference near tRNA genes does not necessarily favor the amplification of position-specific integrating elements to high copy numbers under the repressive conditions that prevail in most host cells.

Dictyostelid LTR retrotransposons comprise related families with different tRNA gene-targeting strategies

As previously noted by Malik et al. [7], IN domains of Ty3/*gypsy*-type retrotransposons frequently contain carboxy-terminal extensions including a distinctive GPY/F motif at the end of the IN core followed by relatively unconserved domains of various sizes that may harbor a chromo domain (CHD). *D. discoideum* DGLT-A has a small IN extension of 32 amino acids, whereas Skipper has a long IN extension of 183 amino acids that contains a CHD. In the analysis of dictyostelid genomes described below, we found that all new identified LTR retrotransposons have the Ty3/*gypsy*-type structure including a conserved GPY/F motif (Additional file 1: Figure S1). For convenience, we call retrotransposons "Skipper" if they contain a CHD in the carboxy-terminal extension of the IN domain and "DGLT-A" if a CHD is lacking.

Twenty insertions of DGLT-A are detectable in the *D. discoideum* genome, eleven of which are solo LTRs that were formerly described as "H3R" elements located upstream of tRNA genes [37]. None of the remaining nine DGLT-A sequences are full-length and refer to the derived consensus of this element (Table 1). This suggests that the DGLT-A population may no longer be able to amplify in the *D. discoideum* genome, even though all

Table 1 Overview of dictyostelid retrotransposon properties and integration preferences

| Name | Consensus length (bp) | LTR length (bp) | Copy number in genome [a] | | | tRNA gene-specific | Distance to tRNA gene (bp) | |
			total	full length [b]	solo LTR		5' of tDNA (%) [c]	3' of tDNA (%) [c]
LTR retrotransposons								
Dd-Skipper-1	6998	390	60	2	10	no	–	–
Dp_Skipper-1	7485	388	12	1	6	no	–	–
Dl_Skipper-1	4763	251	7	2	2	no	–	–
Pp_Skipper-1	5589	226	14	1	10	yes	–	–
Df_Skipper-1	5120 [d]	n.d. [e]	5	0	n.d. [e]	no	–	–
Pf_Skipper-1.1	5296	259	6	1	0	no	–	–
Pf_Skipper-1.2	6983	382	3	0	1	no	–	–
Pf_Skipper-1.3	7081	363	7	1	6	no	–	–
Dd-Skipper-2 [f]	6178	208	8	0	0	no	–	8–23 (4)
Dp_Skipper-2	5676	315	23	3	5	yes	–	7–133 (5)
Pp_Skipper-2	3675 [d]	n.d. [e]	9	0	n.d. [e]	yes	–	54–136 (9)
Df_Skipper-2	5708	312	12	7	5	yes	–	26–97 (11)
Dd_DGLT-A	5054	265	20	0	5	yes	13–33 (18)	–
Dp_DGLT-A.1	5436	492	15	1	9	yes	13–16 (6)	–
Dp_DGLT-A.2	6114	389	9	2	5	yes	15 (1)	–
Dp_DGLT-A.3	5589	563	8	1	6	yes	10–11 (2)	–
Dp_DGLT-A.4 [g]	3447 [d]	206	30	0	20	yes	16–34 (4)	–
Dp_DGLT-A.5 [g]	3440 [d]	354	32	0	32	yes	10–19 (15)	–
Dl_DGLT-A.1	4895	163	10	1	4	yes	63–64 (4)	–
Dl_DGLT-A.2	5112	206	7	1	2	yes	55–65 (2)	–
Pp_DGLT-A.1	7295	601	23	1	13	no	–	–
Pp_DGLT-A.2	6160	393	12	2	6	no	–	–
Pp_DGLT-A.3	5942	212	11	2	3	no	–	–
Pp_DGLT-A.4 [g]	3650 [d]	n.d. [e]	10	0	n.d. [e]	yes	14–24 (3)	–
Pf_DGLT-A	8367	168	2	1	1	no	–	–
Non-LTR retrotransposons								
Dd_TRE3-A	5229	–	67	13	–	yes	–	14–228 (60)
Dd_TRE3-B	5279	–	43	9	–	yes	–	34–188 (39)
Dd_TRE3-C	4734	–	29	2	–	yes	–	14–305 (29)
Dd_TRE3-D	1559 [d]	–	11	0	–	yes	–	49–285 (11)
Dp_TRE3-A	5150	–	56	2	–	yes	–	69–161 (18)
Dp_TRE3-B	5210	–	9	2	–	yes	–	98–154 (2)
Dp_TRE3-C	1620 [d]	–	37	0	–	yes	–	67–450 (15)
Dl_TRE3-A	4386	–	17	2	–	yes	–	23/87 (7)
Pp_TRE3-A	4515	–	35	4	–	yes	–	26–138 (10)
Pp_TRE3-B	4741	–	38	2	–	yes	–	57–151 (11)
Df_TRE3-A	1867 [d]	–	14	0	–	yes	–	29–404 (11)
Dd_TRE5-A	5647	–	102	5	–	yes	37–90 (98)	–
Dd_TRE5-B	5971	–	25	1	–	yes	34–82 (25)	–
Dd_TRE5-C	879 [d]	–	18	0	–	yes	38–95 (18)	–
Dl_TRE5-A [h]	7405	–	30	1	–	no	–	–

Table 1 Overview of dictyostelid retrotransposon properties and integration preferences *(Continued)*

Pp_TRE5-A	1169[d]	–	21	0	–	yes	38–74 (12)	–
Df_TRE5-A.1	2587[d]	–	56	1	–	yes	45–88 (20)	–
Df_TRE5-A.2	1275[d]	–	20	0	–	yes	31–98 (18)	–
Df_TRE5-A.3	2941[d]	–	7	0	–	yes	39–67 (6)	–
Df_TRE5-B	1534[d]	–	27	0	–	yes	44–90 (9)	–
Dp_NLTR-A	7438	–	28	1	–	yes	2–6 (16)	–
Pp_NLTR-B	5550[d]	–	3	0	–	yes	39–64 (3)	–
Pp_NLTR-C	3536[d]	–	12	1	–	no	–	–

[a]Total copy numbers refer to both full-length and partial sequences

[b]Full-length copies with intact open reading frames

[c]Distances are listed only for retrotransposons found in the direct neighborhood of tRNA genes; in cases where other tRNA gene-specific retrotransposons have integrated at the same tRNA gene and therefore upstream of a previously inserted element, distances of the original insertion to the target could not be determined. The number of elements used for determination of target distances are shown in parentheses

[d]No full-length consensus available

[e]No LTR sequences detectable

[f]Previous name DGLT-B (GenBank AF474004) [17]

[g]No ORFs for phylogenetic analysis; classification as DGLT-A according to integration preference

[h]Classified as TRE5 by similarity of RT sequence (compare Fig. 5)

ORF domains are transcribed in growing *D. discoideum* cells (T.W., unpublished observation).

The *D. purpureum* genome contains three related DGLT-A elements, of which each retained at least one retrotransposition-competent copy. *D. purpureum* DGLT-As have the same structure and display the same target preference 13–16 bp upstream of tRNA genes as the prototype DGLT-A of *D. discoideum* (Table 1). Two related full-length DGLT-A elements were detected in the *D. lacteum* genome. These elements also display integration preference upstream of tRNA genes (Table 1). The *P. pallidum* genome contains four related DGLT-A elements. Of these, Pp_DGLT-A.1, Pp_DGLT-A.2, and Pp_DGLT-A.3 comprise a population of elements with intact open reading frames and probable retrotransposition competence. Unlike other DGLT-As, Pp_DGLT-As contain long carboxy-terminal IN extensions of 264–333 amino acids but no detectable CHDs. The IN extensions in *P. pallidum* DGLT-A elements are poorly conserved among each other and do not show similarity with other retrotransposons such as dictyostelid Skipper or yeast Ty1 and Ty3 elements. Notably, Pp_DGLT-A.1, Pp_DGLT-A.2, and Pp_DGLT-A.3 do not show a preference to integrate near tRNA genes. However, we detected a partial sequence of a fourth DGLT-A in the *P. pallidum* genome (Pp_DGLT-A.4) that is related to the other *P. pallidum* DGLT-As by phylogenetic analysis of the intact RT and RNH domains (data not shown) and its preference to integrate 14–25 bp upstream of tRNA genes (Table 1). This suggests that the tRNA gene preference of DGLA-A has also been established in the *P. pallidum* genome but was lost in some DGLT-A lineages. The conclusion from this observation is that tRNA gene targeting by DGLT-As was established in the earliest diverged species of Dictyostelia.

The Skipper-1 element of *D. discoideum* is 34 % identical with DGLT-A in the RT-RNH-IN core domains but does not display integration specificity at tRNA genes. Instead, the approximately 60 Skipper copies are highly enriched in centromeric transposon clusters [21]. Two Skipper copies can be identified in the *D. discoideum* genome that have intact open reading frames and may be retrotransposition-competent.

The *D. purpureum* genome contains two related Skipper elements. Dp_Skipper-1 is highly similar to Dd_Skipper-1 and does not show association with tRNA genes. In contrast, Dp_Skipper-2, of which three intact copies exist in the *D. purpureum* genome, is found within a range of 7–133 bp downstream of tRNA genes (Table 1). This integration preference of an LTR retrotransposon had not been observed before. However, in the course of this study, we re-evaluated the previously described DGLT-P element of *D. discoideum* [17] and detected a CHD in the highly degenerated ORF of this element and surprisingly noticed that 4 of 8 copies of this element are located in a range of 8–23 bp downstream of tRNA genes. We therefore renamed DGLT-P "Dd_Skipper-2". Interestingly, a Skipper-like element with target preference downstream of tRNA genes was also detected in the *D. fasciculatum* genome. The Df_Skipper-2 element was found inserted 26–97 bp downstream of tRNA genes, whereas a related Df_Skipper-1 element does not display target specificity (Table 1). The *P. pallidum* genome also contains two related Skipper-like elements, of which the Skipper-2 is found within a window of 54–136 bp downstream of tRNA genes. The *D. lacteum* genome contains one intact copy of a Skipper element (Dl_Skipper-1) that is not associated with a tRNA gene. In summary, it seems that Skipper elements diverged into two subfamilies, of which one (Skipper-2) developed a previously unnoticed preference to integrate downstream of tRNA genes. This is interesting because integration preference for the same region was also invented by the unrelated non-LTR retrotransposons of the TRE3 family described later.

Phylogenetic analyses based on alignments of the concatenated RT-RNH-IN core domains of all LTR retrotransposons (Additional file 1: Figure S2) support the division of these elements into DGLT-A and Skipper families but also reveal interesting differences in the evolution of these elements (Fig. 3, Additional file 1: Figure S3). For example, DGLT-A elements from *D. discoideum*, *D. purpureum*, and *D. lacteum* form a robust group of elements that share an integration preference upstream of tRNA genes. However, DGLT-A.1, DGLT-A.2, and DGLT-A.3 of *P. pallidum* clustered with Skipper elements, which was unexpected because *P. pallidum* DGLT-A.4 (not included in the phylogenetic analysis shown in Fig. 3) showed the DGLT-A-typical integration preference upstream of tRNA genes. On the other hand, the *P. pallidum* DGLT-As that clustered among Skipper elements have long IN extensions reminiscent of Skipper elements, but they lack a detectable CHD.

The phylogenetic analysis presented in Fig. 3 implies a further separation of Skipper elements into two subfamilies: Skipper-1 without target preference and Skipper-2 that integrate downstream of tRNA genes. Notably, all Skipper elements contain carboxy-terminal extensions of the IN core ranging from 99 to 192 amino acid that include distinctive CHDs. The CHDs of Skipper elements are compared in Fig. 4 with the CHD and chromo shadow domain (CSD) of *D. discoideum* heterochromatin protein 1 (HP1), which is known to bind to heterochromatin via its CHD interacting with methylated lysine-9 of histone H3 (H3K9) while its CSD comprises a dimerization domain [38]. Each Skipper-1 retrotransposon contains a canonical HP1-like CHD that has three conserved aromatic amino acids known to build a "cage" responsible for the binding to methylated H3K9 [39] (Fig. 4). Whether CHDs of Skipper-1 elements indeed bind to methylated histone H3 lysine 9 marks and tether the elements to centromeric regions has not yet been experimentally tested. Gao et al. [24] analyzed CHDs of various LTR retrotransposons and concluded that they can be grouped into "canonical" CHDs (group I CHDs) and derivatives that lack the first and usually also the third of the aromatic cage residues (group II CHDs).

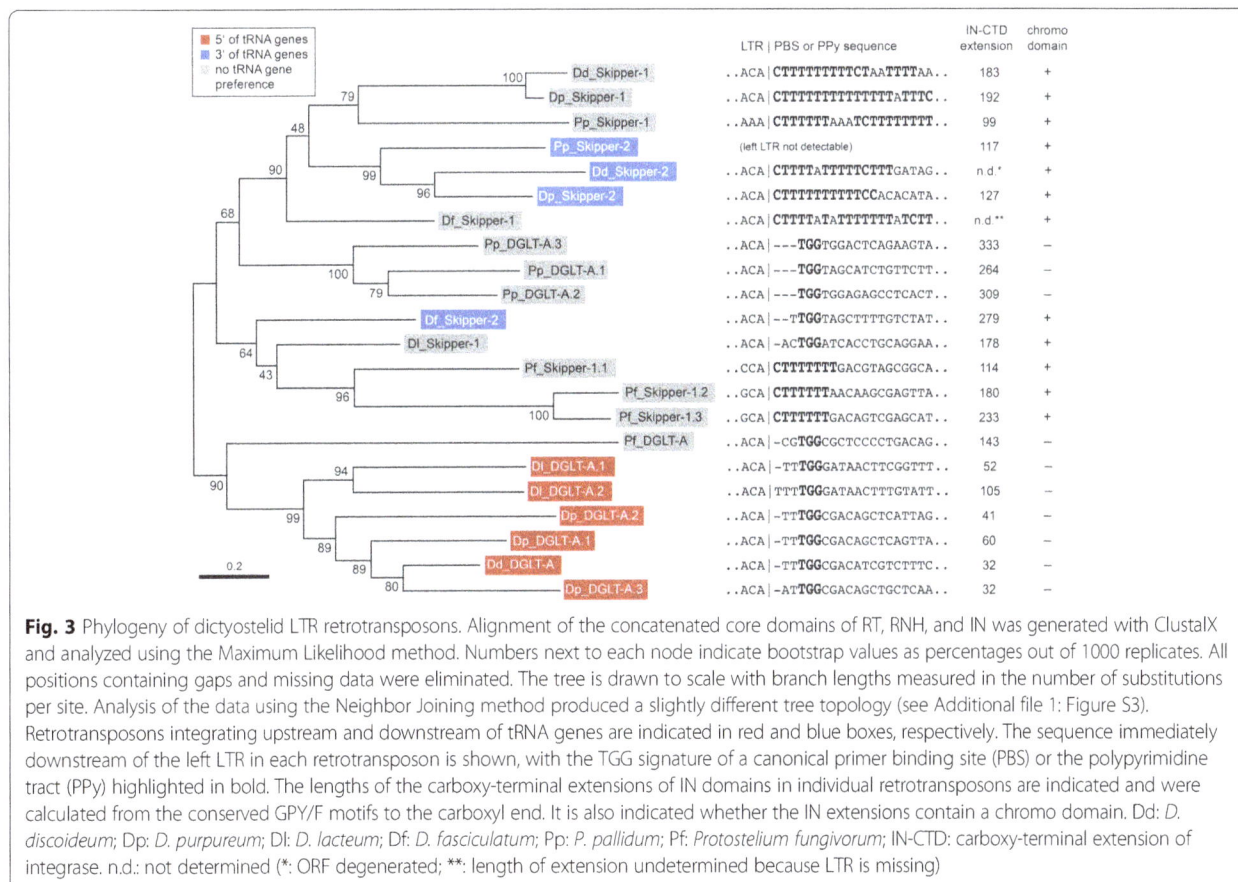

Fig. 3 Phylogeny of dictyostelid LTR retrotransposons. Alignment of the concatenated core domains of RT, RNH, and IN was generated with ClustalX and analyzed using the Maximum Likelihood method. Numbers next to each node indicate bootstrap values as percentages out of 1000 replicates. All positions containing gaps and missing data were eliminated. The tree is drawn to scale with branch lengths measured in the number of substitutions per site. Analysis of the data using the Neighbor Joining method produced a slightly different tree topology (see Additional file 1: Figure S3). Retrotransposons integrating upstream and downstream of tRNA genes are indicated in red and blue boxes, respectively. The sequence immediately downstream of the left LTR in each retrotransposon is shown, with the TGG signature of a canonical primer binding site (PBS) or the polypyrimidine tract (PPy) highlighted in bold. The lengths of the carboxy-terminal extensions of IN domains in individual retrotransposons are indicated and were calculated from the conserved GPY/F motifs to the carboxyl end. It is also indicated whether the IN extensions contain a chromo domain. Dd: *D. discoideum*; Dp: *D. purpureum*; Dl: *D. lacteum*; Df: *D. fasciculatum*; Pp: *P. pallidum*; Pf: *Protostelium fungivorum*; IN-CTD: carboxy-terminal extension of integrase. n.d.: not determined (*: ORF degenerated; **: length of extension undetermined because LTR is missing)

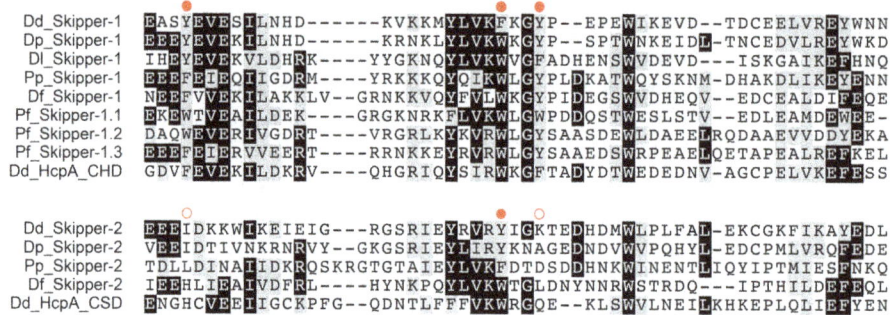

Fig. 4 Alignments of chromo domains in dictyostelid and protostelid Skipper retrotransposons. Alignments were generated with ClustalX. Shading is to a 50 % consensus and was generated with BoxShade. Black boxes indicate invariant amino acids, and gray boxes represent similar amino acids. The corresponding sequences of the chromo domain (CHD) and the "shadow" chromo domain (CSD) of *D. discoideum* heterochromatin protein 1 (HcpA) [38] are shown for comparison. Red dots depict aromatic cage residues present in canonical chromo domains [39]. The alignment is separated into retrotransposons containing canonical (group I) chromo domains (Skipper-1 elements) and group II chromo domains (Skipper-2 elements) in which the first and third aromatic amino acid of the cage are diverged (indicated by open circles). Dd: *Dictyostelium discoideum*; Dp: *D. purpureum*; Dl: *D. lacteum*; Pp: *Polysphondylium pallidum*; Df: *D. fasciculatum*; Pf: *Protostelium fungivorum*

Interestingly, all Skipper-2 elements have diverged exactly the same aromatic cage residues in their CHDs, which in fact resembles the HP1 CSD (Fig. 4). This suggests that CHDs of Skipper-2 elements may be in the process of functional degeneration or, more intriguing, have been modified to shift the integration behavior of these elements to new locations outside of heterochromatin. In this regard, it is of note that Skipper-2 elements apparently evolved a new integration preference downstream of tRNA genes in intergenic regions as described above.

Many Skipper elements have lost the canonical primer binding site to initiate reverse transcription

A primer binding site (PBS) located immediately downstream of the U5 sequence in the left LTR is required to initiate minus-strand strong-stop cDNA synthesis in most Ty3/*gypsy* retrotransposons [40, 41]. The PBS usually presents a TGG trinucleotide signature as a complement of the CCA 3' end of a host tRNA that is used as primer to initiate reverse transcription. In *D. discoideum* DGLT-A, the sequence TGGCGACATCGTCTTTC is located 2 bp downstream of the left LTR (Fig. 3), but no tRNA or any other genomic sequence complementary to the PBS could be identified in the *D. discoideum* genome as a potential primer for reverse transcription of DGLT-A.

In contrast to DGLT-A, most elements classified as Skipper according to the presence of a CHD have apparently replaced the canonical PBS with degenerate polypyrimidine (PPy) sequences (Fig. 3) that suggest a noncanonical mechanism of reverse transcription priming. Interesting exceptions are found in Skipper-like elements from *D. lacteum* and *D. fasciculatum*: Dl_Skipper-1 has a CHD indicative of Skipper, but contains a PBS typical for DGLT-A. Likewise, Df_Skipper-2 contains a DGLT-A-type PBS and a group II CHD. At least seven intact

copies Df_Skipper-2 suggest that the element is retrotransposition-competent; all copies are found within a window of 26–97 bp downstream of tRNA genes (Table 1).

The Skipper and DGLT-A families originated before the evolution of dictyostelds

The long independent evolutionary history of Amoebozoa makes it difficult to trace the origin of DGLT-A- and Skipper-like retrotransposons and the invention of their tRNA gene targeting mechanisms outside the Dictyostelia. The recently obtained genome sequence of a *Protostelium* species (F.H., T.W., G.G., manuscript in preparation) is helpful, because even though Protostelia are polyphyletic [42], they are considered closer related to the monophyletic Dictyostelia than other amoebozoan species sequenced so far such as *Acanthamoeba castellanii* or *Physarum polycephalum*. The genome of the sequenced protostelid, *P. fungivorum*, contains one DGLT-A-like and three Skipper-like elements (Table 1). The Skipper-like elements contain the typical PPy signature downstream of the left LTR (Fig. 3) and a canonical CHD downstream of IN (Fig. 4), supporting the hypothesis that the Skipper-type LTR retrotransposons arose outside the Dictyostelia. Although the gene density of the *P. fungivorum* genome is comparable with the dictyostelids, none of the *P. fungivorum* DGLT-A- or Skipper-like elements has developed integration preferences for tRNA genes. Because the absence of targeting preferences of LTR retrotransposons in this particular *Protostelium* isolate is not an argument for the *de novo* invention of such a specificity in dictyostelids, the origin of tRNA gene targeting in dictyostelid genomes remains a mystery until more amoebozoan genomes have been sequenced.

Dictyostelid non-LTR retrotransposons evolved four different tRNA gene-targeting strategies

In the *D. discoideum* genome, TRE elements can be distinguished between the TRE5 and TRE3 subfamilies according to their exclusive integration behavior [17]. TRE elements comprise 3.7 % of the *D. discoideum* genome, with TRE5-A and TRE3-A contributing the majority of individual copies (Table 1). In *D. discoideum*, 61 % of tRNA genes are associated with at least one TRE element (Additional file 1: Table S2), and 13 % of tRNA genes have been targeted by both TRE3 and TRE5.

We considered newly discovered non-LTR retrotransposons in dictyostelid genomes as TRE5-like and TRE3-like if they were found upstream and downstream of tRNA genes, respectively, at similar distances as in the *D. discoideum* genome. We examined the evolution of TRE5- and TRE3-like elements using the complete ORF2 sequences of *D. discoideum* TREs as query sequences in TBLASTX searches. We identified TRE5- and TRE3-like sequences in *D. lacteum*, *D. fasciculatum* and *P. pallidum*, whereas *D. purpureum* contains only TRE3-like sequences (Table 1). Alignments of the conserved RT domains (Additional file 1: Figure S4) and phylogenetic analyses (Fig. 5) support the evolution of TRE5 and TRE3 in separate subfamilies with the exception of Dd_TRE3-C, which appeared to be more related to TRE5 elements than to TRE3 elements in these analyses. This grouping of Dd_TRE3-C is likely caused by the relatively short RT amino acid sequences used in this analysis because this element clusters robustly with the

other TRE3 elements when examining the complete ORF2 sequences [17]. Phylogenetic analyses on the entire ORF2 proteins across species was not feasible in this study because complete elements could not be reconstructed in all genomes. TRE-like retrotransposons were found to be associated with tRNA genes at locations typical for *D. discoideum* TRE5 and TRE3 elements (Table 1), suggesting that this type of integration behavior is deeply rooted within the dictyostelids. TRE-like elements have not been identified in the genomes of distantly related amoebozoans such as *Physarum polycephalum* and *Acanthamoeba castellanii* and are also absent in the recently sequenced isolate of *Protostelium fungivorum*. Therefore, the origin of the last common ancestor of the TREs (including the evolution of their unique integration preferences) remains to be determined.

We detected new non-LTR retrotransposons in the genomes of *D. purpureum* and *P. pallidum* that we tentatively named "non-LTR" (NLTR) elements because they are only distantly related to TRE elements based on phylogenetic analysis of RT domains (Fig. 5, Additional file 1: Figure S4). *D. purpureum* NLTR-A and *P. pallidum* NLTR-B are 38 % identical to each other in their RT domains and are characterized by an RNH domain located downstream of the RT (Fig. 6). Intriguingly, Dp_NLTR-A and Pp_NLTR-B developed different target preferences upstream of tRNA genes (Table 1). Dp_NLTR-A was found 2–6 bp upstream of the first nucleotide of the mature coding sequence of the targeted tRNA gene, which represents an as-yet unobserved integration specificity, whereas Pp_NLTR-B was found at similar positions as TRE5 elements ~50 bp upstream of tRNA genes. *P. pallidum* NLTR-C was identified as a partial sequence that contains an RT domain. This element is only distantly related to Dp_NLTR-A and Pp_NLTR-B (~26 % sequence identity in the RT domain) and does not show association with tRNA genes. Phylogenetic analysis based on RT domains considering all major subgroups of non-LTR retrotransposons [11] failed to place the Dp_NLTR-A and Pp_NLTR-B elements in any of the subfamilies of non-LTR retrotransposons that are known to harbor an RNH domain (Additional file 1: Figure S5). A phylogenetic evaluation of RNH domains of non-LTR retrotransposons based on alignments previously proposed by Malik et al. [11] confirmed that Dp_NLTR-A and Pp_NLTR-B may form a separate group within the supergroup of non-LTR retrotransposons (Fig. 6; Additional file 1: Figure S6). The Pp_NLTR-C RT sequence aligned best with subgroup R4 elements; however, this grouping could not be evaluated further because no restriction enzyme-like endonuclease domain, which is typically located downstream of RTs in R4-like elements [11], was included in the partial Pp_NLTR-C sequence.

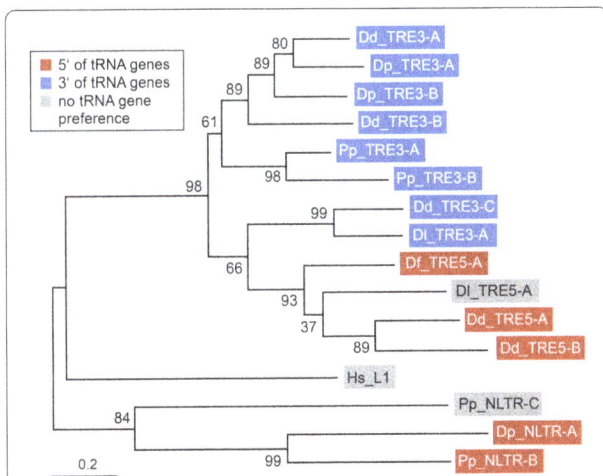

Fig. 5 Phylogenetic analysis of dictyostelid non-LTR retrotransposons. Alignment of RT domains was generated with ClustalX and analyzed using the Maximum Likelihood method. All positions containing gaps and missing data were eliminated. The tree is drawn to scale with branch lengths in the same units as those of the evolutionary distances used to infer the phylogenetic tree. There were a total of 227 amino acid positions in the final dataset. Bootstrap support (percentage from 1000 trials) is indicated next to each node

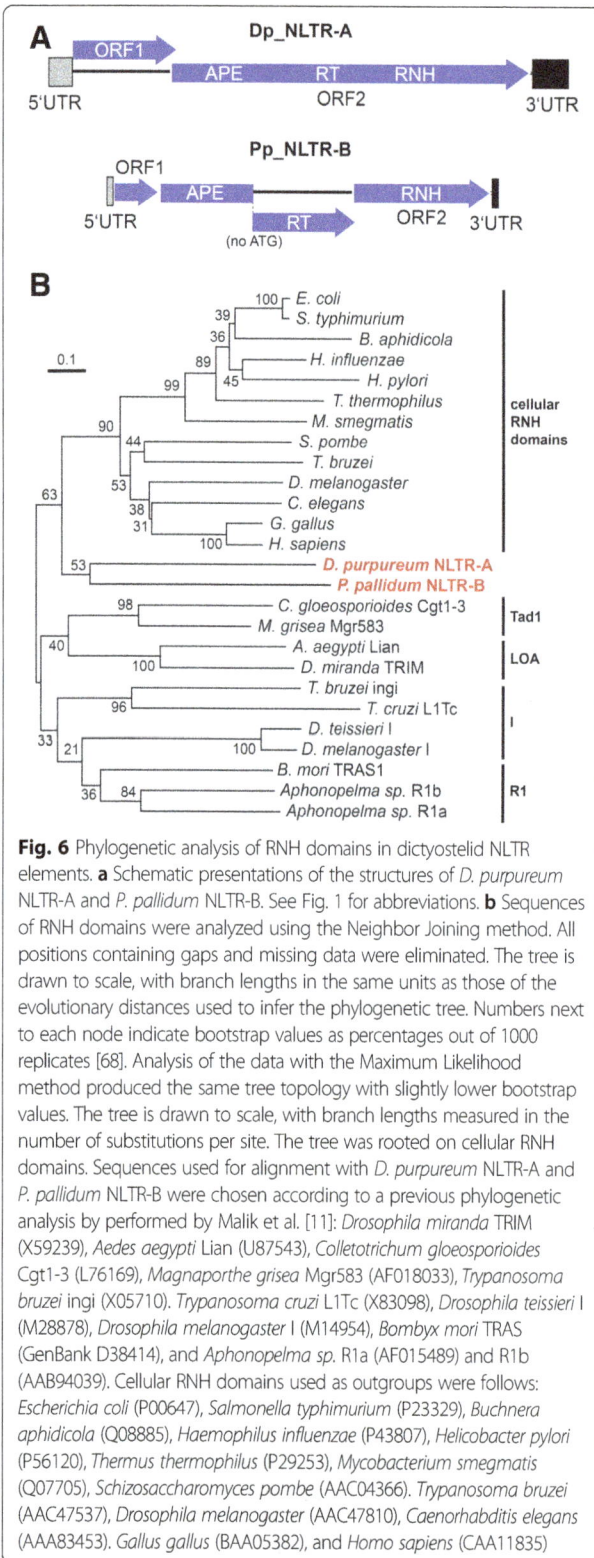

Fig. 6 Phylogenetic analysis of RNH domains in dictyostelid NLTR elements. **a** Schematic presentations of the structures of *D. purpureum* NLTR-A and *P. pallidum* NLTR-B. See Fig. 1 for abbreviations. **b** Sequences of RNH domains were analyzed using the Neighbor Joining method. All positions containing gaps and missing data were eliminated. The tree is drawn to scale, with branch lengths in the same units as those of the evolutionary distances used to infer the phylogenetic tree. Numbers next to each node indicate bootstrap values as percentages out of 1000 replicates [68]. Analysis of the data with the Maximum Likelihood method produced the same tree topology with slightly lower bootstrap values. The tree is drawn to scale, with branch lengths measured in the number of substitutions per site. The tree was rooted on cellular RNH domains. Sequences used for alignment with *D. purpureum* NLTR-A and *P. pallidum* NLTR-B were chosen according to a previous phylogenetic analysis by performed by Malik et al. [11]: *Drosophila miranda* TRIM (X59239), *Aedes aegypti* Lian (U87543), *Colletotrichum gloeosporioides* Cgt1-3 (L76169), *Magnaporthe grisea* Mgr583 (AF018033), *Trypanosoma bruzei* ingi (X05710). *Trypanosoma cruzi* L1Tc (X83098), *Drosophila teissieri* I (M28878), *Drosophila melanogaster* I (M14954), *Bombyx mori* TRAS (GenBank D38414), and *Aphonopelma sp.* R1a (AF015489) and R1b (AAB94039). Cellular RNH domains used as outgroups were follows: *Escherichia coli* (P00647), *Salmonella typhimurium* (P23329), *Buchnera aphidicola* (Q08885), *Haemophilus influenzae* (P43807), *Helicobacter pylori* (P56120), *Thermus thermophilus* (P29253), *Mycobacterium smegmatis* (Q07705), *Schizosaccharomyces pombe* (AAC04366). *Trypanosoma bruzei* (AAC47537), *Drosophila melanogaster* (AAC47810), *Caenorhabditis elegans* (AAA83453). *Gallus gallus* (BAA05382), and *Homo sapiens* (CAA11835)

Discussion

Convergent evolution of integration site selection in compact genomes

Integration behaviors of retrotransposons residing in compact genomes of different organisms show parallels that suggest strong convergent pressures to avoid insertional mutagenesis of genes and to preserve genome stability of the host. The haploid state of dictyostelid genomes may further increase the selection pressure on mobile elements because the disruption of an essential host gene in the absence of a second compensatory allele would ultimately eliminate the parasite along with its host. In dictyostelids, two principally different strategies have emerged to counter this selection pressure: (i) integration in gene-poor regions of centromeric DNA, which restricts mobile elements to certain spots of repetitive DNA in the host genome and (ii) the targeting of tRNA genes, which not only appears to represent the prime "safe sites" to integrate in gene-rich regions but also enables mobile elements to settle anywhere in the genome due to the multicopy nature of their targets and dispersal on all chromosomes.

In *S. cerevisiae*, the Ty1/*copia*-type retrotransposon Ty5 is tethered to regions of silent chromatin via direct protein interactions of Ty5 IN with heterochromatin-associated protein Sir4 [43]. There are no Ty1/*copia*-type retrotransposons found in dictyostelid genomes, but Skipper and DIRS-1 elements accumulate in centromer regions that are organized as heterochromatin. The heterochromatin-targeting mechanisms developed by Skipper and DIRS are different from each other and from Ty5. As we discuss in more detail below, Skipper elements are likely tethered to centromeres via interactions between their chromo domains and histone methylation marks that are characteristic for heterochromatin. The DIRS-1 element is special because it encodes a tyrosine recombinase (YR) instead of a canonical IN and is thought to generate circular retrotransposition intermediates that are probably targeted to centromers via YR-mediated homologous recombination into pre-existing DIRS-1 copies [18, 20].

The targeting of tRNA genes as presumed safe integration sites has been independently developed at least six times by retrotransposons during dictyostelid evolution (summarized in Fig. 7) and at least twice in the yeast *S. cerevisiae*. Ty1 and Ty3 elements, which belong to different classes of LTR retrotransposons, obviously evolved different mechanisms for tRNA gene recognition. Ty1 integrates within a window of ~750 bp upstream of tRNA genes that is defined by nucleosome positioning [44, 45] and direct interactions between Ty1 IN and RNA polymerase III subunits [46, 47]. A Ty1-like integration behavior of retrotransposons has not been observed in dictyostelid genomes. In contrast, there is a

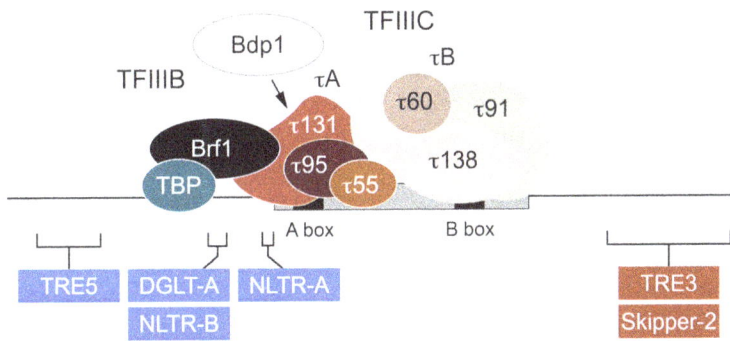

Fig. 7 Summary of integration sites near tRNA genes in dictyostelid genomes. The topology of RNA polymerase III transcription factors TFIIIC and TFIIIB on a tRNA gene is shown as deduced by Male et. al. [69]. The composition of TFIIIB in three subunits is inferred by the presence of orthologs of TBP, Brf1 and Bdp1 in all dictyostelid species. TFIIIC is a six-subunit factor that consists of two subcomplexes, τA and τB [69]. Note that TFIIIB subunit Bdp1 may enter the transcription complex only transiently by displacing TFIIIC subcomplex τB [69]. In dictyostelid genomes only the most conserved TFIIIC subunits τ131 (TFC4) and τ95 (TFC1) can be identified by homology to either yeast or human orthologs. The schematic is not drawn to scale. The tRNA gene, including its internal regulatory sequences (A box and B box), is indicated as a gray bar. Integration windows in tRNA gene-flanking regions of six different dictyostelid retrotransposon families are indicated. Note that DGLT-A and NLTR-B belong to different retrotransposon classes and therefore independently developed a similar integration behavior upstream of tRNA genes. The same is true for TRE3 and Skipper-2 elements, which target similar regions downstream of tRNA genes

striking similarity of integration site selection between Ty3 and dictyostelid DGLT-A elements. Ty3 targets the entire RNA polymerase III transcriptome of *S. cerevisiae* [48], particularly in regions 1–4 bp upstream of the transcription start sites of tRNA genes (that is, ~15 bp upstream of the first nucleotide of the mature tRNA) [25]. This target preference is mediated by an interaction between Ty3 IN and subunits of RNA polymerase III transcription factor TFIIIB [27]. In most dictyostelids evaluated in this study, DGLT-A elements have conserved an integration preference approximately 15 bp upstream of tRNA genes. It would be interesting to determine whether DGLT-A elements use the same molecular interactions to recognize RNA polymerase III-transcribed genes as Ty3 or whether selection pressure to avoid gene mutagenesis has generated other solutions to the problem of targeting tRNA gene-upstream regions in different lineages of retrotransposon evolution.

The targeting of tRNA genes by TRE elements is unique to and deeply rooted in the dictyostelids. Although TRE5 and TRE3 elements evolved from a common ancestor [17] that most likely dates back before dictyostelid evolution, these elements developed strikingly different integration preferences and thus use different molecular mechanisms for target recognition. The integration window preferred by TRE3-A elements strikingly overlaps with the integration profile displayed by the unrelated Skipper-2 elements, suggesting that a region ~100 bp downstream of tRNA genes is accessible for retrotransposons to develop harmless integration strategies in compact genomes. The targeting mechanisms of TRE elements have been investigated experimentally in some detail only in the TRE5-A element,

which requires intact B boxes in targeted tRNA genes and probably DNA-bound RNA polymerase III transcription complexes for integration [49]. In vitro data suggest interaction between TRE5-A ORF1 protein and TFIIIB subunits during the integration process [32], which in turn is a remarkable parallel to target recognition by the otherwise unrelated yeast Ty3 element.

Interestingly, high copy numbers of retrotransposons were only found in *D. discoideum* and not in other dictyostelid genomes. Our data suggest that *D. discoideum* is different from the other investigated dictyostelids in that it was specifically affected by an unknown selection pressure that either demanded or coincidentally enabled a burst of retrotransposon expansion. It seems unlikely that the propagation of the sequenced laboratory strain AX4 for about four decades has caused this retrotransposon expansion, because Southern blot data on genomic DNA of the parent strain NC4 probing for TRE5-A and TRE3-A indicated similarly high copy numbers of both elements (T.W., unpublished data). It is conceivable that *D. discoideum* has evolved to enable DIRS-1 amplification in centromeres to serve the organism as a substrate for kinetochore complex formation. The tRNA gene-targeting retrotransposons may have profited from this selection and, as a consequence, expanded throughout the genome. However, cells affected in such a manner may have been negatively selected even if there was no direct damage to genes because the haploid genome is particularly vulnerable to non-allelic recombinations forced by the accumulation of repetitive DNA. This consideration may explain why the targeting of tRNA genes by TRE elements achieved a steady state at approximately 60 % saturation of tRNA gene loci.

Skipper elements may use unconventional priming of reverse transcription

During the analysis of dictyostelid genomes, the question arose as to whether Skipper elements use a novel mechanism of reverse transcription initiation. Many retroviruses and LTR retrotransposons use cellular tRNAs as primers to initiate minus-strand strong-stop cDNA synthesis [40, 41]. These elements are characterized by a typical TGG trinucleotide signature located a few base pairs downstream of the left LTR that presents the complement of the CCA 3'-end of tRNA primers. All identified DGLT-A elements have this typical TGG motif 2 bp downstream of the left LTR (Fig. 3), but no cellular tRNAs could be identified that may be used as primers for cDNA synthesis. In contrast, most Skipper elements lack the TGG motif and instead contain a degenerate polypyrimidine (PPy) stretch. Although this characteristic feature of Skipper elements could be traced to a *Protostelium* species suggesting a root outside the dictyostelids, it has not been found in other organisms to the best of our knowledge. Some LTR retrotransposons lacking the TGG signature are assumed to use self-priming to initiate reverse transcription [50]. In such elements, RNA sequences located in the left LTR at the 5' ends of the retrotransposon transcripts loop back to the region immediately downstream of the LTR and prime reverse transcription [51]. Regarding the Skipper elements, no such complementary regions in the left LTRs are present, suggesting that a novel type of self-priming may be involved. It is unlikely, however, that a "simple" poly(A) stretch somewhere in the Skipper sequence is used in a self-priming process because the PPy sequences in all Skipper elements bear a characteristic C nucleotide facing the orientation of minus-strand cDNA synthesis (Fig. 3).

Dictyostelid Skipper elements are typical chromoviruses

In *D. discoideum*, DIRS-1 and Skipper elements form large clusters at the nuclear periphery during interphase that splits into six distinct spots during mitosis representative of the centromeric DNA of the six chromosomes [23]. Interestingly, the clustering of retrotransposons in heterochromatic regions has also been reported in fungal genomes such as that in *Magnaporthe grisea*, an organism with a similarly high gene density as dictyostelids [52]. This type of retrotransposon clustering appears to differ from the targeting of yeast Ty5 to heterochromatin and likely involves interactions of chromo domains located downstream of IN domains in Ty3/*gypsy*-type retrotransposons with heterochromatin marks. Similar to DIRS-1, Skipper-1 from *D. discoideum* has been shown to co-localize with sites of H3K9me2 methylation [23] and binding sites of CenH3, a marker for centromeric heterochromatin [53]. DIRS-1 and Skipper also co-localize with heterochromatin protein 1 (HP1; HcpA), which is recruited to centromeric heterochromatin through the binding of its chromo domain (CHD) to H3K9me2 marks [38]. Skipper shows interesting parallels to centromeric Ty3/*gypsy*-type retrotransposons bearing CHDs known as chromoviruses, which are found in plants and fungi. For instance, the MAGGY retrotransposon from *M. grisea* targets heterochromatin via interaction with a "canonical" or group I CHD (CHD_I) with histone marks such as H3K9me2 and H3K9me3 [24]. The CHD of Skipper-1 elements is similar to that of *D. discoideum* HP1 (Fig. 4) and is a representative of group I CHDs (CHD_I); this is consistent with its centromeric accumulation. Some plant chromoviruses contain group II (CHD_II) domains that diverged from CHD_I domains and usually lack the first and third conserved aromatic amino acid that form the "cage" required to interact with methylated histone tails [24, 54] (see Fig. 4). CHD_II motifs can tether retrotransposons to heterochromatin without interacting with histone marks [24], yet many CHD_II-bearing chromoviruses are not heterochromatin-associated but spread on chromosomes [55]. CHD_II domains are notably similar to chromo shadow domains (CSD), which are required to mediate the homo- and heterodimerization of HP1 proteins, for instance, in *D. discoideum* [38]. Thus, CSDs may represent protein interaction platforms that mediate the integration of CHD_II-bearing chromoviruses into heterochromatin by recognizing specific heterochromatin-associated factors [24]. It is tempting to speculate that the divergence of CHD_II domains from canonical CHDs in Skipper-2 elements enabled the development of a new integration preference away from centromeric DNA into intergenic regions downstream of tRNA genes. Interestingly, the transition from CHD_I to CHD_II domains in plant chromoviruses was estimated to date back 500-400 mya [54], which is approximately the time (~600 mya) when the dictyostelids began to evolve from their last common ancestor [33].

Conclusions

In the environments of gene-dense genomes, retrotransposons from organisms as divergent as dictyostelid social amoebae and budding yeast reveal convergent evolution leading to the selection of tRNA gene-flanking sequences as potential safe integration sites. In the evolution of dictyostelids, at least six inventions of targeted integration can be discriminated by the choice of distinct integration windows upstream or downstream of tRNA genes by phylogenetically distinctive retrotransposons. In *D. discoideum*, the strong preference of TRE family retrotransposons to integrate near tRNA genes has likely promoted their expansion to almost 4 % of the genome; however, comparing different dictyostelid genomes suggests that *D. discoideum* is an exception to the rule and may have been affected by an unknown evolutionary force that either demanded or coincidentally enabled a burst of retrotransposon amplification in this particular species. In general, it is evident from our analysis that non-mutagenic retrotransposition is not a license to amplify possibly because host cells keep track of their repetitive sequences to maintain genome stability.

Methods

Annotated genome sequences of *D. discoideum* [16], *D. purpureum* [34], *D. fasciculatum* [33], and *P. pallidum* [33] were accessed at dictyBase (http://dictybase.org/) [56]. A genome sequence of *D. lacteum* was obtained from Genbank (LODT01000000). The genome sequence of *Protostelium fungivorum* will be reported elsewhere (F.H., T.W., G.G., manuscript in preparation).

To identify new retrotransposons in dictyostelid genomes, known retrotransposon sequences from *D. discoideum* [17] were used as queries in TBLASTX searches with a cutoff value of e < 10^{-15}. Found sequences were expanded by 3000 bp upstream and downstream and analyzed using Jemboss [57]. Blast searches were performed to construct consensus sequences from DNA alignments of individual retrotransposon copies. Searches for LTR sequences were performed using LTR_FINDER [58] to determine full-length LTR retrotransposons and to identify solo LTR sequences. Flanking sequences of retrotransposon copies were analyzed for the presence of tRNA genes using tRNAscan-SE [59] and ARAGORN [60]. To specifically search for tRNA gene-associated retrotransposons, tRNA genes were identified genome-wide using tRNAscan-SE [59] and ARAGORN [60] and listed with 3000 bp upstream and downstream flanking regions for BLASTN searches using the aforementioned identified retrotransposon sequences as queries. Consensus sequences of full-length retrotransposons have been deposited in Repbase (http://www.girinst.org/repbase/) [61]. The following elements have alternative names in Repbase: Pp_Skipper-1: Gypsy-1_PPP; Pp_DGLT-A.1: Gypsy-3_PPP; Pp_DGLT-A.2: Gypsy-2_PPP; Pp_DGLT-A.3: Gypsy-4_PPP; Df_Skipper-1: Gypsy-2_DFa; Df_Skipper-2: Gypsy-1_DFa.

For phylogenetic analyses of LTR retrotransposons, the core domains of reverse transcriptase (RT), ribonuclease H (RNH), and integrase (IN) were determined by searching the Conserved Domain Database [62]. Alignments were generated using ClustalX [63], and conserved amino acid positions were highlighted using BoxShade (http://www.ch.embnet.org/software/BOX_-form.html). Shading is to a 50 % consensus with black boxes indicating invariant amino acids and gray boxes representing similar amino acids. Phylogenetic analyses were conducted using the MEGA7 software package

[64]. Phylogenetic trees were analyzed using the Neighbor-Joining [65] or the Maximum Likelihood method [66] as indicated in the figures.

Additional file

Additional file 1: Figure S1. Alignments of GPY/F motifs in the carboxy-terminal regions of integrase domains in dictyostelid LTR retrotransposons. **Figure S2.** Alignments of RT, RNH, and IN domains of dictyostelid LTR retrotransposons. **Figure S3.** Phylogenetic analysis of dictyostelid LTR elements. **Figure S4.** Alignment of RT domains of dictyostelid non-LTR retrotransposons. **Figure S5.** Phylogenetic analysis of RT domains of dictyostelid non-LTR elements. **Figure S6.** Alignment of dictyostelid non-LTR elements bearing RNH domains. **Table S1.** Comparison of retrotransposon in annotated dictyostelid genomes. **Table S2.** Association of tRNA genes with TREs in the *D. discoideum* genome. (PDF 7243 kb)

Acknowledgments

This work was supported by grant WI 1142/10-1 from the Deutsche Forschungsgemeinschaft.

Funding

Grant WI 1142/10-1 (Deutsche Forschungsgemeinschaft) to T.W.

Authors' contributions

TW conceived the study. TW, TS, EK, FH and GG designed and performed the experiments and revised the manuscript critically for important intellectual content. TW, TS, EK, FH and GG analyzed the data and drafted the manuscript. All authors read and approved the final manuscript.

Competing interests

The authors declare that they have no competing interests.

Author details

[1]Institute of Pharmacy, Department of Pharmaceutical Biology, Friedrich Schiller University Jena, Semmelweisstraße 10, Jena 07743, Germany. [2]Institute for Biochemistry I, Medical Faculty, University of Cologne, Berlin, Germany. [3]Institute for Freshwater Ecology and Inland Fisheries, IGB, Berlin, Germany. [4]Junior Research Group Evolution of Microbial Interaction, Leibniz Institute for Natural Product Research and Infection Biology—Hans Knöll Institute, Jena, Germany.

References

1. Deininger PL, Moran JV, Batzer MA, Kazazian HH. Mobile elements and mammalian genome evolution. Curr Opin Genet Dev. 2003;13:651–8.
2. Kazazian HH. Mobile elements: drivers of genome evolution. Science. 2004; 303:1626–32.
3. Cordaux R, Batzer MA. The impact of retrotransposons on human genome evolution. Nat Rev Genet. 2009;10:691–703.
4. Levin HL, Moran JV. Dynamic interactions between transposable elements and their hosts. Nat Rev Genet. 2011;12:615–27.
5. Mita P, Boeke JD. How retrotransposons shape genome regulation. Curr Opin Genet Dev. 2016;37:90–100.
6. Craig NL, Craigie R, Gellert M, Lambowitz AM, editors. Mobile DNA II. Washington DC: ASM Press; 2002.
7. Malik HS, Eickbush TH. Modular evolution of the integrase domain in the Ty3/gypsy class of LTR retrotransposons. J Virol. 1999;73: 5186–90.
8. Marin I, Llorens C. Ty3/Gypsy retrotransposons: description of new *Arabidopsis thaliana* elements and evolutionary perspectives derived from comparative genomic data. Mol Biol Evol. 2000;17:1040–9.

9. Malik HS, Eickbush TH. Phylogenetic analysis of ribonuclease H domains suggests a late, chimeric origin of LTR retrotransposable elements and retroviruses. Genome Res. 2001;11:1187–97.

10. Malik HS. Ribonuclease H, evolution in retrotransposable elements. Cytogen Genome Res. 2005;110:392–401.

11. Malik HS, Burke WD, Eickbush TH. The age and evolution of non-LTR retrotransposable elements. Mol Biol Evol. 1999;16:793–805.

12. Romeralo M, Skiba A, Gonzalez-Voyer A, Schilde C, Lawal HM, Kedziora S, et al. Analysis of phenotypic evolution in Dictyostelia highlights developmental plasticity as a likely consequence of colonial multicellularity. Proc Biol Sci. 2013;280:20130976.

13. Glöckner G, Noegel AA. Comparative genomics in the Amoebozoa clade. Biol Rev. 2013;88:215–25.

14. Loomis WF. Genetic control of morphogenesis in Dictyostelium. Dev Biol. 2015;402:146–61.

15. Du Q, Kawabe Y, Schilde C, Chen ZH, Schaap P. The evolution of aggregative multicellularity and cell-cell communication in the Dictyostelia. J Mol Biol. 2015;427:3722–33.

16. Eichinger L, Pachebat JA, Glöckner G, Rajandream M-A, Sucgang R, Berriman M, et al. The genome of the social amoeba Dictyostelium discoideum. Nature. 2005;435:43–57.

17. Glöckner G, Szafranski K, Winckler T, Dingermann T, Quail M, Cox E, et al. The complex repeats of Dictyostelium discoideum. Genome Res. 2001; 11:585–94.

18. Capello J, Handelsman K, Lodish HF. Sequence of Dictyostelum DIRS-1: An apparent retrotransposon with inverted terminal repeats and an internal circle junction sequence. Cell. 1985;43:105–15.

19. Poulter RTM, Goodwin TJD. DIRS-1 and the other tyrosine recombinase retrotransposons. Cytogenet Genome Res. 2005;110:575–88.

20. Capello J, Cohen SM, Lodish HF. Dictyostelium transposable element DIRS-1 preferentially inserts into DIRS-1 sequences. Mol Cell Biol. 1984;4:2207–13.

21. Glöckner G, Heidel AJ. Centromere sequence and dynamics in Dictyostelium discoideum. Nucleic Acids Res. 2009;37:1809–16.

22. Leng P, Klatte DH, Schumann G, Boeke JD, Steck TL. Skipper, an LTR retrotransposon of Dictyostelium. Nucleic Acids Res. 1998;26:2008–15.

23. Kaller M, Földesi B, Nellen W. Localization and organization of protein factors involved in chromosome inheritance in Dictyostelium discoideum. Biol Chem. 2007;388:355–65.

24. Gao X, Hou Y, Ebina H, Levin HL, Voytas DF. Chromodomains direct integration of retrotransposons to heterochromatin. Genome Res. 2008;18: 359–69.

25. Chalker DL, Sandmeyer SB. Ty3 integrates within the region of RNA polymerase III transcription initiation. Genes Dev. 1992;6:117–28.

26. Hopper AK. Transfer RNA, post-transcriptional processing, turnover, and subcellular dynamics in the yeast Saccharomyces cerevisiae. Genetics. 2013; 194:43–67.

27. Qi X, Sandmeyer S. In vitro targeting of strand transfer by the Ty3 retroelement integrase. J Biol Chem. 2012;287:18589–95.

28. Yieh L, Kassavetis GA, Geiduschek EP, Sandmeyer SB. The Brf and TATA-binding protein subunits of the RNA polymerase III transcription factor IIIB mediate position-specific Integration of the gypsy-like element, Ty3. J Biol Chem. 2000;275:29800–7.

29. Szafranski K, Glöckner G, Dingermann T, Dannat K, Noegel AA, Eichinger L, et al. Non-LTR retrotransposons with unique integration preferences downstream of Dictyostelium discoideum transfer RNA genes. Mol Gen Genet. 1999;262:772–80.

30. Kolosha VO, Martin SL. High affinity, non-sequence-specific RNA binding by the open reading frame 1 (ORF1) protein from long interspersed nuclear element 1 (LINE-1). J Biol Chem. 2003;278:8112–7.

31. Martin SL, Bushman D, Wang F, Li PW-L, Walker A, Cummiskey J, et al. A single amino acid substitution in ORF1 dramatically decreases L1 retrotransposition and provides insight into nucleic acid chaperone activity. Nucleic Acids Res. 2008;18:5845–54.

32. Chung T, Siol O, Dingermann T, Winckler T. Protein interactions involved in tRNA gene-specific integration of Dictyostelium discoideum non-long terminal repeat retrotransposon TRE5-A. Mol Cell Biol. 2007;27:8492–501.

33. Heidel AJ, Lawal HM, Felder M, Schilde C, Helps NR, Tunggal B, et al. Phylogeny-wide analysis of social amoeba genomes highlights ancient originis for complex intercellular communication. Genome Res. 2011;21:1882–91.

34. Sucgang R, Kuo A, Tian X, Salerno W, Parikh A, Feasley CL, et al. Comparative genomics of the social amoebae Dictyostelium discoideum and

35. Goffeau A, Barrell BG, Bussey H, Davis RW, Dujon B, Feldmann H, et al. Life with 6000 genes. Science. 1996;274:546–67.

36. Kim JM, Vanguri S, Boeke JD, Gabriel A, Voytas DF. Transposable elements and genome organization: A comprehensive survey of retrotransposons revealed by the complete Saccharomyces cerevisiae genome sequence. Genome Res. 1998;8:464–78.

37. Hofmann J, Schumann G, Borschet G, Gosseringer R, Bach M, Bertling WM, et al. Transfer RNA genes from Dictyostelium discoideum are frequently associated with repetitive elements and contain consensus boxes in their 5′-flanking and 3′-flanking regions. J Mol Biol. 1991;222:537–52.

38. Kaller M, Euteneuer U, Nellen W. Differential effects of heterochromatin protein 1 isoforms on mitotic chromosome distribution and growth in Dictyostelium discoideum. Eukaryot Cell. 2006;5:530–43.

39. Blus BJ, Wiggins K, Khorasanizadeh S. Epigenetic virtues of chromodomains. Crit Rev Biochem Mol Biol. 2011;46:507–26.

40. Boeke JD, Corces VG. Transcription and reverse transcription of retrotransposons. Ann Rev Microbiol. 1989;43:403–34.

41. Le Grice SFJ. "In the beginning": initiation of minus strand DNA synthesis in retroviruses and LTR-containing retrotransposons. Biochemistry. 2003;42: 14349–55.

42. Shadwick LL, Spiegel FW, Shadwick JD, Brown MW, Silberman JD. Eumycetozoa = Amoebozoa?: SSUrDNA phylogeny of protosteloid slime molds and its significance for the amoebozoan supergroup. PLoS One. 2009;4:e6754.

43. Xie W, Gai X, Zhu Y, Zappulla DC, Sternglanz R, Voytas DF. Targeting of the yeast Ty5 retrotransposon to silent chromatin is mediated by interactions between integrase and Sir4p. Mol Cell Biol. 2001;21:6606–14.

44. Baller JA, Gao J, Stamenova R, Curcio MJ, Voytas DF. A nucleosomal surface defines an integration hotspot for the Saccharomyces cerevisiae Ty1 retrotransposon. Genome Res. 2012;22:704–13.

45. Mularoni L, Zhou Y, Bowen T, Gangadharan S, Wheelan SJ, Boeke JD. Retrotransposon Ty1 intgration targets specifically positioned asymmetric nucleosomal DNA segments in tRNA hotspots. Genome Res. 2012;22:693–703.

46. Bridier-Nahmias A, Tchalikian-Cosson A, Baller JA, Menouni R, Fayol H, Flores A, et al. An RNA polymerase III subunit determines sites of retrotransposon integration. Science. 2015;348:585–8.

47. Cheung S, Ma L, Chan PH, Hu HL, Mayor T, Chen HT, et al. Ty1-Integrase interacts with RNA Polymerase III specific subcomplexes to promote insertion of Ty1 elements upstream of Pol III-transcribed genes. J Biol Chem. 2016;291:6396–411.

48. Qi X, Daily K, Nguyen K, Wang HX, Mayhew D, Rigor P, et al. Retrotransposon profiling of RNA polymerase III initiation sites. Genome Res. 2012;22:681–92.

49. Siol O, Boutliliss M, Chung T, Glöckner G, Dingermann T, Winckler T. Role of RNA polymerase III transcription factors in the selection of integration sites by the Dictyostelium non-long terminal repeat retrotransposon TRE5-A. Mol Cell Biol. 2006;26:8242–51.

50. Levin HL. It's prime time for reverse transcriptase. Cell. 1997;88:5–8.

51. Levin HL. A novel mechanism of self-primed reverse transcription defines a new family of retroelements. Mol Cell Biol. 1995;15:3310–7.

52. Dean RA, Talbot NJ, Ebbole DJ, Farman ML, Mitchell TK, Orbach MJ, et al. The genome sequence of the rice blast fungus Magnaporthe grisea. Nature. 2005;434:980–6.

53. Dubin M, Fuchs J, Gräf R, Schubert I, Nellen W. Dynamics of a novel centromeric histone variant CenH3 reveals the evolutionary ancestral timing of centromere biogenesis. Nuc Acids Res. 2010;38:7526–37.

54. Novikov A, Smyshlyaev G, Novikova O. Evolutionary history of LTR retrotransposon chromodomains in plants. Int J Plant Genomics. 2012; 2012:874743.

55. Neumann P, Navrátilová A, Koblížková A, Kejnovský E, Hřibová E, Hobza R, et al. Plant centromeric retrotransposons: a structural and cytogenetic perspective. Mob DNA. 2011;2:4.

56. Basu S, Fey P, Pandit Y, Dodson R, Kibbe WA, Chisholm RL. dictyBase 2013: integrating multiple Dictyostelid species. Nuc Acids Res. 2013;41(Database issue):D676–D83.

57. Carver T, Bleasby A. The design of Jemboss: a graphical user interface to EMBOSS. Bioinformatics. 2003;19:1837–43.

58. Xu Z, Wang H. LTR_FINDER: an efficient tool for the prediction of full-length LTR retrotransposons. Nucleic Acids Res. 2007;35(Web Server issue):W265–W8.

59. Schattner P, Brooks AN, Lowe TM. The tRNAscan-SE, snoscan and snoGPS

web servers for the detection of tRNAs and snoRNAs. Nucleic Acids Res. 2005;33(Web Server issue):W686–W9.

60. Laslett D, Canback B. ARAGORN, a program to detect tRNA genes and tmRNA genes in nucleotide sequences. Nucleic Acids Res. 2004;32:11–6.

61. Jurka J, Kapitonov VV, Pavlicek A, Klonowski P, Kohany O, Walichiewicz J. Repbase Update, a database of eukaryotic repetitive elements. Cytogen Genome Res. 2005;110:462–7.

62. Marchler-Bauer A, Derbyshire MK, Gonzales NR, Lu S, Chitsaz F, Geer LY, et al. CDD: NCBI's conserved domain database. Nucleic Acids Res. 2015; 43(Database issue):D222–D6.

63. Thompson JD, Gibson TJ, Plewniak F, Jeanmougin F, Higgins DG. The ClustalX windows interface: flexible strategies for multiple sequence alignment aided by quality analysis tools. Nuc Acids Res. 1997;24:4876–82.

64. Kumar S, Stecher G, Tamura K. MEGA7: Molecular Evolutionary Genetics Analysis version 7.0 for bigger datasets. Mol Biol Evol. 2016;Mar 22. pii: msw054. [Epub ahead of print].

65. Saitou N, Nei M. The neighbor-joining method: A new method for reconstructing phylogenetic trees. Mol Biol Evol. 1987;4:406–25.

66. Jones DT, Taylor WR, Thornton JM. The rapid generation of mutation data matrices from protein sequences. Comput Appl Biosci. 1992;8:275–82.

67. Marschalek R, Hofmann J, Schumann G, Gosseringer R, Dingermann T. Structure of DRE, a retrotransposable element which integrates with position specificity upstream of *Dictyostelium discoideum* tRNA genes. Mol Cell Biol. 1992;12:229–39.

68. Felsenstein J. Confidence limits on phylogenies: An approach using the bootstrap. Evolution. 1985;39:783–91.

69. Male G, von Appen A, Glatt S, Taylor NM, Cristovao M, Groetsch H, et al. Architecture of TFIIIC and its role in RNA polymerase III pre-initiation complex assembly. Nat Commun. 2015;6:7387.

12

Non-canonical Helitrons in *Fusarium oxysporum*

Biju Vadakkemukadiyil Chellapan[1,2], Peter van Dam[2], Martijn Rep[2*], Ben J. C. Cornelissen[2] and Like Fokkens[2]

Abstract

Background: Helitrons are eukaryotic rolling circle transposable elements that can have a large impact on host genomes due to their copy-number and their ability to capture and copy genes and regulatory elements. They occur widely in plants and animals, and have thus far been relatively little investigated in fungi.

Results: Here, we comprehensively survey Helitrons in several completely sequenced genomes representing the *F. oxysporum* species complex (FOSC). We thoroughly characterize 5 different Helitron subgroups and determine their impact on genome evolution and assembly in this species complex. FOSC Helitrons resemble members of the Helitron2 variant that includes Helentrons and DINEs. The fact that some Helitrons appeared to be still active in FOSC provided the opportunity to determine whether Helitrons occur as a circular intermediate in FOSC. We present experimental evidence suggesting that at least one Helitron subgroup occurs with joined ends, suggesting a circular intermediate. We extend our analyses to other Pezizomycotina and find that most fungal Helitrons we identified group phylogenetically with Helitron2 and probably have similar characteristics.

Conclusions: FOSC genomes harbour non-canonical Helitrons that are characterized by asymmetric terminal inverted repeats, show hallmarks of recent activity and likely transpose via a circular intermediate. Bioinformatic analyses indicate that they are representative of a large reservoir of fungal Helitrons that thus far has not been characterized.

Keywords: Helitrons, Transposon, Rolling circle, Terminal inverted repeats, Helitron2, Helentrons, *Fusarium oxysporum*

Background

Transposable elements (TEs) are stretches of DNA that are able to copy or move from one site to another in a genome. Autonomous TEs contain one or more sequences coding for proteins that are involved in transposition, combined with TE-specific DNA motifs such as terminal inverted repeats. These motifs are required for transposition. Non-autonomous elements possess the DNA motifs but do not encode a functional transposase. They profit from their autonomous counterparts and often greatly outnumber them.

Helitrons are a family of TEs that encode an Y2-transposase consisting of an N-terminal rolling circle replication initiator (Rep) domain and a C-terminal helicase (Hel) domain. They were first characterized in an *in silico* analysis of the genomes of *A. thaliana*, *O. sativa* and *C. elegans* [1], where they were found to have a 5'-TC and 3'-CTRR (where R stands for A or G) motif and a short hairpin at 10–12 nucleotides distance from the 3' terminus. Recent reports indicate that Helitrons can be divided into two groups: Helitron1 and Helitron2 [2–4]. The motifs that were found upon first discovery of Helitrons are specific to the Helitron1. In contrast Helitron2 TEs are characterized by an asymmetric terminal inverted repeat (ATIR) and a hairpin at both termini. Helentrons cluster phylogenetically with Helitron2 proteins and possess similar termini, but, in addition to the Rep and Hel domains, they possess an endonuclease domain that they obtained through insertion of a retrotransposon [2–6]. DINEs, also known as HINEs, the most abundant TE in Drosophila, are non-autonomous elements derived from Helentrons [3] (see [7] for a recent review).

* Correspondence: M.Rep@uva.nl

[2]Molecular Plant Pathology, Swammerdam Institute for Life Sciences, Faculty of Science, University of Amsterdam, P.O. Box 94215, 1090 Amsterdam, GE, The Netherlands

Full list of author information is available at the end of the article

Recent in-depth analyses of a mobile pathogenicity chromosome of the ascomycete *Fusarium oxysporum* f. sp. *lycopersici* strain Fol4287 revealed 9 nearly identical genes encoding proteins with a Rep-Hel domain architecture [8]. The *Fusarium oxysporum* species complex (FOSC) consists of clonal lines of *Fusarium oxysporum*, a filamentous fungus that colonizes plant roots and occasionally enters the plant's roots and vascular system, causing wilting or root-rot disease symptoms. Individual pathogenic strains are usually pathogenic to only a small number of related host plants, but the species complex as a whole is a versatile pathogen with great economic impact [9]. *Fusarium oxysporum* represents an extreme case of a two-speed genome: its chromosomes can be classified as either 'core' or 'lineage specific' (LS), where core chromosomes are largely syntenic with chromosomes of other *Fusarium* species, while LS chromosomes are largely absent in other *Fusarium* species [10–12]. The LS chromosomes are enriched in TEs and in genes involved in pathogenicity. Genomes of 12 strains of this species complex have been sequenced, assembled and annotated [13], providing an excellent dataset for a thorough study of Helitrons in an ascomycete.

The genomic impact of Helitrons, in terms of copy number as well as in terms of whether Helitrons inserted in or near genes, varies strongly between different species (see [7] for a recent review). This depends on transposition efficiency and effectiveness of TE silencing, but also on whether we are observing a host genome that experienced a recent Helitron outbreak versus the remnants of past activity. In the latter case we expect for example that Helitron copies that adversely affect coding or regulatory regions or gene regulation have been removed from a population through purifying selection. A factor that is often overlooked is the completeness of genome assembly. Within our FOSC dataset, the genomes are assembled up to different levels of completeness, which allows us to assess the impact of incomplete genome assembly on copy number estimates.

A recent study using a reconstructed ancestral bat Helitron1 sequence provided important insights into the mechanisms underlying transposition and gene capture in canonical Helitrons [14]. First of all, the authors could demonstrate that Helitrons transpose as single stranded DNA. This is congruent with the fact that Helitrons do not cause target site duplications that are associated with double stranded, staggered breaks. Recent biochemical studies show that they transpose via copy-paste rather than cut-and-paste, which explains their high copy number [14]. Helitrons can capture (parts of) genes and thus contribute to the emergence of new genes through combining of different coding and non-coding sequence that have been sequentially captured [4, 6, 7, 14–22]. Grabundzija and others confirmed the 'end-bypass' model of

gene capture in Helitrons, in which the transposase skips the 3' terminus and thus includes 3' flanking DNA sequence in the excised Helitron. Finally, Grabundzija and others demonstrated that canonical Helitrons occur as a circular intermediate [14], as has been observed previously for the Insertion Sequence IS91 in *Escherichia coli* [23]. Transposition via a circular intermediate can also explain the presence of multiple tandem insertions of truncated Helitrons that have recently been found in plant centromeres [24]. This indicates that the processes of excision and insertion are decoupled in Helitrons. We extensively survey footprints of past Helitron activity, focussing on putative Helitron self-insertions, to shed light on the transposition process in FOSC Helitrons.

Helitrons are found in a wide range of eukaryotes, including plants, animals, fungi and oomycetes, but have predominantly been described in plants and animals [1, 4–6, 15, 18, 21, 25–37]. We ask whether FOSC Helitrons are relatively unique or whether they represent a larger and relatively unknown reservoir of Pezizomycotina Helitrons. Finally, we study conservation of terminal sequences and ask how the Helitrons we uncovered are related to the two known Helitron families.

Results
FOSC Helitrons divide into two groups and 5 subgroups
Most software designed to identify Helitrons are based on the DNA motifs of the Helitron1 variant and will overlook instances of Helitron2 because these have different termini [5, 18, 20, 38–40]. Moreover, DNA sequence similarity can be hard to recognize over long evolutionary distances and very few ascomycete Helitron sequences were available at the start of our studies. Therefore we selected 35 FOSC proteins with a Rep—Hel domain architecture and used those to search the FOSC genomes for additional, unannotated genes that encode putative Helitron proteins. We found in total 63 proteins in 10 different strains that encode proteins with the typical Helitron domain architecture and named them FoHelis (Fig. 1). Conserved motifs within the Rep as well as the Hel domain are present in most FOSC Helitrons, suggesting that these proteins are functional (Additional file 1: Figure S1 and Figure S2) [41–43]. Like other Helitrons, the putative Helitron proteins we predicted in FOSC have an N-terminal zinc finger-like motif (Additional file 1: Figure S3) [5].

To distinguish different subgroups, we inferred a phylogenetic tree for these 63 protein sequences. We found that they divide into two major groups and five subgroups: FoHeli1 and FoHeli2 in group I, and FoHeli3—FoHeli5 in group II (Fig. 1). FoHeli1 is the subgroup identified earlier [8] and differs from the other subgroups in several respects: (i) they're found only in the genome of *F. oxysporum* f. sp. *lycopersici* Fol4287 (hereafter referred to as

Legend (species colour key):

- F. oxysporum f. sp. *lycopersici* 4287 NRRL 34936
- F. oxysporum f. sp. *vasinfectum* NRRL 25433
- F. oxysporum f. sp. *melonis* NRRL 26406
- F. oxysporum f. sp. *raphani* PHW815 NRRL 54005
- F. oxysporum f. sp. *conglutinans* PHW808 NRRL 54008
- F. oxysporum NRRL 32931 (human)
- F. oxysporum f. sp. *pisi* NRRL 25433
- F. oxysporum Fo5176 (arabidopsis)
- F. oxysporum Fo47 NRRL 54002 (biocontrol)
- F. oxysporum f. sp. *lycopersici* MN25 NRRL 54003

FoHeli1: FoHeli1.1, FoHeli1.2, FoHeli1.3 *, FOXG_22121 *, FoHeli1.4, FoHeli1.5, FoHeli1.6, FoHeli1.7 *, FoHeli1.8, FoHeli1.9, FoHeli1.10, FoHeli1.11, FoHeli1.12 *, FoHeli1.13 *, FoHeli1.14 *

FoHeli2: FOXG_14222, FOWG_18004, FoHeli2.1, FoHeli2.2, FoHeli2.3, FoHeli2.4, FOVG_18253, FoHeli2.5, FoHeli2.6, FoHeli2.7

FoHeli3: FOQG_18559, FOMG_18872, FOVG_19318, FOXB_00617, FOYG_17266, FoHeli3.1 (99), FOVG_18534, FOMG_19038, FOXG_06805, FOXG_06459, FOXG_07365, FOXG_16388, FOYG_16843, FoHeli3.2

FoHeli4: FOQG_16341 (88), FOXG_12542, FOXG_12592, FOXG_13999, FoHeli4.1, FOPG_17871, FOYG_17354, FOXG_14881, FOTG_18615, FOTG_18153 (97), FOVG_18686, FOVG_18938, FoHeli4.2, FoHeli4.3, FoHeli4.4, FOXG_19322, FOZG_17422, FOXG_14988

FoHeli5: FOXG_19766, FOXG_22453, FOXG_06452, FOMG_18625, FOTG_15771, FoHeli5.1

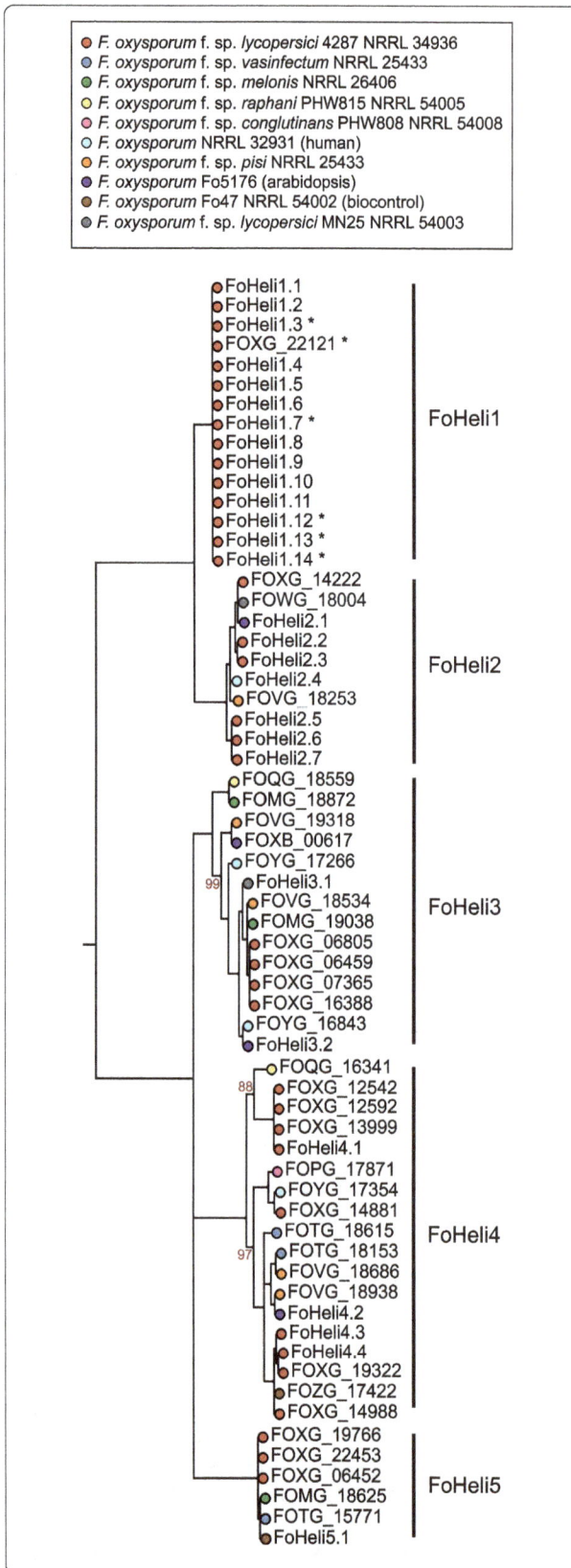

Fig. 1 FOSC Helitrons divide into two groups and 5 subgroups. The phylogeny inferred from a multiple sequence alignment of the (predicted) protein sequences of FOSC Helitrons shows that they are separated into two major groups that can be further subdivided into 5 subgroups in total. These subgroups are designated FoHeli1 – FoHeli5. All except three branches have 100% bootstrap support, the bootstrap support (based on 100 replicates) of those three is indicated in red adjacent to the respective branch. Nodes are coloured according to the fungal strain the Helitron was found in. FoHeli1 is distinct from the other 4 subgroups because the protein sequences are nearly identical, because this subgroup is only found in the most completely assembled genome, that of Fol4287, and because this subgroups is also found on core chromosomes of Fol4287. Copies on core chromosomes are indicated with an *. In the genomes of strains *F. oxysporum* f. sp. *radicis-lycopersici* CL57 and *F. oxysporum* f.sp. *cubense* Il5 no genes encoding proteins with a Rep-Hel domain architecture were detected. This is most likely due to deficiencies in genome assembly and gene annotation as we do find partial Helitron copies in these genomes, albeit in small numbers (Additional file 5: Table S5)

Fol4287) among the 12 strains, (ii) they're found on many different chromosomes, including core chromosomes (Fig. 1, Additional file 2: Table S1) and (iii) there is very little sequence diversity in this subgroup.

FoHeli termini are non-canonical and resemble those of the Helitron2 variant

Using multiple sequence alignments for sets of similar sequences within each subgroup, we identified termini for 48 out of 63 Helitrons, despite the fact that many Helitrons reside on the borders of contigs or supercontigs. More importantly, we found termini for members of each subgroup (Additional file 2: Table S1, Fig. 2a). Interestingly, all FoHeli termini we have identified include asymmetric terminal inverted repeats (ATIRs), like members of the Helitron2 variant. In addition, FoHeli1 and FoHeli2 have hairpins at both termini, as is also observed in some Helitron2 TEs.

Within each of the two major groups, the sequences of the termini are very similar: subgroups FoHeli1 and FoHeli2 have "**TCAGCCGAAGGCTG**AC" and "T[c/a]**AGTCCGAAGGA**CTT", respectively, at the 5' end, where underlined nucleotides indicate the stem of a hairpin. Nucleotides in bold are present as an inverted repeat, that is itself also part of a hairpin, at the 3' end of the element, 38 (FoHeli2) to 51 (FoHeli1) bp upstream from their 3' terminus 'ATATTTT'. The distance between the termini (i.e. the length of the full Helitron transposable element) is quite short: ~6 kb for FoHeli1 and ~5 kb for FoHeli2 (Additional file 2: Table S1). In the other major group, subgroups FoHeli3-FoHeli5 have "TGCCT" and a degenerate hairpin at the 5' end, and "CTCCTGT" at the 3' end, combined with an inverted repeat of between 13–16 bp. The distance between termini is much larger in this group, ranging from ~9 to ~11 kb (Additional file 2: Table S1).

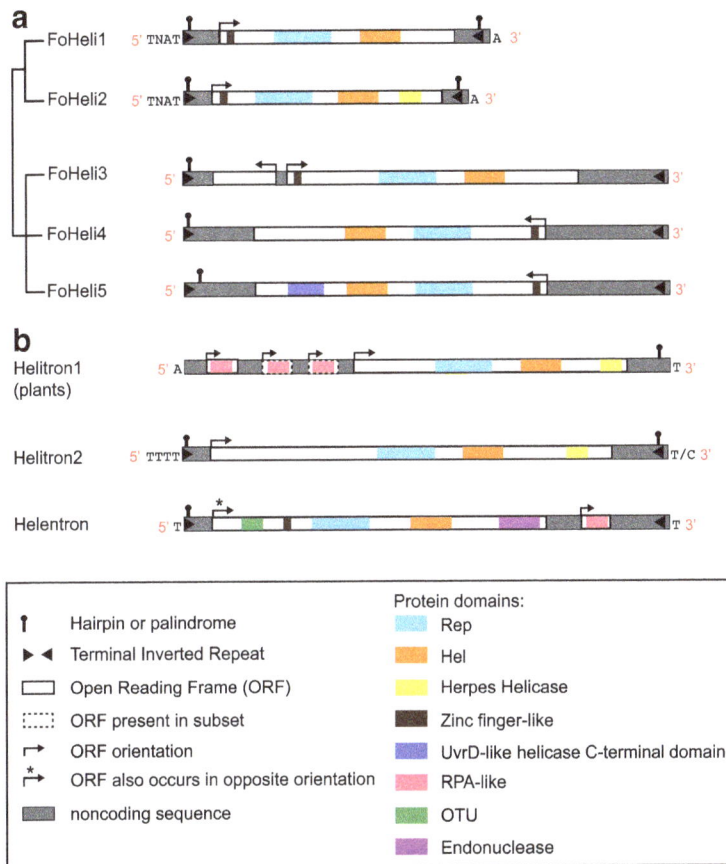

Fig. 2 Terminal features and coding capacity for FOSC and other Helitrons. **a** FoHeli termini are characterized by hairpins and inverted repeats, where the 3′ inverted repeats is ~20–40 bp upstream from the terminal sequence. Within each of the two groups, termini are very similar. FoHeli1 and FoHeli2 have two hairpins, one at each terminus, a 12 bp long inverted repeat, start with 'TCAG' and end with 'ATTTT'. Similar to canonical Helitrons, the 3′ inverted repeat and hairpin are located at ~30-40 bp from the 3′ terminus. In the other group, all FoHelis start with 'TGCCT' and end with 'CTCCTGT'. At the 5′ end, they have a hairpin but they lack a hairpin at the 3′ end. The ORFs in FoHeli4 and FoHeli5 have an opposite orientation when compared to FoHeli3. FoHeli1 and FoHeli2 insert between 'TNAT' and 'T', for the other group we could not establish an insertion preference. **b** When we compare these structural features to those of known Helitrons, we find that FOSC Helitrons resemble Helitron2 transposons. Structural features of Helitron1 (canonical Helitrons), Helitron2 and Helentrons were compiled from [2, 7] and RepBase [51]. Helitron1/canonical Helitrons insert between 'A' and 'T', Helitron2 between 'TTTT' and 'T' or 'C' and Helentrons in a 'TT' dinucleotide. See Additional file 1: Figures S13 and S14 for more detail on Helitron domain composition

Alignment of reconstructed pre-insertion sites confirmed that the termini we found are correct (Fig. 3a). In contrast to what has been reported on Helentrons [3], we have not observed variations in the number of Ts at 3′ ends. Canonical Helitrons insert preferentially into an 'AT' dinucleotide. The preferred insertion site for FoHeli1 and FoHeli2 is between 'TNAT' and 'A', where 'N' denotes any nucleotide (Figs. 2a and 3a). Note that because FoHeli1 and FoHeli2 have a 'T' at the 5′ terminus and preferentially insert between 'T' and 'A', we can not be certain that FoHeli1s and FoHeli2s start with 'TC' like canonical Helitrons, or with 'TTC' (Fig. 3b). From here, we assume that FoHeli1 and FoHeli2 start with 'TC', like canonical Helitrons.

In subgroup FoHeli1, two copies are 100% identical from 5′ to 3′ terminus: FoHeli1.11 on chromosome 14 and FoHeli1.15 (FOXG_22121) on chromosome 8. Within this subgroup all copies are more than 99% identical to each-other, from terminus to terminus. This suggests that FoHeli1 has been active relatively recently and may still be active. The other subgroups do contain identical copies, but these lie in regions that are part of large segmental duplications in Fol4287 and are not the result of recent transposition (Additional file 1: Figure S4, [10]). Only FoHeli2 has two members for which both termini have been identified, that are on the same genome and not interrupted by contigbreaks. FoHeli2.8 (FOXG_14222) and FoHeli2.2 are 98.76% identical from the 5′ to the 3′ terminus. For the other subgroups the period of activity can not be compared based on sequence divergence.

Several FoHelis have multiple predicted Open Reading Frames (ORFs) but most ORFs overlap with the Helitron

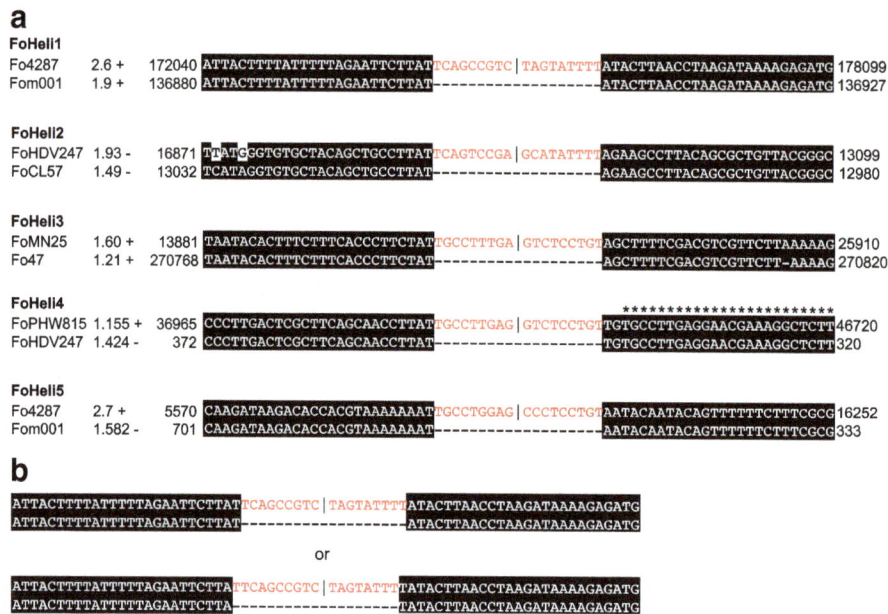

Fig. 3 a Alignment of insertion sites confirms FoHeli termini. For each subgroup, we reconstruct pre-insertion sites by concatenating FoHeli flanking sequences and search for these pre-insertion sequences in our set of FOSC genomes. Alignment of FoHeli flanking sequenes with these pre-insertion sites showed that the termini we had inferred before are correct. Fom001 – *F. oxysporum* f.sp. *melonis* 26406, FoMN25 - *F. oxysporum* f.sp. *lycopersici* MN25, FoHDV247 - *F. oxysporum* f.sp. *pisi* HDV247, FoCL57 - *F. oxysporum* f.sp. *radicis-lycopersici* CL57, FoPHW815 - *F. oxysporum* f.sp. *raphani* PHW815, FoPHW808 - *Fusarium oxysporum* f.sp. *conglutinans* PHW808. In this example a FoHeli4 is inserted 2 bp from another FoHeli4 (indicated with * above the sequence). **b** Because FoHelis have a 'T' at the 5' terminus and preferentially insert between 'T' and 'A', we can not be certain that e.g. FoHeli1s and FoHeli2s start with 'TC' like canonical Helitrons, or with 'TTC'.

transposase and are probably the result of gene prediction errors (Fig. 2, Additional file 3: Table S2). FoHeli3 is the only subgroup with a predicted ORF that does not overlap with the gene encoding the transposase. This additional ORF is located upstream from the transposase gene and has an opposite orientation. It has no known domains and only occurs in Helitron TEs (Additional file 3: Table S2). Several plant Helitrons contain one or more genes encoding an RPA-like protein; we found no RPA-like genes in FoHelis. Interestingly, the transposase ORFs in FoHeli4 and FoHeli5 have an inverted orientation when compared to FoHeli3 in the same major group (Fig. 2). This phenomenon has been observed before in Helitron2-like elements: in a Helentron in the fish *Danio rerio*, a Helentron in the fruit fly *Drosophila ananassae* and in a Helitron2 in the green alga *Chlamydomonas reinhardtii* [2, 3].

FOSC genomes contain non-autonomous FoHelis

In plant and animal genomes, the most abundant Helitrons are non-autonomous; they possess the structural terminal features that are needed for transposition, but do not encode a functional transposase. They are typically much shorter than autonomous Helitrons. The fact that we have terminus-to-terminus sequences for each subgroup allowed us to query the 12 FOSC genomes for non-autonomous elements. We found two types of non-autonomous elements in which (part of) the Helitron coding sequence was deleted. Interestingly, these non-autonomous elements all appear to have derived from FoHeli1, and we found them only in genomes in which we could not find a putative autonomous FoHeli1 copy. Moreover, their high sequence similarity and distinct termini suggest they have recently transposed.

The shortest element of the two, named FoHeliNA1, is 830 bp in size. We found this element in low copy number in the genomes of *F. oxysporum* f. sp. *raphani* PHW815, *F. oxysporum* f. sp. *vasinfectum*, *F. oxysporum* f. sp. *conglutinans* PHW808 and *F. oxysporum* Fo5176 (Additional file 4: Table S3). Its first 27 bp and last 166 bp are, respectively, ~92.5% identical to the 5' and ~78.2% identical to the 3' terminus of FoHeli1. The 637 bp between the termini are not similar to any of the Helitrons we had identified before (Additional file 1: Figure S5). The second type of non-autonomous element, named FoHeliNA2, is 1929 bp in size and was found in the genomes of *F. oxysporum* f.sp. *raphani* PHW815, *F. oxysporum* f.sp. *vasinfectum*, *F. oxysporum* NRRL 32931 and *F. oxysporum* f.sp. *pisi* HDV247, again in low copy number. Its first 1092 and last 837 bp are ~90% identical to FoHeli1 termini (Additional file 1: Figure S6). One copy of FoHeliNA2 has inserted into a putative autonomous FoHeli, namely FoHeli3.3 (FOQG_18559) in *Fusarium oxysporum* f.sp. *raphani* PHW815.

Increasing the maximum distance between matching termini allowed us to detect a few full-length Helitrons that were previously unrecognized, mostly because no or an incomplete ORF was predicted. Possibly, these Helitrons have pseudogenized, or the presence of assembly gaps in the coding sequence has hampered the correct prediction of the ORF. We also identified a few cases in which a hAT or a Hornet TE was inserted into a Helitron, truncating the ORF (Additional file 2: Table S1, Additional file 4: Table S3), but found no evidence that these 'chimeric' TEs have transposed (Additional file 1: Figure S4).

FoHeli copy number is underestimated due to genome assembly being hampered by the presence of identical FoHeli copies

The presence of non-autonomous Helitrons in genomes that do not have an autonomous version suggests that we may have failed to identify the putative autonomous copies in these genomes. Most FOSC genome sequences are based on short reads generated by second-generation sequencing. The occurrence of multiple, highly similar copies of a long sequence, due to recent gene duplications or recent transposition of TEs, greatly impacts these assemblies. Single reads only cover a small section of the repeated sequence and for those reads that do not contain a portion of unique flanking sequence, it is impossible to infer to which copy they belong. Most assemblers introduce a contig break and assemble all reads that fall completely within the repeated sequence into a single contig with very high coverage [44, 45].

If incomplete genome assembly hampered the detection of Helitrons, we should find partial Helitron copies at the borders of contigs and supercontigs, and some contigs that consist entirely of a Helitron sequence. Indeed, when we query the 12 FOSC genomes with DNA sequences of full-length (terminus-to-terminus) elements, we find that for FoHeli1, FoHeli2 and FoHeli4, most partial copies are located near the edge of a (super)contig (Additional file 5: Table S5). Especially the presence of FoHeli1 and FoHeli2 copies seem to have impaired genome assembly: respectively 82% and 96% of partial copies are located near contig borders, or span entire contigs, compared to 32% to 68% percent of FoHeli3—FoHeli5. Notably, a large fraction of these partial copies are between 80 and 150 bp long, which is what is expected given the read length that was achieved on Illumina platforms at the time these genomes were sequenced.

Conversely, due to incomplete genome assembly, the copy number of Helitrons in FOSC is potentially severely underestimated. If we assume that every Helitron 'start' is actually an unrecognized complete (potentially non-autonomous) copy, counting multiple termini as one, we arrive at an upper-bound copy number estimate that is almost ten-fold higher than the number of Helitrons we identified in our initial search (Additional file 5: Table S5). In total we then predict 559 copies in the FOSC, where FoHeli1 and FoHeli2 are most abundant with 115 and 327 copies in all 12 strains, respectively. Notably, FoHeli2 is particularly abundant in strains that are able to infect Arabidopsis (*F. oxysporum* f. sp. *conglutinans* PHW808: 95, *F. oxysporum* Fo5176:147 and *F. oxysporum* f. sp. *raphani* PHW815: 54), whereas other subgroups are more evenly distributed among the different strains.

Amplicons with the sequence of FoHeli1 with joined ends suggest presence of a circular intermediate

A recent study demonstrated that canonical Helitrons transpose via a circular intermediate [14]. We tested for the presence of a FoHeli circle in Fol4287 by trying to amplify the junction sequence of FoHeli with joined ends by PCR, using primers that anneal close to termini of FoHeli and are directed outwards, and genomic DNA from Fol4287 as template (Additional file 4: Table S3, Fig. 4). Interestingly, using FoHeli1-specific primers, a PCR product of 800 bp was amplified. The sequence of this PCR product corresponds to a FoHeli1 with joined ends (Fig. 4d) and does not occur in the assembled genome. Moreover, the intensity of the PCR product obtained using this primer pair is low compared to that of the PCR products obtained using the other, 'genomic' primer pairs (Fig. 4c), which is to be expected if its template is low-abundance, extra-chromosomal circular DNA. Notably, no PCR products corresponding to a FoHeli with joined ends were obtained using outward directed primer pairs specific for the subgroups that were more diverged in sequence, and therefore predicted to be non-active, FoHeli2—FoHeli5.

We tried to confirm the presence of circular Helitrons through multiply-primed Rolling Circle Amplification (RCA) [46] in which circular templates are overamplified with respect to the linear 'background' genome into concatemers. These concatemers can then be digested with an enzyme and run on a gel to produce bands corresponding to the size of the circle. In our experiments we could not detect overamplification of FoHeli1 (Additional file 1: Figure S8), rather we observed bands that most likely correspond to mitochondrial DNA. This can be explained by the extremely low abundance of FoHeli1 circles—caught in the act during DNA isolation- in the genomic DNA. They could easily have been outcompeted by the large amount of mitochondrial DNA during RCA and thus not have been amplified to such an extent that it would result in observable bands. However, when we isolated ~6–7 kb fragments from the gel (corresponding to the size of FoHeli1), we were able to obtain amplicons that correspond to FoHeli1 with

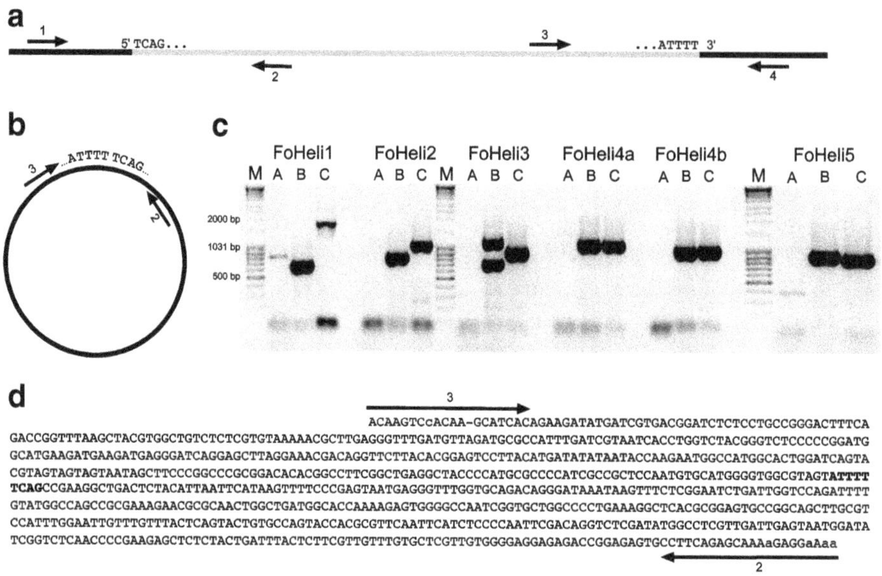

Fig. 4 Putative circular Helitrons detected by PCR. **a** Schematic representation of FoHeli1 in the genome. The grey line represents FoHeli1 and the 5′ and 3′ terminal sequences are indicated above. The black thick lines represent the flanking genomic region. The arrows indicate the positions of the primers. For each subgroup, FoHeli1 to FoHeli5, we designed four specific PCR primers (Additional file 1: Table S4). Primer pairs 1 + 2 and 3 + 4 are specific to FoHeli 5′ and 3′ ends and their flanking sequences, respectively. Primers 2 + 3 anneal close to FoHeli ends and are directed outwards; these are expected to amplify a PCR product only from molecules that contain nearby or joined FoHeli ends. **b** Schematic representation of a FoHeli1 circle with joined ends (possible template for the amplification of a PCR product using primers 2 + 3) **c** PCR experiment showing amplification of PCR products using primer pair A (primers 2 + 3), B (primers 1 + 2) and C (primers 3 + 4) specific for FoHeli1 – FoHeli5. The template for the PCR reaction was genomic DNA isolated from Fol4287. We used two sets of primers for FoHeli4, because this subgroup is more divergent than the others. Note that there is ~400 bp PCR product of FoHeli5 using outward directed primers. However, the sequence of FoHeli5 with joined ends between these primers is 570 bp. Moreover, the sequence of this amplicon did not show any similarity to a FoHeli. Hence we concluded that this amplicon does not correspond to a FoHeli5 with joined ends. **d** Structure of FoHeli1 joined ends. The terminal sequences are shown in bold

closed ends (Additional file 1: Figure S8), thus confirming our previous result.

In *M. lucifugus*, *Drosophila*, Rice and Maize, multiple tandem insertions of Helitrons or Helitron-derived elements have been reported [3, 6, 7, 24]. We observed one case in which a FoHeli4 was inserted 2 bp upstream of the 5′ terminal partial sequence another FoHeli4 element (Fig. 3). We considered the possibility that the PCR product was amplified from a tandem insertion of FoHeli1 in the Fol4287 genome that was not assembled correctly, rather than a circular intermediate.

We mapped Illumina sequencing reads of Fol4287 from three different libraries with distinct insert sizes to a constructed sequence corresponding to a tandem insertion of FoHeli1 (see Additional file 1: Figures S9 and S10 for more detail). The mate-pair library, with the largest insert size (5 kb), contained one read that spanned the junction of the two FoHeli1 copies, and a few paired reads that were mapped on either side of this junction. However, mate-pair libraries tend to suffer from contamination with paired-end and overlapping reads and we found no reads either spanning the junction or crossing the junction as pairs in the other two libraries. Hence we conclude that it is unlikely that FoHeli1 occurs as a

tandem insertion in Fol4287 (Additional file 1: Figure S9 and Figure S10).

Some FoHelis have multiple 5′ termini

Some Helitrons, including non-autonomous Helitrons and partial Helitron copies, possess multiple termini (Additional file 6: Table S6). Interestingly, different genomes harbor different 'versions' of multiple termini. For example, *F. oxysporum* f.sp. *vasinfectum* contains partial copies in which the first 73 nucleotides of FoHeli1 are repeated once, whereas copies in *F. oxysporum* f.sp. *conglutinans* PHW808 repeat the first 85 nucleotides (Additional file 6: Table S6). *F. oxysporum* f.sp. *cubense* II5 contains partial copies of FoHeli1 that contain the first 31 or 65 bp of the 5′ terminus, and combinations thereof. Two tomato infecting strains, *F. oxysporum* f.sp. *lycopersici* MN25 and Fol4287, contain partial Helitron copies in which the first 31 nucleotides are duplicated. Helitrons with two or more 5′ termini are found in different locations in the genome. Multiple sequence alignments of these termini, including flanking genomic sequences, show a sharp decline in similarity at Helitron borders, indicating that these copies arose via transposition rather than via segmental duplication (Additional file 1: Figure S11).

Helitrons are found in close proximity to pathogenicity-related genes

As mentioned above, Helitrons are potentially able to capture (parts of) genes and combine them into new transcripts transcripts [6, 18, 20, 25, 26]. Gene capture by Helitrons occurs very frequently in maize, but has rarely been observed to that extent in other species. Hence, pervasive gene capture is not a universal property of Helitrons. We investigated whether genes could have been captured by FoHelis. To this end, we compared all full-length putative autonomous and non-autonomous elements to NCBI's non-redundant nucleotide database, removing hits that were likely to be misannotated Helitrons rather than captured host genes. This resulted in a list of 27 putative gene capture events, most of which are hypothetical proteins identified in the fungus *Metarhizium* (Additional file 7: Table S7).

Although we didn't find evidence that gene capture by FoHelis plays an important role in FOSC evolution, we did note that some Helitrons are located in very close proximity to genes that have been implicated in pathogenicity in FOSC. For example, in the Arabidopsis-infecting strain *Fusarium oxysporum* Fo5176, a Helitron is found upstream of both *SIX9a* and *SIX9b*, homologs of the effector gene *SIX9* (Secreted In Xylem 9) encoding a protein identified in the xylem sap of tomato plants infected with Fol4287 [47]. A partial copy of a FoHeli2 is found 167 bp upstream from SIX9a, and a partial copy (the last 34 residues) of FoHeli1 is found 412 bp from SIX9b. Moreover, we find in the same strain a partial copy of a FoHeli1 or FoHeliNA2 located ~2 kb from a homolog of *SIX1* (Secreted In Xylem 1) of Fol4287 [47, 48]. Additionally, in the reference strain Fol4287, FoHeli1.6 is located 251 bp upstream from *SIX6* (Secreted In Xylem 6). In a race 1 tomato-infecting strain (Fol004) a FoHeli1 is located 156 bp upstream from a gene for a secreted oxido-reductase (ORX1-like) protein (AKC01502.1). Finally, in a melon-infecting isolate (Mel02010), we find a partial copy of a FoHeli1 located 476 bp upstream from a predicted argininosuccinate lyase gene (*ARG1*, AB045736.1). Deletion of *ARG1* leads to a reduction in virulence [49]. All partial copies in these examples lie on the border of the sequence that was submitted to GenBank, hence they could very well be complete copies that have either not been sequenced or not correctly assembled. Ectopic recombination between (almost) identical Helitron sequences can result in deletion of genomic regions. If these regions contain genes that are involved in infection, this may contribute to changes in virulence [50].

FoHeli elements cluster phylogenetically with Helitron2 proteins

The termini of FoHelis suggest that they belong to the Helitron2 variant [2]. To test this, we compiled a set of Helitron protein sequences extracted from RepBase [51] and Helitron2 sequences described in [2]. We also wanted to know how FoHelis relate to Helitrons found in relatively closely related fungi, hence we searched 102 Pezizomycotina proteomes for proteins with a Rep-Hel domain architecture. We predicted 45 proteins in 16 Pezizomycotina species to be putative Helitrons and added those to our dataset. We inferred a phylogeny and find that FoHelis and most fungal Helitrons group with Helentrons and other Helitron2 elements with high bootstrap support (Fig. 5, Additional file 1: Figure S12, Figure S13). This suggests that many fungal Helitrons have non-canonical termini. We find no fungal putative Helitrons that contain an endonuclease domain, the hallmark domain of Helentrons.

Conservation of terminal features: FoHeli-like termini in other fungi

FoHelis share several features with members of the Helitron2 variant, but none of these members have the exact same terminal sequences as FoHelis [2]. To determine to what extent the exact termini are FOSC-specific we searched a database of 102 Pezizomycotina genomes for Helitrons with FoHeli-like termini (Additional file 8: Table S8). For each subgroup, we find at least one sequence outside FOSC that possesses FoHeli termini (Additional file 9: Table S9). The species in which we find completely conserved FoHelis (i.e. including termini) corresponds to what we would expect given the tree presented above: FoHeli1 is present in *Metarhizium anisopliae* ARSEF 23 (currently corrected to *Metarhizium robertsii*), FoHeli4 in *Verticillium dahliae VdLs.17* and FoHeli5 in *Chaetomium globosum. Fusarium solani* has all FoHeli subgroups except FoHeli5. In *Metarhizium acridum*, we only find 3' termini, except for one case in which we observe three Helitron copies in tandem. Either a Helitron was inserted into the 5' end of another Helitron twice (MAC_03224 and MAC_3225 in Additional file 1: Figures S12 and S13), or this is the result of rolling circle replication of single stranded circular DNA. Finally, we find FoHeli2 in *F. acuminatum*, and FoHeli2, FoHeli3 and FoHeli4 in *F. virguliforme*, genome sequences for which annotations are not publicly available.

Interestingly, FoHeli1 sequences in *F. solani* bear hallmarks of Repeat Induced Point (RIP) mutation with a more than 3-fold increase in CpA to TpA and TpG to TpA mutations compared to other G->A and C->T mutations (Additional file 1: Figure S14). RIP is hypothesized to function as a genome defence mechanism against duplicated genes and TEs and RIP can at least partially explain why we do not find a large number of proteins with a Hel-Rep domain architecture in *F. solani* [52, 53].

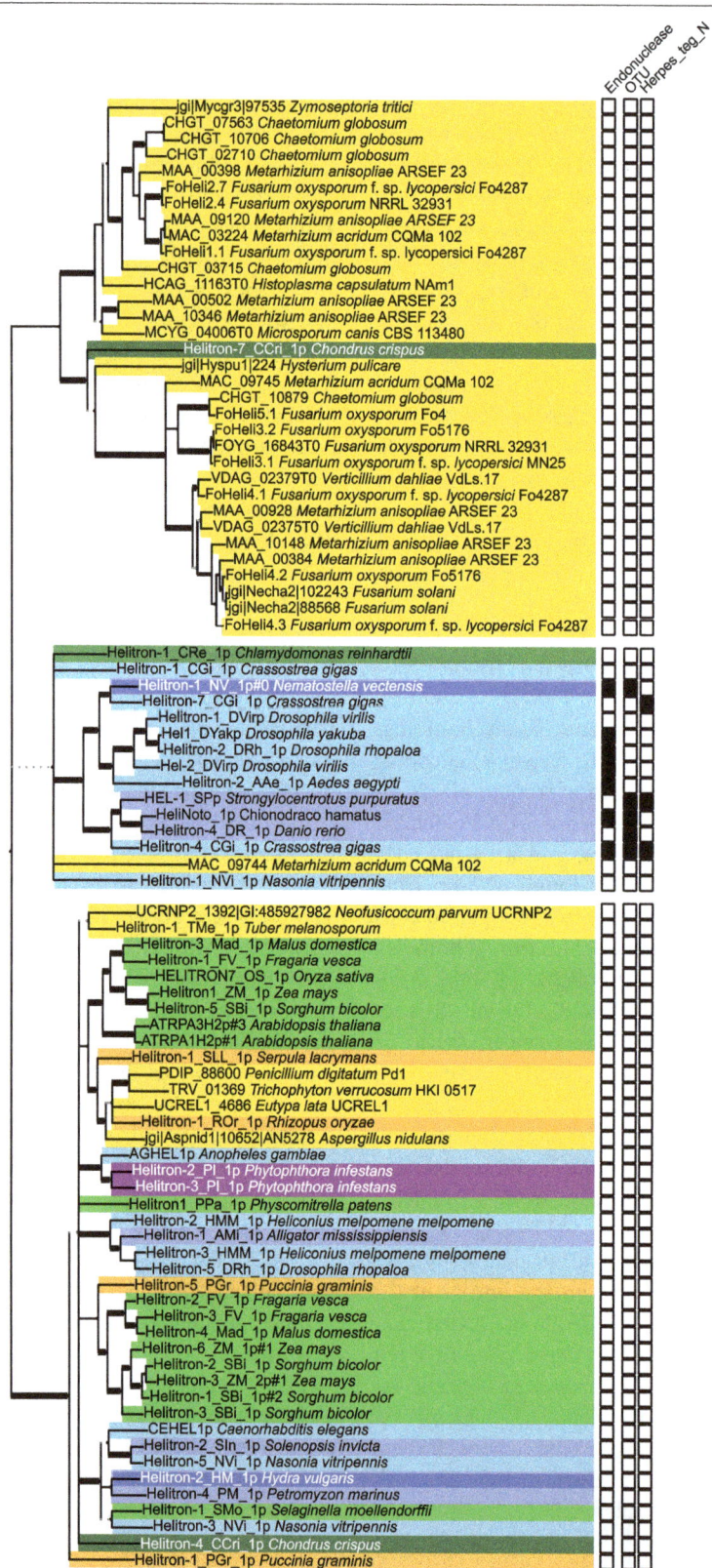

Fig. 5 (See legend on next page.)

(See figure on previous page.)
Fig. 5 Phylogenetic relationships and domain architecture of FOSC and other fungal Helitrons, Helentrons and canonical Helitrons. Phylogeny based on a multiple sequence alignment of known Helitrons and Helentrons from RepBase, and a set of fungal Helitrons detected by scanning fungal proteomes for proteins with a Hel-Rep domain architecture. Due to space constraints, we do not show all proteins included in the phylogeny but selected a subset that represents the full tree as depicted in Additional file 1: Figure S13. We inferred a 100 bootstrap replicates: thick branches have a bootstrap value of 100 and grey branches a bootstrap value < 70. Branches with bootstrap values < 50 have been removed. Background color of leaves indicates taxonomy: yellow - Fungi, light-blue insects, blue – other Animals, green - Plants, dark green - Red Algae, purple – Oomycetes. All proteins included in this tree have a Hel and a Rep domain. The Helentrons that posses an endonuclease domain, often combined with a OTU domain, form a distinct group

In the trees of Helitron sequences presented in Fig. 5, Additional file 1: Figure S13 and Figure S14, we find two clades of very similar Helitron sequences in *Chaetomium globosum* that neighbour the FoHeli1 and FoHeli2 clades. Yet we did not find FoHeli1- or FoHeli2-like termini in *Chaetomium globosum* using our blastn. To determine the termini for these Helitrons, we took the same approach as we did originally for FOSC Helitron sequences: we aligned the gene sequences including a large up- and downstream region and inspected these alignments to find termini for the Helitrons. We find that *C. globosum* Helitrons possess the 3' terminus of FOSC Helitrons, including the 'ATTTT' and the inverted repeat, but do not have a hairpin at the 5' end (Fig. 6). The 3' terminus is more conserved than the 5' terminus. Finally, *Chaetomium globosum* Helitrons, like FoHeli1 and FoHeli2, appear to insert between 'TNAT' and 'A', where 'N' denotes any nucleotide.

Discussion
Detection of non-canonical Helitrons
FoHelis likely represent a large reservoir of Pezizomycotina Helitrons that group phylogenetically with Helitron2 transposons, suggesting that most fungal Helitrons have non-canonical termini (Fig. 5). Indeed, we were able to confirm that Helitrons with FoHeli-like termini also occur in other fungi (Additional file 9: Table S9, Fig. 6). In the case of the FOSC, we would not have detected any Helitrons using conventional approaches based on termini or DNA sequences of canonical Helitrons [5, 18, 20, 39]. Our analyses of predicted putative

Helitrons in other fungi suggests that the same may hold true for many other species [38, 40].

Another factor that hampered detection of FoHelis is their size. Genome assemblies based on second generation sequencing data are unlikely to include recently transposed elements of more than 5 kb [44, 45, 54]. Hence the repeat content of genomes that are assembled to different levels of completeness cannot be directly compared [40, 54]. Similarly, non-autonomous elements are often more abundant than their autonomous counterparts [1, 3], which can be explained by the intuitive assumption that shorter sequences are more efficiently transposed. On the other hand, non-autonomous elements are more likely to be assembled in one piece and therefore more easily detected. Hence we may have been overestimating their success as a parasite's parasite. Improvements in genome assembly through the use of third and fourth generation sequencing technologies will allow us to better estimate and compare the TE repertoires of different genomes, to reconstruct the influence of transposons on genome evolution, but also to gain understanding on (co-) evolution of selfish elements in and across host genomes [19, 55–57].

Self-insertion may have led to composite FoHelis
Self-insertion can lead to nested, composite or chimeric Helitrons [14, 24, 30]. In this study, we've found one example of a non-autonomous FoHeli nested into a putative autonomous one. Moreover, we've found a number of FoHelis in which multiple 5' termini were combined with a single 3' terminus. Typically, the 5' sequence that

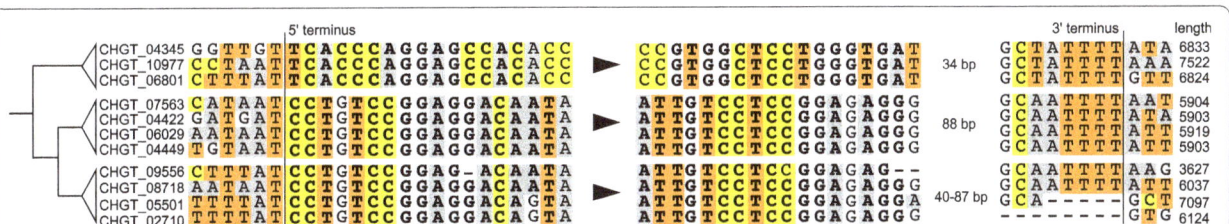

Fig. 6 Termini of Helitrons in *Chaetomium globosum* exemplify conservation of 3' terminal sequences. We determined the termini for two groups of *Chaetomium globosum* Helitrons that group together with FoHeli1 and FoHeli2 in the tree in Fig. 5, Additional file 1: Figure S12 and Figure S13. Terminal inverted repeats (TIRs) are in bold. In contrast to the ATIRs in FoHeli1 and FoHeli2, the ATIRs of these Helitrons are not hairpins. The sequence of the 3' termini closely resembles those of FoHeli1 and FoHeli2, as they also end in 'ATTTT'. Moreover, the bottom two subgroups possess imperfect palindromes overlapping their ATIRS

is duplicated is short (<200 bp). These 5' duplications may stem from nested FoHelis that result from self-insertion. If, during transposition, the transposase nicks the leftmost 5' terminus of the nested Helitron, and continues to unwind the DNA until it encounters the first 3' terminus, where it stops, it may transpose a FoHeli with two 5' termini (Fig. 7). Reversal of ORF orientation may also stem from a composite or nested Helitron, in which one copy is inserted into the other in opposite orientation, after which the innermost set of termini is deleted or mutated and only the extreme termini are preserved (Fig. 8).

Detection of circular intermediates

Results from this study indicate that FoHelis, like canonical Helitrons [14] transpose via a circular intermediate. However, we failed to amplify circular Helitrons using Rolling Circle Amplification, suggesting that we need additional preprocessing steps to enrich our genomic DNA samples for circular DNA other than from mitochondria to find circular Helitrons via this approach (e.g. as in [58, 59]). DNA isolation provides a snapshot of DNA content of a large number of cells and for a Helitron circle to be present, it has to transpose at that exact time. Therefore we expect very few circles to be present in one DNA sample and need extremely sensitive methods to detect them.

Which FoHelis are still active in the FOSC?

In Fol4287, we've found two identical copies of FoHeli1 that, judging from their flanking sequences, arose through transposition rather than segmental duplication. Moreover, FoHeli1 is the subgroup we have found most at contig borders in Fol4287 and for which we found a PCR amplicon that could stem from a circular intermediate. This suggests that FoHeli1 is still active in the genome of Fol4287. The other subgroup that appeared to have had a strong impact on genome assembly is

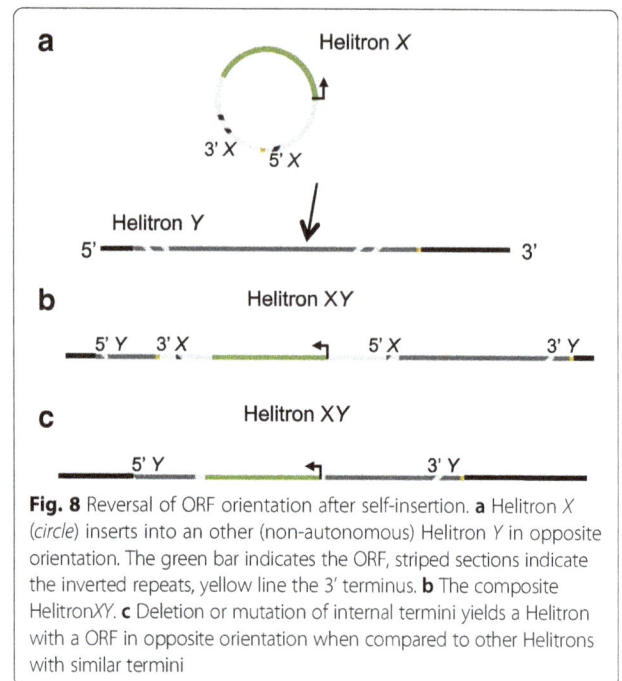

Fig. 8 Reversal of ORF orientation after self-insertion. **a** Helitron X (*circle*) inserts into an other (non-autonomous) Helitron Y in opposite orientation. The green bar indicates the ORF, striped sections indicate the inverted repeats, yellow line the 3' terminus. **b** The composite HelitronXY. **c** Deletion or mutation of internal termini yields a Helitron with a ORF in opposite orientation when compared to other Helitrons with similar termini

FoHeli2 that is predicted to occur in high copynumber in brassicaceae-infecting isolates. In contrast to the genome of Fol4287, the genomes of these isolates have not been assembled with the aid of an optical map. Improved assemblies, combined with detection of putative circular intermediates, may shed light on when FoHeli2 was active in these isolates.

Conclusions

Helitrons have been studied for more than a decade, where the main focus has been on canonical Helitrons, or Helitron1, in plants and animals. Here we present the first study of non-canonical Helitron transposons in Pezizomycotina, shedding light on a Helitron variant in

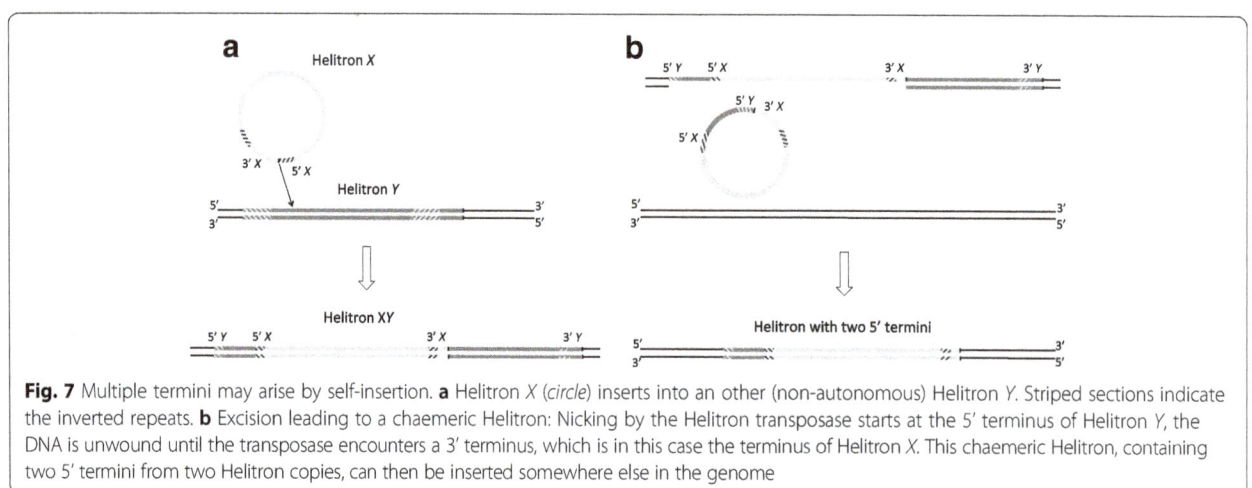

Fig. 7 Multiple termini may arise by self-insertion. **a** Helitron X (*circle*) inserts into an other (non-autonomous) Helitron Y. Striped sections indicate the inverted repeats. **b** Excision leading to a chaemeric Helitron: Nicking by the Helitron transposase starts at the 5' terminus of Helitron Y, the DNA is unwound until the transposase encounters a 3' terminus, which is in this case the terminus of Helitron X. This chaemeric Helitron, containing two 5' termini from two Helitron copies, can then be inserted somewhere else in the genome

a subphylum that both have been relatively underrepresented in scientific literature on Helitrons. In FOSC, we've identified 2 groups with distinct terminal sequences. We presented data suggesting that FOSC Helitrons transpose via a circular intermediate, which has been shown for canonical Helitrons very recently [14]. Importantly, we found that most Pezizomycotina Helitrons are probably non-canonical. The information we provide here will aid in future identifications of Helitrons and thus contribute to a more accurate characterization of transposon repertoires, especially in Pezizomycotina.

Methods
Identification of putative autonomous Helitrons in FOSC
We select 35 genes from 10 different strains encoding proteins with a Rep (PF14214) and a Hel (PF05790) domain based on Pfam annotation for the 12 FOSC genomes provided by the Broad Institute [10, 13, 47, 60]. To detect additional copies that were excluded from the gene annotation, we used these 35 proteins as a query in a tblastn search to find homologous regions in the 12 FOSC genomes [61]: sequences were included if the alignment returned by BLAST covered at least 80% of the query with > = 80% identity. These sequences were extended up to 10 kb in each direction and annotated by FgenesH [62], an online program for gene prediction, using parameters of *Fusarium graminearum*. We determined the domain architecture for the proteins encoded in these predicted ORFs using hmmscan and the PfamA database, applying default inclusion thresholds. The genes that encode proteins with a Rep (PF14214) and a Hel (PF05790) domain, were considered putative autonomous Helitrons. In this way, we found 28 more Helitrons, bringing our total to 63 (Additional file 2: Table S1). These were subsequently used as queries to search for additional copies using blastn. We found no additional full-length copies. In total, we retrieved 63 Helitron protein sequences in the FOSC (Fig. 1, Additional file 2: Table S1).

Phylogenetic analyses of FOSC Helitrons
To assess how these 63 FOSC Helitrons are clustered into subgroups, we aligned protein sequences using prank [63] with default settings, trimmed the multiple sequence alignment with trimAl −strictplus [64] and inferred a tree using PhyML v3.0 [65] with 4 substitution rate categories, estimated proportion of invariable sites and gamma distribution. We run PhyML once to produce bootstrap support (100 bootstraps) and once with aLRT branch support (SH-like). For the tree depicted in Fig. 1, branches that have aLRT-support < 0.9 and/or bootstrap support < 80 were collapsed using a custom python script implementing ete2 [66]. We found that FOSC Helitrons can be divided into 5 subgroups, here designated FoHeli1-FoHeli5 (Fig. 1).

Identification of Helitron termini
If different copies of a transposable element arose through transposition (as opposed to segmental duplication), sequence similarity between the copies extends up to the termini of the transposable elements, but not further. We use this to identify termini for FOSC Helitrons. For each FoHeli subgroup, we add 1–7 kb of flanking sequences to the predicted gene sequences, if possible, i.e. if the Helitron is not to close to the border of a (super)contig. We align these sequences using Clustal Omega [47] and manually inspect alignments to find the regions where the sequences change from dissimilar to very similar (5' terminus) or from very similar to dissimilar (3' terminus). We use this approach to identify termini in *Chaetomium globosum* as well (Fig. 6). To identify FOSC termini in other fungal species we queried a database of 102 Pezizomycotina genomes with the DNA sequences of FoHeli full elements. We combined all partial hits of the same FoHeli query that are located within close distance (<3 kb), aligned the corresponding region with the query and inspected the alignment to determine whether FoHeli termini were indeed present.

For each subgroup, we reconstructed pre-insertion sites by concatenating 500 bp 5' flanking sequence of the FoHeli with 500 bp of 3' flanking sequence of the FoHeli. In some cases the FoHeli resides closer than 500 bp to a supercontig border, then we took as much flanking sequence as we could. We use blastn to search for these pre-insertion sites within the 12 FOSC genomes. We used a custom python script to extract the sequences of BLAST hits that bridge the two flanking sequences, write these sequences to a fastafile and align these with Clustal Omega to confirm the termini we inferred are correct.

Estimation of FoHeli copynumber from partial hits
We expected that the number of Helitrons we found in our initial survey [63] is an underestimate of the real copynumber as a result of e.g. gaps in the genome assembly or regions of high divergence within Helitrons. We search the 12 FOSC genomes using megablast, with the 41 FOSC terminus-to-terminus Helitron sequences as queries, each with 100 bp of flanking sequence. The resulting blast output was parsed using a Python script. We only considered hits that start after the first 90 and end before the last 90 bp.

Due to low complexity or gaps between the contigs that are represented by 'N's, BLAST may produce multiple alignments of a query (sub)sequence and a subject (sub)sequence. To avoid overestimating the number of partial hits because of this, we first merged hits that were less than 200 bp apart in the query, but for which the overlap in the query was <50 bp (to ensure that individual hits represent different parts of the query), and less than 2000 bp apart in the subject (scaffold) sequence, assuming that these multiple hits represent one putative Helitron sequence on the supercontig. Moreover, we merged hits that represented multiple termini.

Identification of putative gene capture events
Helitrons are well-known for their ability to capture (parts of) genes [4, 6, 7, 14–22]. To determine the extent of gene capture in FOSC Helitrons, we search NCBIs non-redundant nucleotide database (nr/nt) using 48 full-length FOSC Helitrons. We use a custom python script to query the Entrez database with the Genbank Identifiers returned by the BLAST search. We select hits that contain a coding sequence and find the corresponding protein sequence. We infer domain architectures for these protein sequences using hmmscan from the hmmer3 package [67] and the PfamA database (Pfam 27.0) [68] and select proteins that do not contain a Helitron-like_N (Rep) or PIF1 (Hel) domain. We thus obtain a list of 27 genes that have been (partially) captured by a FOSC Helitron.

DNA isolation, PCR analysis and sequencing
We use PCR to detect circular intermediates of FOSC Helitrons (Fig. 4). Fungal genomic DNA (gDNA) was extracted using the following method: a patch of mycelium was scraped from the margin of a colony and suspended in 400 µl Tris-EDTA buffer (1 M Tris pH 8, 0.5 M EDTA pH 8) together with 300 µl phenol:chloroform (1:1) and approximately 300 µl glass beads (212–300 µm). Cells were mechanically disrupted in a tissuelyser for 30 s. The supernatant (150 µl) was collected after centrifugation (5 min) at maximum speed and mixed with equal volume of chloroform. Again, the supernatant (100 µl) was collected after vortexing and centrifugation (5 min) and kept in -20 °C for further use. 1 µl of genomic DNA was used for PCR experiments. Primers used for amplification of the FoHeli joined ends are listed in Additional file 1: Table S4. The amplified products were resolved electrophoretically in a 1% agarose gel. PCR products were sequenced and analyzed using Seqbuilder.

Rolling circle amplification and downstream analyses
Rolling circle amplification was performed on 80 ng Fol4287, 250 ng Fol4287, 80 ng Fol029, 80 ng Fo5176, 80 ng Fo47, 80 ng of Fom001 genomic DNA and a 5169 bp plasmid spiked into 80 ng of Fo47 genomic DNA, as described by [46] using phi29 DNA polymerase (#EP0091, Thermo Scientific), inorganic pyrophosphatase (#EF0221, Thermo Scientific) and exo-resistant random primers (#S0181, Thermo Scientific) in a 12.5 h, 20 µL reaction at 30 °C (Additional file 1: Figure S7). The reaction was stopped by elevating the temperature to 65 °C for 10 min. Subsequently, 5 µL of the amplification product was digested with Acc65I, XhoI or

EcoRV for 3 h and run on a 1% agarose gel. A band of the expected size (~6–7 kb) was observed and extracted from the gel using a QIAquick Gel Extraction Kit according to the manufacturer's protocol. 1 µL of the 6–7 kb fragment was used for a regular PCR using primer pairs distributed over the length of FoHeli1 (Additional file 1: Figure S8).

Phylogenetic analyses of FOSC, pezizomycotina and known Helitrons
For the phylogenetic analyses including known Helitron1 and Helitron2 from RepBase (version 19.11), we used custom Python scripts to parse RepBase files for protein sequences of Helitrons. In addition, we obtained Helitron2 sequences described in [2] from the authors. These include all proteins that reside within Helitron termini, hence also e.g. replication protein A (RPA)-like proteins. We predicted domain architecture for these proteins using hmmscan from the hmmer3 package [67] and the PfamA database (Pfam 27.0) [68]. We used custom Python scripts and manual curation to determine the final domain architecture of individual proteins: in case of overlapping domain predictions (mostly PIF1 domains that also matched AAA domains), we kept the domain with the highest score (PIF1), or, in cases in which predictions likely correspond to the same domain, we merged overlapping regions that also overlapped in a similar fashion (e.g. no inversions) in the hmm model. In further analyses, we only include protein sequences that contain a Rep and a Hel domain (PF14214 and PF05970). We use hmmsearch from the hmmer3 package [67] with PF14214 and PF05970 to scan all Pezizomycotina proteomes in our dataset (Additional file 8: Table S8) for proteins that contain both these domains. We constructed two different multiple sequence alignments for this set of proteins. First, we cut out Rep and Hel domains from each protein, removed identical sequences, aligned the domain sequences using hmmalign and concatenated the alignments of both domains. Second, we aligned full protein sequences using Clustal Omega with default settings [69]. We then trimmed this alignment with trimAl (–gappyout), removed identical sequences and used RaxML to infer the phylogeny (options: –f a -N 100 -m PROTGAMMAIWAG -x 1234567 -p 123 (Additional file 1: Figure S12 and Figure S13 and Fig. 5) [70]. Figure 5 shows the Clustal Omega tree, where branches with a bootstrap support of less than 50 trees were collapsed.

For all the putative Helitron proteins in the these trees we predicted whether any domain other than the Rep and Hel domain was present using hmmscan from the hmmer3 package and the PfamA database (Pfam 27.0) [67, 68]. The domain architecture of proteins is summarized in Fig. 5, and shown more elaborately in Additional file 1: Figure S12 and Figure S13. Many putative Helitrons have an N-terminal Helicase domain that is classified in Pfam as either a Herpes_Helicase, an UvrD_C_2 or a Viral_helicase_1 domain (Additional file 1: Figure S12 and Figure S13). However, when we overlay the conserved Hel motifs I-VI in FoHelis (Additional file 1: Figure S2) to the automated Pfam domain prediction, we find that only motifs I until IV/V lie in the predicted Hel domain, whereas the other two motifs (V and VI) lie in the predicted N-terminal Helicase. This means that the automated prediction of both the Hel and the N-terminal Helicase is probably incorrect and that the predicted N-terminal Helicase domains are actually part of the Hel domain.

Additional files

Additional file 1: Supplemental file with Figure S1,S2 and S3 that show conserved motifs in FoHeli protein sequences, Figure S4 that shows a dotplot of chromosome 3 and chromosome 6, with FoHelis on the diagonal, Table S4 that contains the primers in this study, Figure S5 that shows an alignment of non-autonomous FoHeliNA1 with FoHeli1, Figure S6 that shows an alignment of non-autonomous FoHeliNA2 with FoHeli1, Figure S7 that shows a gel with the product of RCAs, digested with different enzymes, Figure S8 that shows results of PCR amplification of a 6-7 kb band from Figure S7, Figure S9 and S10 shows the results of mapping unpaired and paired reads on a constructed tandem FoHeli1, Figure S11 shows alignments of FoHelis with multiple termini and their flanking regions, Figure S12 and S13 show phylogenetic trees of FoHelis with other predicted Helitrons and Figure S14 shows putative RIP mutations in FoHeli1 homologs in *F. solani*. (PDF 17380 kb)

Additional file 2: Table S1: FoHelis found in this study: position on the genome, start and end sequences, position of the predicted ORFs. (XLSX 46 kb)

Additional file 3: Table S2, additional ORFs predicted within FoHeli termini. (XLSX 38 kb)

Additional file 4: Table S3: non-autonomous FoHelis. (XLSX 47 kb)

Additional file 5: Table S5: Copynumber estimate based on partial hits. (XLSX 272 kb)

Additional file 6: Table S6: FoHelis with multiple 5' termini. (XLSX 65 kb)

Additional file 7: Table S7: Putative gene capture events. (XLSX 123 kb)

Additional file 8: Table S8: Number of protiens with a Rep and/or Hel domain found in queried species. (XLSX 22 kb)

Additional file 9: Table S9: FoHeli-like sequences found in other species. (XLSX 39 kb)

Acknowledgements
The autors are very grateful to Frank Takken for useful discussions. The genome sequencing and annotation for 11 strains were supported by the National Research Initiative Competitive Grants Program Grant no. 2008-35604-18800 and MASR-2009-04374 from the USDA National Institute of Food and Agriculture.

Funding
Biju V.C. is supported by the Erasmus Mundus External Cooperation Window 15 (EMECW15). Like Fokkens is supported by a Horizon grant from the Netherlands Genomics Iniative. The genome sequencing and annotation for 11 strains were supported by the National Research Initiative Competitive Grants Program Grant no. 2008-35604-18800 and MASR-2009-04374 from the USDA National Institute of Food and Agriculture.

Availability of data and materials
FOSC Helitron sequences have been submitted to RepBase. FOSC genome sequences can be downloaded from GenBank: *Fusarium oxysporum* f. sp. *lycopersici* 4287, accession number GCA_000149955.2; *Fusarium oxysporum* f. sp. *lycopersici* MN25, accession number GCA_000259975.2; *Fusarium oxysporum* f. sp. *pisi* HDV247, accession number GCA_000260075.2; *Fusarium oxysporum* f. sp. *radicis-lycopersici* 26381, accession number GCA_000260155.3; *Fusarium oxysporum* f. sp. *vasinfectum* 25433, accession number GCA_000260175.2; *Fusarium oxysporum* f. sp. *cubense* tropical race 4 54006, accession number GCA_000260195.2; *Fusarium oxysporum* f. sp. *conglutinans* race 2 54008, accession number GCA_000260215.2; *Fusarium oxysporum* f. sp. *raphani* 54005, accession number GCA_000260235.2; *Fusarium oxysporum* f. sp. *melonis* 26406, accession number GCA_000260495.2; *Fusarium oxysporum* Fo47, accession number GCA_000271705.2; *Fusarium oxysporum* FOSC 3-a, accession number GCA_000271745.2; *Fusarium oxysporum* Fo5176, accession number GCA_000222805.1.

Authors' contributions
BVC designed and performed PCR experiments and bioinformatic analyses and assisted in writing the manuscript, PvD performed RCA experiments, LF designed and performed bioinformatic analyses and wrote the manuscript, MR and BJC Cornelissen assisted in experimental design and writing the manuscript. All authors read and approved the final manuscript.

Competing interests
The authors declare that they have no competing interests.

Author details
[1]Department of Computational Biology and Bioinformatics, University of Kerala, Karyavattom Campus, Karyavattom PO, Trivandrum, Kerala, India. [2]Molecular Plant Pathology, Swammerdam Institute for Life Sciences, Faculty of Science, University of Amsterdam, P.O. Box 94215, 1090 Amsterdam, GE, The Netherlands.

References
1. Kapitonov VV, Jurka J. Rolling-circle transposons in eukaryotes. Proc Natl Acad Sci U S A. 2001;98(15):8714–9.
2. Bao W, Jurka J. Homologues of bacterial TnpB_IS605 are widespread in diverse eukaryotic transposable elements. Mob DNA. 2013;4(1):12. -8753-4-12.
3. Thomas J, Vadnagara K, Pritham EJ. DINE-1, the highest copy number repeats in Drosophila melanogaster are non-autonomous endonuclease-encoding rolling-circle transposable elements (Helentrons). Mob DNA. 2014;5:18. 8753-5-18. eCollection 2014.
4. Castanera R, Perez G, Lopez L, Sancho R, Santoyo F, Alfaro M, Gabaldon T, Pisabarro AG, Oguiza JA, Ramirez L. Highly expressed captured genes and cross-kingdom domains present in Helitrons create novel diversity in Pleurotus ostreatus and other fungi. BMC Genomics. 2014;15:1071. 2164-15-1071.
5. Poulter RT, Goodwin TJ, Butler MI. Vertebrate helentrons and other novel Helitrons. Gene. 2003;313:201–12.
6. Thomas J, Phillips CD, Baker RJ, Pritham EJ. Rolling-circle transposons catalyze genomic innovation in a Mammalian lineage. Genome Biol Evol. 2014;6(10):2595–610.
7. Thomas J, Pritham EJ: Helitrons, the Eukaryotic Rolling-circle Transposable Elements. Microbiol Spectr 2015, 3(4):10.1128/microbiolspec.MDNA3-0049-2014.
8. Schmidt SM, Houterman PM, Schreiver I, Ma L, Amyotte S, Chellappan B, Boeren S, Takken FL, Rep M. MITEs in the promoters of effector genes allow prediction of novel virulence genes in Fusarium oxysporum. BMC Genomics. 2013;14:119. 2164-14-119.
9. Ma LJ, Geiser DM, Proctor RH, Rooney AP, O'Donnell K, Trail F, Gardiner DM, Manners JM, Kazan K. Fusarium pathogenomics. Annu Rev Microbiol. 2013;67:399–416.
10. Ma LJ, van der Does HC, Borkovich KA, Coleman JJ, Daboussi MJ, Di Pietro A, Dufresne M, Freitag M, Grabherr M, Henrissat B, Houterman PM, Kang S, Shim WB, Woloshuk C, Xie X, Xu JR, Antoniw J, Baker SE, Bluhm BH, Breakspear A, Brown DW, Butchko RA, Chapman S, Coulson R, Coutinho PM, Danchin EG, Diener A, Gale LR, Gardiner DM, Goff S, Hammond-Kosack KE, Hilburn K, Hua-Van A, Jonkers W, Kazan K, Kodira CD, Koehrsen M, Kumar L,

Lee YH, Li L, Manners JM, Miranda-Saavedra D, Mukherjee M, Park G, Park J, Park SY, Proctor RH, Regev A, Ruiz-Roldan MC, Sain D, Sakthikumar S, Sykes S, Schwartz DC, Turgeon BG, Wapinski I, Yoder O, Young S, Zeng Q, Zhou S, Galagan J, Cuomo CA, Kistler HC, Rep M. Comparative genomics reveals mobile pathogenicity chromosomes in Fusarium. Nature. 2010;464(7287): 367–73.

11. Raffaele S, Farrer RA, Cano LM, Studholme DJ, MacLean D, Thines M, Jiang RH, Zody MC, Kunjeti SG, Donofrio NM, Meyers BC, Nusbaum C, Kamoun S. Genome evolution following host jumps in the Irish potato famine pathogen lineage. Science. 2010;330(6010):1540–3.

12. Croll D, McDonald BA. The accessory genome as a cradle for adaptive evolution in pathogens. PLoS Pathog. 2012;8(4):e1002608.

13. Fusarium Comparative Sequencing Project, Broad Institute of Harvard and MIT, accessed 2014 [http://www.broadinstitute.org/]

14. Grabundzija I, Messing SA, Thomas J, Cosby RL, Bilic I, Miskey C, Gogol-Doring A, Kapitonov V, Diem T, Dalda A, Jurka J, Pritham EJ, Dyda F, Izsvak Z, Ivics Z. A Helitron transposon reconstructed from bats reveals a novel mechanism of genome shuffling in eukaryotes. Nat Commun. 2016;7:10716.

15. Lai J, Li Y, Messing J, Dooner HK. Gene movement by Helitron transposons contributes to the haplotype variability of maize. Proc Natl Acad Sci U S A. 2005;102(25):9068–73.

16. Bennetzen JL. Transposable elements, gene creation and genome rearrangement in flowering plants. Curr Opin Genet Dev. 2005;15(6):621–7.

17. Feschotte C, Pritham EJ. DNA transposons and the evolution of eukaryotic genomes. Annu Rev Genet. 2007;41:331–68.

18. Sweredoski M, DeRose-Wilson L, Gaut BS. A comparative computational analysis of nonautonomous helitron elements between maize and rice. BMC Genomics. 2008;9:467. 2164-9-467.

19. Feschotte C, Pritham EJ. A cornucopia of Helitrons shapes the maize genome. Proc Natl Acad Sci U S A. 2009;106(47):19747–8.

20. Dong Y, Lu X, Song W, Shi L, Zhang M, Zhao H, Jiao Y, Lai J. Structural characterization of helitrons and their stepwise capturing of gene fragments in the maize genome. BMC Genomics. 2011;12:609. 2164-12-609.

21. Barbaglia AM, Klusman KM, Higgins J, Shaw JR, Hannah LC, Lal SK. Gene capture by Helitron transposons reshuffles the transcriptome of maize. Genetics. 2012;190(3):965–75.

22. Han MJ, Shen YH, Xu MS, Liang HY, Zhang HH, Zhang Z. Identification and evolution of the silkworm helitrons and their contribution to transcripts. DNA Res. 2013;20(5):471–84.

23. del Pilar Garcillan-Barcia M, Bernales I, Mendiola MV, de la Cruz F. Single-stranded DNA intermediates in IS91 rolling-circle transposition. Mol Microbiol. 2001;39(2):494–501.

24. Xiong W, Dooner HK, Du C. Rolling-circle amplification of centromeric Helitrons in plant genomes. Plant J. 2016. doi:10.1111/tpj.13314.

25. Gupta S, Gallavotti A, Stryker GA, Schmidt RJ, Lal SK. A novel class of Helitron-related transposable elements in maize contain portions of multiple pseudogenes. Plant Mol Biol. 2005;57(1):115–27.

26. Morgante M, Brunner S, Pea G, Fengler K, Zuccolo A, Rafalski A. Gene duplication and exon shuffling by helitron-like transposons generate intraspecies diversity in maize. Nat Genet. 2005;37(9):997–1002.

27. Pritham EJ, Feschotte C. Massive amplification of rolling-circle transposons in the lineage of the bat Myotis lucifugus. Proc Natl Acad Sci U S A. 2007;104(6):1895–900.

28. Hollister JD, Gaut BS. Population and evolutionary dynamics of Helitron transposable elements in Arabidopsis thaliana. Mol Biol Evol. 2007;24(11):2515–24.

29. Cultrone A, Dominguez YR, Drevet C, Scazzocchio C, Fernandez-Martin R. The tightly regulated promoter of the xanA gene of Aspergillus nidulans is included in a helitron. Mol Microbiol. 2007;63(6):1577–87.

30. Tempel S, Nicolas J, El Amrani A, Couee I. Model-based identification of Helitrons results in a new classification of their families in Arabidopsis thaliana. Gene. 2007;403(1–2):18–28.

31. Du C, Caronna J, He L, Dooner HK. Computational prediction and molecular confirmation of Helitron transposons in the maize genome. BMC Genomics. 2008;9:51. 2164-9-51.

32. Yang L, Bennetzen JL. Distribution, diversity, evolution, and survival of Helitrons in the maize genome. Proc Natl Acad Sci U S A. 2009;106(47): 19922–7.

33. Yang L, Bennetzen JL. Structure-based discovery and description of plant and animal Helitrons. Proc Natl Acad Sci U S A. 2009;106(31):12832–7.

34. Haas BJ, Kamoun S, Zody MC, Jiang RH, Handsaker RE, Cano LM, Grabherr

M, Kodira CD, Raffaele S, Torto-Alalibo T, Bozkurt TO, Ah-Fong AM, Alvarado L, Anderson VL, Armstrong MR, Avrova A, Baxter L, Beynon J, Boevink PC, Bollmann SR, Bos JI, Bulone V, Cai G, Cakir C, Carrington JC, Chawner M, Conti L, Costanzo S, Ewan R, Fahlgren N, Fischbach MA, Fugelstad J, Gilroy EM, Gnerre S, Green PJ, Grenville-Briggs LJ, Griffith J, Grunwald NJ, Horn K, Horner NR, Hu CH, Huitema E, Jeong DH, Jones AM, Jones JD, Jones RW, Karlsson EK, Kunjeti SG, Lamour K, Liu Z, Ma L, Maclean D, Chibucos MC, McDonald H, McWalters J, Meijer HJ, Morgan W, Morris PF, Munro CA, O'Neill K, Ospina-Giraldo M, Pinzon A, Pritchard L, Ramsahoye B, Ren Q, Restrepo S, Roy S, Sadanandom A, Savidor A, Schornack S, Schwartz DC, Schumann UD, Schwessinger B, Seyer L, Sharpe T, Silvar C, Song J, Studholme DJ, Sykes S, Thines M, van de Vondervoort PJ, Phuntumart V, Wawra S, Weide R, Win J, Young C, Zhou S, Fry W, Meyers BC, van West P, Ristaino J, Govers F, Birch PR, Whisson SC, Judelson HS, Nusbaum C. Genome sequence and analysis of the Irish potato famine pathogen Phytophthora infestans. Nature. 2009;461(7262):393–8.

35. Langdon T, Thomas A, Huang L, Farrar K, King J, Armstead I. Fragments of the key flowering gene GIGANTEA are associated with helitron-type sequences in the Pooideae grass Lolium perenne. BMC Plant Biol. 2009;9:70. 2229-9-70.

36. Cantu D, Govindarajulu M, Kozik A, Wang M, Chen X, Kojima KK, Jurka J, Michelmore RW, Dubcovsky J. Next generation sequencing provides rapid access to the genome of Puccinia striiformis f. sp. tritici, the causal agent of wheat stripe rust. PLoS One. 2011;6(8):e24230.

37. Fu D, Wei L, Xiao M, Hayward A. New insights into helitron transposable elements in the mesopolyploid species Brassica rapa. Gene. 2013;532(2): 236–45.

38. Curcio MJ, Derbyshire KM. The outs and ins of transposition: from mu to kangaroo. Nat Rev Mol Cell Biol. 2003;4(11):865–77.

39. Xiong W, He L, Lai J, Dooner HK, Du C. HelitronScanner uncovers a large overlooked cache of Helitron transposons in many plant genomes. Proc Natl Acad Sci U S A. 2014;111(28):10263–8.

40. Platt 2nd RN, Blanco-Berdugo L, Ray DA. Accurate transposable element annotation is vital when analyzing new genome assemblies. Genome Biol Evol. 2016;8(2):403–10.

41. Koonin EV, Ilyina TV. Computer-assisted dissection of rolling circle DNA replication. BioSystems. 1993;30(1–3):241–68.

42. Koonin EV, Corbalenya AE. Helicases: amino acid sequence comparisons and structure-function relationships. Curr Opin Struct Biol. 1993;3:419–29.

43. Fairman-Williams ME, Guenther UP, Jankowsky E. SF1 and SF2 helicases: family matters. Curr Opin Struct Biol. 2010;20(3):313–24.

44. Alkan C, Sajjadian S, Eichler EE. Limitations of next-generation genome sequence assembly. Nat Methods. 2011;8(1):61–5.

45. Treangen TJ, Salzberg SL. Repetitive DNA and next-generation sequencing: computational challenges and solutions. Nat Rev Genet. 2011;13(1):36–46.

46. Dean FB, Nelson JR, Giesler TL, Lasken RS. Rapid amplification of plasmid and phage DNA using Phi 29 DNA polymerase and multiply-primed rolling circle amplification. Genome Res. 2001;11(6):1095–9.

47. Thatcher LF, Gardiner DM, Kazan K, Manners JM. A highly conserved effector in Fusarium oxysporum is required for full virulence on Arabidopsis. Mol Plant Microbe Interact. 2012;25(2):180–90.

48. Rep M, van der Does HC, Meijer M, van Wijk R, Houterman PM, Dekker HL, de Koster CG, Cornelissen BJ. A small, cysteine-rich protein secreted by Fusarium oxysporum during colonization of xylem vessels is required for I-3-mediated resistance in tomato. Mol Microbiol. 2004;53(5):1373–83.

49. Namiki F, Matsunaga M, Okuda M, Inoue I, Nishi K, Fujita Y, Tsuge T. Mutation of an arginine biosynthesis gene causes reduced pathogenicity in Fusarium oxysporum f. sp. melonis. Mol Plant Microbe Interact. 2001;14(4):580–4.

50. Biju VC, Fokkens L, Houterman P, Rep M, Cornelissen BJC: Multiple evolutionary trajectories have led to the emergence of races in Fusarium oxysporum f. sp. lycopersici.(in press). Applied and Environmental Microbiology 2016,.

51. Jurka J. Repbase update: a database and an electronic journal of repetitive elements. Trends Genet. 2000;16(9):418–20.

52. Coleman JJ, Rounsley SD, Rodriguez-Carres M, Kuo A, Wasmann CC, Grimwood J, Schmutz J, Taga M, White GJ, Zhou S, Schwartz DC, Freitag M, Ma LJ, Danchin EG, Henrissat B, Coutinho PM, Nelson DR, Straney D, Napoli CA, Barker BM, Gribskov M, Rep M, Kroken S, Molnar I, Rensing C, Kennell JC, Zamora J, Farman ML, Selker EU, Salamov A, Shapiro H, Pangilinan J, Lindquist E, Lamers C, Grigoriev IV, Geiser DM, Covert SF, Temporini E,

Vanetten HD. The genome of Nectria haematococca: contribution of supernumerary chromosomes to gene expansion. PLoS Genet. 2009;5(8): e1000618.

53. Hane JK, Oliver RP. RIPCAL: a tool for alignment-based analysis of repeat-induced point mutations in fungal genomic sequences. BMC Bioinformatics. 2008;9:478. 2105-9-478.

54. Rius N, Guillen Y, Delprat A, Kapusta A, Feschotte C, Ruiz A. Exploration of the Drosophila buzzatii transposable element content suggests underestimation of repeats in Drosophila genomes. BMC Genomics. 2016;17(1):344. 016-2648-8.

55. Venner S, Feschotte C, Biemont C. Dynamics of transposable elements: towards a community ecology of the genome. Trends Genet. 2009;25(7):317–23.

56. Schaack S, Gilbert C, Feschotte C. Promiscuous DNA: horizontal transfer of transposable elements and why it matters for eukaryotic evolution. Trends Ecol Evol. 2010;25(9):537–46.

57. Arkhipova IR, Batzer MA, Brosius J, Feschotte C, Moran JV, Schmitz J, Jurka J. Genomic impact of eukaryotic transposable elements. Mob DNA. 2012;3(1):19. -8753-3-19.

58. Moller HD, Parsons L, Jorgensen TS, Botstein D, Regenberg B. Extrachromosomal circular DNA is common in yeast. Proc Natl Acad Sci U S A. 2015;112(24): E3114–22.

59. Moller HD, Bojsen RK, Tachibana C, Parsons L, Botstein D, Regenberg B: Genome-wide Purification of Extrachromosomal Circular DNA from Eukaryotic Cells. J Vis Exp 2016, (110). doi(110):10.3791/54239.

60. Ma LJ, Shea T, Young S, Zeng Q, Kistler HC: Genome Sequence of Fusarium oxysporum f. sp. melonis Strain NRRL 26406, a Fungus Causing Wilt Disease on Melon. Genome Announc 2014, 2(4):10.1128/genomeA.00730-14.

61. Altschul SF, Gish W, Miller W, Myers EW, Lipman DJ. Basic local alignment search tool. J Mol Biol. 1990;215(3):403–10.

62. Salamov AA, Solovyev VV. Ab initio gene finding in Drosophila genomic DNA. Genome Res. 2000;10(4):516–22.

63. Loytynoja A, Goldman N. Phylogeny-aware gap placement prevents errors in sequence alignment and evolutionary analysis. Science. 2008;320(5883):1632–5.

64. Capella-Gutierrez S, Silla-Martinez JM, Gabaldon T. trimAl: a tool for automated alignment trimming in large-scale phylogenetic analyses. Bioinformatics. 2009;25(15):1972–3.

65. Guindon S, Dufayard JF, Lefort V, Anisimova M, Hordijk W, Gascuel O. New algorithms and methods to estimate maximum-likelihood phylogenies: assessing the performance of PhyML 3.0. Syst Biol. 2010;59(3):307–21.

66. Huerta-Cepas J, Dopazo J, Gabaldon T. ETE: a python Environment for Tree Exploration. BMC Bioinformatics. 2010;11:24. 2105-11-24.

67. Eddy SR. A new generation of homology search tools based on probabilistic inference. Genome Inform. 2009;23(1):205–11.

68. Finn RD, Bateman A, Clements J, Coggill P, Eberhardt RY, Eddy SR, Heger A, Hetherington K, Holm L, Mistry J, Sonnhammer EL, Tate J, Punta M. Pfam: the protein families database. Nucleic Acids Res. 2014;42(Database issue): D222–30.

69. Sievers F, Wilm A, Dineen D, Gibson TJ, Karplus K, Li W, Lopez R, McWilliam H, Remmert M, Soding J, Thompson JD, Higgins DG. Fast, scalable generation of high-quality protein multiple sequence alignments using Clustal Omega. Mol Syst Biol. 2011;7:539.

70. Stamatakis A. RAxML version 8: a tool for phylogenetic analysis and post-analysis of large phylogenies. Bioinformatics. 2014;30(9):1312–3.

13

Convergence of retrotransposons in oomycetes and plants

Kirill Ustyantsev[1]* (ID), Alexandr Blinov[1] and Georgy Smyshlyaev[2]

Abstract

Background: Retrotransposons comprise a ubiquitous and abundant class of eukaryotic transposable elements. All members of this class rely on reverse transcriptase activity to produce a DNA copy of the element from the RNA template. However, other activities of the retrotransposon-encoded polyprotein may differ between diverse retrotransposons. The polyprotein domains corresponding to each of these activities may have their own evolutionary history independent from that of the reverse transcriptase, thus underlying the modular view on the evolution of retrotransposons. Furthermore, some transposable elements can independently evolve similar domain architectures by acquiring functionally similar but phylogenetically distinct modules. This convergent evolution of retrotransposons may ultimately suggest similar regulatory pathways underlying the lifecycle of the elements.

Results: Here, we provide new examples of the convergent evolution of retrotransposons of species from two unrelated taxa: green plants and parasitic protozoan oomycetes. In the present study we first analyzed the available genomic sequences of oomycete species and characterized two groups of Ty3/Gypsy long terminal repeat retrotransposons, namely Chronos and Archon, and a subgroup of L1 non-long terminal repeat retrotransposons. The results demonstrated that the retroelements from these three groups each have independently acquired plant-related ribonuclease H domains. This process closely resembles the evolution of retrotransposons in the genomes of green plants. In addition, we showed that Chronos elements captured a chromodomain, mimicking the process of chromodomain acquisition by Chromoviruses, another group of Ty3/Gypsy retrotransposons of plants, fungi, and vertebrates.

Conclusions: Repeated and strikingly similar acquisitions of ribonuclease H domains and chromodomains by different retrotransposon groups from unrelated taxa indicate similar selection pressure acting on these elements. Thus, there are some major trends in the evolution of the structural composition of retrotransposons, and characterizing these trends may enhance the current understanding of the retrotransposon life cycle.

Keywords: Convergent evolution, Retrotransposons, Plants, Oomycetes, Ribonuclease H, Chromodomain

Background

Retrotransposons are "copy-and-paste" mobile elements transferred via an RNA intermediate through the process of reverse transcription. Generally, retrotransposons are further subdivided in two major groups: long terminal repeat retrotransposons (LTR-RTs), with their viral descendants (retroviruses), and non-LTR retrotransposons (non-LTR-RTs). The only general structural feature shared between autonomous elements from both groups is the reverse transcriptase (RT) domain, a key enzyme

responsible for reverse transcription. In contrast, the set of other encoded activities could largely vary and rely on the life cycle organization and insertion strategy of the retrotransposon [1–3]. Each of these additional domains can have an evolutionary history independent from that of the RT domain. There are multiple examples of independent acquisitions of domains with the same enzymatic activity by the diverse retrotransposons, suggesting the importance of the domain-encoded function for the performance of each element [4–10]. One of these examples is the ribonuclease H (RNH) domain, which has been captured by diverse retrotransposons on different occasions [4–6, 8, 11–14].

* Correspondence: ustyantsev@bionet.nsc.ru
[1]Institute of Cytology and Genetics, Laboratory of Molecular Genetic Systems, Prospekt Lavrentyeva 10, 630090 Novosibirsk, Russia
Full list of author information is available at the end of the article

RNH activity is required for the removal of an RNA template from a cDNA/RNA hybrid generated during reverse transcription. Retrotransposons rely on either the host genome-encoded RNH enzyme or encode their own RNH domains [4]. For example, non-LTR-RTs often rely on host genome-encoded RNH activity, as the reverse transcription of these transposons occurs directly in the nucleus where the host cellular RNH enzyme is naturally present [4, 15]. Nevertheless, some non-LTR-RTs encode their own RNH. For example, some non-LTR-RTs of oomycetes and plants have acquired RNH closely related to the Archaea-like RNHs (aRNH). Interestingly, these two groups of non-LTR-RTs independently acquired aRNHs [6, 11]. In case of the LTR-RTs, the presence of the element-encoded RNH is obligatory, as reverse transcription occurs in the cytoplasm where no host-encoded enzyme is available [4]. Accordingly, the RNH domain has been detected in all LTR-RTs, and the evolution of the domains follows that of the RT [5]. However, some retroelements, such as retroviruses, have captured additional RNH domains, resulting in a 'dual' RNH [4, 5, 16]. Strikingly similar to retroviruses, the Tat LTR-RTs of green plants have acquired an additional RNH domain, aRNH, indicating structural and functional convergence between plant Tat LTR-RTs and vertebrate retroviruses [5].

In the present study, we mined all aRNH-containing retrotransposons from oomycete genomes and provided new examples of convergence in retrotransposons between plants and oomycetes. We identified and characterized two groups of Ty3/Gypsy LTR-RTs, Chronos and Archon, and a subgroup of L1 non-LTR-RTs in the genomes of oomycetes, which to our knowledge has not previously been described. These retrotransposons captured aRNH in the same manner as plant retrotransposons. In addition, we showed that Chronos LTR-RTs also captured a chromodomain (CHD), resembling the evolution of plant Chromoviruses and Ty1/Copia CoDi-I LTR-RTs from the free-living Stramenopiles *Phaeodactylum tricornutum* [7, 17–19].

Results

Diversity of aRNH-containing retrotransposons in oomycete genomes

aRNH is a subgroup of the type I RNH, which also includes Fungi/Metazoa-like RNHs (fmRNH) and LTR-RT RNH. While fmRNHs and aRNHs are characterized by the presence of histidine or arginine residues respectively in the active site, LTR-RTs RNHs lack any conserved residues in that position [4, 16]. aRNHs were originally described in the archaeal genomes and were also identified as cellular genes in the genomes of plants and some bacteria [20]. Furthermore, RNH domains that were found in Ty3/Gypsy Tat LTR-RTs and Ta11 L1

non-LTR-RTs of higher plants [12–14] were shown to be phylogenetically related to cellular-like aRNHs [5, 6]. In addition, Kojima and Jurka [11] identified a subgroup of aRNH-containing non-LTR-RTs of the Utopia group in oomycete genomes.

To determine the presence of the aRNH in other retroelements, we screened for aRNH sequences in Repbase Update (RU, v. 20.08), the database of eukaryotic transposable elements [21, 22]. Consistent with previous data, all retrotransposons predicted to have an aRNH domain (see Methods for details) were detected in either the genomes of higher plants or the parasitic protozoans oomycetes. Surprisingly, in addition to the previously described Utopia non-LTR-RTs [11], some oomycete Ty3/Gypsy LTR-RTs and L1 non-LTR-RTs also encode aRNH (for the RU accession numbers see Additional file 1: Table S1).

Since the variability of the oomycete retrotransposons annotated and deposited in RU 20.08 was restricted only to retrotransposons from seven species, of which retrotransposons from only four species contained aRNH (Additional file 1: Table S1), to provide comprehensive insight into the diversity of the identified elements, we further analyzed oomycete genomic sequences for the presence of aRNH-containing retrotransposons. This mining resulted in an overall set of 2899 distinct retrotransposon sequences from 21 out of 25 analyzed oomycete genomes. We initially classified the identified elements into the three groups, Ty3/Gypsy, L1 and Utopia, based on homology to the ORF2 amino acid sequences of aRNH-containing retrotransposons identified in RU. When possible, full-length copies were retrieved as representatives for each genome, and their structure and domain composition were analyzed (Fig. 1a, Additional file 1: Table S2).

Based on the RT phylogeny and comparative structural analysis, we identified two groups of aRNH-containing Ty3/Gypsy LTR-RTs in oomycetes. The first group, designated here as Archon, is specific for Saprolegniales genomes, and its members have an aRNH next to the original Ty3/Gypsy RNH domain. Interestingly, this RNH-aRNH junction resembles the 'dual' RNH domains of Tat LTR-RTs and retroviruses [5]. The second group, named Chronos, comprises elements detected in the Peronosporales and Pythiales genomes. In addition, a single copy of a Chronos element was identified in *Aphanomyces astaci* (Saprolegniales). These retrotransposons also have 'dual' RNH domains. However, in contrast to all other known aRNH-containing elements, these transposons possess a CHD in the 3′ end of their *pol* next to the INT domain (Fig. 1b, Additional file 2: Figure S1, Additional file 1: Table S2). Previously, the presence of a CHD was shown only for two groups of LTR-RTs: Chromoviruses (a group of Ty3/Gypsy LTR-RTs [7, 9, 18, 23]) and CoDi-I elements (a group of Ty1/Copia LTR-RTs from the free-living Stramenopiles, pennate diatom, *Phaeodactylum tricornutum*

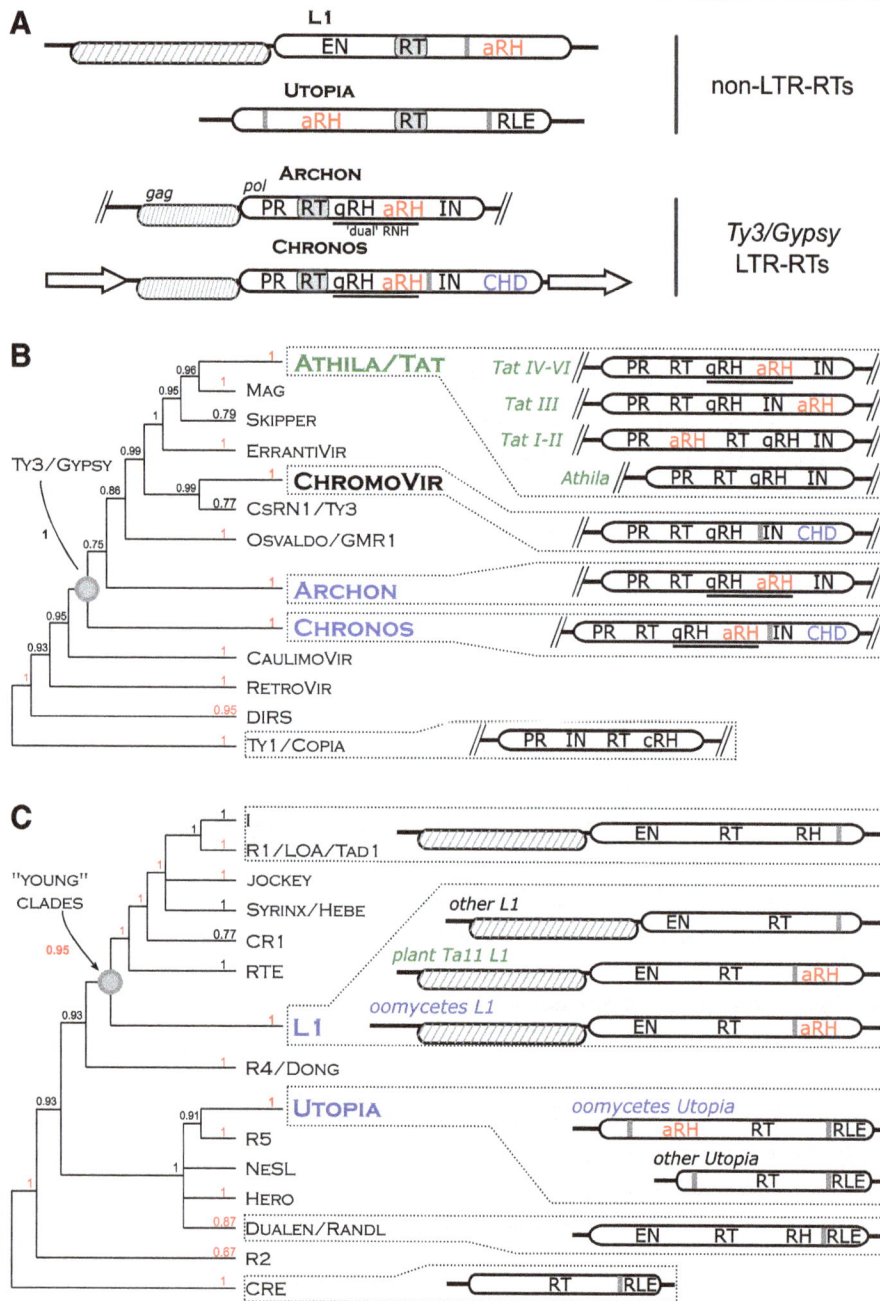

Fig. 1 (See legend on next page.)

(See figure on previous page.)
Fig. 1 Diversity of aRNH-containing retrotransposons in oomycetes. **a** Schematic structural composition of the elements from the identified groups: ORFs are shown as horizontal ovals (ORFs 1 are shaded); PR – protease; gRH – RNH of Ty3/Gypsy LTR-RTs; aRH – aRNH (in red); IN – integrase; CHD – chromodomain (in blue); EN – apurinic/apyrimidinic endonuclease-like endonuclease, RLE – restriction-like endonuclease; CCHC Zn finger motif indicated as vertical gray line; gray arrows, LTRs – long terminal repeats. **b** Consensus of Maximum-likelihood and Bayesian trees based on the amino acid sequences of RT domain of LTR-RTs. Approximate likelihood-ratio test (aLRT) statistical support values (unit fractions) are shown at the corresponding nodes of the tree; the values are highlighted in red if the corresponding node was additionally supported by more than 60 of 100 bootstrap replicates. Groups of aRNH-containing retrotransposons of oomycetes and plants are emphasized in bold and highlighted in blue and green, respectively. CHD-containing retrotransposons without aRNH of Chromoviruses (ChromoVir) group are emphasized in bold. On the right from the tree schemes of the consensus structures of the ORF2 of the corresponding groups are shown; cRH – RNH of Tc1/Copia LTR-RTs. The complete Maximum-likelihood and Bayesian phylogenetic trees with accession numbers, the names of the elements, and all the statistical support values are presented in Additional file 2: Figure S1. **c** Consensus of Maximum-likelihood and Bayesian trees based on the amino acid sequences of RT domain of non-LTR-RTs. Approximate likelihood-ratio test (aLRT) statistical support values (unit fractions) are shown at the corresponding nodes of the tree; the values are highlighted in red if the corresponding node was additionally supported by more than 60 of 100 bootstrap replicates. On the right of the tree, the schemes of the consensus structures of the corresponding groups are shown; RH – RNH domain of non-LTR-RTs. The complete Maximum-likelihood and Bayesian phylogenetic trees with accession numbers, the names of the elements, and all the statistical support values are presented in Additional file 2: Figure S1

[17]). Although Archon and Chronos LTR-RTs share similar structural organization with Tat LTR-RTs and Chromoviruses, they seem to be only distantly related to these elements (Fig. 1b, Additional file 2: Figure S1).

Identified in most of the Peronosporales and Pythiales genomes and undetectable in the Saprolegniales genomes (Additional file 1: Table S2), oomycete aRNH-containing L1 elements are similar in general organization to aRNH-containing Ta11 L1 of plants (Fig. 1c). In both groups, the aRNH domain is positioned at the C-terminal end of ORF2. Notably, both groups are also characterized by a CCHC cysteine motif located upstream of the aRNH. In other non-LTR-RTs harboring an RNH, the CCHC is positioned downstream of the RNH in ORF2 [24]. However, despite the similarities in the general organization of ORF2 (Fig. 1c, Additional file 3: Figure S2), oomycete and plant L1s do not form a monophyletic clade within the L1 group.

Oomycete Utopia elements were identified in most Peronosporales and Pythiales genomes, while only one copy was detected in *Saprolegnia diclina* (Saprolegniales) (Additional file 1: Table S2). Utopia is one of the "old" clades of non-LTR-RTs (such as R2, R5, and CRE) and its elements have sequence-specific restriction-like endonuclease domain (RLE), which guides their insertion to U2 small nuclear RNA genes [11]. The Utopia elements identified in our study did not differ in organization from the original Utopias identified by Kojima and Jurka [11] (Fig. 1c, Additional file 3: Figure S2).

The distinct positions of the oomycete Chronos, Archon, L1, and Utopia groups on the RT phylogenetic trees from all previously known aRNH-containing retrotransposons and from each other suggested that aRNH was independently acquired by each of these groups. However, to further elaborate on this idea, we performed a comparative analysis of the aRNHs from genomes of oomycetes, plants and other organisms.

Diversity of aRNH in oomycetes

After screening the oomycete genomic sequences, we detected aRNHs that were not associated with RT (individual aRNHs) and could therefore represent potential cellular genes. To obtain reference cellular RNH sequences, we additionally screened for fmRNHs using a set of sequences from a previous study [5]. Table 1 summarizes the results of the analysis comparing the distribution of individual aRNHs and fmRNHs to that of the RT-associated aRNH domains. We identified individual aRNHs in 21 out of 25 oomycete genomes. Notably, we previously identified aRNH-containing retrotransposons in these same 21 genomes. In contrast, fmRNH was identified in all studied genomes. For a majority of the genomes there was only single copy of an individual aRNH, while other genomes contained up to eleven copies of an individual aRNH. The copy number of fmRNHs per genome was also relatively low, varying from one to seven (Table 1), suggesting that due to its ubiquity and low copy number, fmRNH is the most likely candidate for the cellular RNH gene in oomycetes. However, the functions and origins of the individual aRNHs in oomycetes remain elusive.

To unveil the origin of both RT-associated aRNHs and individual aRNHs in oomycetes we performed a comparative analysis of RNH genes and domains from various sources (Figs. 2 and 3, Additional file 4: Figure S3, Table 1). L1, Archon, Chronos, and Utopia oomycete aRNH domains and aRNHs of plant retrotransposons form distinct clades on the tree (Fig. 2). The identified individual aRNHs were split into three clades on the tree: aRNH 1, aRNH 2, and aRNH 3. Two clades, aRNH 1 and aRNH 3, clustered together with the aRNH domains from oomycete retrotransposons Archon and L1, respectively, although this clustering was not supported by the bootstrap. aRNH 2 formed a distinct clade that did not show any significant clustering with any RT-associated aRNHs (Fig. 2, Additional file 4: Figure S3).

Table 1 Diversity, distribution, and the number of aRNH and fmRNH domains in the studied oomycete species

Taxonomic position according to the NCBI taxonomy				RT-associated RNHs number				Individual RNHs number			
Order	Family	Genus	Species	Chronos	Archon	L1	Utopia	aRNH			fmRNH
								aRNH 1	aRNH 2	aRNH 3	
Albuginales	Albuginaceae	Albugo	A. candida	-	-	-	-	-	-	-	2
			A. laibachii	-	-	-	-	-	-	-	1
Peronosporales	Peronosporaceae	Hyaloperonospora	H. arabidopsidis Emoy2	57	-	88	-	-	1	-	1
		Phytophthora	P. alni	106	-	3	2	4	-	-	7
			P. capsici	64	-	14	16	1	1	-	1
			P. cinnamomi var cinnamomi	92	-	3	5	-	1	-	1
			P. infestans T30-4	1555	-	43	25	-	1	-	2
			P. kernovia 00238/432	3	-	-	1	-	1	-	1
			P. lateralis MPF4	30	-	7	5	-	1	-	2
			P. parasitica P1569	11	-	-	1	1	1	-	4
			P. pinifolia CBS 122922	75	-	26	2	8	2	-	2
			P. ramorum	120	-	11	14	-	1	-	3
			P. sojae	271	-	35	31	2	1	-	4
			Average	233	-	18	10	3	1	-	3
		Phytopythium	P. vexans	1	-	-	1	-	1	-	2
		Pseudoperonospora	P. cubensis MSU-1	-	-	-	-	-	-	-	1
Pythiales	Pythiaceae	Pythium	P. aphanidermatum	5	-	3	-	-	1	-	1
			P. arrhenomanes	5	-	5	2	-	1	-	1
			P. insidiosum	33	-	43	2	-	1	-	1
			P. irregulare	1	-	-	1	-	1	-	2
			P. iwayamai	1	-	3	-	-	1	-	2
			P. ultimum var. ultimum	-	-	74	7	-	1	-	2
			Average	9	-	26	3	-	1	-	2
Saprolegniales	Saprolegniaceae	Aphanomyces	A. astaci APO3.2	1	-	-	-	-	-	3	1
			A. invadans 9901.2	-	-	-	-	-	-	-	1
			Average	1	-	-	-	-	-	3	1
		Saprolegnia	S. diclina VS20	-	3	-	1	-	1	10	1
			S. parasitica CBS 223.65	-	2	-	-	-	1	4	1
			Average	-	3	-	1	-	1	7	1

Notably, multiple copies of both aRNH 1 and aRNH 3 were detected in the studied oomycete genomes (Table 1). Thus, together with the potential relationship between the two aRNH groups and the RT-associated aRNHs of oomycetes, these results may suggest that aRNH 1 and aRNH 3 may represent remnants of Archon and L1 retrotransposons. In contrast, aRNH 2 was not related to RT-associated aRNHs (Fig. 2, Additional file 4: Figure S3). Therefore, it is likely that aRNH 2, in addition to fmRNH, could be a cellular RNH gene in oomycetes. This finding is also supported by the wide distribution and low copy number of aRNH 2 (Table 1).

To shed more light on the evolution of both aRNH and fmRNH in oomycetes, we mined aRNH and fmRNH homologs from the free-living Stramenopiles taxa, the closest relatives of oomycetes available in databases (Additional file 1: Table S3) using a tBLASTn search against NCBI WGS and TSA databases with oomycete aRNH and fmRNH amino acid domain sequences as queries (Fig. 2, Additional file 4: Figure S3) [25]. The results revealed aRNHs in the Stramenopiles genomes but did not detect fmRNHs (Additional file 1: Table S3). The aRNH domains of free-living Stramenopiles form a monophyletic clade on the Maximum-likelihood RNH tree (only weakly supported by the bootstrap) and a

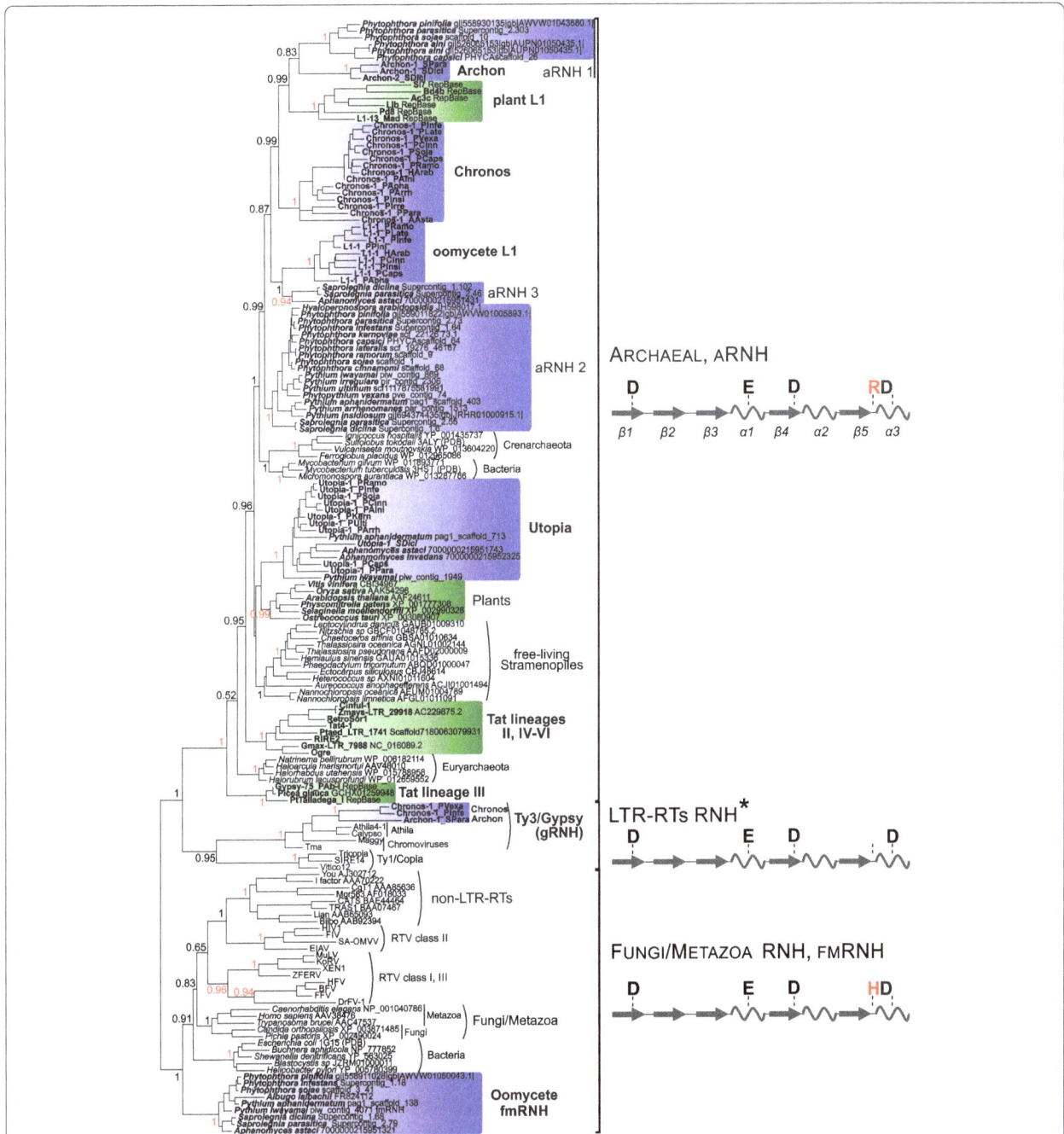

Fig. 2 Maximum-likelihood representative tree based on the amino acid sequences of different types of type I RNHs. Approximate likelihood-ratio test (aLRT) statistical support values (unit fractions) are shown at the corresponding nodes of the tree; the values are highlighted in red if the corresponding node was additionally supported by more than 60 of 100 bootstrap replicates. Comparison of Maximum-likelihood and Bayesian reconstructions and bootstrap values are presented in Additional file 4: Figure S3. RNH lineages specific for oomycetes and plants are highlighted in blue and green gradient blocks, respectively. RTV – retroviruses. The names of oomycete non-LTR-RT and LTR-RT RNH sequences identified in the present study correspond to those in Additional file 1: Table S2. Names of RNHs of other LTR-RTs and non-LTR-RTs correspond to those in GyDB [39] and Repbase Update [21], respectively. NCBI accession numbers are indicated to the right of other RNH sequences. Schemes of the secondary structures of three subtypes of RNH with the corresponding active site residues are shown at the right of the tree. The α-helices are depicted as helices, and the β-sheets are shown as arrows. The conserved R/H residue of the active site, which varies between different RNH subtypes, is highlighted in red. *The D-E-D-D catalytic residues are not conserved in the gRNHs of Archon, Chronos and Tat LTR-RTs

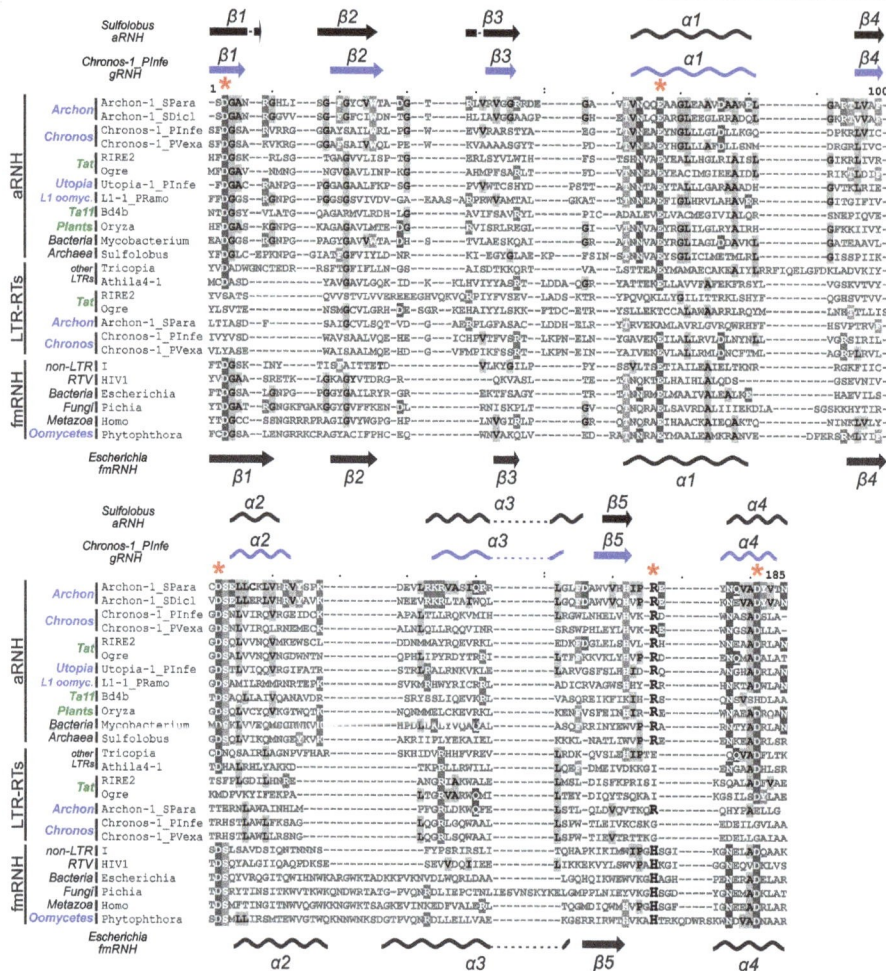

Fig. 3 Multiple amino acid sequence alignment of different types of RNHs. The names of RNH sequences corresponding to oomycete and plant lineages are emphasized in bold and highlighted in blue and green, respectively. Archaeal RNHs, Fungi/Metazoa RNHs, and original RNHs of LTR retrotransposons are designated as aRNH, fmRNH, and LTR-RTs, respectively. Apart from RIRE2 and Ogre gRNH that were retrieved from GyDB, all the sequences are available in the Additional file 8. Conserved catalytic residues (D-E-D-R/H-D) are indicated by asterisks at the top of the alignment. The semiconservative (R/H)-residue varying between the aRNH and fmRNH is additionally denoted by the bigger font at position 166 of the alignment. The conserved residues are highlighted in shades of gray. The secondary structure of *Escherichia coli* fmRNH (PDB: 1g15_A) is shown at the bottom of the alignment. The secondary structures of oomycete Chronos-1_PInfe LTR gRNH (predicted, this study) and *Sulfolobus tokodaii* aRNH (PDB: 3aly_A) are shown at the top of the alignment. The α-helices are depicted as helices, and the β-sheets are shown as arrows

paraphyletic clade on the Bayesian tree. In addition, these RNH sequences did not show any significant clustering with other studied aRNHs (Additional file 4: Figure S3).

Discussion

Potential origin of aRNH and fmRNH in oomycetes

While searching for homologs of aRNH and fmRNH in oomycete genomes, we identified aRNH in both free-living Stramenopiles and oomycete taxa, while fmRNH was detected only in oomycetes (Table 1). In addition, aRNH is absent in some groups of oomycetes, likely reflecting its loss in small genome parasitic lineages, such as Albuginales [26]. One possibility is that aRNH was present in the ancestor of the Stramenopiles lineage

and was vertically transmitted to oomycetes. Alternatively, aRNH might have been horizontally transferred from green plants, onto which most of the oomycete taxa examined in the present study typically parasitize [25, 27, 28]. The lack of aRNH in some oomycete genomes can be explained by the redundancy of aRNH and fmRNH functions.

The lack of fmRNH in the free-living Stramenopiles most likely indicates that oomycetes acquired this gene after the divergence from the Stramenopiles stem. The horizontal transfer of genes from fungi to oomycetes as an adaptation to parasitism on algae and plants has been previously proposed [27, 28]. Fungal genomes encode fmRNHs, which are responsible for the precise removal of RNA primers of Okazaki fragments during DNA replication and are critical for the maintenance of genome integrity (Fig. 2, Additional file 4: Figure S3) [29, 30]. Thus, it could be hypothesized that oomycetes might have acquired fmRNH through horizontal transfer together with other genes from ancient fungal lineages. However, in our phylogenetic reconstruction oomycete fmRNHs are only distantly related to fungal fmRNHs, which contradicts this hypothesis (Additional file 4: Figure S3).

Convergence between oomycete and plant retrotransposons

In the present study we showed that based on RT phylogeny, the identified aRNH-containing oomycete L1 non-LTR-RTs, and Chronos and Archon LTR-RTs are only distantly related to the previously described aRNH-containing Ta11 L1 non-LTR-RTs and Tat LTR-RTs of green plants (Fig. 1, Additional file 2: Figure S1, and Additional file 3: Figure S2). The distinct phylogenetic positions of the elements contradict the possibility of a single origin of all aRNH-containing LTR and non-LTR retroelement from plants and oomycetes. We therefore suggest that presence of aRNH in Tat, Chronos, and Archon LTR-RTs and Ta11 L1 and oomycete L1 non-LTR-RTs could be the best explained by series of independent aRNH acquisitions by ancestors of these elements, reflecting their convergent evolution to the similar structural compositions. However, the single origin of all aRNH-containing LTR and non-LTR retrotransposons from plants and oomycetes could not be completely rejected by the phylogenetic reconstructions due to the low bootstrap support values (in contrast to the aLRT and Bayesian posterior probabilities supports) that we obtained for the paraphyletic origin of the aRNH-containing retrotransposons (Fig. 1, Additional file 2: Figure S1, and Additional file 3: Figure S2), leaving the alternative to convergent evolution still open for discussion.

The repeated sequestration and fixation of some functional domains during the evolution by diverse members of a certain genetic lineage may reflect a beneficial effect on the selection in the environment that this lineage inhabits. Previously, we proposed that the 'dual' RNH domains of plant Tat LTR-RTs reflected convergent evolution with vertebrate retroviruses [5]. With the discovery of Chronos and Archon LTR-RTs in oomycetes, 'dual' RNH acquisition may indicate a more general evolutionary tendency in all LTR-RTs. Indeed, the loss of the conserved catalytic residues (D-E-D-R/H-D) in the original Ty3/Gypsy RNH domain and their complete set in aRNH of Chronos and Archon representatives (Fig. 3) is similar to what was shown for Tat LTR-RTs [5], and resembles transformation of the original retroviral RNH to the connection (tether) RNH domain after

the acquisition of new eukaryotic fmRNH in retroviruses [16] that is supported by the structural study of Ty3 reverse transcriptase [31]. Intriguingly, this evolutionary pathway may resemble an early stage in the transition of a Ty3/Gypsy retrotransposon into a retrovirus, preceding the acquisition of the infection-mediating envelope domain.

The beneficial effect from the RNH acquisition for non-LTR-RTs, however, is still poorly understood, as these elements typically rely on the host-encoded RNH activity. Furthermore, RNH could also be lost within some non-LTR-RT groups [32]. The finding of multiple examples of RNH acquisition in non-LTR-RTs therefore remains enigmatic.

The structural analysis of Chronos LTR-RTs revealed that apart from the aRNH domain, these elements also harbor CHD on the C-terminal end of the ORF2 next to the INT domain (INT-CHD), similar to the Ty3/Gypsy Chromoviruses from plants, fungi, and vertebrates [7, 9, 18, 19, 33]. Based on RT phylogeny, we showed that Chronos LTR-RTs and Chromoviruses are evolutionarily distinct from each other, thereby suggesting the convergent acquisition of the CHD by both groups. Interestingly, apart from Chromoviruses and Chronos LTR-RTs the INT-CHD domain was also reported for phylogenetically distant Ty1/Copia CoDi-I elements observed in the free-living Stramenopiles, pennate diatom, *Phaeodactylum tricornutum* [17]. See Additional file 5: Figure S4 for the multiple sequence alignment of CHDs from Chronos, Chromoviruses, and CoDI-I LTR-RTs. CHDs are widespread domains involved in chromatin remodeling in eukaryotes [34, 35]. The fusion of the CHD to the INT in LTR-RTs likely targets retrotransposon integration to the heterochromatin away from gene-rich regions [36]. Thus, multiple acquisitions of the CHD reflect the evolutionary tendency in LTR-RTs to minimize the damage to the host, while "quietly hitchhiking" its cellular machinery for retrotransposon propagation within the genome.

Conclusions

The current understanding of the diversity of retrotransposons and other mobile elements increases with an increasing number of sequenced genomes from a broad taxa range. In the present study, we identified and characterized several groups of retrotransposons from oomycete genomes, which to our knowledge has not previously been described. Importantly, the similar patterns of acquisitions of aRNH and CHD by unrelated retrotransposon groups from oomycetes and plants suggest that these events may represent a major evolutionary trend in retroelement evolution. This trend is likely independent of the retrotransposon host genome and may reflect similarities in the fundamental organization of retrotransposon life cycle, suggesting a beneficial role for the acquired domains in this cycle.

Methods

Computational mining for aRNH-containing repeats in Repbase update

The complete database of prototypic repetitive sequences Repbase Update (RU, v. 20.08) [21] was downloaded and analyzed for the presence of aRNH-containing repeats. Based on a hidden Markov model profile (HMM profile), aRNH domains were mapped using hmmsearch tool of the HMMER package [37] in translations of the retrieved RU sequences. The HMM profile was constructed from the amino acid alignment of aRNH sequences from the Ustyantsev et al. [5]. Repeats without the predicted similarity to aRNH were filtered out. The remained RU repeats were initially grouped according to the taxon of origin and subsequently grouped according to repeat type.

Computational mining for aRNH-containing retrotransposons, individual aRNH and fmRNH domains in oomycete genomes

The oomycete genomic sequences used in the present study were retrieved from public databases, as listed in Additional file 1: Table S2. To identify all retrotransposons harboring aRNH, the following algorithm was implemented using the UGENE workflow designer [38]. First, based on the aRNH HMM profile, aRNH domains were mapped using the hmmsearch tool of the HMMER [37] package in translations of the genomic DNA sequences. Second, sequences surrounding the regions of significant similarity to the aRNH profile were expanded, when possible, to 10,000 bp in both directions. Third, the enlarged sequences were screened for the presence of significant similarity to RT domains of non-LTR-RTs and LTR-RTs HMM profiles using hmmsearch. The non-LTR-RTs HMM profile was generated from the RT alignment of Repbase [21] non-LTR-RTs amino acid sequences available in the RTclass1 [12] server output. The corresponding HMM profile for LTR-RTs was constructed from the RT alignment of LTR-RTs amino acid sequences available in Gypsy Database [39]. Fourth, RT-positive sequences were divided into two groups corresponding to either non-LTR-RTs or LTR-RTs, and RT-negative sequences were filtered out, and identified aRNH sequences were retained for a further separate analysis as individual aRNHs. For each dataset, representative sequences were retrieved, and the number of elements belonging to each group (Ty3/Gypsy, L1, and Utopia) was counted by repeated BLAST [40], using ORF2 amino acid sequences of the previously identified RU aRNH-containing retrotransposons of oomycetes (Gypsy_18_PIT_I Ty3/Gypsy LTR-RT, L1-5_PI L1 non-LTR-RT, and R2I-1_PI Utopia non-LTR-RT) as seeding quires in the tBLASTn search.

Fungi/Metazoa RNHs (fmRNH) were mined using the HMM profile reconstructed based on the alignment of fmRNH amino acid sequences from Ustyantsev et al. [5] with hmmsearch, and the flanking sequences were expanded 1,000 bp in both directions.

Characterization of the structural composition of aRNH-containing retrotransposons

For each of the identified representative retrotransposons, a detailed analysis of the structural composition was performed. We used NCBI ORFfinder [41] to identify ORFs and NCBI CD-search [42] and HHpred [43] for a subsequent homology-based mining of conserved retrotransposon-specific domains. For LTR-RT representatives, when possible, the sequences of their LTRs were predicted by aligning 5′ upstream and 3′ downstream sequences flanking ORF1 and ORF2 using BLAST [40]. Secondary structure prediction for Chronos-1_PInfe aRNH was performed using Quick2D from the MPI bioinformatics toolkit [44].

Comparative and phylogenetic analysis

The RT amino acid sequences of the LTR-RT and non-LTR-RT representatives were aligned using hmmalign tool from the HMMER package to the corresponding HMM profiles [37]. The amino acid sequences of RNH are less conservative than RT, and a profile multiple alignment with the predicted local structures and 3D constraints (PROMALS3D) server was used to produce the alignment [45]. The alignments (refer to Additional files 6, 7, and 8 for corresponding LTR-RTs RT, non-LTR-RTs RT, and RNH alignments) were manually curated, and the phylogenetic trees were reconstructed using the maximum-likelihood and Bayesian algorithms implemented in the PhyML [46] and MrBayes [47] program tools. The best model for phylogenetic reconstruction, LG + G, was suggested using the ProtTest stand-alone tool [48] based on the Akaike Information Criterion (AIC) and Bayesian Information Criterion (BIC) for each of the alignments. In PhyML, an optimal tree topology was searched among 100 random starting trees under the subtree pruning and regrafting (SPR) algorithm, from which the tree with the largest log-likelihood value was taken, and its robustness was estimated using a Bayesian-like transformation of approximate likelihood-ratio test (aLRT, aBayes) and 100 bootstrap replicates [49]. In MrBayes, 10 split Markov chain Monte Carlo (MCMC) chains were run for 2,500,000 generations with sampling each 250 generations and discarding the first 5000 samples prior to consensus tree estimation.

Additional files

Additional file 1: Table S1. Diversity and distribution of aRNH-containing repetitive elements identified in the Repbase Update v. 20.08 (08-30-2015) database [21]. **Table S2.** Diversity, distribution and selected representatives of identified aRNH-containing retrotransposons in the studied oomycete genomes. **Table S3.** Individual aRNHs identified in the free-living Stramenopiles species. (XLSX 40 kb)

Additional file 2: Figure S1. The complete Maximum-likelihood and Bayesian phylogenetic trees reconstructed based on the amino acid sequences of RT domain of LTR-RTs (see Additional file 6 for the alignment). Statistical support was evaluated using aBayes aLRT (unit fractions) and 100 bootstrap replicates (% after a slash), and MCMC runs (%) in Maximum-likelihood and Bayesian reconstructions, respectively, and are shown at the corresponding nodes of the tree. Bootstrap values are shown only for the main indicated clusters. Chromodomain-containing clade names are underlined, and the names of the aRNH-containing clades are indicated in blue and green for plant and oomycete LTR-RTs, respectively. The names of the oomycete LTR-RT sequences identified in the present study correspond to those in Additional file 1: Table S2. Unless otherwise stated, the names of other LTR-RTs correspond to those in GyDB [39]. (PDF 779 kb)

Additional file 3: Figure S2. The complete Maximum-likelihood and Bayesian phylogenetic trees reconstructed based on the amino acid sequences of RT domain of non-LTR-RTs (see Additional file 7 for the alignment). Statistical support was evaluated using aBayes aLRT (unit fractions) and 100 bootstrap replicates (% after a slash), and MCMC runs (%) in Maximum-likelihood and Bayesian reconstructions, respectively, and the results are shown at the corresponding nodes of the tree. Bootstrap values are shown only for the main indicated clusters. The names of the aRNH-containing clades are indicated in blue and green for plant and oomycete non-LTR-RTs, respectively. The names of oomycete non-LTR-RT sequences identified in the present study correspond to those in Additional file 1: Table S2. The names of other non-LTR-RTs correspond to those in Repbase Update [21]. (PDF 366 kb)

Additional file 4: Figure S3. The complete Maximum-likelihood and Bayesian trees reconstructed based on different type I RNH amino acid sequences (see Additional file 8 for the alignment). Statistical support was evaluated using aBayes aLRT (unit fractions) and 100 bootstrap replicates (% after a slash), and MCMC runs (%) in Maximum-likelihood and Bayesian reconstructions, respectively, and the results are shown at the corresponding nodes of the tree. Bootstrap values are shown only for the main indicated clusters. The names of the RNH clades from plant and oomycete genomes are highlighted in green and blue, respectively. The names of oomycete non-LTR-RT and LTR-RT RNH sequences identified in the present study correspond to those in Additional file 1: Table S2. Names of RNHs of other LTR-RTs and non-LTR-RTs correspond to those in GyDB [39] and Repbase Update [21], respectively. NCBI accession numbers are indicated to the right of other RNH sequences. (PDF 863 kb)

Additional file 5: Figure S4. Multiple amino acid sequence alignment of CHDs from LTR-RTs and human Chromodomain Protein Y-Like 2 (PDB accession number 5JJZ_A). Additional information about the amino acid conservation is shown as a sequence Logo generated from the alignment, which is positioned at the bottom. (PDF 1096 kb)

Additional file 6: Multiple amino acid sequence alignment of RT domains from diverse LTR-RTs constructed and used for the phylogenetic reconstruction in the present study. (TXT 58 kb)

Additional file 7: Multiple amino acid sequence alignment of RT domains from diverse non-LTR-RTs constructed and used for the phylogenetic reconstruction in the present study. (TXT 83 kb)

Additional file 8: Multiple amino acid sequence alignment of RNH genes and domains from diverse taxa constructed and used for the phylogenetic reconstruction in the present study. (TXT 45 kb)

Abbreviations

aRNH: RNH of archaeal and plant origin; CHD: Chromodomain; fmRNH: RNH of Fungi/Metazoa origin; INT: Integrase; LTR-RTs: Long terminal repeat retrotransposons; non-LTR-RTs: Non-long terminal repeat retrotransposons;

ORF: Open reading frame; RLE: Restriction-like endonuclease; RNH: Ribonuclease H; RT: Reverse transcriptase; RU: Repbase update database

Acknowledgments

The authors are grateful to everyone who made the data freely available for the present study. The authors would also like to thank the American Journal Experts (AJE) for English language editing.

Funding

This work was financially supported by the Russian Foundation for Basic Research (Project No. 14-04-01498a) and the State scientific project (Project No. 0324-2016-0008).

Authors' contributions

KU performed all the bioinformatics assays and data analyses. KU and GS conceived and directed the study. AB provided computational resources and helped with the manuscript editing and writing. All authors contributed to the manuscript review. All authors read and approved the final manuscript.

Competing interests

The authors declare that they have no competing interests.

Author details

[1]Institute of Cytology and Genetics, Laboratory of Molecular Genetic Systems, Prospekt Lavrentyeva 10, 630090 Novosibirsk, Russia. [2]Structural and Computational Biology Unit, European Molecular Biology Laboratory, 69117 Heidelberg, Germany.

References

1. Xiong Y, Eickbush TH. Origin and evolution of retroelements based upon their reverse transcriptase sequences. EMBO J. 1990;9:3353–62.
2. Kazazian HH. Mobile elements: drivers of genome evolution. Science (New York, NY). 2004;303:1626–32.
3. Eickbush TH, Jamburuthugoda VK. The diversity of retrotransposons and the properties of their reverse transcriptases. Virus Res. 2008;134:221–34.
4. Malik HS. Ribonuclease H, evolution in retrotransposable elements. Cytogenetic Genome Res. 2005;110:392–401.
5. Ustyantsev K, Novikova O, Blinov A, Smyshlyaev G. Convergent evolution of ribonuclease H in LTR retrotransposons and retroviruses. Mol Biol Evol. 2015; 32:1197–207.
6. Smyshlyaev G, Voigt F, Blinov A, Barabas O, Novikova O. Acquisition of an Archaea-like ribonuclease H domain by plant L1 retrotransposons supports modular evolution. Proc Natl Acad Sci. 2013;110:20140–5.
7. Novikova O, Smyshlyaev G, Blinov A. Evolutionary genomics revealed interkingdom distribution of Tcn1-like chromodomain-containing Gypsy LTR retrotransposons among fungi and plants. BMC Genomics. 2010;11:231.
8. Kojima KK, Fujiwara H. An extraordinary retrotransposon family encoding dual endonucleases. Genome Res. 2005;15:1106–17.
9. Malik HS, Eickbush TH. Modular evolution of the integrase domain in the Ty3/Gypsy class of LTR retrotransposons. J Virol. 1999;73:5186–90.
10. Malik HS, Henikoff S, Eickbush TH. Poised for contagion: evolutionary origins of the infectious abilities of invertebrate retroviruses. Genome Res. 2000;10:1307–18.
11. Kojima KK, Jurka J. Ancient Origin of the U2 Small Nuclear RNA Gene-Targeting Non-LTR Retrotransposons Utopia. Schmitz J, editor. PLOS ONE. Public Library of Science; 2015;10:e0140084.
12. Kapitonov VV, Tempel S, Jurka J. Simple and fast classification of non-LTR retrotransposons based on phylogeny of their RT domain protein sequences. Gene. 2009;448:207–13.
13. Heitkam T, Schmidt T. BNR - a LINE family from Beta vulgaris - contains a RRM domain in open reading frame 1 and defines a L1 sub-clade present in diverse plant genomes. Plant J. 2009;59:872–82.
14. Wenke T, Holtgräwe D, Horn AV, Weisshaar B, Schmidt T. An abundant and heavily truncated non-LTR retrotransposon (LINE) family in Beta vulgaris. Plant Mol Biol. 2009;71:585–97.

15. Han JS. Non-long terminal repeat (non-LTR) retrotransposons: mechanisms, recent developments, and unanswered questions. Mob DNA. 2010;1:15.

16. Malik HS, Eickbush TH. Phylogenetic analysis of ribonuclease H domains suggests a late, chimeric origin of LTR retrotransposable elements and retroviruses. Genome Res. 2001;11:1187–97.

17. Llorens C, Muñoz-Pomer A, Bernad L, Botella H, Moya A. Network dynamics of eukaryotic LTR retroelements beyond phylogenetic trees. Biol Direct. 2009;4:41.

18. Novikov A, Smyshlyaev G, Novikova O. Evolutionary History of LTR Retrotransposon Chromodomains in Plants. Int J Plant Genomics. 2012;2012: 1–17. Hindawi Publishing Corporation.

19. Marín I, Lloréns C. Ty3/Gypsy retrotransposons: description of new Arabidopsis thaliana elements and evolutionary perspectives derived from comparative genomic data. Molecular biology and evolution. 2000;17:1040–9. Oxford University Press.

20. Ohtani N, Yanagawa H, Tomita M, Itaya M. Identification of the first archaeal Type 1 RNase H gene from Halobacterium sp. NRC-1: archaeal RNase HI can cleave an RNA-DNA junction. Biochem J. 2004;381:795–802. Portland Press Ltd.

21. Jurka J, Kapitonov VV, Pavlicek A, Klonowski P, Kohany O, Walichiewicz J. Repbase Update, a database of eukaryotic repetitive elements. Cytogenetic Genome Res. 2005;110:462–7.

22. Bao W, Kojima KK, Kohany O. Repbase Update, a database of repetitive elements in eukaryotic genomes. Mob DNA. 2015;6:11.

23. Novikova O. Chromodomains and LTR retrotransposons in plants. Commun Integr Biol. 2009;2:158–62.

24. Smyshlyaev GA, Blinov AG. Evolution and biodiversity of L1 retrotransposons in angiosperm genomes. Russian J Genetics. 2012;2:72–8.

25. Beakes GW, Glockling SL, Sekimoto S. The evolutionary phylogeny of the oomycete "fungi". Protoplasma. 2012;249:3–19.

26. Links MG, Holub E, Jiang RHY, Sharpe AG, Hegedus D, Beynon E, et al. De novo sequence assembly of Albugo candida reveals a small genome relative to other biotrophic oomycetes. BMC Genomics. 2011;12:1–12.

27. Richards TA, Soanes DM, Jones MDM, Vasieva O, Leonard G, Paszkiewicz K, et al. Horizontal gene transfer facilitated the evolution of plant parasitic mechanisms in the oomycetes. Proc Natl Acad Sci. 2011;108:15258–63.

28. Soanes D, Richards TA. Horizontal Gene Transfer in Eukaryotic Plant Pathogens. Annu Rev Phytopathol. 2014;52:583–614.

29. Qiu J, Qian Y, Frank P, Wintersberger U, Shen B. Saccharomyces cerevisiae RNase H(35) functions in RNA primer removal during lagging-strand DNA synthesis, most efficiently in cooperation with Rad27 nuclease. Mol Cell Biol. 1999;19:8361–71.

30. Cerritelli SM, Crouch RJ. Ribonuclease H: the enzymes in eukaryotes. FEBS J. 2009;276:1494–505.

31. Nowak E, Miller JT, Bona MK, Studnicka J, Szczepanowski RH, Jurkowski J, et al. Ty3 reverse transcriptase complexed with an RNA-DNA hybrid shows structural and functional asymmetry. Nat Struct Mol Biol. 2014;21:389–96. Nature Research.

32. Malik HS, Burke WD, Eickbush TH. The age and evolution of non-LTR retrotransposable elements. Mol Biol Evol. 1999;16:793–805.

33. Gorinsek B, Gubensek F, Kordis D. Evolutionary genomics of chromoviruses in eukaryotes. Mol Biol Evol. 2004;21:781–98. Oxford University Press.

34. Platero JS, Hartnett T, Eissenberg JC. Functional analysis of the chromo domain of HP1. EMBO J. 1995;14:3977–86.

35. Eissenberg JC. Structural biology of the chromodomain: Form and function. Gene. 2012;496:69–78.

36. Gao X, Hou Y, Ebina H, Levin HL, Voytas DF. Chromodomains direct integration of retrotransposons to heterochromatin. Genome Res. 2008;18:359–69.

37. Eddy SR. Accelerated Profile HMM Searches. Pearson WR, editor. PLoS computational biology. Public Library of Science; 2011;7:e1002195.

38. Okonechnikov K, Golosova O, Fursov M. Unipro UGENE: a unified bioinformatics toolkit. Bioinformatics (Oxford, England). 2012;28:1166–7.

39. Llorens C, Futami R, Covelli L, Domínguez-Escribá L, Viu JM, Tamarit D, et al. The Gypsy Database (GyDB) of mobile genetic elements: release 2.0. Nucleic Acids Res. 2011;39:D70–4.

40. Altschul SF, Gish W, Miller W, Myers EW, Lipman DJ. Basic local alignment search tool. J Mol Biol. 1990;215:403–10.

41. NCBI Open Reading Frame finder. https://www.ncbi.nlm.nih.gov/orffinder/. Accessed 10 Dec 2016.

42. Marchler-Bauer A, Derbyshire MK, Gonzales NR, Lu S, Chitsaz F, Geer LY, et al. CDD: NCBI's conserved domain database. Nucleic Acids Res. 2015;43:D222–6.

43. Söding J, Biegert A, Lupas AN. The HHpred interactive server for protein homology detection and structure prediction. Nucleic Acids Res. 2005;33.

44. Alva V, Nam S-Z, Söding J, Lupas AN. The MPI bioinformatics Toolkit as an integrative platform for advanced protein sequence and structure analysis. Nucleic Acids Res. 2016;44:W410–5.

45. Pei J, Kim BH, Grishin NV. PROMALS3D: A tool for multiple protein sequence and structure alignments. Nucleic Acids Res. 2008;36:2295–300.

46. Guindon S, Dufayard J-F, Lefort V, Anisimova M, Hordijk W, Gascuel O. New algorithms and methods to estimate maximum-likelihood phylogenies: assessing the performance of PhyML 3.0. Syst Biol. 2010;59:307–21.

47. Ronquist F, Teslenko M, van der Mark P, Ayres DL, Darling A, Höhna S, et al. MrBayes 3.2: efficient Bayesian phylogenetic inference and model choice across a large model space. Syst Biol. 2012;61:539–42.

48. Darriba D, Taboada GL, Doallo R, Posada D. ProtTest 3: fast selection of best-fit models of protein evolution. Bioinformatics (Oxford, England). 2011; 27:1164–5. Oxford University Press.

49. Anisimova M, Gil M, Dufayard JF, Dessimoz C, Gascuel O. Survey of branch support methods demonstrates accuracy, power, and robustness of fast likelihood-based approximation schemes. Syst Biol. 2011;60:685–99.

14

TGTT and *AACA*: two transcriptionally active LTR retrotransposon subfamilies with a specific LTR structure and horizontal transfer in four Rosaceae species

Hao Yin[1,2], Xiao Wu[1,2], Dongqing Shi[1,2], Yangyang Chen[1,2], Kaijie Qi[1,2], Zhengqiang Ma[2,3] and Shaoling Zhang[1,2]*

Abstract

Background: Long terminal repeat retrotransposons (LTR-RTs) are major components of plant genomes. Common LTR-RTs contain the palindromic dinucleotide 5′-'TG'–'CA'-3′ motif at the ends. Thus, further analyses of non-canonical LTR-RTs with non-palindromic motifs will enhance our understanding of their structures and evolutionary history.

Results: Here, we report two new LTR-RT subfamilies (*TGTT* and *AACA*) with atypical dinucleotide ends of 5′-'TG'–'TT'-3′, and 5′-'AA'–'CA'-3′ in pear, apple, peach and mei. In total, 91 intact LTR-RTs were identified and classified into four *TGTT* and four *AACA* families. A structural annotation analysis showed that the four *TGTT* families, together with *AACA1* and *AACA2*, belong to the *Copia*-like superfamily, whereas *AACA3* and *AACA4* appeared to be TRIM elements. The average amplification time frames for the eight families ranged from 0.05 to 2.32 million years. Phylogenetics coupled with sequence analyses revealed that the *TGTT1* elements of peach were horizontally transferred from apple. In addition, 32 elements from two *TGTT* and three *AACA* families had detectable transcriptional activation, and a qRT-PCR analysis indicated that their expression levels varied dramatically in different species, organs and stress treatments.

Conclusions: Two novel LTR-RT subfamilies that terminated with non-palindromic dinucleotides at the ends of their LTRs were identified in four Rosaceae species, and a deep analysis showed their recent activity, horizontal transfer and varied transcriptional levels in different species, organs and stress treatments. This work enhances our understanding of the structural variation and evolutionary history of LTR-RTs in plants and also provides a valuable resource for future investigations of LTR-RTs having specific structures in other species.

Keywords: LTR retrotransposon, Horizontal transfer, Transcription activity, Pear, Rosaceae

Background

Long terminal repeat retrotransposons (LTR-RTs) are major components that are widespread in flower plant genomes [1]. They are capable of propagating to reach thousands of copies in a genome using RNA as an intermediate [2, 3]. LTR-RTs are the most significant contributor to genome size, representing 43% of the nuclear DNA in pear [4], 38% in apple [5], 19% in peach [6], 53% in cotton [7] and over 70% in maize genomes [8]. A representative intact LTR-RT usually contains two highly identical LTRs, which are typically flanked by 2-bp palindromic motifs, commonly 5′-TG–CA-3′. The internal region of an autonomous LTR should contain a primer-binding site (PBS), a polypurine tract (PPT) and two functional genes (*gag*, and *pol*) [9]. Based on the order of Reverse transcriptase (*rt*) and Integrase (*int*) in *pol*, LTR-RTs can be further classed into *Gypsy* and *Copia* superfamilies [9]. In addition, the LTR-RTs also contain two types of non-autonomous groups, large retrotransposon derivatives (LARDs) [10] and terminal-repeat retrotransposons in

---

* Correspondence: slzhang@njau.edu.cn
[1]Center of Pear Engineering Technology Research, College of Horticulture, Nanjing Agricultural University, Nanjing 210095, China
[2]State Key Laboratory of Crop Genetics and Germplasm Enhancement, Nanjing Agricultural University, Nanjing, China
Full list of author information is available at the end of the article

miniature (TRIMs) [11]. The insertion of an LTR-RT is accompanied by the duplication of a 4–6-bp sequence immediately flanking with the 5′ and 3′ ends of the element, called target site duplication (TSD).

The most common dinucleotide motif flanking the direct LTR-RT repeat regions is the palindromic 5′-TG–CA-3′ motif. However, several LTR-RT families with non-TGCA motifs have been reported. For example, *Tos17*, a rice LTR-RT that can be activated by tissue culture, has a non-canonical motif of 5′-TG–GA-3′ [12] and *TARE1*, which was identified as intensively amplified in the tomato genome, ends with 5′-TA–CA-3′ motifs [13]. In addition, *AcCOPIA1* that terminated with 5′-'TG'–'TA'-3′ at both ends of the LTRs was identified in onion [14]. However, no such non-canonical elements have been identified in the Rosaceae species.

Horizontal transfers (HTs) indicate the transmission of genetic material among sexually isolated species. As a possible dissemination mechanism of transposable elements (TEs) in eukaryotes, the horizontal transfer of TEs (HTTs) into a new organism is an important step for the TE to escape from the silencing machinery of their host genome and obtain a new 'life cycle' [15]. The first case of horizontal TE transfer (HTT) was the *P* TE identified between *Drosophila willistoni* and *Drosophila melanogaster* [16]. Recently, with the availability of many plant genome sequences, several HTT cases have been reported mainly through comparative genomic approaches. For example, multiple HTs of the LTR-RT *RIRE1* were identified within the genus *Oryza* [17], and another LTR-RT family *Route66* were found and proven to be HTs among the rice, maize and sorghum genomes through a comparative genomics analysis [18]. In addition, 32 HTs of LTR-RTs were discovered by whole genome surveys and comparative analyses in 46 sequenced plant genomes [19].

The propensity of LTR-RTs not only contributed to genome size but also resulted in byproducts of gene disruption, expression level alterations and genomic rearrangements by inserting themselves into genes or their promoter regions [20, 21]. In plants, LTR-RTs are usually silent under normal conditions, but some show transcriptional activities and increased accumulations while under stress, potentially triggering the genetic diversity required to evolve adaptations [21, 22]. For example, salt (*AtCopeg1* in *Arabidopsis* [23]), drought (*BARE1* in barley [24]), heat (*ONSEN* in *Arabidopsis* [25, 26]), cold (*mPing* in rice [27, 28]) and wounding (*Corky* from Quercus [29]; *CLCoy1* in lemon [30], *OARE1* in oat [31] and *Tnt1* in tobacco [32]). Recently, several LTR-RTs were identified as being expressed in the fruits and buds of pear in the RNA-seq databases [33]. However, their study did not focus on the LTR-RTs' transcription activities under stress in pear.

The Rosaceae family is an economically important angiosperm lineage, containing over 3000 distinct species with chromosome's numbering from 7 to 17 pairs [34]. Some genera with higher economic values that are widely cultivated have had their whole genomes sequenced in the last decade, including pear (*Pyrus bretschneideri*, n = 17, 527 Mb) [4], apple (*Malus domestica*, n = 17, 743 Mb) [5], peach (*Prunus persica*, n = 8, 265 Mb) [6], mei (*Prunus mume*, n = 8, 280 Mb) [35] and woodland strawberry (*Fragria vesca*, n = 7) [36] (Additional file 1: Table S1). Based on DNA sequence data, *Fragaria* belongs to the Rosoideae, supertribe Rosadea, tribe Potentilleae, *Malus* and *Pyrus* occur in the Spiraeoideae, supertribe Pyrodeae, tribe Pyreae and *Prunus* is in the Spiraeoideae, tribe Amygdaleae [37]. The availability of the five Rosaceae genomic sequences provided opportunities to undertake comparative analyses of LTR-RTs in pear and four other genomes [3, 38]. In this study, a genome wide identification of non-typical LTR-RTs in pear genome was conducted. Two new subfamilies of LTR-RTs, *TGTT* and *AACA*, were identified in pear, apple, peach and mei, but not in strawberry. Their structures, abundance levels, insertion time frames, evolution and transcription activities have been comprehensively analyzed between the four Rosaceae species. *TGTT* and *AACA* elements terminate in short inverted repeat dinucleotides, such as 'TG' and 'TT', 'AA' and 'CA', and the *AACA1* elements in peach may have been horizontally transferred from apple. In addition, multiple elements from the two subfamilies present differential expression levels in different pear organs and also show different expression levels under heat, cold and salt stress treatments. Our study reveals novel structures, horizontal transfer and the transcription activation of two new LTR-RT subfamilies, providing additional information on, and knowledge of, the structure, evolution and activity of TEs in plants.

Results

Identification, structural characterization and sequence analysis of *TGTT* and *AACA* TEs in the pear genome

We started our analyses by focusing on a class of atypical LTR-RTs identified in the pear (*P. bretschneideri*) genome. Initially, 12 intact TEs with atypical characteristics were identified using the LTR_STRUC program [39]. The LTRs of the 12 intact TEs terminate in the dinucleotide 5′-TG–TT-3′ or 5′-AA–CA-3′ (Figs. 1a,b and 2) instead of 5′-TG–CA-3′ usually found in typical LTR-RTs. Thus, these TEs were classified into two subfamilies, named *TGTT* and *AACA*, based on their terminal dinucleotides. In total, 66 intact TEs with two clearly defined boundaries and TSDs were identified using combined homology-based approaches as previously described [3, 13, 38]. Using the unified classification for eukaryotic

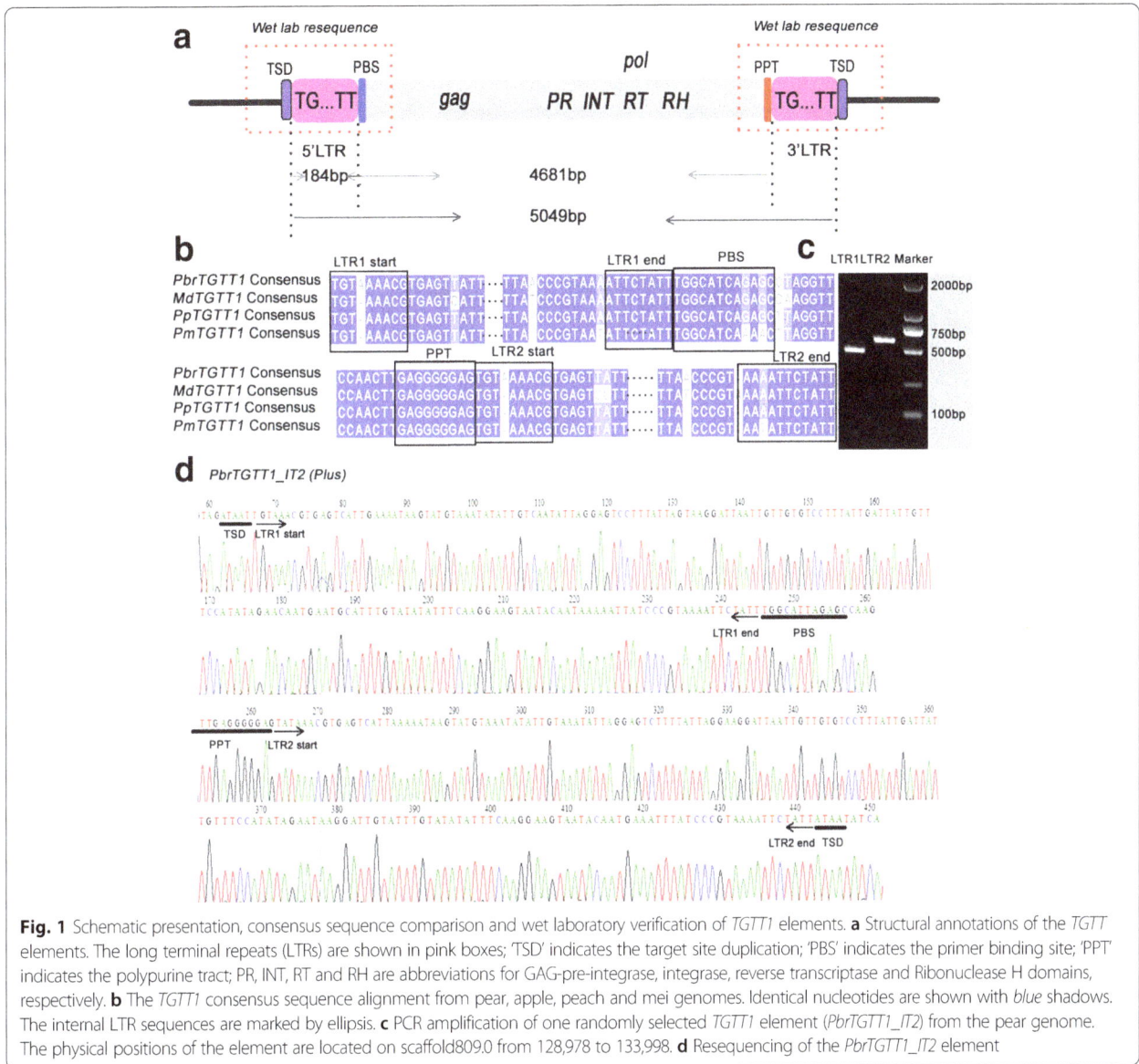

Fig. 1 Schematic presentation, consensus sequence comparison and wet laboratory verification of *TGTT1* elements. **a** Structural annotations of the *TGTT* elements. The long terminal repeats (LTRs) are shown in pink boxes; 'TSD' indicates the target site duplication; 'PBS' indicates the primer binding site; 'PPT' indicates the polypurine tract; PR, INT, RT and RH are abbreviations for GAG-pre-integrase, integrase, reverse transcriptase and Ribonuclease H domains, respectively. **b** The *TGTT1* consensus sequence alignment from pear, apple, peach and mei genomes. Identical nucleotides are shown with *blue* shadows. The internal LTR sequences are marked by ellipsis. **c** PCR amplification of one randomly selected *TGTT1* element (*PbrTGTT1_IT2*) from the pear genome. The physical positions of the element are located on scaffold809.0 from 128,978 to 133,998. **d** Resequencing of the *PbrTGTT1_IT2* element

TEs [40], the 66 TEs were grouped into eight distinct families based on an over 80% identity in at least 80% of their LTR regions (Table 1, Additional file 1: Tables S2 and S3 and Additional file 2: Figure S1). Four families, containing 35 TEs, belonged to the *TGTT* subfamily and the other four families, containing 31 TEs, belonged to the *AACA* subfamily. We randomly selected nine elements and confirmed the existence of 5′-TG–TT-3′ and 5′-AA–CA-3′ terminals in the LTR sequences by PCR and Sanger sequencing (see Methods, Fig. 1c and d, Additional file 2: Figure S2 and Additional file 1: Table S4).

The consensus sequence sizes of the eight families ranged from 2129 (*PbrAACA3*) to 5114 bp (*PbrTGTT3*), and the LTR sequence sizes ranged from 152 (*PbrTGTT3*) to 266 bp (*PbrAACA1*, Table 1). The coding sequences of the 66 elements indicated that all of the *PbrTGTT* elements contained the *Gag* and *Pol* genes, including the protease (PR), integrase (INT), reverse transcriptase (RT), and RNase H (RH) domains. The *PbrAACA1* and *PbrAACA2* TEs also contain the *Gag* and *Pol* genes, but the PR domain was absent in their *Pols*. The order of *int*, *rt*. and *rh* defined the six families (46 elements) as *Copia*-like elements (Fig. 2). Interestingly, no coding sequences were identified in the short internal sequences between the two LTRs (1641–2042-bp) of *PbrAACA3* and *PbrAACA4* (20 elements), indicating that these two *AACA* families were TRIM families (Fig. 2). Notably, the TSD sizes of the *PbrAACA* elements were 6 bp, while those of the *PbrTGTT* family varied from 4 to 6 bp (Table 1).

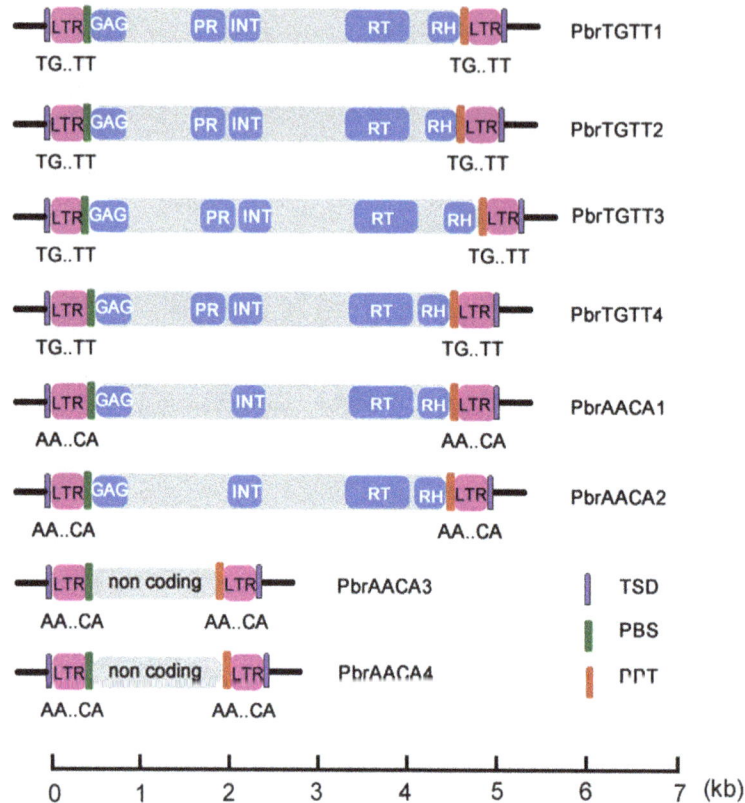

Fig. 2 Schematic representation of *TGTT* and *AACA* structures in pear. The *black* lines at the ends represent the DNA sequences. The scale below measures the lengths of the elements

Table 1 Summary of *TGTT* and *AACA* elements in four Rosaceae genomes

Family	Superfamily Lineage	No. of intact element	Length of LTR (bp)	Length of intact element (bp)	Start and end of LTR (Plus)	Length of TSDs	Ave. age (mys)
PbrTGTT1	Copia/Ale	8	184	5049	TG..TT	5	2.32
MdTGTT1	Copia/Ale	6	179	5113	TG..TT	5	0.91
PpTGTT1	Copia/Ale	4	179	5009	TG..TT	5	0.16
PmTGTT1	Copia/ Ale	2	180	5055	TG..TT	5	0.65
PbrTGTT2	Copia/ Ale	22	180	5039	TG..TT	5	0.3
MdTGTT2	Copia/ Ale	3	169	5046	TG..TT	5	0.92
PbrTGTT3	Copia/ Ale	2	152	5114	TG..TT	4	0.38
PbrTGTT4	Copia/ Ale	3	229	4999	TG..TT	6	0.28
Subtotal/average		50	180	5053	TG..TT	5	0.74
PbrAACA1	Copia/ Ale	7	266	4924	AA..CA	6	0.94
PmAACA1	Copia/ Ale	2	251	4883	AA..CA	6	0.15
PbrAACA2	Copia/ Ale	4	201	4857	AA..CA	6	0.05
PbrAACA3	TRIM	4	244	2129	AA..CA	6	1.85
PbrAACA4	TRIM	16	203/242	2364/2522	AA..CA	6	1.84
MdAACA4	TRIM	8	207/242	2606/2735	AA..CA	6	1.03
Subtotal/average		41	232	3378	AA..CA	6	1.26

Note: *PbrAACA4* and *MdAACA4* elements can be separated into two sub-groups based on their sequence length, since their sequence identity and sequence length are still over 80%, the two sub-groups were still classified into one family

TGTT and AACA TEs are also present in other Rosaceae genomes

To detect whether the *TGTT* and *AACA* elements are specific to the pear genome, these elements were annotated in other published plant genomes at pyhtozome (http://www.phytozome.net) using the same strategies as described above. Only four families of *TGTT* and *AACA* were identified in three other closely related Rosaceae genomes, apple (*M. domestica*) [5], peach (*P. persica*) [6] and mei (*P. mume*) [35] (Additional file 1: Table S1). To distinguish these TEs in different genomes, we have named them *MdTGTT1*, *PpTGTT1*, *PmTGTT1*, *MdTGTT2*, *PmAACA1* and *MdAACA4* (Table 1, Additional file 1: Table S3). In total, six *MdTGTT1* copies, four *PpTGTT1* copies, two *PmTGTT1* copies, three *MdTGTT2* copies, two *PmAACA1* and eight *MdAACA4* copies, which are all less than the number in pear, were identified. No *TGTT* or *AACA* TEs were identified in the closely related Rosaceae species, woodland strawberry (*F. vesca*) [36] or other published plant genomes.

Variable spectra of activities for amplification of *TGTT* and *AACA* elements over evolutionary time

To compare the activities and amplification time frames of *TGTT* and *AACA* elements among the four Rosaceae species, the full-length TEs with TSDs were dated using a previously described approach [41, 42]. Even though the two LTR sequences of an intact LTR-RT element are identical at the time of insertion, both LTRs accumulate nucleotide substitutions independently over evolutionary time. Thus, when an evolutionary rate is applied to the LTR-RT element, the sequence divergence of two LTRs can be roughly converted into the insertion time.

Although the evolution rate of LTR-RTs varies among different loci, families, and lineages [43], an estimation of 1.3×10^{-8} per site per year has been applied in many studies [13, 42, 44]. Using this rate, the insertion times of the 50 *TGTT* and 41 *AACA* intact copies with TSDs from the four Rosaceae species were estimated. The following was observed: 1) the average insertion times of *TGTT* and *AACA* subfamilies are 0.74 and 1.26 million years (Mys), respectively; 2) the average insertion times of the eight families in the four Rosaceae species ranged from 0.05 (*PbrAACA2*) to 2.32 Mys (*PbrTGTT1*). Most of these elements (65, 71.43%) inserted into the genome <1.0 million years ago (Mya), and 21 copies (23.08%) integrated into the genome within 1–3 Mya. In addition, only five copies (5.49%) have been dated >3 Mya; 3) over one third of these *TGTT* and *AACA* elements (31, 34.07%) have two identical LTRs, and the ratio of *TGTT* to *AACA* TEs is almost 1:1 (15:16, Fig. 3, Additional file 1: Table S3); and 4) the average insertion times of *TGTT* and *AACA* TEs varied among pear, apple, peach and mei, at 1.00, 0.95, 0.16 and 0.40 Mya, respectively (Fig. 3).

The evolutionary relationship between *TGTT* and *AACA* TEs

To understand the evolutionary relationships among *TGTT* and *AACA* TEs in the four Rosaceae species, a phylogenetic tree using the 5′ LTR sequences was constructed (Fig. 4a). The 91 *TGTT* and *AACA* TEs can be successfully separated into eight clades. The four *TGTT* families clustered together, and the four *AACA* families were closer to each other, indicating that the *TGTT* and *AACA* TEs evolved independently. Although *TGTT* and *AACA* TEs are separated from each other, the *MdTGTT2*

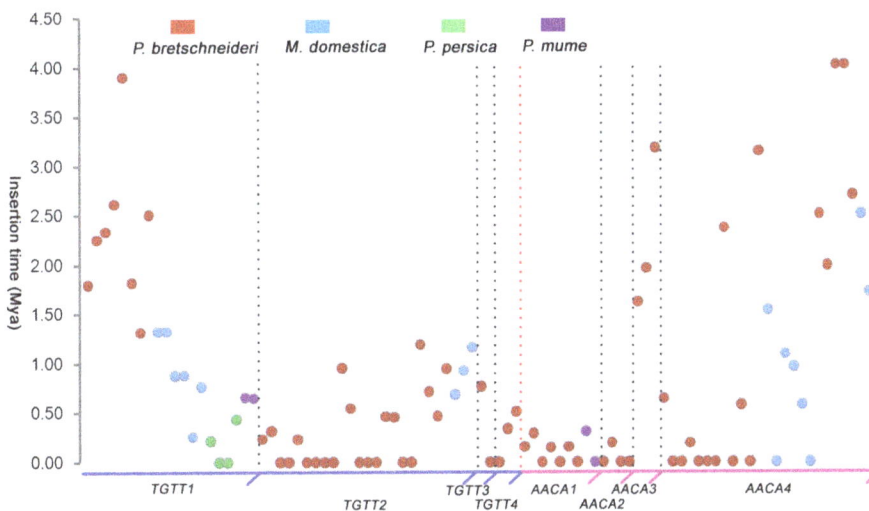

Fig. 3 Insertion times of *TGTT* and *AACA* elements in the four Rosaceae species. The y-axis represents the insertion time. Each *TGTT* and *AACA* family is separated by dotted lines. Elements from different species are represented by red (*Pyrus bretschneideri*), blue (*Malus domestica*), green (*Prunus persica*) and purple (*Prunus mume*) circles, respectively

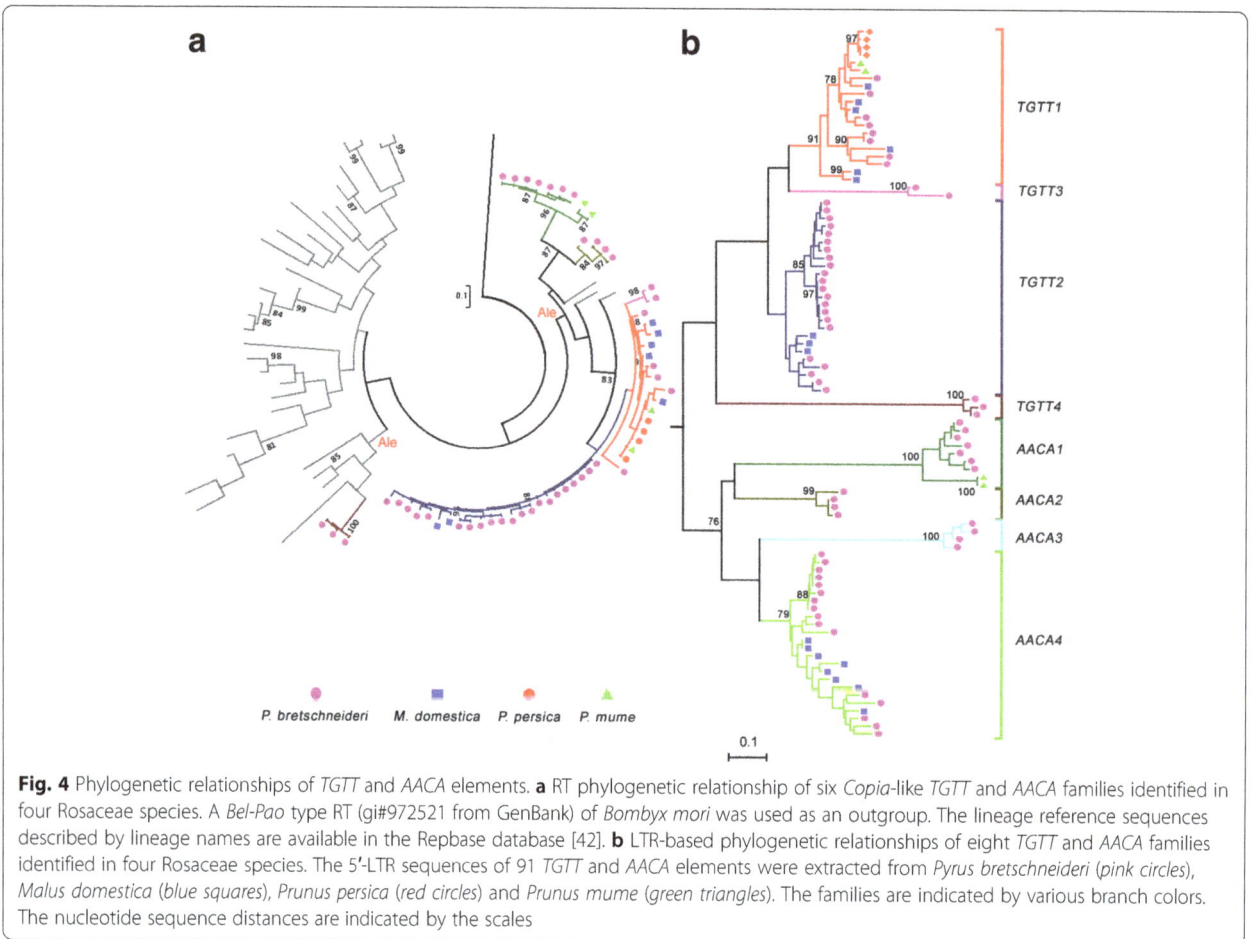

Fig. 4 Phylogenetic relationships of *TGTT* and *AACA* elements. **a** RT phylogenetic relationship of six *Copia*-like *TGTT* and *AACA* families identified in four Rosaceae species. A *Bel-Pao* type RT (gi#972521 from GenBank) of *Bombyx mori* was used as an outgroup. The lineage reference sequences described by lineage names are available in the Repbase database [42]. **b** LTR-based phylogenetic relationships of eight *TGTT* and *AACA* families identified in four Rosaceae species. The 5′-LTR sequences of 91 *TGTT* and *AACA* elements were extracted from *Pyrus bretschneideri* (*pink circles*), *Malus domestica* (*blue squares*), *Prunus persica* (*red circles*) and *Prunus mume* (*green triangles*). The families are indicated by various branch colors. The nucleotide sequence distances are indicated by the scales

and *MdAACA4* elements were mixed with *PbrTGTT2* and *PbrAACA4* elements, respectively. In addition, the *PpTGTT1* and *PmTGTT1* were also found with *MdTGTT1* and *PbrTGTT1*, indicating that the species may have experienced some introgression in early stages of their evolution or HT events after their divergence.

The individual *Copia*-like LTR-RT families can be separated into six major evolutionary lineages, *Angela*, *Ale*, *Bianca*, *Ivana*, *Maximus* and *TAR*. To discern the evolutionary history and phylogenetic relationships among the four *TGTT* and two *AACA Copia*-like families and the major evolutionary lineages, the conserved RT DNA sequences from each of the *TGTT* and *AACA* elements, as well as *Copia*-like LTR-RTs in *Arabidopsis*, rice and soybean, which were previously identified [42], were used to construct a Maximum Likelihood (ML) phylogenetic tree. As shown in Fig. 4a, the six *Copia*-like *TGTT* and *AACA* families all belong to the *Ale* lineage but formed three distinct sublineages. The two *AACA* families were separated into a sublineage, three *TGTT* families (*TGTT1–3*) formed another sublineage, with two sublineages being closer to soybean *Ale* elements, while the *TGTT4* elements, together with the *Arabidopsis Ale*

elements, grouped into a distinct sublineage. Because the two *AACA* TRIM families have no coding genes inside the internal regions, their PBS sites were used to make multiple alignments with those of other elements. Interestingly, the PBS sites were highly conserved with those of other *Ale* lineage elements (Additional file 2: Figure S3). Thus, the two *AACA* TRIM families may also originate from the *Ale* lineage.

HT of *TGTT1* elements between apple and peach genomes

TGTT1 is the only family identified in all four Rosaceae species (pear, apple, peach and mei) but not in the woodland strawberry and other species, indicating that these TEs arose after the divergence of strawberry (*F. vesca*, $n = 7$) and the ancestors of pear (*P. bretschneideri*, $n = 17$), apple (*M. domestica*, n = 17), peach (*P. persica*, $n = 8$) and mei (*P. mume*, n = 8). Previously, the 20 *TGTT1* elements from four Rosaceae species were shown to be mixed in an ML phylogenetic tree based on LTR sequences (Fig. 4b). To understand the evolutionary history of the *TGTT1* elements, the ML phylogenetic tree of *TGTT1* was rebuilt using the whole complete sequences of the 20 TEs, and the data indicated that the *TGTT1* TEs had a patchy distribution in the phylogenies (Fig. 5a). Specifically, four *PpTGTT1* and two *PmTGTT1* TEs were phylogenetically closer to the apple *TGTT1* element (*MdTGTT1_IT4*). The phylogenetic relationships among these *TGTT1* TEs were not fully congruent with their host species (Fig. 5b), which prompted an investigation into the possibility of HT occurring between distantly related Prunus (peach and mei) and Maloideae (apple and pear) species.

To test this hypothesis, the sequence identities between pairs of *TGTT1* elements were initially analyzed. A higher sequence similarity (92.26%, gaps were excluded) was identified between peach (*PpTGTT1_IT3*) and apple (*MdTGTT1_IT4*), and even when taking into account the gaps between the two sequences, the sequence similarity was 89.05% (Additional file 1: Table S5). Additionally, to make a comparison, sequence identities of single orthologous genes were calculated between peach and apple based on their synonymous substitutions per site (Ks; average = 61.29% ± 12.85%) and sequence similarities (gaps calculated average = 69.80% ± 15.56%; gaps excluded average = 86.57% ± 3.59%). Sequence identity at the peak of the distribution should be a good indicator of overall genomic divergence [19]. Here, the sequence identities of *PpTGTT1_IT3* and *MdTGTT1_IT4* were always higher

than the sequence identity peak values of 822 orthologous single genes (Fig. 6, Additional file 1: Tables S6 and S7). This hypothesis was also supported by the Ks values of *TGTT1* family integrases, which are much lower for *PpTGTT1* (0.02 ± 0.01) than for *MdTGTT1* (0.78 ± 0.29) (Table 2), suggesting that it recently entered the peach genome. The presence of the two elements with higher sequence identities in apple and peach were tested by PCR amplification of the LTRs and Sanger sequencing (Fig. 6 and Additional file 2: Figure S2). Thus, the HT of *TGTT1* might have occurred between the distantly related apple and peach.

Transcriptional activities of *TGTT* and *AACA* TEs in different organs and under stress treatments in pear

Because over 70% of *TGTT* and *AACA* TEs were inserted into the four genomes <1.0 Mya, and over one third of these elements (34.07%) contained two identical LTRs, these TEs may still be transcriptionally or even transpositionally active. To detect the transcriptional activity of *TGTT* and *AACA* TEs in the four Rosaceae species, HISAT alignments for each were constructed using the RNA-seq data from Sequence Read Archive (SRA) database of NCBI. A total of 12 *AACA* TEs from pear, including 7 *PbrAACA1*, 1 *PbrAACA3* and 4 *PbrAACA4*, were transcriptionally active in fruit and buds of *P.*

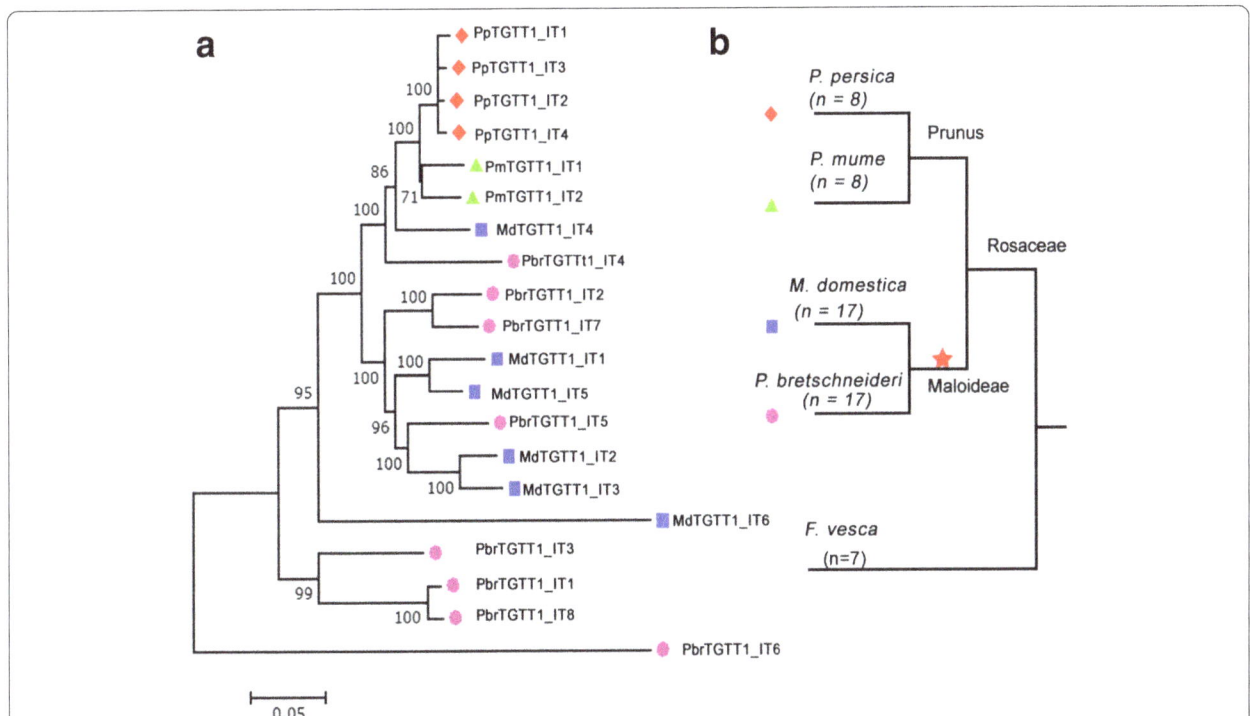

Fig. 5 Phylogenetic incongruences between horizontally transferred *TGTT1* elements and trees of four Rosaceae species. **a** *TGTT1* tree; the ML phylogenetic tree was based on intact sequences of 20 *TGTT1* elements from 8 *PbrTGTT1* (*pink circles*), 6 *MdTGTT1* (*blue squares*), 4 *PpTGTT1* (*red circles*) and 2 *PmTGTT1* (*green triangles*). **b** Species tree; The red star indicates the recent whole-genome duplication event. The nucleotide sequence distances are indicated by the scales

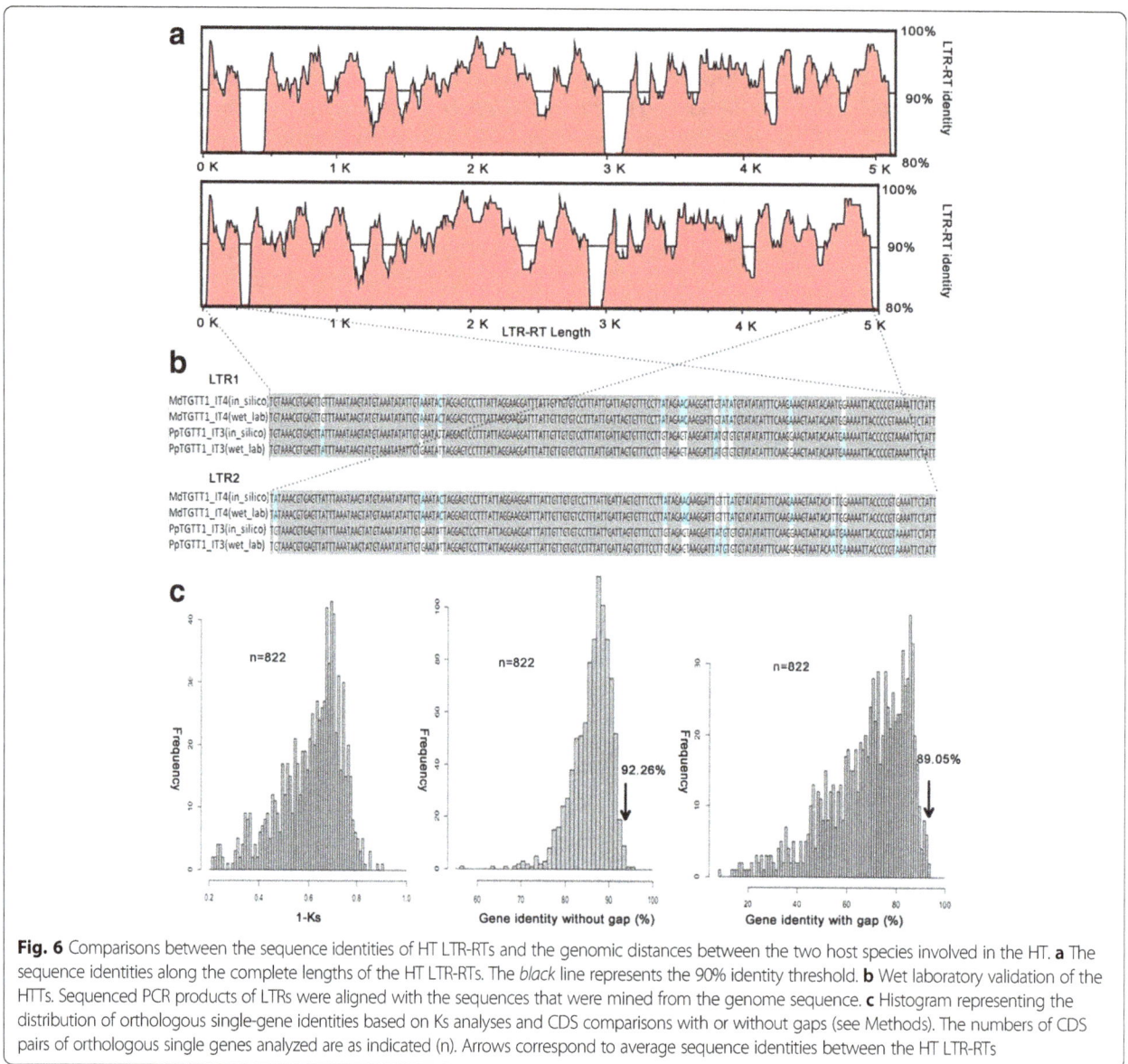

Fig. 6 Comparisons between the sequence identities of HT LTR-RTs and the genomic distances between the two host species involved in the HT. **a** The sequence identities along the complete lengths of the HT LTR-RTs. The *black* line represents the 90% identity threshold. **b** Wet laboratory validation of the HTTs. Sequenced PCR products of LTRs were aligned with the sequences that were mined from the genome sequence. **c** Histogram representing the distribution of orthologous single-gene identities based on Ks analyses and CDS comparisons with or without gaps (see Methods). The numbers of CDS pairs of orthologous single genes analyzed are as indicated (n). Arrows correspond to average sequence identities between the HT LTR-RTs

bretshneideri ('Dangshansuli'; Additional file 1: Table S3 and Additional file 2: Figure S5), whereas none of the *TGTT* TEs were active according to the RNA-seq data. To gain insight into the expression levels of these active TEs, three pair of primers (Additional file 1: Table S8) that corresponded to the transcribed U5 of the 5′ LTR or U3 of the 3′ LTR and the partial internal regions of three randomly selected elements (*PbrAACA1_IT5*, *PbrAACA4_IT11* and *PbrAACA4_IT14*) were used for qRT-PCR in different pear ('Dangshansuli') organs, including fruit flesh at four developmental periods (32, 65, 99 and 143 d after flower bloom), leaves, pericarp, pollen and stylet. Transcripts of the three elements could be detected in all eight samples (Fig. 7a–c, Additional file 2: Figure S5). The transcriptional levels varied for the three elements in the different organs. For example, the

transcriptional levels of *PbrAACA1_IT5* (Fig. 7a), *PbrAA-CA4_IT11* (Fig. 7b) and *PbrAACA4_IT14* (Fig. 7c) were highest at 32, 99 and 143 d after flower bloom in fruit flesh, respectively, and they all expressed at their lowest levels in the pollen.

The transcriptional activities of LTR-RTs are usually repressed in plant tissues during normal development, as well as in response to a variety of biotic and abiotic stresses [45]. Thus, the expression patterns of the former three elements in leaves of pear ('Dangshansuli') were examined under various stress treatments, including cold, heat and salt (see Methods). *PbrAACA1_IT5* was significantly up-regulated in heat (6 h, Fig. 7g) and salt (6 h, Fig. 7j) stress treatments (two-tailed *t*-test, $P < 0.01$), but down-regulated in the cold stress treatments (two-tailed *t*-test, $P < 0.01$, Fig. 7d). Elevated

Table 2 Ka, Ks and Ka/Ks values of integrases in the *TGTT* and *AACA* families

Family	Ka	Ks	Ka/Ks
PbrTGTT1	0.73 ± 0.42	0.56 ± 0.18	1.33 ± 0.65
MdTGTT1	0.74 ± 0.39	0.78 ± 0.29	0.99 ± 0.48
PpTGTT1	0.01 ± 0.00	0.02 ± 0.01	0.51 ± 0.36
PmTGTT1	0.51 ± 0.00	0.50 ± 0.00	1.01 ± 0.00
PbrTGTT2	0.15 ± 0.17	0.32 ± 0.21	0.42 ± 0.38
MdTGTT2	0.42 ± 0.24	0.44 ± 0.20	0.85 ± 0.24
PbrTGTT3	0.12 ± 0.00	0.09 ± 0.00	1.25 ± 0.00
PbrTGTT4	0.58 ± 0.00	0.61 ± 0.00	0.96 ± 0.00
PbrAACA1	0.49 ± 0.73	0.33 ± 0.42	0.63 ± 0.67
PmAACA1	0.00 ± 0.00	0.01 ± 0.00	0.29 ± 0.00
PbrAACA2	0.03 ± 0.00	0.11 ± 0.00	0.25 ± 0.00

levels of *PbrAACA4_IT11* transcripts were observed in cold (1 h, Fig. 7e) and salt (6 h, Fig. 7k) stress treatments (two-tailed t-test, $P < 0.01$) but not in heat. For *PbrAACA4_IT14*, the up-regulation was also significant in cold (two-tailed t-test, $P < 0.01$, 1 h, Fig. 7f) and salt stress treatments (two-tailed t-test, $P < 0.01$, 24 h, Fig. 7l). These data may indicate the constitutive expression of *TGTT* and *AACA* TEs in different pear tissues and the elevation of their expression levels under various stresses.

Discussion

We have isolated two novel LTR-RT subfamilies called *TGTT* and *AACA* in four Rosaceae genomes, and these can be classified into eight families according to the TE classification system proposed by Wicker et al. 2007 [40]. Six *Copia*-like families were classified into the *Ale* lineage using a phylogenetic analysis (Fig. 4a), while the other two *AACA* families were *TRIM* elements owing to their short non-coding internal regions (Fig. 2). However, *TGTT* and *AACA* TEs were restricted to four Rosacese species (pear, apple, peach and mei), and not even truncated fragments were detected in the closest woodland strawberry genome and other published plant genomes. This suggested that the two subfamilies evolved specifically after the divergence between *F. vesca* and the ancestor of the four Rosacese species.

Specific dinucleotide termini of two new subfamilies of non-canonical LTR-RTs in the four Rosaceae species

LTR sequences start with 'TG' and end with 'CA' in typical LTR-RTs [9]. In a previous study, our group described a systematic survey of LTR-RTs in the sequenced pear (*P. bretschneideri*) genome, and 3221 full-length LTR-RTs have been found to terminate with 5′-TG–CA-3′ [3, 38]. Here, although these new TEs contained most of the typical

features of LTR-RTs, a salient difference was identified in the dinucleotide positions at both ends of the LTRs. When these TEs with their two flanking sequences (500-bp for each site) where combined and aligned, they showed an accurate insertion site terminating with 5′-TG–TT-3′ or 5′-AA–CA-3′, and flanked by perfect 4 to 6-bp TSDs. Thus, these atypical LTR-RTs were defined as two new subfamilies (*TGTT* and *AACA*). Previously, an atypical LTR-RT family (*TARE1*) in tomato [13] that contained 5′-TA–CA-3′ at both ends of the LTRs was reported, and a plausible mutation model explaining the creation of such atypical dinucleotides in the LTRs was proposed. In addition, another exception, *AcCOPIA1*, which terminates with 'TG' and 'TA' at both ends of the LTRs, was identified in onion [14]. Similar to *TARE1*, only one nucleotide changed from 'CA' to 'TA' in *AcCOPIA1*. Compared with the *TARE1* and *AcCOPIA1* TEs, the 'CA' has been changed to 'TT' in the *TGTT* elements and 'TG' turned into 'AA' in the *AACA* elements. Interestingly, the 5′-TG–TT-'3′ of *TGTT* is the reverse complement of the 5′-AA–CA-3′ of *AACA*. The simultaneous mutation of the dinucleotides is a low probability event that cannot be easily explained by the mutation model [13]. A global annotation, structural analysis and phylogenetic study of all the non-TGCA TEs within the eukaryotes is worth performing in the future, to unravel the scale and frequencies of non-canonical LTR-RTs that terminate with non-TGCA motifs and whether they exist naturally or derived from the normal TGCA-containing LTR-RT elements.

Although most of LTR-RTs carry the palindromic dinucleotide motif (5′-TG–CA-3′) flanking each LTR, the importance of this conserved motif is still poorly understood. Studies of retrovirus integration indicate that the 3′ CA terminal sequences of retroviral LTRs are essential for viral integration [46, 47]. The close relationship between retroviruses and LTR-RT TEs, with an additional envelope protein [46, 47], may explain why most LTR-RTs have the conserved 5′-TG–CA-3′ motif. Here, based on the phylogenetic tree of *TGTT*, *AACA* and typical TGCA integrases, most *TGTT* integrases (*TGTT1*, *TGTT2* and *TGTT3* elements) can be differentiated from typical integrases (Additional file 2: Figure S4), whereas the *AACA* elements could not be differentiated. The Ka/Ks values of the INT domains of *TGTT1* and *TGTT3* families (>1) are significantly greater than those of *AACA1* and *AACA2* families (<1) (Table 2). Thus, the functional divergence of the integrase active sites from these *3′-TT ends* LTR-RTs has occurred and might result in a novel integration mechanism. Based on sequence comparisons, structural and phylogenetic analyses, the newly identified *TGTT* and *AACA* TEs should provide a valuable resource to study the non-canonical LTR-RTs integration mechanism.

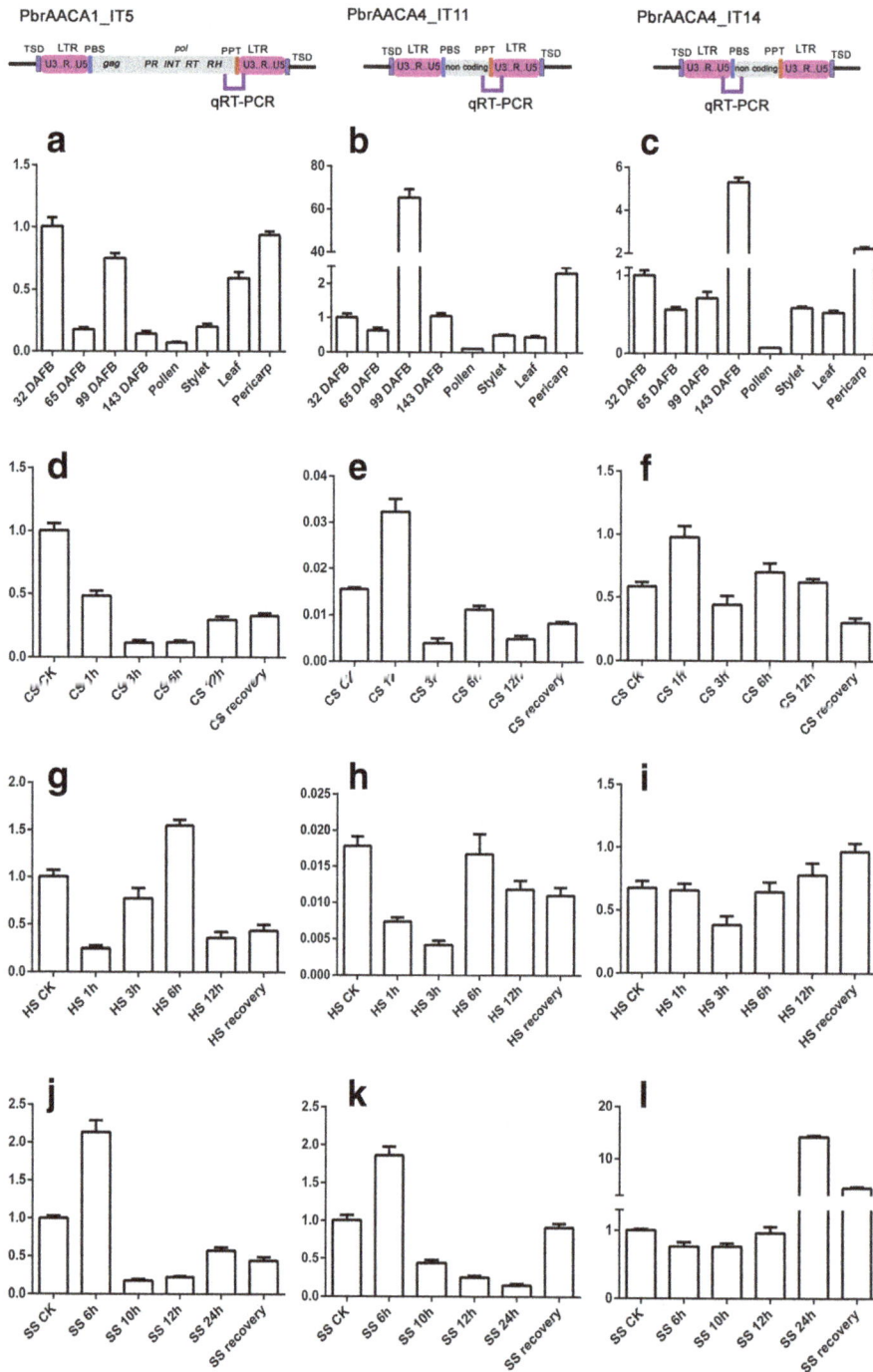

Fig. 7 Time-course expression levels of active *AACA* elements in *Pyrus bretschneideri*. The positions of the primers used for transcriptional validation are indicated in the schematic of each element in qRT-PCR region. The expression levels were detected in the fruit flesh of four developmental stages, pollen, stylet, leaf and pericarp. **a–c**, leaves under cold (**d–f**), heat (**g–i**), and salt (**j–l**) treatments for *PbrAACA1_IT5* (**a, d, g, j**), *PbrAACA4_IT11* (**b, e, h, k**) and *PbrAACA4_IT14* (**c, f, i, l**). Error bars indicate the standard deviations of three biological replicates

The *PpTGTT1* elements originated from *MdTGTT1* through HT

The transmission of genetic materials among sexually isolated species is usually defined as HTs. HTTs were first proposed as a possible dissemination mechanism of TEs in eukaryotes [15]. Because TEs could undergo epigenetic-mediated silencing by the host genome [48], HTTs could be the mechanism of escaping the silencing and ensuring the long-term survival of TEs among eukaryotic lineages. Based on this model, most of the active TEs found in plant and animal genomes may originate from other species through HTT [49]. Owing to the availability of more released eukaryotic genome sequences and standard comparative genomics approaches [50], hundreds of cases of HTTs have been reported over the past years [18, 19, 51]. Recently, Fawcet and Innan (2016) [52] proposed a method to differentiate the HTT and vertical transmission scenarios, which involves testing whether the hypothetical HTT copies are present in the orthologous regions of the two species [52]. If the two species acquired the copies independently by HTs, then the two species should not share any copies and each copy should be species-specific. This theory is reasonable because TE-based recombination and loss occur frequently in host genomes. The same analysis was conducted for the *TGTT1* TEs among the four Rosaceae genomes, however, no shared intact *TGTT1* TEs or degenerated fragments were detected between pairs of the Rosaceae genomes. Combined with the high similarity between *MdTGTT1_IT4* and *PpTGTT1_IT3*, the deep divergence time between Maloideae and Prunus (>45 Mys), and their patchy distribution in Rosaceae, indicates that the HT of the *TGTT1* TEs occurred between apple and peach. Recently, 32 clear cases of recent HTTs of LTR-RTs, including 5 HTTs between apple and peach, were detected among 46 sequenced plant genomes [19]. As expected, the HTT of *TGTT1* TEs was not included in the five reported HTTs between apple and peach, possibly because the *TGTT1* TEs were not identified initially through their LTR-RTs annotation method. Thus, the estimation of millions of HTTs occurring among the angiosperms in the recent evolutionary past may have been an underestimation.

The insertion time of the HTT elements have facilitated us to speculate the HTT history and time frame. Based on the sequence divergence of two LTRs of *MdTGTT1_IT4* and *PpTGTT1_IT3*, we propose that the presence of *TGTT1* in peach was resulted from HT of *MdTGTT1* between 0.43 Mys and 0.88 Mys. First, the average insertion time of the four *PpTGTT1* elements (0.16 Mys) is much younger than that of the six *MdTGTT1* elements (0.91 Mys, Additional file 1: Table S3), especially, *PpTGTT1_IT3* (0.43 Mys) aged much younger than *MdTGTT1_IT4* (0.88 Mys); Second, *PpTGTT1_IT3* is the oldest element of the four *PpTGTT1* elements (Additional file 1: Table S3); Third, four *MdTGTT1* elements including *MdTGTT1_IT4* are still transcriptionally active (Additional file 2: Figure S5 and Additional file 1: Table S3). Last, the cluster of the horizontally transferred *PpTGTT1* copies is included in the larger cluster of copies from apple and pear. All of these evidences suggest that peach is the recipient species of the HT event. Although several studies suggest that "host-vector species" interactions may favor HTTs in animals [53, 54]. However, evidence of "host-vector-driven" HTTs has not been provided in plants, and no experimental evidence of this

process has been reported yet. Thus it is unclear how the transfer of *TGTT1* may have occurred between apple and peach. Considering that apple and peach belong to Rosaceae fruit crops with higher economic values, the most plausible explanation is that the *TGTT1* was transmitted by their common pathogen, such as bacteria, fungi and virus often believed to be the vectors of HT [55, 56], and perhaps with the help of a plant cell-piercing insect [54, 55]. Thus *TGTT1* should be an attractive candidate for testing whether similar mechanisms of HTTs exist in plants.

Varied transcription activities of these non-canonical TEs in pear and apple

Although the LTR-RTs are less likely to be actively expressed in plant tissues during normal development, several exceptions have been reported in various organs belonging to different species, such as *Ogre* elements in leaves, roots and flowers of pea [57], *Grande* elements in leaves of *Zea* and *Tripsacum* [58], eight LTR-RT families in leaves, stalks and roots of *Eucalyptus* genus [59], and *EARE-1* elements in roots, staminate flowers, pistillate flowers, leaves and seeds of *Excoecaria agallocha*, which were all detected as transcriptionally active [60]. In our study, three families (12 *AACA* TEs) from pear were initially detected with transcriptional activity using the published RNA-seq data in the SRA database of NCBI, whereas no transcripts of the six families from peach (*PpTGTT1*), mei (*PmTGTT1* and *PmAACA1*) and apple (*MdTGTT1*, *MdTGTT2* and *MdAACA1*) were identified. In particular, all of the *TGTT* TEs were silenced in the four species, indicating that the transcriptional activities of the *TGTT* and *AACA* TEs varied in different species. The qRT-PCR analyses of three randomly selected elements from pear (*PbrAACA1_IT5*, *PbrAACA4_IT11* and *PbrAACA4_IT14*) also proved that these *TGTT* and *AACA* TEs are transcriptionally active at different levels in different organs. In addition, various elevated transcript levels of the three pear elements were observed following heat, cold and salt treatments, indicating that the *TGTT* and *AACA* TEs could be activated by stresses. This is coincident with the discovery that several other LTR-RTs are frequently activated under stress conditions [28, 60, 61], and also conforms to McClintock's theory of genome shock in which the enhanced activities of TEs under stress might represent an evolutionary strategy for plant species to increase the chances of survival under unfavorable conditions [62]. Although the transcriptional activities of *TGTT* and *AACA* TEs were not detected in peach and mei using the SRA and EST databases from NCBI, the *PpTGTT1* family has proliferated into four copies since the HT of *PpTGTT1_IT3* from apple, indicating that the element was active for a short time after invasion and then the life cycle of *PpTGTT1* may have been firmly controlled at the post-transcriptional level. Further studies will be conducted in the future.

Conclusions

TGTT and *AACA* are two new types of LTR-RT subfamilies isolated from pear, apple, peach and mei that terminate with atypical dinucleotide structures. Their family and element copy numbers, proliferation time frames and transcriptional activities varied among the four Rosaceae species. HT might have played a significant role in the life cycle of *TGTT1*. These newly identified TEs should be valuable materials for the further investigation of atypical LTR-RTs in other sequenced plant species and will provide interesting insights into their structural evolution and TE-driven genomic evolution.

Methods

Genome sequence resources, annotation and classification of *TGTT* and *AACA* LTR-RTs

The genome sequence data for the four Rosaceae species are available in Additional file 1: Table S1. The annotation method of *TGTT* and *AACA* LTR-RT elements has been widely used in previous studies [3, 13, 38]. First, based on the structural analysis, several intact elements were identified by the LTR_STRUC program [39]. Then, all of the identified LTR sequences of the intact elements with clearly defined boundaries were used as queries to detect additional intact elements through sequence homology searches using CROSS_MATCH and CLUSTALW programs with default parameters. Finally, the structures and boundaries of all of the identified LTR-RTs were manually inspected and confirmed, and the TSD sites were defined with one mismatch allowed. Fragments and truncated elements were not analyzed in this study. The *TGTT* and *AACA* LTR-RTs were classified into superfamilies based on the conserved functional domains detected using the BLASTX tool. The queried domains included GAG (for UBN2 superfamily domain, pfam14223), PR (for GAG-pre-integrase domain, pfam13976), INT (for integrase core domain, pfam00665), RT (for reverse transcriptase domain, pfam07727) and RNase H (for Ribonuclease H domain, cd09272). Each individual family was classified using sequence homology comparisons according to the criteria described previously [40].

Estimation of insertion time

The insertion time of intact elements with TSD sites was estimated by comparing the divergence of their 5′ and 3′ LTR sequences because both LTR sequences of a newly proliferated LTR-RT were believed to be identical at the time of integration [41]. To investigate the nucleotide substitution rate for each element, the two LTR sequences were aligned using the MUSCLE program with default parameters [63]. The insertion time (T) for each intact element was calculated using the formula: T = K/2r, in which the average number of substitutions per aligned site (K) was corrected using the Jukes–Cantor method [64], and 1.3×10^{-8} substitutions per site per year was used as the average LTR substitution rate (r) [44].

Phylogenetic analysis

For each *TGTT* and *AACA* TE, the 5′ LTR sequence, and RT and INT domains were extracted from the intact sequence using a perl script. Sequence alignments were performed by the MAFFT version 7 program with default options [65]. The MEGA 5.2 program implemented with Jukes–Cantor model was employed for building the Maximum Likelihood trees based on 1000 bootstrap replicates [66]. The taxonomic tree was built using the common tree tool on the NCBI website (http://www.ncbi.nlm.nih.gov/Taxonomy/CommonTree/wwwcmt.cgi, last accessed).

Identification of orthologous single genes and estimation of genomic sequence divergence

The strategy to identify orthologous single genes between the apple and peach genomes has been used in previous studies [4, 38]. First, the genomic protein and CDS sequences of apple and peach were downloaded from the Phytozome website (http://www.phytozome.net) and set as a database. Then, the BLASTP and orthoMCL software [67] were employed to identify all the orthologous single genes in the two genomes using the same parameters in the previous study [38]. All of the identified single-copy orthologous genes were manually inspected, and gene sequences that contained frame-shift mutations or stop codons were excluded from further analysis.

The Ka, Ks, and Ka/Ks ratio of orthologous single genes and the intra-family INT domains of *AACA* and *TGTT* TEs were calculated using the YN00 program implemented in the PAML software package [68].

The CDS sequence identities of orthologous single genes in apple and peach were computed using an in-house perl script, which ran in the following three steps: (1) All of the identified orthologous single-gene pairs were separately aligned using MUSCLE software; (2) For each orthologous single-gene pair, the numbers of identical nucleotides (I), mismatches (M) and gaps (G) were counted. (3) Gene identities without gaps were calculated using the formula: I/(I + M) × 100, and gene identities with gaps were calculated using the formula: I/(I + M + G) × 100 (Additional file 1: Table S7). The sequence identity analysis between the *TGTT1* TEs were also conducted using the same strategy.

PCR and sequencing analysis

The total genomic DNA of the four Rosaceae species were extracted from the young leaves using the improved cetyltrimethyl ammonium bromide method. In total, 11 *TGTT* and 5 *AACA* TEs from eight families were randomly selected for validation. For each element, 300-bp 5′-flanking sequences and 300-bp 3′-flanking sequences of both LTR sequences were extracted and used to design primers (Additional file 2: Figure S2 and Additional file 1: Table S5). Polymerase chain reactions (PCR) were performed in a total volume of 25 ml, containing 1 ml of 50 ng/ml genomic DNA template, 2.5 ml of 10× buffer (without MgCl2), 2.5 ml of 2.5 mM dNTP mixture, 2.5 ml of 25 mM $MgCl_2$, 0.8 ml each of forward and reverse primer (10 pmol/ml) and 0.2 ml of 5 U/ml Taq polymerase (Takara Biotechnology Company, Dalian, China). The reactions were performed with the following conditions: 94 °C for 3 min, then 35 cycles of 94 °C for 30 s, 57 °C for 40 s and 72 °C for 2 min, and a final step at 72 °C for 10 min. The PCR products were resolved on 1% agarose and detected by ethidium bromide staining. The analyses were performed three times and loaded on independent gels. All of the specific PCR products were isolated with the DNA Gel Extraction kit AxyPrep (Axygen Inc.). The fragments were cloned into the pMD19-T vector (Takara, China), and the plasmids were sequenced by Invitrogen (Shanghai, China).

Transcriptional activity analysis of *TGTT* and *AACA* elements

The Illumina RNA-Seq data of four samples from the SRA database of NCBI (https://www.ncbi.nlm.nih.gov/), including pear fruit peel (*P. bretshneideri* 'Dangshansuli', SRX298075), pear bud (*P. bretshneideri* 'Dangshansuli', SRX147917) and apple leaves (*M. domestica* 'Gala', SRX1150925), were used to identify the transcriptional patterns. For each element, the whole-nucleotide sequences were used as queries to construct HISAT alignments using the default parameters [69].

Stress treatments and LTR-RT expression analysis by quantitative RT-PCR

The scions of 'Dangshansuli' (*P. bretschneideri*) were grafted to 1-year-old *Pyrus betulifolia* plants and grown in a culture room at 25 °C under long-day conditions (16 h light/8 h dark) for 30 d prior to stresses. For the cold treatment, seedlings were placed in a growth incubator set at 4 °C for 0, 1, 3, 6 and 12 h. For heat stress treatments, plants were transferred to 40 °C for 0, 1, 3, 6 and 12 h. Salt stress was carried out by watering the plants with 1600 mM NaCl solution for 0, 1, 3 and 12 h. All of the samples were recovered for 24 h.

The total RNA from fruit flesh at four developmental periods (32, 65, 99 and 143 d after flower bloom), leaves, pericarp, pollen and stylet were extracted using a cetyltrimethyl ammonium bromide-based method and digested with RNase-free DNase I (Thermo) to remove DNA contamination. According to the manufacturer's instructions, 1 mg of total RNA was reverse transcribed into cDNA using the ReverTra Ace qPCR RT Kit (Toyobo, Shanghai, China). Specific primers for *PbrAACA1_IT5*, *PbrAACA4_IT11* and *PbrAACA4_IT14* were designed using the Primer 5 software (Additional file 1: Table S7). Quantitative real-time RT-PCR (qRT-PCR) was used for measuring the transcript levels of the three LTR-RTs. The PCR solution (20 ml) contained 10 ml of SYBR-Green PCR Master Mix (SYBR Premix EX TaqTM, TaKaRa), 0.25 mM each of forward and reverse primer, 100 ng of cDNA template, and nuclease free water. The qRT-PCR analysis with a SYBR Green PCR kit was performed in a Light Cycler 480 (Roche, USA) Real-Time System. The reactions were conducted under the following conditions: 95 °C for 5 min, then 45 cycles of 94°C4for 10 s, 60° C0for 30 s and 72 °C for 30 s, followed by a final extension at 72 °C for 3 min. The $2^{-\Delta\Delta CT}$ method [70] was used to calculate the relative expression levels of each gene. Each sample was analyzed for three replicates. The mRNA capping enzyme gene (*Pbr035952.1*) and cytochrome B561 gene (*Pbr013721.1*) were used as internal controls for five different tissue and three stress treatments, respectively, and to normalize the relative expression levels of each LTR-RT. The expression analysis of each time point was repeated three times.

Additional files

Additional file 1: Table S1. List of four species and their genomic sequence and DNA material source information used in this study. **Table S2.** Sequence identity between all TGTT and AACA consensus sequences in the four Rosaceae genomes. **Table S3.** Summary of *TGTT* and *AACA* elements identified in *P. bretschneideri*, *M. domestcia*, *P. persica* and *P. mume*. **Table S4.** Primers used for wet lab validations. **Table S5.** Sequence identity between all TGTT1 elements identified in the four Rosaceae genomes. **Table S6.** Sequence divergence of orthologous singletons between apple and peach. **Table S7.** Sequence identity of the orthologous singletons between apple and peach. **Table S8.** Primers used for Real time Quantitative PCR. (XLSX 142 kb)

Additional file 2: Figure S1. Sequence alignment of the *TGTT1* elements. **Figure S2.** Wet laboratory validation of the *TGTT* and *AACA* elements. **Figure S3.** Multiple alignments of PBS sites from *TGTT*, *AACA* and normal *TGCA* elements of the *Ale* lineage. **Figure S4.** INT phylogenetic relationships among *TGTT*, *AACA* and normal *TGCA* elements. **Figure S5.** Evidence of transcriptional activity in five *TGTT* and *AACA* elements. (PDF 2824 kb)

Abbreviations

HTs: Horizontal transfers; HTT: Horizontal TE transfer; HTTEs: Horizontal transfer of transposable elements; int/INT: Integrase; LARDs: Large retrotransposon derivatives; LTR-RTs: Long terminal repeat retrotransposons; Mya: Million years ago; Mys: Million years; PBS: Primer-binding site; PPT: Polypurine tract; PR: Protease; qRT-PCR: Quantitative real time polymerase chain reaction; rh/RH: RNase H; rt/RT: Reverse transcriptase; SRA: Sequence read archive; TEs: Transposable elements; TRIMs: Terminal-repeat retrotransposons in miniature; TSD: Target site duplication

Acknowledgements

We would like to thank many genome data contributors. The high-quality genome sequences generated by them were instrumental in conducting this study.

Funding

This study was supported by the Natural Science Foundation of Jiangsu Province in China (BK20160715), China Postdoctoral Science Foundation funded project (2015 M570456) and the National Natural Science Foundation of China (31701890).

Author's contributions

Conceived and designed the experiments: HY SZ. Performed the experiments: HY XW. Analyzed the data: HY XW DS YC KQ ZM. Contributed reagents/materials/analysis tools: HY XW ZM. Wrote the paper: HY SZ. All authors read and approved the final maunscript.

Competing interests

The authors have declared that no competing interests exist.

Author details

[1]Center of Pear Engineering Technology Research, College of Horticulture, Nanjing Agricultural University, Nanjing 210095, China. [2]State Key Laboratory of Crop Genetics and Germplasm Enhancement, Nanjing Agricultural University, Nanjing, China. [3]College of Agricultural Sciences, Nanjing Agricultural University, Nanjing, China.

References

1. Bennetzen JL, Ma J, Devos KM. Mechanisms of recent genome size variation in flowering plants. Annu Bot. 2005;95(1):127–32.
2. GT V, Schmutzer T, Bull F, Cao HX, Fuchs J, Tran TD, et al. Comparative genome analysis reveals divergent genome size evolution in a carnivorous plant genus. Plant Genome. 2015;8(3)
3. Yin H, Du J, Li L, Jin C, Fan L, Li M, et al. Comparative genomic analysis reveals multiple long terminal repeats, lineage-specific amplification, and frequent Interelement recombination for Cassandra Retrotransposon in pear (*Pyrus bretschneideri* Rehd.). Genome Biol Evol. 2014;6(6):1423–36.
4. Wu J, Wang Z, Shi Z, Zhang S, Ming R, Zhu S, et al. The genome of the pear (*Pyrus bretschneideri* Rehd.). Genome Res. 2013;23(2):396–408.
5. Velasco R, Zharkikh A, Affourtit J, Dhingra A, Cestaro A, Kalyanaraman A, et al. The genome of the domesticated apple (*Malus* × *domestica* Borkh.). Nature Genet. 2010;42(10):833–9.
6. Verde I, Abbott AG, Scalabrin S, Jung S, Shu S, Marroni F, et al. The high-quality draft genome of peach (*Prunus persica*) identifies unique patterns of genetic diversity, domestication and genome evolution. Nature Genet. 2013;45(5):487–94.
7. Paterson AH, Wendel JF, Gundlach H, Guo H, Jenkins J, Jin D, et al. Repeated polyploidization of *Gossypium* genomes and the evolution of spinnable cotton fibres. Nature. 2012;492(7429):423–7.
8. Schnable PS, Ware D, Fulton RS, Stein JC, Wei F, Pasternak S, et al. The B73 maize genome: complexity, diversity, and dynamics. Science. 2009; 326(5956):1112–5.
9. Kumar A, Bennetzen JL. Plant retrotransposons. Annu Rev Genet. 1999;33(1): 479–532.
10. Kalendar R, Vicient CM, Peleg O, Anamthawat-Jonsson K, Bolshoy A, Schulman AH. Large retrotransposon derivatives: abundant, conserved but nonautonomous retroelements of barley and related genomes. Genetics. 2004;166(3):1437–50.
11. Witte C-P, Le QH, Bureau T, Kumar A. Terminal-repeat retrotransposons in miniature (TRIM) are involved in restructuring plant genomes. P Natl Acad Sci USA. 2001;98(24):13778–83.
12. Hirochika H, Sugimoto K, Otsuki Y, Tsugawa H, Kanda M. Retrotransposons of rice involved in mutations induced by tissue culture. P Natl Acad Sci USA. 1996;93(15):7783–8.
13. Yin H, Liu J, Xu Y, Liu X, Zhang S, Ma J, et al. *TARE1*, a mutated *Copia*-like LTR retrotransposon followed by recent massive amplification in tomato. PLoS One. 2013;8(7):e68587.
14. Kim S, Park JY, Yang T-J. Characterization of three active transposable elements recently inserted in three independent *DFR*-A alleles and one high-copy DNA transposon isolated from the pink allele of the ANS gene in onion (*Allium cepa* L.). Mol Gen Genomics. 2015; 290(3):1027–37.
15. Hartl DL, Lohe AR, Lozovskaya ER. Regulation of the transposable element *mariner*. Genetica. 1997;100(1):177–84.
16. Daniels SB, Peterson KR, Strausbaugh LD, Kidwell MG, Chovnick A. Evidence for horizontal transmission of the P transposable element between *Drosophila* species. Genetics. 1990;124(2):339–55.
17. Roulin A, Piegu B, Wing RA, Panaud O. Evidence of multiple horizontal transfers of the long terminal repeat retrotransposon *RIRE1* within the genus *Oryza*. Plant J. 2008;53(6):950–9.
18. Roulin A, Piegu B, Fortune PM, Sabot F, D'Hont A, Manicacci D, et al. Whole genome surveys of rice, maize and sorghum reveal multiple horizontal transfers of the LTR-retrotransposon *Route66* in Poaceae. BMC Evol Biol. 2009;9(1):1–10.
19. El Baidouri M, Carpentier M-C, Cooke R, Gao D, Lasserre E, Llauro C, et al. Widespread and frequent horizontal transfers of transposable elements in plants. Genome Res. 2014;24(5):831–8.
20. Feschotte C, Jiang N, Wessler SR. Plant transposable elements: where genetics meets genomics. Nat Rev Genet. 2002;3(5):329–41.
21. Paszkowski J. Controlled activation of retrotransposition for plant breeding. Curr Opin Biotech. 2015;32:200–6.
22. Grandbastien M-A. Activation of plant retrotransposons under stress conditions. Trends Plant Sci. 1998;3(5):181–7.
23. Duan K, Ding X, Zhang Q, Zhu H, Pan A, Huang J. *AtCopeg1*, the unique gene originated from *AtCopia95* retrotransposon family, is sensitive to external hormones and abiotic stresses. Plant Cell Rep. 2008;27(6):1065–73.
24. Manninen O, Kalendar R, Robinson J, Schulman A. Application of *BARE-1* retrotransposon markers to the mapping of a major resistance gene for net blotch in barley. Mol Gen Genet. 2000;264(3):325–34.
25. Ito H, Gaubert H, Bucher E, Mirouze M, Vaillant I, Paszkowski J. An siRNA pathway prevents transgenerational retrotransposition in plants subjected to stress. Nature. 2011;472(7341):115–9.
26. Ito H, Yoshida T, Tsukahara S, Kawabe A. Evolution of the *ONSEN* retrotransposon family activated upon heat stress in Brassicaceae. Gene. 2013;518(2):256–61.
27. Jiang N, Bao Z, Zhang X, Hirochika H, Eddy SR, McCouch SR, et al. An active DNA transposon family in rice. Nature. 2003;421(6919):163–7.
28. Butelli E, Licciardello C, Zhang Y, Liu J, Mackay S, Bailey P, et al. Retrotransposons control fruit-specific, cold-dependent accumulation of anthocyanins in blood oranges. Plant Cell. 2012;24(3):1242–55.
29. Rocheta M, Carvalho L, Viegas W, Morais-Cecílio L. *Corky*, a *gypsy*-like retrotransposon is differentially transcribed in *Quercus* Suber tissues. BMC Res Notes. 2012;5(1):432.
30. Felice B, Wilson RR, Argenziano C, Kafantaris I, Conicella C. A transcriptionally active *copia*-like retroelement in *Citrus* limon. Cell Mol Biol Lett. 2008;14(2):289.
31. Kimura Y, Tosa Y, Shimada S, Sogo R, Kusaba M, Sunaga T, et al. *OARE-1*, a *Ty1-copia* retrotransposon in oat activated by abiotic and biotic stresses. Plant Cell Physiol. 2001;42(12):1345–54.
32. Mhiri C, Morel J-B, Vernhettes S, Casacuberta JM, Lucas H, Grandbastien M-A. The promoter of the tobacco *Tnt1* retrotransposon is induced by wounding and by abiotic stress. Plant Mol Biol. 1997;33(2):257–66.
33. Jiang S, Cai D, Sun Y, Teng Y. Isolation and characterization of putative functional long terminal repeat retrotransposons in the *Pyrus* genome. Mob DNA. 2016;7(1):1.
34. Rosaceae KC. In:Kubitzki K. (eds) Flowering Plants · Dicotyledons. The Families and Genera of Vascular Plants. Berlin, Heidelberg: Springer; 2004. p. 343–86.
35. Zhang Q, Chen W, Sun L, Zhao F, Huang B, Yang W, et al. The genome of *Prunus mume*. Nat Commun. 2012;3(4):1318.
36. Shulaev V, Sargent DJ, Crowhurst RN, Mockler TC, Folkerts O, Delcher AL, et al. The genome of woodland strawberry (*Fragaria vesca*). Nat Genet. 2011; 43(2):109–16.
37. Potter D, Eriksson T, Evans RC, Oh S, Smedmark J, Morgan DR, et al. Phylogeny and classification of Rosaceae. Plant Syst Evol. 2007;266(1-2):5–43.
38. Yin H, Du J, Wu J, Wei S, Xu Y, Tao S, et al. Genome-wide annotation and comparative analysis of long terminal repeat Retrotransposons between pear species of P. *bretschneideri* and *P. Communis*. Sci Rep. 2015;5
39. McCarthy EM, JF MD. LTR_STRUC: a novel search and identification program

for LTR retrotransposons. Bioinformatics. 2003;19(3):362–7.

40. Wicker T, Sabot F, Hua-Van A, Bennetzen JL, Capy P, Chalhoub B, et al. A unified classification system for eukaryotic transposable elements. Nat Rev Genet. 2007;8(12):973–82.

41. SanMiguel P, Gaut BS, Tikhonov A, Nakajima Y, Bennetzen JL. The paleontology of intergene retrotransposons of maize. Nat Genet. 1998;20(1):43–5.

42. Du J, Tian Z, Hans CS, Laten HM, Cannon SB, Jackson SA, et al. Evolutionary conservation, diversity and specificity of LTR-retrotransposons in flowering plants: insights from genome-wide analysis and multi-specific comparison. Plant J. 2010;63(4):584–98.

43. Zhao M, Ma J. Co-evolution of plant LTR-retrotransposons and their host genomes. Protein Cell. 2013;4(7):493–501.

44. Ma J, Bennetzen JL. Recombination, rearrangement, reshuffling, and divergence in a centromeric region of rice. P Natl Acad Sci USA. 2006;103(2):383–8.

45. Wessler SR. Plant retrotransposons: turned on by stress. Curr Biol. 1996; 6(8):959–61.

46. Hobaika Z, Zargarian L, Boulard Y, Maroun RG, Mauffret O, Fermandjian S. Specificity of LTR DNA recognition by a peptide mimicking the HIV-1 integrase α4 helix. Nucleic Acids Res. 2009;37(22):7691–700.

47. Zhou H, Rainey GJ, Wong S-K, Coffin JM. Substrate sequence selection by retroviral integrase. J Virol. 2001;75(3):1359–70.

48. Schaack S, Gilbert C, Feschotte C. Promiscuous DNA: horizontal transfer of transposable elements and why it matters for eukaryotic evolution. Trends Ecol Evol. 2010;25(9):537–46.

49. El Baidouri M, Panaud O. Horizontal transfers and the new model of TE-driven genome evolution in eukaryotes. In:Pontarotti P. (eds) Evolutionary biology: biodiversification from genotype to phenotype: Springer Cham. 2015. p. 77–92.

50. Walsh AM, Kortschak RD, Gardner MG, Bertozzi T, Adelson DL. Widespread horizontal transfer of retrotransposons. P Natl Acad Sci USA. 2013;110(3):1012–6.

51. Wallau GL, Ortiz MF, Loreto ELS. Horizontal transposon transfer in eukarya: detection, bias, and perspectives. Genome Biol Evol. 2012;4(8):689–99.

52. Fawcett JA, Innan H. High similarity between distantly related species of a plant SINE family is consistent with a scenario of vertical transmission without horizontal transfers. Mol Biol Evol. 2016;33(10):2593–604.

53. Gilbert C, Chateigner A, Ernenwein L, Barbe V, Bézier A, Herniou EA, et al. Population genomics supports baculoviruses as vectors of horizontal transfer of insect transposons. Nat Commun. 2014;5

54. Gilbert C, Schaack S, Pace IIJK, Brindley PJ, Feschotte C. A role for host-parasite interactions in the horizontal transfer of transposons across phyla. Nature. 2010;464(7293):1347–50.

55. Won H, Renner SS. Horizontal gene transfer from flowering plants to Gnetum. P Natl Acad Sci USA. 2003;100(19):10824–9.

56. Dupuy C, Periquet G, Serbielle C, Bézier A, Louis F, Drezen J-M. Transfer of a chromosomal maverick to endogenous bracovirus in a parasitoid wasp. Genetica. 2011;139(4):489–96.

57. Neumann P, Požárková D, Macas J. Highly abundant pea LTR retrotransposon Ogre is constitutively transcribed and partially spliced. Plant Mol Biol. 2003;53(3):399–410.

58. Gómez E, Schulman AH, Martínez-Izquierdo JA, Vicient CM. Integrase diversity and transcription of the maize retrotransposon Grande. Genome. 2006;49(5):558–62.

59. Marcon HS, Domingues DS, Silva JC, Borges RJ, Matioli FF, de Mattos Fontes MR, et al. Transcriptionally active LTR retrotransposons in Eucalyptus genus are differentially expressed and insertionally polymorphic. BMC Plant Biol. 2015;15(1):198.

60. Huang J, Wang Y, Liu W, Shen X, Fan Q, Jian S, et al. EARE-1, a transcriptionally active Ty1/copia-like retrotransposon has colonized the genome of Excoecaria agallocha through horizontal transfer. Front Plant Sci. 2017;8

61. Pietzenuk B, Markus C, Gaubert H, Bagwan N, Merotto A, Bucher E, et al. Recurrent evolution of heat-responsiveness in Brassicaceae COPIA elements. Genome Biol. 2016;17(1):209.

62. McClintock B. The significance of responses of the genome to challenge. Singapore: World Scientific Pub. Co; 1993.

63. Edgar RC. MUSCLE: multiple sequence alignment with high accuracy and high throughput. Nucleic Acids Res. 2004;32(5):1792–7.

64. Jukes TH, Cantor CR. Evolution of protein molecules. Mamm Protein Metab. 1969;3(21):132.

65. Katoh K, Standley DM. MAFFT multiple sequence alignment software version 7: improvements in performance and usability. Mol Biol Evol. 2013;30(4):772–80.

66. Tamura K, Peterson D, Peterson N, Stecher G, Nei M, Kumar S. MEGA5: molecular evolutionary genetics analysis using maximum likelihood, evolutionary distance, and maximum parsimony methods. Mol Biol Evol. 2011;28(10):2731–9.

67. Li L, Stoeckert CJ, Roos DS. OrthoMCL: identification of ortholog groups for eukaryotic genomes. Genome Res. 2003;13(9):2178–89.

68. Yang Z. PAML: a program package for phylogenetic analysis by maximum likelihood. Bioinformatics. 1997;13(5):555–6.

69. Kim D, Langmead B, Salzberg SL. HISAT: a fast spliced aligner with low memory requirements. Nat Methods. 2015;12(4):357–60.

70. Livak KJ, Schmittgen TD. Analysis of relative gene expression data using real-time quantitative PCR and the $2^{-\Delta\Delta CT}$ method. Methods. 2001;25(4):402–8.

15

Genome-wide comparison of Asian and African rice reveals high recent activity of DNA transposons

Stefan Roffler and Thomas Wicker[*]

Abstract

Background: DNA (Class II) transposons are ubiquitous in plant genomes. However, unlike for (Class I) retrotransposons, only little is known about their proliferation mechanisms, activity, and impact on genomes. Asian and African rice (*Oryza sativa* and *O. glaberrima*) diverged approximately 600,000 years ago. Their fully sequenced genomes therefore provide an excellent opportunity to study polymorphisms introduced from recent transposon activity.

Results: We manually analyzed 1,821 transposon related polymorphisms among which we identified 487 loci which clearly resulted from DNA transposon insertions and excisions. In total, we estimate about 4,000 (3.5% of all DNA transposons) to be polymorphic between the two species, indicating a high level of transposable element (TE) activity. The vast majority of the recently active elements are non-autonomous. Nevertheless, we identified multiple potentially functional autonomous elements. Furthermore, we quantified the impacts of insertions and excisions on the adjacent sequences. Transposon insertions were found to be generally precise, creating simple target site duplications. In contrast, excisions almost always go along with the deletion of flanking sequences and/or the insertion of foreign 'filler' segments. Some of the excision-triggered deletions ranged from hundreds to thousands of bp flanking the excision site. Furthermore, we found in some superfamilies unexpectedly low numbers of excisions. This suggests that some excisions might cause such large-scale rearrangements so that they cannot be detected anymore.

Conclusions: We conclude that the activity of DNA transposons (particularly the excision process) is a major evolutionary force driving the generation of genetic diversity.

Keywords: DNA transposon activity, Rice, Proliferation mechanism

Background

Transposable elements (TEs) are found in practically all eukaryotes and are thought to have co-evolved with cellular life. Due to their virus-like lifestyle, TEs are considered 'parasitic' or 'selfish' DNA. However, recent studies revealed more detail about their role as potent genome shapers [1-4]. Most generally, TEs can be divided into two major classes such as: Class I (retrotransposons) and Class II (DNA transposons). Each class is further subdivided into several superfamilies [5]. For this study, we used the proposed classification system where each superfamily was assigned a 3-letter code [5] which will be given in parenthesis.

Retrotransposons use a mRNA intermediate that is reverse transcribed and integrated somewhere else in the genome. Therefore, each successful transposition produces an additional copy, which can lead to massive genome expansions [2,6]. In contrast, DNA transposons use a *cut and paste* mechanism to transpose and multiply. Because DNA transposons of the terminal inverted repeat (TIR) order [5] are the main focus of this study, we will describe their characteristics in more detail. In most TIR superfamilies, the pivotal transposase is flanked by TIRs and is transcribed and translated by the host machinery. *Mariner (DTT)* elements are, in copy numbers, the most abundant DNA transposons in rice and other grasses such as Sorghum [7] or Brachypodium [8] and usually encode a single transposase protein containing a catalytic DDD/E motif as do elements of the *hAT (DTA)* and *Mutator (DTM)* superfamilies. In contrast, *Harbinger*

* Correspondence: wicker@botinst.uzh.ch
Institute for Plant Biology, University of Zürich, Zollikerstrasse 107, CH-8008 Zürich, Switzerland

(DTH) and *CACTA (DTC)* elements also encode a second open-reading frame (ORF) of yet unknown function. Additionally, *CACTA* elements often contain complex arrays of subterminal repeats and large arrays of low complexity repeats which make them difficult to assemble and annotate [9].

Many transposons have lost their ability to transpose on their own. These non-autonomous elements usually lack protein-coding domains and transpose by recruiting enzymes of active, full size 'mother' elements. Such trans-acting systems have been described in both DNA [10] and retrotransposons [11]. Often, the non-autonomous elements, by far, outnumber their full-length counterparts and can represent a substantial amount of DNA in some genomes [8,12]. For example, in *Brachypodium*, 20,994 non-autonomous *Mariner (DTT)* elements were found whereas only 50 putative mother elements were identified [3]. Small non-autonomous DNA transposons are often referred to as 'Miniature Inverted Transposable Elements' (MITEs, [13,14]). Since we found non-autonomous elements of various sizes in multiple superfamilies, we prefer not to use the term MITE but rather refer to them simply as non-autonomous elements.

So far, several active DNA transposons have been found and documented in rice. The first described element was the non-autonomous, low copy element *mPing* of the *Harbinger (DTH)* superfamily [15-17]. One study identified *mPing* through mutability of a slender mutation of the glume which was caused by the insertion of *mPing* into the *slg* locus [15]. Kikuchi *et al.* identified *mPing* by a computational approach and presented a putative corresponding autonomous element which they named *Ping* [16]. Moreover, they showed experimentally that the transposition of both *mPing* and *Ping* preferentially occurs in cells derived from germ-line cells. Jiang *et al.* identified an additional, more distantly related autonomous element *(Pong)* which can activate *mPing in trans* [17]. Moreover, they could show experimentally that *mPing* preferably inserts in single-copy sequences. The *mPing/Pong* system has later been shown to transpose when introduced in heterologous systems such as in yeast [18] or Arabidopsis [19]. In 2005, Fujino *et al.* identified a non-autonomous element of the *hAT (DTA)* superfamily, *nDart*, that causes an *albino* phenotype and its putative autonomous mother element, *Dart*, which shared identical TIRs and similar subterminal sequences [20]. Finally, another member of the *hAT (DTA)* superfamily, *dTok*, was found to have inserted into the kinase domain of *FON1* during the molecular analysis of the *fon1/mp2* mutant [21]. Also here, they propose a putative autonomous element providing the necessary enzymes for the mobility of *dTok*. Interestingly, also in this study, transposon activity was found only in regenerative tissue.

Upon insertion, the host's DNA is cut similar to a restriction enzyme, generating 3′ overhangs. After the transposable element (TE) has been inserted, these overhangs get complemented by the host's repair system on both sides of the TE which leads to a duplication of the original target site. The length of this target site duplication (TSD) is an important diagnostic feature to classify DNA transposons, especially non-autonomous ones which do not encode any proteins (Table 1).

The current model of transposon excision proposes initial binding of the transposase to the TIR sequences followed by sequential cleavage of the two DNA strands. Thereon, dimerization of the paired-end complex brings the two strands in close proximity and links them by a clamp-loop protein [22]. Most likely, at least two subunits of the transposase (one binding to each TIR) are required for cleavage at the border of the element. When DNA transposons excise, they leave a double-strand break (DSB) with small 3′ overhangs which are derived from the TIRs of the element [22] (Additional file 1: Figure S1). Since DSBs are lethal for dividing cells, they need to be repaired by the host's DSB repair systems. The applied repair pathways and therefore the footprint of the excision can vary substantially between species. There are two main groups of DNA repair pathways [23-25]. Which of the different pathways is applied depends on the cell-cycle phase and the nature of breakpoint ends. The simplest way of DSB repair is that the 3′ overhangs get denatured by exonucleases. This generates blunt ends which allow direct ligation of the two strands, called non-homologous end joining (NHEJ). These cases result in what is referred to as 'perfect excision' where only the TSD remains as a footprint [3] (Additional file 1: Figure S1A). The second major pathway uses short homologous sequences as templates to connect the two strands. These processes employ exonucleases to produce 5′ overhangs which resect until the newly exposed strands find a homologous region of a few bp between each other, allowing annealing of the overhangs. This is referred to as microhomology-mediated end joining (MMEJ) or single strand annealing (SSA). As a consequence, the sequence downstream of the homology will be lost resulting in a deletion (Additional file 1: Figure S1B). In some cases, if the homologous pattern that re-ligates the two

Table 1 Target site characteristics of DNA transposon superfamilies

TE superfamily	Target site motif	Target site size	TIR consensus
Mariner (DTT)	TA	2	CTCCCTC
Harbinger DTH)	TAA/TTA	3	GG(G/C)CC
Mutator (DTM)	Variable	9	GAG
CACTA (DTC)	Variable	3	CACT(A/G)
hAT (DTA)	Variable	8	CA

strands corresponds exactly the complementary target site, this can lead to a restoration of the initial, 'empty-site' situation even before insertion of the TE (Additional file 1: Figure S1D). Such 'precise' excisions have been described to occur frequently when introducing the *mPing/Pong* system into Arabidopsis [19]. Thus, it is important to note that precise excisions are indistinguishable from insertions purely by means of comparative analysis. Alternatively, ectopic recombination can be initiated, which is referred to as synthesis-dependent strand annealing (*SDSA*). This can lead to the introduction of copies of foreign segments as 'filler' DNA (Additional file 1: Figure S1C). SDSA is also the mechanism underlying gene conversions [26]. In some cases, combinations of SSA and SDSA are utilized at the breakpoint leading to chimeric repair patterns [25]. While TE insertions or precise excisions are relatively easy to identify (*via* TSD), in some cases, it can be very difficult to precisely decipher excision footprints. Buchmann *et al.* [3] suggested that excisions of DNA transposons often cause extensive deletions which may also be combined with the introduction of foreign filler DNA.

In this work, we compared the genome sequences of Asian rice, *Oryza sativa ssp. japonica*, and African rice, *Oryza glaberrima*, whose genome sequence recently became available [27]. They diverged only about 600,000 years ago, providing an excellent opportunity to study recent TE activity and fixation. Moreover, it provided insight into the insertion and excision footprints, allowing inferring of qualitative and quantitative differences between TE superfamilies populating the two rice genomes. We aligned more than 63% of the two genomes and investigated 1,821 polymorphisms manually. Among these, we identified 487 loci with polymorphic DNA transposons that either inserted or excised since the divergence of the two species. We therefore estimate that the two rice genomes contain approximately 4,000 such polymorphisms. Moreover, we found differences in the excisions between different TE superfamilies. These seem to cause a multitude of rearrangements; some may be so dramatic that they cannot be detected at all anymore.

Results

TE families are unequally distributed within superfamilies
The assembled genome sizes (excluding Ns) are 372 Mbp for *O. sativa* (Version 5) [28] and 303 Mbp for *O. glaberrima* [27]. We were able to align approximately 63% of the two genomes (see below). We focused this study on DNA transposons of the TIR order (that is, elements are flanked by terminal inverted repeats and move with the help of a transposase enzyme). To obtain an overview of the abundance of DNA transposon families that had been active since the divergence of *O. sativa* and *O. glaberrima*, we used a database that was created based on an iterative search of insertion/excision polymorphic

sequences in the alignments of the two genomes (see below and 'Methods'). Thus, the results of our survey do not reflect the total content of DNA transposons in rice which, in fact, might be much higher [27]. We identified 64,645 Class II transposons of the TIR order in *O. sativa* and 54,280 in *O. glaberrima*, occupying approximately 20.4 Mbp and 12.6 Mbp of the two genomes, respectively (Additional file 2: Table S1). The average sizes of 316 bp and 230 bp reflect the strong outnumbering of autonomous by non-autonomous elements. A closer investigation of the substantial number of unclear sites (Ns) indicated that many sequence gaps in the *O. glaberrima* assembly are caused by Class II transposons (see 'Methods'). We estimate that at least 5,100 sequence gaps actually correspond to Class II TIR elements, resulting in an estimated total of approximately 59,500 elements (approximately 16 Mbp) in *O. glaberrima*. Therefore, the overall DNA transposon content in *O. glaberrima* is probably slightly lower than in *O. sativa*.

In both species, the highest copy numbers were found for *Mariner (DTT)* elements, followed by elements of the *Harbinger (DTH)* superfamily. CACTA elements are on average larger than the other superfamilies (938 bp in *O. sativa* and 600 bp in *O. glaberrima*) and thus they occupy the most space. These findings are also consistent on the family level (that is, among elements that can be aligned at the DNA level [5]). We found strong over representation of a few families that dominate each of the superfamilies (Figure 1 and Additional file 2: Table S1). With the exception of the *DTA_Coraline* family (which was only found in *O. sativa*), all TE families are represented at similar numbers in both genomes (Additional file 2: Table S1).

Identification of transposon polymorphisms
We were able to align 63.3% (235.6 Mbp) of the *O. sativa* and *O. glaberrima* genomes with the Smith-Waterman algorithm in sliding windows of 12 kb (see 'Methods') and examined the presence/absence of polymorphisms larger than 50 bp. We identified 23,709 polymorphisms in the *O. sativa* genome of which 7,542 showed homology to DNA transposons. In *O. glaberrima*, we found 22,003 polymorphisms whereof 4,816 had homology to DNA transposons. Upon visual inspection, we noticed that many of the polymorphisms in *O. glaberrima* that showed homology to TEs also contained large stretches of Ns, indicating that TEs are often problematic to assemble completely. Thus, in an independent approach, we estimated how many of the 'presence' polymorphisms in *O. glaberrima* which are comprised mostly of Ns actually correspond to TE sequences (see 'Methods'). We estimate that there are approximately 1,750 polymorphisms in *O. glaberrima* which can be attributed to DNA transposons but which are not identifiable because the sequence assembly is incomplete

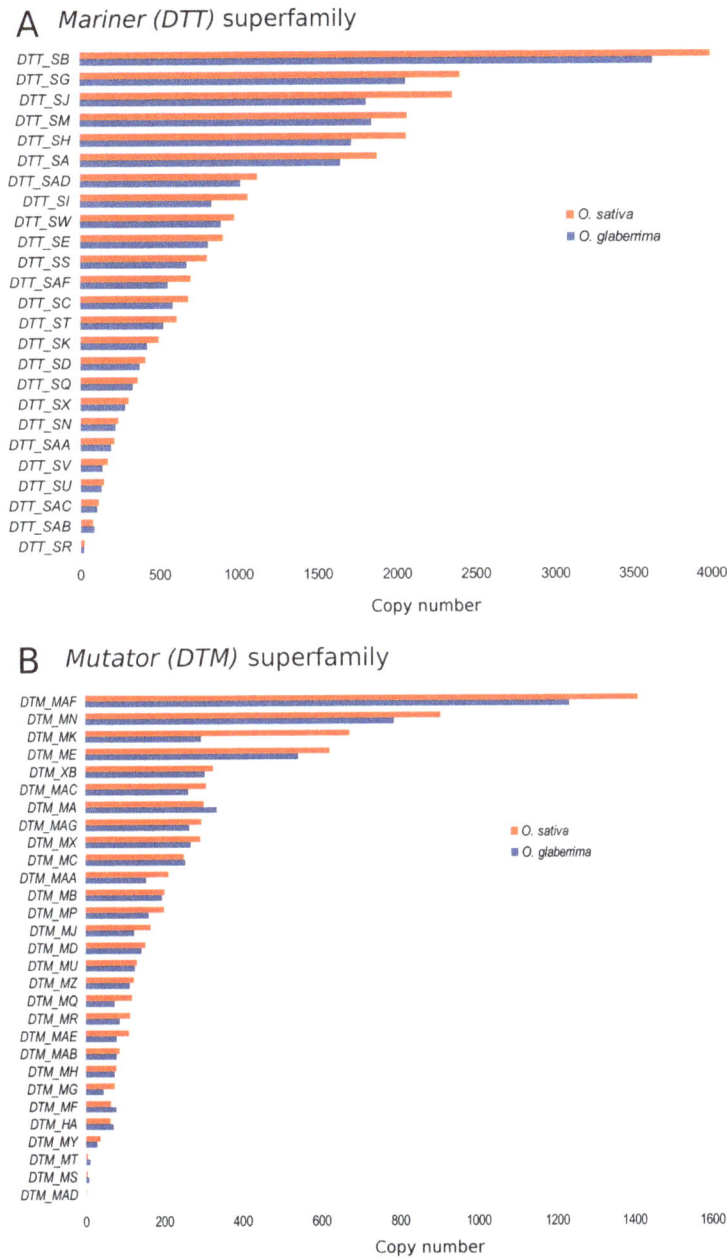

Figure 1 The abundance of *Mariner (DTT)* and *Mutator (DTM)* families in *O. sativa* and *O. glaberrima*. **(A)** Overview of *Mariner (DTT)* abundance. Copy numbers of individual families show large differences within the *Mariner (DTT)* superfamily. For example, in *O. sativa*, the most successful *DTT_SB* is represented 4'702 times while we only identified 25 copies of the *DTT_SR* family. **(B)** Overview of *Mutator (DTM)* superfamily. Despite an overall similar distribution, we found one exception for the *Mutator (DTM)* superfamily *DTM_MA*, where we found slightly more elements in the *O. glaberrima* genome (302 copies in *O. sativa* and 336 copies in *O. glaberrima*).

at these sites. Thus we extrapolate that there are approximately 6,500 presence polymorphisms in *O. glaberrima*, slightly fewer than the 7,542 presence polymorphisms in *O. sativa* (see 'Methods').

Here, it should also be noted that sequence alignments of large genomic regions often contain misalignments caused for example by the presence on non-homologous segments or by sequence gaps in one of the species.

Automated examination of sequence alignments can therefore yield very noisy data. Thus, we decided to manually analyze a subset of the identified polymorphisms. In total, we manually analyzed 1,821 cases which showed homology to TEs, 844 from *O. sativa* and 977 from *O. glaberrima*, representing approximately 15% of all TE-related polymorphisms. Most of them turned out not to be directly associated with TE activity because many represent internal or partial deletions within the elements, which means that the missing sequence obviously was not caused by an insertion or excision of the respective transposon but by a mechanism unrelated to its activity (for example, template slippage, Additional file 3: Figure S2).

In *O. sativa*, we found 238 and in *O. glaberrima* 249 TE polymorphisms that were most likely caused by DNA transposon activity. A complete overview of all active transposons is provided in Additional file 4: Table S2. Thus, 28% and 25% of all presence/absence polymorphisms examined represent likely transposition events. For *O. sativa*, we therefore extrapolate that about 2,100 of the TE-related polymorphisms are actually caused by TE activity. The value for *O. glaberrima* is probably similar. Considering the estimated, unknown part of approximately 450 additional TE-related polymorphisms, we expect a slightly lower overall activity of 1,650 transposition events in *O. glaberrima*.

In *O. sativa*, the most abundant were *Harbinger (DTH)* and *Mariner (DTT)* elements with 95 and 90 transpositions, respectively. Moreover, we identified 33 *Mutator (DTM)* and 15 *CACTA (DTC)* elements to have transposed recently. Finally, we found five elements of the *hAT (DTA)* superfamily. Also in *O. glaberrima*, *Harbinger (DTH)* and *Mariner (DTT)* were the most prominent superfamilies with 110 and 102 transpositions, respectively. Additionally, we identified 32 *Mutator (DTM)*, four *hAT (DTA)*, and a single *CACTA (DTC)* transposition (Table 2).

Distinguishing insertions and excisions

We defined TE insertions in the classic way as follows: one species contains the TE flanked by the two direct repeats created upon insertion (the TSD) while in the other species, the TE is absent and only one copy of the TSD is present (example in Figure 2A). Of the total 487 TE polymorphisms we identified in *O. sativa* and *O. glaberrima*, we classified 393 as insertions (192 in *O. sativa* and 201 in *O. glaberrima*). It is important to note that a precise excision (that is, one that removes the TE plus one target site) cannot be distinguished from an insertion with these criteria.

Table 2 Overview of recently active DNA transposons in *O. sativa* and *O. glaberrima*

TE superfamily	O. sativa	O. glaberrima
Mariner (DTT)	90	102
Harbinger (DTH)	95	110
Mutator (DTM)	33	32
CACTA (DTC)	15	1
hAT (DTA)	5	4

Excisions are much more complex to identify and show various patterns of DSB repair. The simplest case, a perfect excision, was defined as an event where the TE excises and exactly leaves the two copies of the TSD as a footprint [3]. Of a total of 94 putative excision events, we identified only eight perfect excisions (example in Figure 2B). In all other cases, the excision went along with deletions of flanking sequences or the insertion of filler sequences, or both. In 43 excisions, we found that sequences flanking the element were deleted. Excluding one extreme case (see below), on average approximately 18 bp of flanking sequences were deleted per excision event (example in Figure 2C). On the other hand, in 58 cases, excision also went along with the introduction of foreign DNA segments. On average these fillers had a size of 13 bp, ranging in size from 1 to 123 bp (example in Figure 2D). Nine cases showed both deletions and introduced filler segments. The cumulative length of all deleted sequences is 926 bp while the combined length of all filler segments is 880 bp.

The most extreme case was a putative excision of a *Mariner* element of the *DTT_SC* family. Its excision went along with the deletion of a 2,479 bp fragment on one side of the element (Figure 3). We are confident that this deletion was indeed the result of the excision because the left border of the excised fragment coincides precisely with the left end of the *DTT_SC* element (Figure 3). It is highly unlikely (however, not impossible) that a random deletion would have its one breakpoint exactly at the terminus of the TE. If this case is included in the overall calculation, a total of 3,405 bp were deleted in the 94 excision events.

The ratio of insertions and excisions indicates differences in transposition between superfamilies

Inferring recent activity of DNA transposons from the numbers of excisions and insertions is not trivial. Intuitively, one would assume that the ratio of insertions and excisions is 1:1, because each excising element would simply insert somewhere else in the genome. The current hypothesis is that DNA transposons can excise during DNA replication and transpose in front of the replication fork to create an additional copy. This results in two different gametes, one with one copy and one with two copies. If the number of transposition events is large, overall equal numbers of loci derived from the two gamete types should be passed to offspring. Thus, if all the observed insertion/excision ratios in a cross-species comparison such as the one presented is considered, the ratio is actually expected to be 2:1 (Additional file 5: Figure S3). However, it is also possible that transposition happens at other points during the cell cycle, which would not lead to a replication of the respective TE. Considering this, one would therefore expect a ratio somewhere between 1:1 and 2:1 but

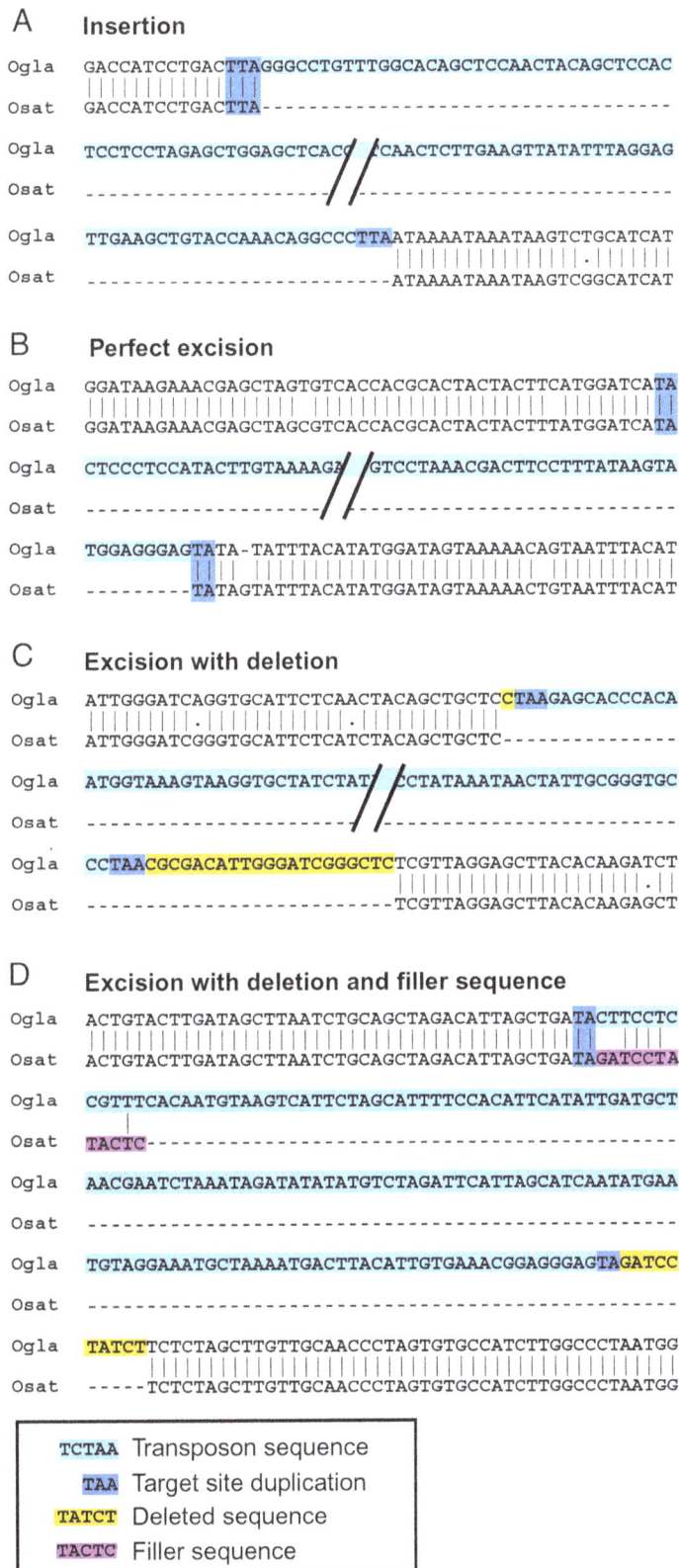

Figure 2 Examples of DNA transposon polymorphisms in *O. sativa (Osat)* and *O. glaberrima (Ogla)*. The alignments show the polymorphic TE plus some of the genomic flanking sequences. Diagnostic sequence motifs are *highlighted* with colors. **(A)** Insertion. **(B)** Perfect excision. **(C)** Excision with deletion. **(D)** Excision with deletion and filler sequence.

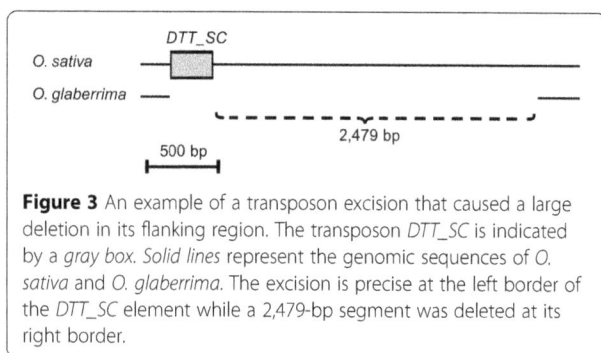

Figure 3 An example of a transposon excision that caused a large deletion in its flanking region. The transposon *DTT_SC* is indicated by a *gray box*. *Solid lines* represent the genomic sequences of *O. sativa* and *O. glaberrima*. The excision is precise at the left border of the *DTT_SC* element while a 2,479-bp segment was deleted at its right border.

Table 3 Overview of TE insertions and excisions by superfamily and species

TE superfamily	Species	Insertions	Excisions	Ratio	P value
DTT	*O. sativa*	59	21	2.81	0.19
DTH	*O. sativa*	84	19	4.42	0.01*
DTM	*O. sativa*	30	7	4.29	0.14
DTC	*O. sativa*	14	1	14	-
DTA	*O. sativa*	5	0	-	-
DTT	*O. glaberrima*	81	31	2.61	0.38
DTH	*O. glaberrima*	90	11	8.18	0.00008*
DTM	*O. glaberrima*	26	3	8.67	0.028*
DTC	*O. glaberrima*	0	1	-	-
DTA	*O. glaberrima*	4	0	-	-

*Significantly different from expected 2:1 ratio.

not higher than 2:1. This 2:1 ratio is only expected if a given transposon family is active at similar levels in both species being compared. Deviation from the 2:1 ratio in the inter-species comparison would therefore indicate different levels of activity of that transposon family in the two species (Additional file 6: Figure S4). For example, a ratio much lower than 2:1 in *O. sativa* and much higher than 2:1 in *O. glaberrima* could indicate that a given transposon family was more active in *O. glaberrima* (Additional file 6: Figure S4).

When comparing the insertion/excision ratio for the different superfamilies, we observed almost the expected 2:1 ratio for the *Mariner (DTT)* superfamily. In both datasets, we found ratios which are not significantly different from 2:1 (2.6:1 in the *O. glaberrima* dataset and 2.8:1 in the *O. sativa* dataset), indicating that the proposed proliferation mechanism is sufficient to explain the observations on *Mariner* elements. It also indicates that precise excisions (removal of the TE plus one target site) are rare in the *Mariner* superfamily. Interestingly, for the *Harbinger (DTH)* superfamily, the ratio differs from what we expected. The ratio in the *O. glaberrima* dataset was 8.2:1 (Fisher's exact test, $P = 0.0001$), and in the *O. sativa* dataset, we found 4.4:1 ($P = 0.013$). These differing ratios between *O. glaberrima* and *O. sativa* could indicate a higher level of *Harbinger* activity in *O. glaberrima*. Finally, we found a significant difference for the *Mutator* superfamily in the *O. glaberrima* dataset where we observed a ratio of 8.7:1 with 26 insertions and only three excisions ($P = 0.03$). The ratio in the *O. sativa* dataset (4.3:1), where we found 30 insertions and seven excisions, did not reach significance level ($P = 0.14$) (Table 3). It is not easy to explain why *Harbingers (DTH)* and *Mutator (DTM)* elements deviate so strongly from the expected 2:1 ratio in both species (see 'Discussion').

High abundance does not necessarily correlate with strong activity

To estimate activities of individual TE families, we had to consider that an additional sequence in *O. sativa* could mean that the TE inserted in *O. sativa* or excised in *O. glaberrima* and *vice versa*. Therefore, we had to combine data from both datasets. We defined the relative activity as the number of copies that moved in relation to the total copies in a particular TE family. As relative abundance we defined the total copy number of the respective family divided by the total number of DNA transposons of the investigated genome. We divided the families into three categories (Figure 4). In the first group, we grouped TE families with high overall copy numbers and also high numbers of insertion polymorphisms. Members of the *Mariner (DTT)* and *Harbinger (DTH)* superfamilies are most prominent in this category. Moreover, the most abundant *CACTA* family, *DTC_Calvin* (4,868 copies), turned out to be also very active with nine identified insertions and one excision.

The families in the second group show a high number of insertions relative to their abundance. Most noticeably, for the *DTH_TR* family, of which we found only 581 copies in the whole *O. sativa* genome, we identified 17 insertions and one excision, more than for any other family overall. Other highly active families in this class are the *Harbinger* family *DTH_TAA* and the *Mutator* family *DTM_MA* which both inserted eight times and excised once while we found 288 and 302 copies, respectively. Furthermore, we found several other *Harbinger (DTH)* and two *hAT (DTA)* families with less than 50 copies and one insertion. The most extreme case here is the *Mutator* family *DTM_MAD* where we only found two copies in the whole genome, one of them inserted recently.

The third group contains families with high abundance but with only little or even no activity. Here, we find the most numerous families, again of the *Harbinger (DTH)* and *Mariner (DTT)* superfamilies, where we found several families with more than 3,000 copies but only five or less polymorphisms. For the most numerous, *Mariner* family *DTT_SB* (3,995 copies), and the most abundant, *Mutator* family *DTM_MAF* (1,408 copies), we did not

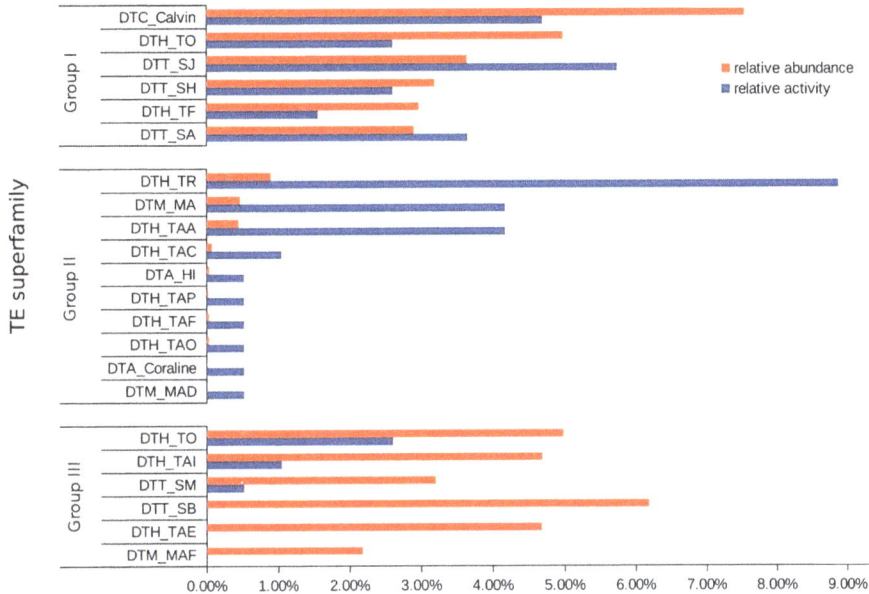

Figure 4 The relative activity and relative abundance of the TE families in *O. sativa*. We compared the relative activity with the relative abundance of all TE families in *O. sativa*. Group I consists of families with high activity and high abundance. The *CACTA* family *DTC_Calvin*, which is the overall most abundant family, also shows remarkable activity. Group II contains elements with high activity but low copy numbers. We found that *Mutator (DTM)* and *hAT (DTA)* families are relatively active despite their poor abundance. Finally, Group III consists of families with high abundance but relatively little activity. This class is dominated by families of the *Harbinger (DTH)* and *Mariner (DTT)* superfamilies. The *Harbinger* family *DTH_TO* seems to be still relatively active despite its high abundance, whereas the most abundant *Mariner* and *Mutator* families *DTT_SB* and *DTM_MAF*, respectively, show no activity at all.

find any polymorphic elements at all (Additional file 2: Table S1 and Additional file 4: Table S2).

Most potentially active and autonomous elements are of the *Mutator* superfamily

As mentioned above, the majority the elements that transposed since the divergence of the two rice species are non-autonomous elements which do not code for any proteins. We found a total of 17 elements which contain at least parts of transposase ORFs and have moved since species divergence. Interestingly, twelve of them belong to the *Mutator (DTM)* superfamily, which had, overall, relatively few active elements (see above). We found eight families where all polymorphic copies contain at least parts of coding sequences (CDS) for transposases (*DTM_MAF, DTM_MS, DTM_MAG, DTA_Coraline, DTA_HL, DTH_TAG, DTH_TAH,* and *DTH_Blip*). For the *Mutator* family *DTM_MU,* we found one CDS-containing element and one non-autonomous deletion derivative. However, besides the *DTA_Coraline* insertion, all the transposase ORFs of the above families have either stop codons or frameshifts, suggesting that they are not functional autonomous elements.

The most interesting *Mutator* family is *DTM_MK* which contains 14 elements that have moved since species divergence (Figure 5). We found a total of nine

insertions and one excision of *DTM_MK* elements in *O. sativa* and four excisions but no insertion in *O. glaberrima*, indicating that they had been active in both species. Here, we found three elements with apparently intact transposase ORFs which we consider potentially active mother elements. The largest among these (6,721 bp) contains an intact transposase ORF and an additional ORF that encodes a 'TE-associated' protein. Interestingly, we found large parts of the same second ORF in three other putative full-length elements that all have disrupted transposase ORFs. Furthermore, two of the disrupted elements acquired an additional sequence which has no homology in any of the other family members (Figure 5). Intriguingly, we also found a subpopulation of six non-autonomous elements that had moved. These elements are very similar to each other in size (604 bp to 684 bp) and share the TIRs of around 120 bp with the other larger elements. The approximate 400 bps between their TIRs is not homologous to any of the larger elements but is highly similar in the six small copies. This indicates that these six copies originated from a single deletion event and multiplied after (Figure 5).

TE activity mainly influences regions close to genes but not coding sequence

We investigated if and to what extend TE activity affects genes by using the coding sequence provided by the Rice Genome Annotation Project [29]. We included all TE

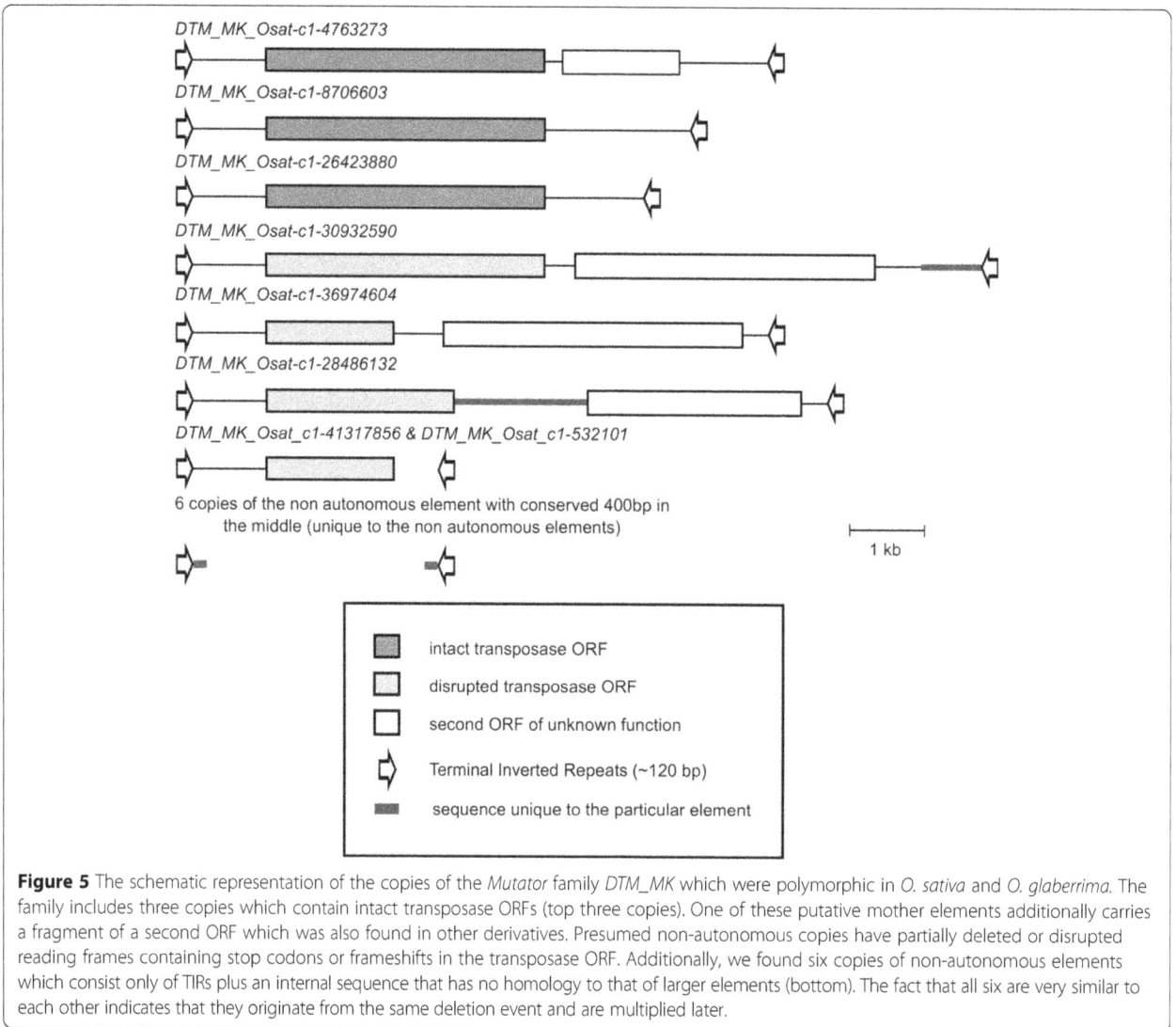

Figure 5 The schematic representation of the copies of the *Mutator* family *DTM_MK* which were polymorphic in *O. sativa* and *O. glaberrima*. The family includes three copies which contain intact transposase ORFs (top three copies). One of these putative mother elements additionally carries a fragment of a second ORF which was also found in other derivatives. Presumed non-autonomous copies have partially deleted or disrupted reading frames containing stop codons or frameshifts in the transposase ORF. Additionally, we found six copies of non-autonomous elements which consist only of TIRs plus an internal sequence that has no homology to that of larger elements (bottom). The fact that all six are very similar to each other indicates that they originate from the same deletion event and are multiplied later.

polymorphisms (insertions and excisions) found in exons as well as those in introns, 1,000 bp upstream and 500 bp downstream of the coding sequence.

Of the 487 investigated TE polymorphisms, 160 matched our criteria. We found 74 insertions or excisions in upstream regions and 24 downstream of genes. Moreover, 61 polymorphisms were identified in introns. Interestingly, only one, a *Mariner* element of the *DTT_SG* family, actually disrupted a gene in *O. sativa*. The element inserted into the second exon of a glutathione S-transferase homolog (LOC_Os01g72120), a protein assumed to be involved in detoxification. We furthermore performed a gene ontology analysis. This revealed that genes involved in nucleoside metabolic and biosynthetic processes, protein dephosphorylation and SRP-dependent proteins targeting the membrane, are affected disproportionately high ($P < 0.01$/not shown).

Discussion

In this study, we conducted a genome-wide analysis of the activity of DNA transposons in the two closely related rice species *O. sativa* and *O. glaberrima*. Numerous studies have described the activity of (Class I) retrotransposons in plants [2,6,30,31], but only very few have focused on DNA (Class II) transposons. To our knowledge, this is the first study that characterizes DNA transposons and assesses their activity at a genome-wide scale. Here, the recently released sequence of the *O. glaberrima* genome [27] provided a unique opportunity for comparative analysis because it is phylogenetically close enough to *O. sativa* to allow reliable sequence alignments of large parts of the genome and yet distant enough to have accumulated numerous TE polymorphisms. Both the *O. sativa* and the *O. glaberrima* genome were sequenced with Sanger technology and assembled independently. This has the important advantage over simple re-sequencing and subsequent mapping onto a reference so that large insertions and deletions in both species can be easily identified and characterized in much detail. A very important part of our work was that we manually inspected over 1,800 TE related polymorphisms in the two species because transposon excisions, especially, can produce complex sequence patterns which are extremely difficult to characterize on an automated basis. Furthermore, it was important to distinguish actual insertions and excisions from random deletions that by chance affected parts of the TEs. The result of this study is a chromosome-scale catalog of TEs that were recently active in rice as well as information on how transposon insertions and excisions affect the genome. Our data allow conclusions and hypotheses on transposon activity. These will be discussed below.

Estimates of frequencies of transposition events in *O. sativa* and *O. glaberrima*

It has been known for several years now that grass genomes contain tens of thousands of DNA transposons [7,12]. However, it was not clear how often these elements actually move. *Wessler et al.* [10] suggested that some families may be active in bursts, creating thousands of copies within only a few generations before they are silenced by the host. The data from our study now allow some conclusions on the actual level of transposition activity such that:. *O. sativa* and *O. glaberrima* were estimated to have diverged approximately 600,000 years ago [27]. Overall, our data indicate that DNA transposons were active at similar levels in both rice species since their divergence. Based on the manual analysis of insertions, we estimated that *O. sativa* contains roughly 2,000 polymorphic elements and *O. glaberrima* approximately 1,600. Assuming that most of these polymorphic transposons are actually fixed in the two species, we can estimate that since species divergence, a DNA transposon polymorphism (insertion or excision) became fixed approximately every 250 years to 300 years in both *O. sativa* and *O. glaberrima*. In other words, about 2.5% to 3.5% of the DNA transposons in the two species have moved within the last 600,000 years.

For the following calculations, we assume that all identified transposition events were selectively neutral as deleterious transpositions would have been selected against. However, fixed polymorphisms only represent a small part of actual TE activity. A measure for actual transposon activity can be defined analogous to a mutation rate (m) as the number of transposition per generation per individual. The total number of transposition events per generation would therefore be the effective population size $N(e)$ times the mutation rate (that is, $N(e)m$). According to Kimura [32], fixation rates are inversely proportional to population sizes. Thus, if all transposition events are neutral, the

probability of fixation of an event is $1/N(e)$. The rate of fixation is therefore $N(e)m \times 1/N(e) = m$. Thus, population size is irrelevant, and the fixation rate is equal to m [32]. In the case of *O. sativa*, fixation rate would therefore be the number of identified transposition events (assuming all of them are fixed) divided by the number of generations since divergence from *O. glaberrima* (2,300/600,000 = 0.004). This would mean that in each generation, 1 out of 250 individuals contains a transposition event.

Most polymorphic DNA transposons are non-autonomous, except in two superfamilies

The vast majority of the polymorphic transposons were small non-autonomous elements (MITEs) of the *Mariner (DTT)* and *Harbinger (DTH)* superfamilies. Interestingly, we did not find any polymorphic potentially autonomous elements for either of the two superfamilies. This could indicate that the required transposase genes may still be expressed, but the mother elements themselves have lost the ability to move. It was previously reported that non-autonomous *Mariner* and *Harbinger* elements could also be cross-activated by even distantly related mother elements and even in heterologous systems when non-autonomous rice elements are introduced into yeast and Arabidopsis [10,18,19].

We found polymorphic putative full-size elements of twelve *Mutator*, three *Harbinger* and two *hAT* families. However, even among these large elements, most carried defective transposase ORFs which contained frameshifts or stop codons. The only exceptions were an insertion of the *hAT* element *DTA_Coraline* and several members of the *Mutator* family *DTM_MK*. Here, we found multiple copies that contain intact transposase ORFs. The *DTM_MK* family is particularly interesting because it illustrates how TEs can diverge into multiple sub-families. The *DTM_MK* family consists of multiple large elements that each contains a unique pattern of internal deletions of additionally acquired sequence fragments. Furthermore, it contains a sub-population of six small deletion derivatives that obviously originated from a single deletion event since they all have a very similar structure. These elements may represent the first steps in the evolution of a population of non-autonomous TEs.

DNA transposon excisions have a large potential to shape the genome

Of particular interest to us was a broad assessment of what types of footprints DNA transposons produce. We found that TE excisions can produce very complex patterns. Previous studies already suggested that excisions may produce a variety of outcomes and that the perfect footprint (that is, the precisely duplicated target site) might actually be rare [3]. Furthermore, it was shown that excisions may lead to large deletions and/or insertions of copies of foreign DNA fragments when 'filler' DNA is inserted in the process of DSB repair [3,16,33]. Our data indeed show that perfect excisions which leave exactly two copies of the TSD are extremely rare, as only 8 out of 94 excisions showed this pattern. In all other cases, excisions lead to the deletion and/or introduction of foreign DNA fragments. Our large dataset allowed us to quantify that on average, 18 bp of the flanking region are deleted while 13 bp of the new sequence are introduced at the excision site. These numbers do not include the most extreme case wherein an excision apparently went along with the deletion of a 2.4 kb fragment. Furthermore, our dataset does not include possible cases where large segments on both sides of the element were deleted upon excision (such events would be indistinguishable from random deletions that by chance removed a large segment containing the TE). Also data from insertion/excision ratios of some superfamilies suggest that many excisions may have 'catastrophic' outcomes (see below). Thus, we conclude that excisions of DNA transposons are a major driving force in genome evolution as they can cause relatively large-scale rearrangements such as deletions and integrations of new sequences surrounding the excision site.

Why do *Harbinger* and *Mutator* elements show more insertions than expected?

The current model of proliferation during DNA replication postulates that one would find a ratio of insertions to excisions that lies somewhere between 1:1 and 2:1 (see Additional file 5: Figure S3 and Additional file 6: Figure S4) when comparing two closely related genomes. Interestingly, in all DNA transposon superfamilies, we found insertion/excision ratios higher than 2:1 in both species. Only for the *Mariner (DTT)* superfamily did we find a ratio of insertions to excisions that was only slightly above 2:1 in both *O. sativa* and *O. glaberrima*. In contrast, the insertion/excision ratios of the *Harbinger (DTH)* and *Mutator (DTM)* superfamilies are clearly higher than 2:1 (that is, they show a much higher number of insertions than expected). The same is also probably true for *CACTA (DTC)* elements, but there, the sample size is smaller and the insertion/excision ratio does not significantly deviate from 2:1.

One explanation for the distorted ratio is that, for some reason, we can simply not see excisions in our sequence alignments. Buchmann *et al.* [3] suggested that some excisions go along with deletions of several kb of the flanking regions. Indeed, for example for *Harbinger* elements we identified the highest proportion of "unclear" events. These comprise large sequence gaps which we could not clearly classify as excisions because too much sequence was deleted or rearranged surrounding the

element. Thus, our hypothesis is that *Harbinger* and *Mutator* (and possibly *CACTA*) elements frequently cause large rearrangements (mostly large deletions) upon excision, so that the orthologous regions of the two species cannot be aligned easily anymore. If such deletions are in the size range of 3 kb to 5 kb, it would undermine our initial mapping of homologous loci that was based on blast searches of 5 kb segments. Additionally, if the fitness of a gamete carrying the excision is reduced or even lethal, this would also contribute to raising the ratio above the 2:1. One possible reason for frequent large deletions could be the size of the elements, simply because excisions of large elements may be more difficult to repair. Indeed, *Mariner* elements are on average the smallest of all the elements studied, and there, we find an insertion/excision ratio to be the closest to 2:1. With increasing average size of elements, we also see an increasing insertions/excision ratio.

A second explanation why we find fewer excisions than expected is that the DSB is repaired by using the sister chromatid as a template *via* the SDSA mechanism [34] analogous to what happens during gene conversion. In this case, the excision would be undetectable because it was repaired perfectly with a copy of the sister chromatid that still contains the insertion. Such reversion of excision sites has been described in *Drosophila melanogaster* [35] and *Caenorhabditis elegans* [36]. However, it is not clear why this repair mechanism would preferably be used in certain superfamilies such as *Harbinger* and *Mutator*.

Finally, it is possible that many excisions are precise (that is, the TE and one target site is removed) and thus could not be distinguished from insertions. This could, for example, explain our findings of the high ratios of 4.4:1 for *O. sativa* and 8.2:1 for *O. glaberrima* in the *Harbinger* superfamily. However, previous studies produced conflicting results on the frequency of precise excisions. Yang *et al.* [19] described that 83% of approximately all 30 excisions were precise for the *Harbinger* element *mPing* when expressing it in *A. thaliana*. In contrast, Kikuchi *et al.*, who worked with the same element in rice anther cultures, stated that only one case out of approximately 70 excision sites showed the footprint of a precise excision [16]. Thus, it is possible that the frequency of precise excisions depends on the conditions under which the transposition occurs. Additionally, the frequency of precise excisions could also differ between TE superfamilies. Indeed, for Mariner elements, we found a ratio close to 2:1, indicating that we were able to distinguish insertions and excisions well.

Conclusions

We conclude that the activity of DNA transposons (particularly the excision process) is a major evolutionary force driving the generation of genetic diversity. Additionally, our data indicate that some DNA transposon excisions might cause such large-scale rearrangements so that they cannot be detected anymore. It is therefore likely that our study still under-estimates the impact of DNA transposon excisions on genome evolution. However, it will require further and more detailed studies of these transposable elements in multiple species to conclusively answer this question.

Methods

Genome-wide sequence alignments

The genome of *O. sativa* was split into fragments of 5 kb. Each of these fragments were then used in BLASTN searches against the *O. glaberrima* genome to identify the orthologous regions. As a primary filter criteria, we considered only fragments in the same orientation on the same chromosome with an identity of at least 96%. Then, 12 kb of sequence from both species (5 kb fragment + 7 kb adjacent 3′ sequence to create an overlap with the following fragment) were excised for pair-wise alignment. Here, we used the EMBOSS (emboss.sourceforge.net) program Water which implements the Smith-Waterman algorithm. We used a gap opening penalty of 30 and gap extension penalty of 0.1 to obtain alignments that preferably contain fewer but larger gaps.

Each of these pairs was scanned for alignment quality. We included all sequences that were embedded between at least 200 continuous bases that could be aligned with more than 90% perfect matches. The corresponding positions in the *O. sativa* genome were determined, and the overlapping individual alignments were re-assembled into one global alignment per chromosome. The consistency of the global alignment with the original assembly of *O. sativa* was tested extensively by manual comparisons of positions of randomly chosen sequences in and across the breakpoints of the overlaps. The global alignments were scanned for insertions or deletions (InDels) larger than 50 bp. InDels only separated by less than 4 bp were considered as one event. Additionally, InDels that bordered to sequence gaps (stretches of Ns) were discarded.

The remaining InDels were scanned for homologies to Class II TIR-order transposons from our in-house database that is derived from the TREP database (http://wheat.pw.usda.gov/ITMI/Repeats/) with the following BLASTN parameters: minimum alignment size of 50 bp and identity of at least 70%. The InDels that could not be associated to known TEs were used as a query for an iterative BLASTN search against the whole genome in which the InDel was found. Sequences with at least 15 copies and a minimum identity of 85% were considered putative TEs. The top 15 hits were extracted from the genome including a few hundred bp of flanking sequences. These were aligned with ClustalW to determine the precise borders of the element and to generate a consensus sequence. Consensus sequences were curated manually and added to the repeat database. Like this, we were able to expand the existing dataset for rice repeats at TREP from 59 sequences to 235 sequences. All scripts were written in PERL and are available upon request.

The data for this analysis were retrieved from Wang *et al.* [27] for *O. glaberrima* and the International Rice Genome Sequencing Project (IRGSP) for *O. sativa Nipponbare cultivar* [20] (plantbiology.msu.edu/pub/data/), respectively. We retrieved the annotation of the *O. sativa* genome from the Rice Genome Annotation Project [29] (Version 6/plantbiology.msu.edu). We removed all entries that included the word 'transpos' in the description line as well as putative genes (which also mostly correspond to TE sequences) and mapped the remaining genes on our version (version 5) of the genome using GMAP [37] (research-pub.gene.com/gmap/). For the gene ontology analysis, we used the online platform 'Rice Oligonucleotide Array Database' [38] (ricearray.org/analysis/) at default settings. We included the genes found to be affected by active TEs in exons, introns, 1,000 bp upstream, or 500 bp downstream of the CDS to check if they are often involved in certain biological processes disproportionately.

Estimate of the number of sequence gaps caused by DNA transposons

To assess the different assembly qualities, we first counted all Ns in both genomes. With a total of 93,930 Ns, the *O. sativa* assembly contains a low number of sequence gaps. In contrast, in *O. glaberrima*, we identified 20,080 gaps consisting of more than 50 Ns (total N count, 12,768,901). To study the cause of these sequence gaps, we extracted 500 bp up and downstream of these regions and identified the orthologous position in *O. sativa*. We identified 7,301 cases (4,413,818 Ns), where both flanking sequences mapped within 10 kb from each other in the same orientation (blast hits with a minimum of 400 bp length and 95% identity). We then screened the segment in *O. sativa* that corresponds to the gap in *O. glaberrima* for TE sequence. Of these orthologous loci, 25.6% (1,871 cases) showed homology to TIR DNA transposons. From this number, we extrapolated that proximately 5,150 sequence gaps in the *O. glaberrima* genome correspond to TIR DNA transposon sequences.

In the alignment of the two genomes, we identified 1,745 insertions (that is, additional sequence) in *O. glaberrima* larger than 50 bp which consist of more than 80% Ns. Assuming that about 25% of these loci correspond to DNA transposons, we expect 447 additional DNA transposon-related polymorphisms in *O. glaberrima*.

Additional data files

The following additional data are available with the online version of this paper. Additional file 1 is an illustration of the different DSB repair mechanisms. Additional file 2 is a table listing all annotated DNA transposons in the genomes of *O. sativa* and *O. glaberrima*. Additional file 3 is a figure explaining different mechanisms that lead to InDels. Additional file 4 is a table listing all described TE polymorphisms. Additional file 5 is a figure explaining the inheritance of insertion and excision patterns of DNA transposons. Additional file 6 is a figure explaining that the differences in the ratio of insertions and excisions is an indicator for differential TE activity between species.

Additional files

Additional file 1: Figure S1. Overview of the different mechanisms of DSB repair following DNA transposon excision. A.) Non-homologous end joining (NHEJ) which leads to perfect excisions. B.) Single stranded annealing (SSA) which leads to a deletion of adjacent sequences. C.) Synthesis-dependent strand annealing (SDSA) which can lead to introduction of 'filler' DNA segments. D.) A special case of SSA which leads to precise excisions.

Additional file 2: Table S1. Overview of TE abundance in *O. sativa* and *O. glaberrima*.

Additional file 3: Figure S2. Summary of causes for insertions in rice species. Many of the identified insertions showed homology with DNA transposons but were not caused directly by their activity (for example, partial deletions of TEs). Therefore, we divided the remaining insertions into three classes based on their presumed molecular mechanism as follows: (i) repeat slippage, (ii) partial deletion, and (iii) unknown. Repeat slippage happens if DNA polymerase loses its template while synthesizing the new strand during replication and then re-adopts at a similar template close by. We found 149 insertions in *O. glaberrima* and 51 in *O. sativa* which represent differences in the number of tandem repeats between the two species. Template lengths ranged from simple dinucleotides to more than 20 bp. In two cases, entire TEs served as templates for slippage, deleting several kb between two elements. In these cases, unequal homologous cross over (similar to the mechanism that produces solo LTRs of retrotransposons) could be an alternative interpretation. Another 68 insertions in *O. glaberrima* and 94 in *O. sativa* resulted from partial deletions of TEs. These were deletions of apparently random segments within or close to TEs. Finally, 35 insertions in *O. glaberrima* and 66 in *O. sativa* could not be clearly classified. These InDels are often larger than the average InDel. These include cases where it was not possible to deduce the original, ancient state because, for example, multiple TEs were nested in these positions. Also included here are cases where a TE was found in the middle if a large insertion. These could potentially represent excisions which went along with deletions of large segments of the flanking sequence.

Additional file 4: Table S2. Overview of all transpositions.

Additional file 5: Figure S3. Inheritance of transposon insertion/excision patterns. For this model we assume that all transposition effects are selectively neutral. It is commonly accepted that a mechanism of multiplication is for DNA transposons to excise during DNA replication and to reinsert in front of the replication fork. This leads to one daughter strand with one copy of the element (A-type gamete) and one with two copies (B-type gamete). If a large number of transposons are active in many different loci in a species (this may be spread out over many generations), the offspring genome will be a mosaic of loci derived from A- and B-type gametes. When comparing that genome to that of a closely related species, loci resulting from A-type gametes will identify an excision and an insertion, while loci resulting from A-type gametes will only identify insertions. Thus the observed overall ratio of insertions to excisions from a given transposon family will be 2:1.

Additional file 6: Figure S4. Detection of differences in transposon activity in different species. This model assumes that a given transposon family was present in many copies in the ancestor species. After species divergence, the transposon family is active at different levels in the two species (100 transpositions in one and 200 in the other species). As

described in Additional file 1: Figure S1, A- and B-type gametes are passed on to offspring in a 1:1 ratio. In a cross-species comparison which identifies transposons (additional sequences) which are present in one but absent in the other species, insertions in one species and excisions in the other will be detected. If a transposon family had different levels of activity in the two species since their divergence, insertion/excision ratios will deviate from the 2:1 ratio.

Abbreviations
TE: Transposable element; TIR: Terminal inverted repeat; MITE: Miniature inverted transposable element; TSD: Target site duplication; ORF: Open reading frame; DSB: Double-strand break; NHEJ: Non-homologous end joining; MMEJ: Microhomology-mediated end joining; SSA: Single strand annealing; SDSA: Synthesis-dependent strand annealing; InDel: Insertion or deletion.

Competing interests
The authors declare that they have no competing interests.

Authors' contributions
SR performed the TE annotation and analysis and wrote the paper. TW designed the study and wrote the paper. Both authors have read and approved the final version of the paper.

Acknowledgements
We would like to thank R. A. Wing for granting early access to the *O. glaberrima* data and G. Treier for support in the statistical analysis. This study was supported by the Swiss National Foundation grant # 31003A_138505/1.

References
1.   Feschotte C, Pritham EJ. DNA transposons and the evolution of eukaryotic genomes. Annu Rev Genet. 2007;41:331.
2.   Piegu B, Guyot R, Picault N, Roulin A, Saniyal A, Kim H, et al. Doubling genome size without polyploidization: dynamics of retrotransposition-driven genomic expansions in Oryza australiensis, a wild relative of rice. Genome Res. 2006;16:1262–9.
3.   Buchmann JP, Matsumoto T, Stein N, Keller B, Wicker T. Inter-species sequence comparison of Brachypodium reveals how transposon activity corrodes genome colinearity. Plant J. 2012;71(4):550–63.
4.   Wicker T, Buchmann JP, Keller B. Patching gaps in plant genomes results in gene movement and erosion of colinearity. Genome Res. 2010;20:1229–37.
5.   Wicker T, Sabot F, Hua-Van A, Bennetzen JL, Capy P, Chalhoub B, et al. A unified classification system for eukaryotic transposable elements. Nat Rev Genet. 2007;8(12):973–82.
6.   SanMiguel P, Gaut BS, Tikhonov A, Nakajima Y, Bennetzen JL. The paleontology of intergene retrotransposons of maize. Nat Genet. 1998;20:43–5.
7.   Paterson AH, Bowers JE, Bruggmann R, Dubchak I, Grimwood J, Gundlach H, et al. The Sorghum bicolor genome and the diversification of grasses. Nature. 2009;457:551–6.
8.   The International Brachypodium Initiative. Genome sequencing and analysis of the model grass Brachypodium distachyon. Nature. 2010;463:763–8.
9.   Wicker T, Guyot R, Yahiaoui N, Keller B. CACTA transposons in Triticeae. A diverse family of high-copy repetitive elements. Plant Physiologist. 2003;132:52–63.
10. Yang G, Holligan Nagel D, Feschotte C, Hancock CN, Wessler SR. Tuned for transposition: molecular determinants underlying the hyperactivity of a Stowaway MITE. Science. 2009;325:1391–4.
11. International Human Genome Sequencing Consortium. Initial sequencing and analysis of the human genome. Nature. 2001;409:860–921.
12. Sabot F, Picault N, El-Baidouri M, Llauro C, Chaparro C, Piegu B, et al. Transpositional landscape of the rice genome revealed by paired-end mapping of high-throughput re-sequencing data. Plant J. 2011;66(2):241–6.
13. Bureau TE, Wessler SR. Stowaway: a new family of inverted-repeat elements associated with genes of both monocotyledonous and dicotyledonous plants. Plant Cell. 1994;6:907–16.

14. Bureau TE, Ronald PC, Wessler SR. A computer-based systematic survey reveals the predominance of small inverted-repeat elements in wild-type rice genes. Proc Natl Acad Sci. 1996;93:8524–9.

15. Nakazaki T, Okumoto Y, Horibata A, Yamahira S, Teraishi M, Nishida H, et al. Mobilization of a transposon in the rice genome. Nature. 2003;421:170–2.

16. Kikuchi K, Terauchi K, Wada M, Hirano HY. The plant MITE *mPing* is mobilized in anther culture. Nature. 2003;421:167–70.

17. Jiang N, Bao Z, Zhang X, Hirochika H, Eddy SR, McCouch SR, et al. An active DNA transposon family in rice. Nature. 2003;421:163–7.

18. Yang G, Weil CF, Wessler SR. A rice Tc1/mariner-like element transposes in yeast. Plant Cell. 2006;18:2469–78.

19. Yang G, Zhang F, Hancock CN, Wessler SR. Transposition of the rice miniature inverted repeat transposable elemten *mPing* in *Arabidopsis thaliana*. Proc Natl Acad Sci U S A. 2007;104(26):10962–7.

20. Fujino K, Sekiguchi H, Kiguchi T. Identification of an active transposon in intact rice plants. Mol Gen Genomics. 2005;273:150–7.

21. Moon S, Jung KH, Lee DE, Jiang WZ, Koh HJ, Heu MH, et al. Identification of active transposon *dTok*, a member of the *hAT* family, in rice. Plant Cell Physiologist. 2006;47(11):1473–83.

22. Richardson JM, Colloms SD, Finnegan DJ, Walkinshaw MD. Molecular architecture of the Mos1 paired-end complex: the structural basis of DNA transposition in a eukaryote. Cell. 2009;138(6):1096–108.

23. Symington LS, Gautier J. Double-strand break end resection and repair pathway choice. Annu Rev Genet. 2011;45:247–71.

24. Edlinger B, Schlögelhofer P. Have a break: determinants of meiotic DNA double strand break (DSB) formation and processing in plants. J Exp Bot. 2011;62(5):1545–63.

25. Vu GTH, Cao HX, Watanabe K, Hensel G, Blattner FR, Kumlehn J. Schubert I: repair of site-specific DNA double-strand breaks in barley occurs via diverse pathways primarily involving the sister chromatid. Plant Cell. 2014;26(5):2156–67.

26. Duret L, Galtier N. Biased gene conversion and the evolution of mammalian genomic landscapes. Annu Rev Genomics Hum Genet. 2009;10:285–311.

27. Wang M, Yu Y, Haberer G, Marri PR, Fan C, Goicoechea JL, et al. The genome sequence of African rice (Oryza glaberrima) and evidence for independent domestication. Nat Genet. 2014;46:982–8.

28. International Rice Genome Sequencing Project. The map-based sequence of the rice genome. Nature. 2005;436:793–800.

29. Ouyang S, Zhu W, Hamilton J, Haining L, Campbell M, Childs K, et al. The TIGR Rice Genome Annotation Resource: improvements and new features. Nucleic Acids Res. 2007;35:D883–7.

30. Wicker T, Keller B. Genome-wide comparative analysis of copia retrotransposons in Triticeae, rice, and Arabidopsis reveals conserved ancient evolutionary lineages and distinct dynamics of individual copia families. Genome Res. 2007;17:1072–81.

31. Kalendar R, Tanskanen J, Immonen S, Nevo E, Schulman AH. Genome evolution of wild barley (Hordeum spontaneum) by BARE-1 retrotransposon dynamics in response to sharp microclimatic divergence. Proc Natl Acad Sci. 2000;97:6603–7.

32. Kimura M. On the probability of fixation of mutant genes in a population. Genetics. 1962;47:713–9.

33. Robert V, Bessereau JL. Targeted engineering of the Caenorhabditis elegans genome following Mos1-triggered chromosomal breaks. EMBO J. 2007;26:170–83.

34. Hartlerode AJ, Scully R. Mechanisms of double-strand break repair in somatic mammalian cells. Biochem J. 2009;423:157–68.

35. Engels WR, Johnson-Schlitz DM, Eggleston WB, Sved J. High-frequency P element loss in Drosophila is homolog dependent. Cell. 1990;62:515–25.

36. Plasterk RH. The origin of footprints of the Tc1 transposon of Caenorhabditis elegans. EMBO J. 1991;10:1919–25.

37. Wu TD, Watanabe CK. GMAP: a genomic mapping and alignment program for mRNA and EST sequences. Bioinformatics. 2005;21:1859–75.

38. Cao P, Jung KH, Choi D, Hwang D, Zhu J, Ronald PC. The Rice Oligonucleotide Array Database: an atlas of rice gene expression. Rice. 2012;5:17.

16

Analysis of *CACTA* transposases reveals intron loss as major factor influencing their exon/intron structure in monocotyledonous and eudicotyledonous hosts

Jan P Buchmann[1,4*], Ari Löytynoja[1], Thomas Wicker[2] and Alan H Schulman[1,3]

Abstract

Background: *CACTA* elements are DNA transposons and are found in numerous organisms. Despite their low activity, several thousand copies can be identified in many genomes. *CACTA* elements transpose using a 'cut-and-paste' mechanism, which is facilitated by a DDE transposase. DDE transposases from *CACTA* elements contain, despite their conserved function, different exon numbers among various *CACTA* families. While earlier studies analyzed the ancestral history of the DDE transposases, no studies have examined exon loss and gain with a view of mechanisms that could drive the changes.

Results: We analyzed 64 transposases from different *CACTA* families among monocotyledonous and eudicotyledonous host species. The annotation of the exon/intron boundaries showed a range from one to six exons. A robust multiple sequence alignment of the 64 transposases based on their protein sequences was created and used for phylogenetic analysis, which revealed eight different clades. We observed that the exon numbers in *CACTA* transposases are not specific for a host genome. We found that ancient *CACTA* lineages diverged before the divergence of monocotyledons and eudicotyledons. Most exon/intron boundaries were found in three distinct regions among all the transposases, grouping 63 conserved intron/exon boundaries.

Conclusions: We propose a model for the ancestral *CACTA* transposase gene, which consists of four exons, that predates the divergence of the monocotyledons and eudicotyledons. Based on this model, we propose pathways of intron loss or gain to explain the observed variation in exon numbers. While intron loss appears to have prevailed, a putative case of intron gain was nevertheless observed.

Keywords: Transposases, Intron loss, Molecular evolution, DNA transposons, Plants

Background

CACTA elements are DNA transposons found in genomes across the phylogenetic spectrum, from algae [1] to vascular plants [2-6] to animals [7,8]. The first *CACTA* element described at the molecular level was *En-1* in *Zea mays* [2]; since then, they have been well documented in the grasses. Although *CACTA* elements usually do not account for the large genome sizes found in grasses, *CACTA* families nevertheless can be highly abundant. In a few cases, however, including *Tpo1* in Lolium *perenne* (ryegrass) and *Caspar* in the Triticeae, *CACTA* elements are known to have contributed considerably to the expansion of the genome size of their host [9-12]. Moreover, *CACTA*s can influence the evolution of the host genome in other ways [12]. In *Glycine max* (soybean), *CACTA* elements can affect flower color and capture host genes [13-16]. *CACTA* elements are sometimes associated with regulatory elements of genes, therefore possibly influencing gene expression [10,17]. Despite their prevalence and impact, evolutionary studies about

* Correspondence: jan.buchmann@sydney.edu.au
[1]Institute of Biotechnology, Viikki Biocenter, University of Helsinki, PO Box 65, FIN-00014 Helsinki, Finland
[4]Present address: Marie Bashir Institute for Infectious Diseases and Biosecurity, Charles Perkins Center, University of Sydney, Sydney NSW 2006, Australia
Full list of author information is available at the end of the article

CACTA elements, or DNA transposons in general, are scarce.

The *CACTA* superfamily belongs to the Class II of transposable elements, proliferating by a 'cut and paste' mechanism. In contrast to Class I elements, which transpose via an RNA intermediate and therefore copy the original element, *CACTAs* transpose the original element itself. *CACTA* elements constitute approximately 2 to 5% of a grass genome [16,18]. However, only few active *CACTA* elements have been identified in plants [2-6,19]. In addition, only seven putative transcribed transposases have been identified in the Triticeae [10].

A full-length *CACTA* element consists of two terminal inverted repeats (TIRs) bordering two open reading frames(ORFs), one encoding a transposase and the other, called ORF2, a protein of unknown function. The first and last 5 bp of the TIRs consist of the highly conserved CACTA and TAGTG motifs, respectively, hence the name of the element. The function of the ORF2 protein has been determined in specific *CACTA* families to support excision and transposition [20]. However, the transposase is the key transposition enzyme. It binds to the TIR during excision, creating a 3-bp target site duplication (TSD) [21]. The catalytic center of the transposase is the acidic triad known as the 'DDD/E' motif, which is highly conserved [22].

The presence of *CACTA* elements across the phylogenetic spectrum and the highly conserved catalytic core of their transposases indicate an ancient presence. Interestingly, the number of exons in transposases among *CACTA* transposons differs even among the grasses. Transposases in rice were found that have four exons [23], while studies in maize reported up to eleven exons for *CACTA* transposases [2,24]. In the recently sequenced grass *Brachypodium distachyon*, the exon number for transposases among *CACTA* superfamilies ranges from one to three. Therefore, the analysis of the exon/intron configuration of *CACTA* transposases offers an excellent opportunity to study the evolutionary mechanisms of intron gain and loss in DNA transposons. In addition, analyzing exon number variations in such a highly conserved and ancient gene as the *CACTA* transposase can offer a perspective on the 'intron-early' and 'intron-late' models [25,26].

The goal of this study was to analyze the differences in exon numbers in *CACTA* transposases in monocotyledonous and eudicotyledonous plants and to identify an evolutionary mechanism to explain those differences. This was accomplished using phylogenetic and comparative analyses, which required a solid and robust multiple sequence alignment (MSA). We constructed such an MSA based on protein consensus sequences of 64 transposases from *CACTA* families annotated in ten monocotyledonous and eudicotyledonous species.

Our phylogenetic analysis revealed that ancient *CACTA* lineages diverged before the divergence of the monocotyledons and eudicotyledons, supporting an intron-early model for *CACTA* transposases. The analysis of the MSA identified conserved exon/intron boundaries and putative intron gain among the transposases examined. Combining these analyses lead to a model for a putative ancient *CACTA* transposase, in which intron loss was the main mechanism shaping the exon/intron configurations of current transposases found in monocotyledonous and eudicotyledonous plants.

Results

We analyzed 64 autonomous *CACTA* transposases from ten different monocotyledonous and eudicotyledonous species. All analyzed transposases are derived from consensus sequences from distinctive *CACTA* families. Because families of transposable elements (TEs) differ from each other based on the 80-80-80 rule, they were considered orthologous [27]. Therefore, the name of the family, for example, Calvin, will indicate the consensus sequence of the transposase and not the consensus of the whole element. We refer to the plant in which a *CACTA* family and its transposase were annotated as its host. Except for transposases identified in *B. distachyon*, we searched the PTREP [28] and Repbase [29] databases for *CACTA* families with annotated transposases (see Materials and Methods). The selection was based on two criteria: i) the annotation had to clearly state 'transposase', that is annotations without ORFs described as transposases were omitted because *CACTA* elements have two ORFs, the transposase and ORF2; ii) the presence of two ORFs was expected, thereby avoiding selection of annotations having a predicted transposase that spans most of a consensus sequence, such as ATENSPM10 in Repbase, where the consensus is 8,272 bp and the predicted transposase covers positions 1,201 to 7,766. We selected nine transposases from *Sorghum bicolor*, eight transposases from *Z. mays*, five transposases from *Triticum aestivum*, 13 from *Oryza sativa*, and 11 from *B.distachyon* (Additional file 1). This resulted in a total of 46 transposases from monocotyledonous hosts. For the eudicotyledonous dataset, we selected all transposases from eudicotyledonous hosts in Repbase fitting our criteria, totaling in eighteen elements: seven transposases from elements annotated in *Arabidopsis thaliana*, five from *Fragaria vesca*, three from *Vitis vinifera*, and one each from *Petunia hybrida*, *Malus domestica*, and *G. max* (Additional file 1).

Annotation of exon/intron boundaries on *CACTA* transposases

For simplicity, the term 'boundary' will indicate exon/intron boundaries in this study. Except for transposases in *B. distachyon*, boundaries were extracted from the respective PTREP and Repbase entries (Table 1, Material and Methods). The eleven *Brachypodium distachyon* transposases were derived from consensus sequences of the autonomous families in this genome [18]. We manually annotated the transposases and boundaries by aligning to the most similar BLASTX hit within the PTREP database. Additional alignments against transcription databases from rice and *B. distachyon* did not increase the quality of the boundary predictions, because transcriptome data is scarce for *CACTA* transposases. *De novo* gene prediction did not return significant results.

Our final dataset consisted of 64 transposases with 86 annotated boundaries on the 40 transposases that contained more than one exon (Table 1). Out of the 64 annotated transposases, 24 contained only one exon and therefore no boundaries. On the remaining 40 transposases, we annotated between two and six exons (Additional file 1). The length of the transposases ranged from 552 amino acids (amino acids; PSL, 1 exon) to 4,785 amino acids (EnSpm4_Fves, 4 exons), and averaged 1,163 amino acids. The six transposases Isidor, Rufus, Sandro, Radon, Ivan, and Isaac were annotated on the 3' end of the corresponding *CACTA* consensus sequence (Additional file 1).

Generation of a robust multiple sequence alignment using confidence scores

Our phylogenetic and comparative analyses were based on an MSA derived from the selected 64 consensus transposase protein sequences. Due to the possibly ancient origin of certain *CACTA* transposases and their generally low activity, we assumed that some parts of sequences might be more evolutionarily diverged than others. In addition, the formation of consensus sequences can introduce weak regions into an MSA. A robust MSA is therefore crucial because errors or uncertainties can influence the downstream analysis. In addition, identifying weakly aligned regions or positions in an MSA and then removing them may improve downstream phylogenetic analysis [30].

GUIDANCE is a method to infer unreliable regions in an MSA and remove the potentially erroneous signal from subsequent analyses ([31]; Materials and Methods). The final MSA was 2,516 residues long and contained five unstable regions placed between positions 120 to 186, 196 to 251, 381 to 416, 728 to 766, and in the 3' end, starting from position 1,665 (Additional file 2). GUIDANCE scores range from 0 (low confidence) to 1 (high confidence) and are calculated for single residues as well as for whole columns. Because there is no recommended confidence score for residues and columns in an MSA, a trade-off between sensitivity and specificity is required. High sensitivity (low cutoff value) retains as many columns as possible while high specificity

(high cutoff value) keeps only columns of very high confidence.

The default GUIDANCE cutoff of 0.93 removed 638 columns (approximately 25%) from the alignment, including the badly aligned regions and 34 annotated boundaries. However, GUIDANCE kept columns with only one residue, for example, most of the badly aligned 3' end. To retain as many boundaries as possible for the analysis we applied our own trimming: we removed columns containing only residues with scores below 0.804 (keeping boundaries) and columns with only one residue (not comparable and/or bad aligned). This approach removed 1,398 columns (approximately 44%): the badly aligned regions but only 13 annotated boundaries. This final MSA was 1,118 residues long and contained 73 annotated boundaries in 64 transposases (Figure 1). Because the first boundary is also the beginning of the first intron, introns were named in the 5' to 3' direction and designated as subscripts to the name of the transposase, for example, the first intron and boundary of transposase Baron is described as $Baron_1$. We mapped conserved DDE motifs [22] onto the MSA, which were all in positions with high confidence values (Figure 1). This MSA was used for all further analysis.

Exon numbers in *CACTA* transposases are not specific to a host genome

RAxML [32] was used to calculate the phylogenetic tree (Figure 2). A maximum likelihood (ML) tree was generated based on 200 distinct, randomized, maximum parsimony trees and its robustness assessed by using 1,000 bootstrap replicates and by testing the influence of several outgroups (Additional file 3, Material and Methods). The resulting tree shows the relation between individual transposases but not their evolution over time; that is the branch lengths do not indicate the time when transposases diverged from each other but how close they are on the molecular level (Figure 2). We identified eight clades, designated α to θ (Figure 2). Crucially, the transposases grouped primarily by their exon numbers rather than by their hosts and the analysis of the clusters found no host-specific exon numbers for *CACTA* transposases (Figure 2).

Ancient *CACTA* lineages diverged before the divergence of monocotyledons and eudicotyledons

We identified three clades in which monocotyledonous and eudicotyledonous transposases clustered together. EnSpm2_Gmax from soybean grouped in Clade α with transposases from several monocotyledonous hosts, analogous to EnSpm3_Fves and EnSpm4_Fves from strawberry in Clade ζ. Clade δ grouped transposases from strawberry, apple, and several grasses. The other clades contained only transposases from either eudicotyledonous or

Table 1 Exon/intron boundaries of the 34 analyzed *CACTA* transposases with more than one exon.

	1	2	3	4	5
EnSpm12_Fves	462 | 564[G]				
C	718 | 771[I]				
EnSpm10_Fves	826 | 846				
Joey	842 | 893[II]				
Janus	837 | 894[II]				
F	846 | 894[II]				
G	847 | 894[II]				
Norman	879 | 921[III]				
En1	879 | 925[III]				
Alfred	885 | 925[III]				
H	838 | 972				
EnSpm3_Vvin	821 | 894[II]	856 | 925[III]			
EnSpm8_Sbic	754 | 783[I]	*976 | 0*			
Storm	827 | 782[I]	*951 | 0*			
Sherman	831 | 782[I]	*954 | 0*			
J	495 | 521	750 | 885			
EnSpm2_Mdom	755 | 782[I]	886 | 893[II]			
Baldur	731 | 782[I]	837 | 895[II]			
I	834 | 782[I]	954 | 910			
Isidor	857 | 894[II]	892 | 920[III]			
Radon	841 | 894[II]	877 | 921[III]			
Rufus	851 | 894[II]	887 | 921[III]			
EnSpm13_Vvin	821 | 894[II]	856 | 925[III]			
EnSpm5_Vvin	824 | 894[II]	859 | 925[III]			
Isaac	861 | 894[II]	900 | 925[III]			
Sandro	744 | 782[I]	851 | 928[III]			
Balduin	850 | 895[II]	890 | 930[III]			
DOPPIA	843 | 894[II]	890 | 936[III]			
K	744 | 7,82I	850 | 936[III]			
Horace	712 | 711	981 | 1,054			
EnSpm4_Fves	*812 | 0*	*992 | 0*	*1,244 | 0*		
EnSpm3_Fves	*681 | 0*	770 | 781[I]	*919 | 0*		
Seamus	730 | 782[I]	833 | 892[II]	878 | 925[III]		
Dario	726 | 711	842 | 839	895 | 890[III]		
Aron	851 | 833	899 | 879[II]	1,013 | 1,060		
Korbin	510 | 567[G]	718 | 782[I]	814 | 894[II]	*853 | 0*	
Chester	520 | 563[G]	728 | 777[I]	823 | 889[II]	858 | 920[III]	
Baron	522 | 568[G]	730 | 781[I]	825 | 893[II]	861 | 925[III]	
EnSpm8_Fves	158 | 163	830 | 893[II]	*975 | 0*	*1,219 | 0*	*1,500 | 0*
ATENSPM6_Athal	802 | 809	918 | 922[III]	978 | 981	1,011 | 1,012	*1,141 | 0*

The positions are relative to the beginning of the transcription start and given as follows:
On the protein sequence | on the trimmed multiple sequence alignment (MSA). 0 and numbers in italic indicate boundaries with GUIDANCE scores below 0.804 and removed in the final MSA. Superscripts indicate Regions I to III and G cluster, respectively (Figure 1).

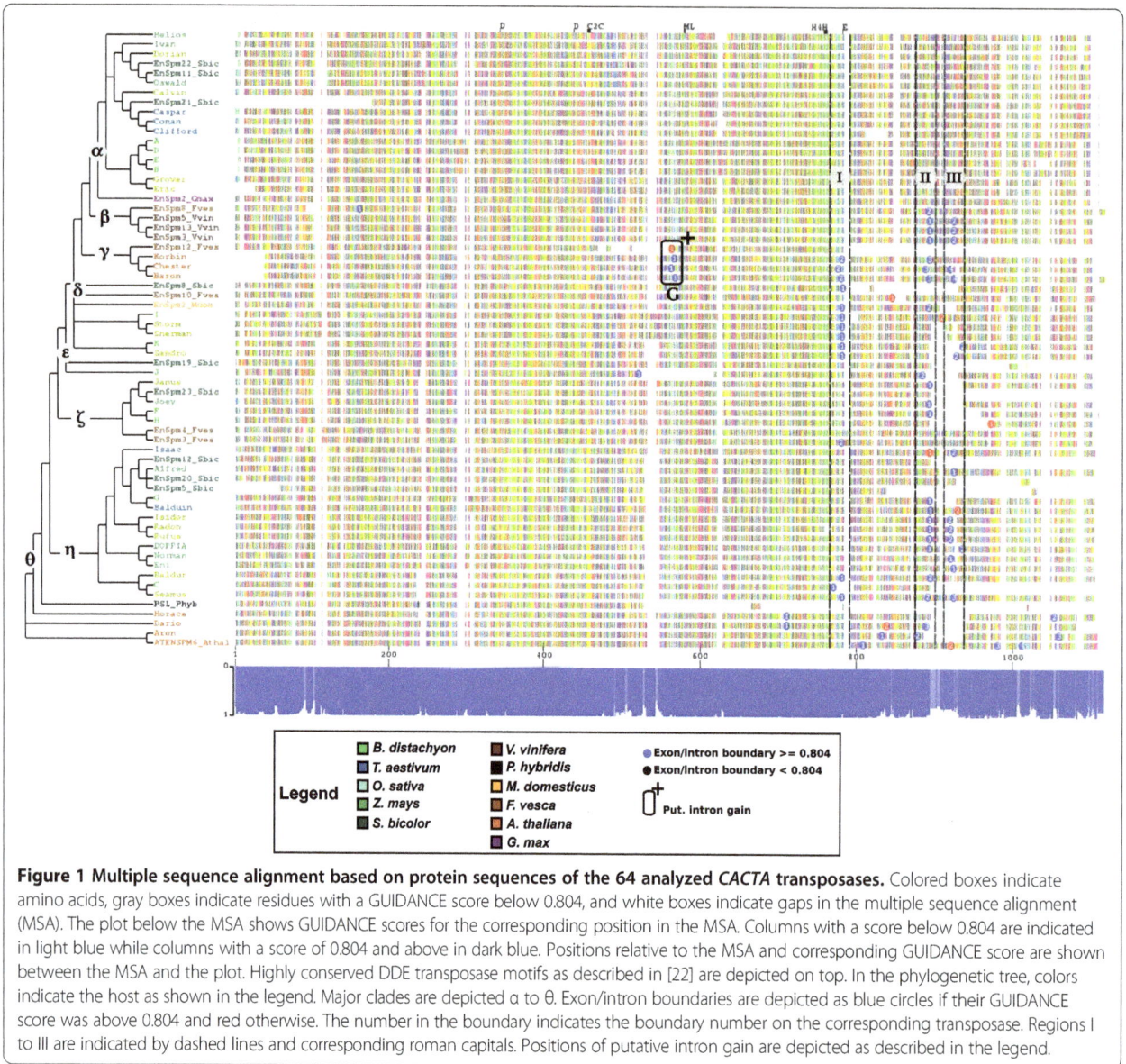

Figure 1 Multiple sequence alignment based on protein sequences of the 64 analyzed *CACTA* transposases. Colored boxes indicate amino acids, gray boxes indicate residues with a GUIDANCE score below 0.804, and white boxes indicate gaps in the multiple sequence alignment (MSA). The plot below the MSA shows GUIDANCE scores for the corresponding position in the MSA. Columns with a score below 0.804 are indicated in light blue while columns with a score of 0.804 and above in dark blue. Positions relative to the MSA and corresponding GUIDANCE score are shown between the MSA and the plot. Highly conserved DDE transposase motifs as described in [22] are depicted on top. In the phylogenetic tree, colors indicate the host as shown in the legend. Major clades are depicted α to θ. Exon/intron boundaries are depicted as blue circles if their GUIDANCE score was above 0.804 and red otherwise. The number in the boundary indicates the boundary number on the corresponding transposase. Regions I to III are indicated by dashed lines and corresponding roman capitals. Positions of putative intron gain are depicted as described in the legend.

monocotyledonous hosts (Figure 2). Despite the long evolutionary time separating monocotyledonous and eudicotyledonous hosts, the presence of mixed clades and the close relation of clades with only monocotyledonous or eudicotyledonous hosts suggests that the *CACTA* transposase phylogeny rather than the host phylogeny is primary, that is that the main transposase branches diverged already before the divergence of monocotyledons and eudicotyledons. Indeed, a closer look at the phylogenetic tree revealed that transposases within clades tend to have the same number of exons (Figure 2).

The majority of *CACTA* transposase boundaries are found in three regions on the MSA

To analyze the evolution of exon/intron arrangements in *CACTA* transposases, we compared the boundaries from

the 33 transposases containing 73 introns that were not removed in the trimming process (Table 1, Figure 1). We identified 3 regions, labeled I to III, in the MSA, which contain 63 out of the 73 boundaries (Figure 1). Outside those regions, we identified eight boundaries inside the DDE motif, four boundaries between Regions I and II, one boundary between Regions II and III and five boundaries downstream of Region III. Most boundaries are close to each other but not in the same position on the alignment. This can be due to small errors introduced by calculating the MSA or consensus sequences. Therefore, we analyzed the distances between boundaries to identify which were shared among transposases.

We analyzed the boundaries by clustering them based on their positions on the MSA. We set the maximal distance between boundaries still considered to be in the

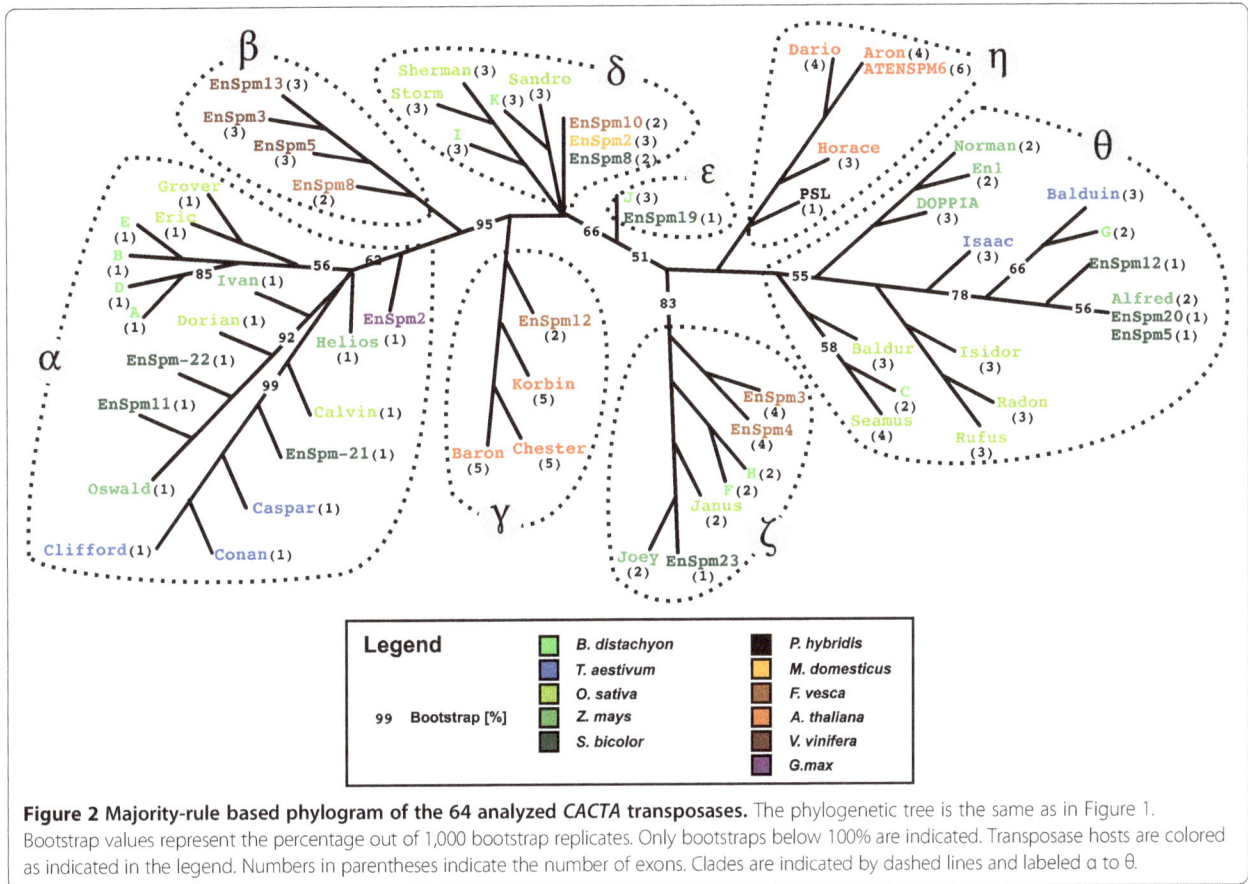

Figure 2 Majority-rule based phylogram of the 64 analyzed *CACTA* transposases. The phylogenetic tree is the same as in Figure 1. Bootstrap values represent the percentage out of 1,000 bootstrap replicates. Only bootstraps below 100% are indicated. Transposase hosts are colored as indicated in the legend. Numbers in parentheses indicate the number of exons. Clades are indicated by dashed lines and labeled α to θ.

same region to 16 residues, which is half the length of the shortest intron annotated (33 amino acids in ATENSPM_Athal$_3$). Boundaries that were closer than 16 residues to each other were grouped together. No boundaries within a region were further than 16 residues apart (Tables 2, 3, Additional files 4, 5, 6). The distances between the closest boundaries of Regions I and II is 98 residues (Additional file 7), but 30 residues between Region II and III (Additional file 7). The closest boundary upstream of Region I is 60 residues away, whereas the closest boundary downstream of Region III is 36 residues away. This clustering confirmed the previously identified regions as clearly distinct. The four boundaries EnSpm10_Fves$_1$, Dario$_2$, Aron$_1$, and ATENSPM6_Athal$_1$ between Region I

Table 2 Distances between exon/intron boundaries within Region I

	Baldur$_1$											
Baron$_2$	1	Baron$_2$										
C$_1$	11	10	C$_1$									
Chester$_2$	5	4	6	Chester$_2$								
EnSpm2_Mdom$_1$	0	1	11	5	EnSpm2_Mdom$_1$							
EnSpm3_Fves$_2$	1	0	10	4	1	EnSpm3_Fves$_2$						
I$_1$	0	1	11	5	0	1	I$_1$					
K$_1$	0	1	11	5	0	1	0	K$_1$				
Korbin$_2$	0	1	11	5	0	1	0	0	Korbin$_2$			
Sandro$_1$	0	1	11	5	0	1	0	0	0	Sandro$_1$		
Seamus$_1$	0	1	11	5	0	1	0	0	0	0	Seamus$_1$	
Sherman$_1$	0	1	11	5	0	1	0	0	0	0	0	Sherman$_1$
Storm$_1$	0	1	11	5	0	1	0	0	0	0	0	0

Distances between exon/intron boundaries in the MSA within Region I (depicted in Figure 1). The distances are given in residues in the alignment.

and II, as well as I_2 between Region II and III could not be clustered in those Regions. We identified only one additional cluster containing four boundaries outside Regions I to III. It groups the first introns from all members of Clade γ and was therefore named Region G.

Based on these analyses of distances between all boundaries, we established that Regions I to III and G in the MSA were clearly separated from each other as well as from all other boundaries. Given the distinctness of the four boundary regions, we examined if the boundaries themselves were conserved among the analyzed transposases.

Boundaries in Regions I to III are conserved among most transposases while Region G represents putative intron gain

Due to the proximity of boundaries in Regions I to III and their clear separation from other boundaries, we established that boundaries within a region are shared between the different transposases. The clustering of boundaries within Regions I to III indicates that the boundaries are conserved among the analyzed transposases. This is supported by the phylogenetic tree, in which purely monocotyledonous or eudicotyledonous clades share boundaries (Figure 1). Boundaries in Region I are on, or close to, the position of the conserved E from the DDE motif, supporting the claim that Region I represents conserved boundaries among the transposases (Figure 1). Therefore, we considered the 63 boundaries in Regions I to III as conserved within each region. All transposases in Clade γ share their first introns with a maximum distance of five residues (Figure 1, Table 4). This is a unique cluster in the whole tree, indicating intron gain since all members of Clade γ share this intron but none of its ancestor nodes and transposases in other clades.

Only two boundaries from a monocotyledonous host are found outside Regions I to III

We identified 17 boundaries outside Regions I to III (Figure 1). Only J_1 and H1 are from a monocotyledonous host, whereas the remaining 15 boundaries were annotated in transposases from eudicotyledonous hosts. Boundaries I_1 and $ATENSPM6_{1,2,3}$ cannot be clustered and therefore were not further characterized. The transposases Horace, Dario, and Aron have three separate boundaries which are not farther apart than six residues: $Horace_1$ and $Dario_1$, $Daron_2$ and $Aron_1$, $Horace_2$ and $Aron_3$. While this appears as another case of intron gain, their relation in the phylogenetic tree is not properly resolved and does not support this interpretation.

Our analysis of the boundaries identified 63 conserved boundaries and 4 cases of putative intron gain in Region G. Most conserved introns were identified in transposases from monocotyledonous hosts. In contrast, all unique boundaries except two were identified in eudicotyledonous hosts. We decided to combine the results of the phylogenetic and boundary analyses to develop a model to understand how the observed exon/intron configuration evolved.

Defining consensus exon numbers for each phylogenetic clade

A comparison of the phylogenetic tree and the conserved boundaries revealed a high consistency between clades and boundary positions. Based on the majority of exons per clade, we constructed a loose consensus to represent the exon number for transposases in the corresponding clade. For example, Clade ζ groups together seven transposases of which four, the majority, have two exons. Therefore, a representative transposase from Clade ζ has two exons and one consensus boundary. We used this approach for each clade (Figure 3). Our approach resulted in following exon numbers for representative transposases: one exon for Clade α; Clades β, δ, and θ three exons each; Clade η four exons; Clade γ five exons. Designating consensus exon numbers for each clade simplified further the analysis to develop a model for the loss and gain of boundaries in *CACTA* transposases.

A model for loss and gain of exon/intron boundaries in *CACTA* transposases

Because it had the largest number of confirmed exons, we compared all consensus boundaries to Clade γ (Figure 3). Clade α has no annotated introns. The second, third, and fourth intron of Clade γ can be found throughout the phylogenetic tree, whereby the third intron of Clade γ is the most conserved, followed by its fourth and second intron. The fourth intron of Clade γ is found among Clades β, θ, ι, and in Isaac. The third intron is missing in the Clades EnSpm8, δ, and θ, but otherwise is found in all clades containing introns. The second intron of Clade γ is present in Clades δ, EnSpm8, and η. This comparison indicates that *CACTA* transposases were as a whole losing rather than gaining introns. However, Clades γ and ζ have introns that are not found in other clades (Figure 3), the first intron in Clade γ representing an intron gain. The unique introns in Clade ζ cannot be classified as losses or gains because the phylogenetic tree does not allow a definitive classification.

We propose that the consensus transposase in Clade γ represents the most likely exon/intron configuration of an ancient transposase, containing at least four exons and three introns (Figure 3). The three boundaries correspond to those identified in Regions I to III in the MSA (Figures 1, 3). Using the putative ancestor model transposase, we can infer the emergence of the known transposases through intron loss and gain (Figure 3).

Discussion

In sum, we analyzed 64 *CACTA* transposases from 11 monocotyledonous and eudicotyledonous hosts. Our

Table 3 Distances between exon/intron boundaries within Region III

	Alfred$_1$	Balduin$_2$	Baron$_4$	Chester$_4$	En1$_1$	EnSpm13_Vvin$_2$	EnSpm3_Vvin$_2$	EnSpm5_Vvin$_2$	Isaac$_2$	Isidor$_2$	Norman$_1$	Radon$_2$	Rufus$_2$	Sandro$_2$
Balduin$_2$	5													
Baron$_4$	0	5												
Chester$_4$	5	10	5											
En1$_1$	0	5	5	5										
EnSpm13_Vvin$_2$	0	5	0	5	0									
EnSpm3_Vvin$_2$	0	5	0	5	0	0								
EnSpm5_Vvin$_2$	0	5	0	5	0	0	0							
Isaac$_2$	0	5	0	5	0	0	0	0						
Isidor$_2$	5	10	5	0	5	5	5	5	5					
Norman$_1$	4	9	4	1	4	4	4	4	4	1				
Radon$_2$	4	9	4	1	4	4	4	4	4	1	0			
Rufus$_2$	4	9	4	1	4	4	4	4	4	1	0	0		
Sandro$_2$	3	2	3	8	3	3	3	3	3	8	7	7	7	
Seamus$_3$	0	5	0	5	0	0	0	0	0	5	4	4	4	3

Distances between exon/intron boundaries in the MSA within Region III (depicted in Figure 1). The distances are given in residues in the alignment.

Table 4 Distances between exon/intron boundaries within Cluster G

	Baron₁		
Chester₁	5	Chester₁	
EnSpm12_Fves₁	4	1	EnSpm12_Fves₁
Korbin₁	1	4	3

Distances between exon/intron boundaries in the MSA within Cluster G (depicted in Figure 1). The distances are given in residues in the alignment.

phylogenetic analysis indicates divergence of ancient *CACTA* lineages already before the divergence of the monocotyledons and eudicotyledons. The analysis of 73 boundaries across 33 transposases with more than one exon identified 55 conserved exon/intron boundaries and allowed us to reconstruct the exon/intron configuration of a *CACTA* transposase representing the ancestral state before the divergence of monocotyledonous and eudicotyledonous plants. The model consists of at least four exons. We propose a mechanism for the evolution of the extant *CACTA* transposases in which they were shaped mainly by intron loss, although one case of putative intron gain was found.

Potential for greater regulation of *CACTA* elements in eudicotyledons

Studies of the *P* Element in *Drosophila* and *Ac/Ds* in maize have shown that alternative splicing can regulate tissue-specific transposition of elements. For example, the *P* element retains its third intron in somatic cells, inhibiting transposition [33,34]. Should this occur with *CATCA* transposases as well, our data suggests that elements in dicotyledonous hosts have more possibilities for regulation. Interestingly, most non-clustered boundaries and the putative intron gain cluster were found in transposases from dicotyledonous hosts, whereas the majority of boundaries in Regions I to III were found in transposases from monocotyledonous hosts. The number of transposable elements in eudicotyledonous genomes is generally lower than in monocotyledonous genomes, consistent with a tighter control of transposable elements in eudicotyledonous hosts. Therefore, the large number of unique boundaries found outside Regions I to III could be associated with more control of expression of *CACTA* elements in eudicotyledons than in monocotyledons.

Differences in intron gain and loss among TE transposases

Previously, intron gain and loss in transposases of DNA transposable elements was studied for *Mariner*-like elements in flowering plants [35]. In that study, degenerate primers were used to extract fragments of DDE transposases from 54 plant species for phylogenetic analysis. The results were consistent with vertical transmission

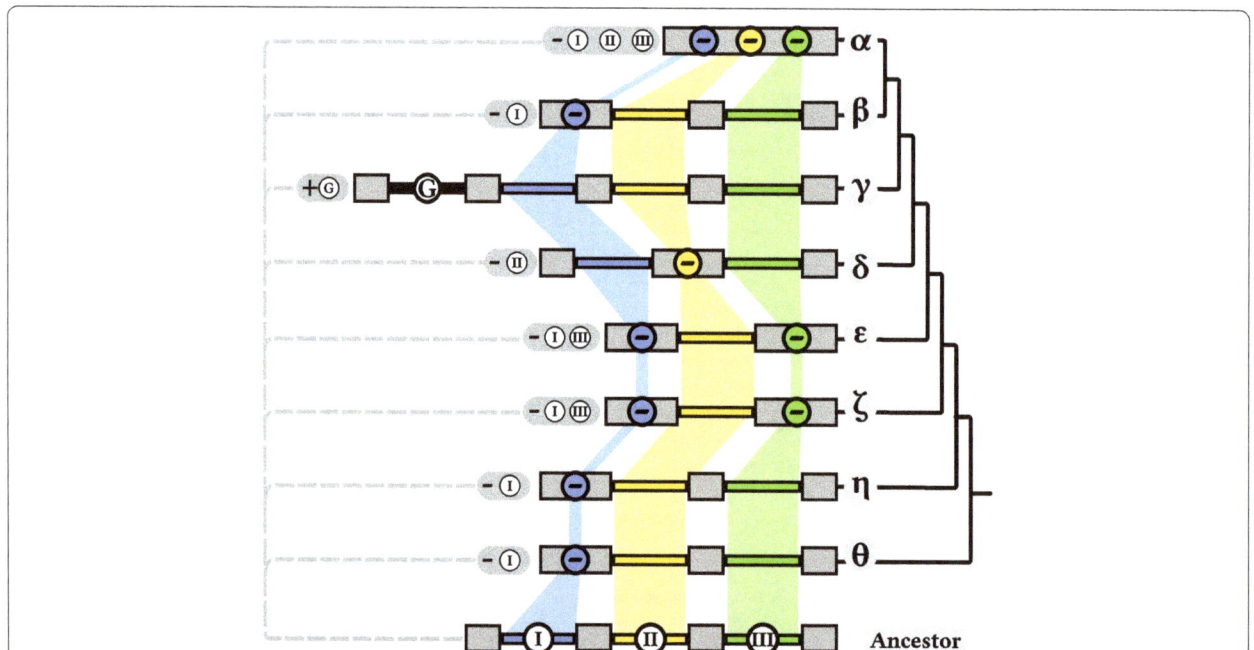

Figure 3 Model for the loss and gain of introns in *CACTA* transposases. Simplified phylogenetic tree based on the consensus exon numbers per clade as described in the text. Below the tree the putative ancestor transposase with four exons is depicted. Exons are depicted as gray rectangles with introns as colored lines. Blue, red and green depict introns conserved in Regions I to III, G indicates cluster G with the putative intron gain. Conserved introns share the same color band. Intron loss is depicted by its corresponding color and circled −, intron gain by an encircled +. Gray balloons indicate how the observed configuration arose from the putative ancestor.

and rapid diversification and indicated a gain of introns in grasses in a localized region of the transposase gene. This may indicate that *Mariner*-like elements generally tend to gain introns, while *CACTA* elements tend to intron loss. However, the *Mariner* fragments analyzed were mainly located within the DDE motif, where exon/intron boundaries have been predicted, whereas our data suggests that most exon/intron boundaries in DDE transposases from *CATCA* elements are downstream of that motif.

Horizontal transfer of *CACTA* elements

We observed several transposases from distinct species grouping in the same clade such as EnSpm2_Fves in Clade α and EnSpm3_Fves and EnSpm4_Fves in Clade ζ. This raises the question of a possible horizontal mode of inheritance, which has been proposed to drive genomic variation in eukaryotic genomes and has been shown for the *Mu*-like elements in plants [36,37]. Experiments that introduced the *Ac/Ds* element from maize into *A. thaliana* and sugar beet found reduced levels of correctly spliced *Ac* transposase transcripts in those distant heterologous host species. Therefore, it has been proposed that intron loss in the transposases of DNA transposons is an adaptation to ease horizontal transfer [36]. Although the ML tree from our analysis clusters transposases from different host together, the closest relations are mainly from the same host (Additional file 3). Some exceptions are found, mostly where transposases from maize, sorghum, wheat and *B. distachyon* are found as closest neighbors. Interestingly, those close neighbors have a very similar exon/intron boundary configuration, for example, G and Balduin in Clade η, Sandro and K in Clade δ, and Oswald and EnSpm11_Sbic in Clade α. Because we did our analysis on consensus protein sequences, analysis on the DNA level as performed earlier [37] was not possible. Therefore, although horizontal transposon transfer for *CACTA* elements cannot be ruled out, our dataset does not provide support for this mechanism.

Using several data sources increases fidelity of the annotated exon/intron boundaries

To counter the various influences of consensus sequences, we used GUIDANCE. The identification of weak regions and residues in the MSA using confidence scores improves subsequent analysis [30]. We decided to apply a threshold lower than the default, 0.804 compared to 0.93, because the boundary annotations are based on predictions and modeling approaches. Certain boundaries may have been wrongly predicted or modeled because transcription data for *CACTA* transposases is scarce. Analyses for the Triticeae have shown only seven putative transcribed transposases out of 41 identified

CACTA elements [10]. Nevertheless, the range of annotated exons in the transposases is similar for the previously published *CACTA* transposons. OsESI1 and Hipa in rice have four exons [23], although studies in maize indicate transposases with up to eleven exons [2,24].

We used three sources to collect transposes: PTREP, Repbase, and our own models for the transposases annotated in *B. distachyon*. The majority of annotated boundaries were found in three Regions, I to III. In several cases, the boundary predictions overlapped. Annotated boundaries in Region II were derived from Repbase, our own modeling and from PTREP. This overlap strongly supports the proper annotation of an exon/intron boundary at those positions. The unique boundaries are missing such support and have, therefore, not been classified because there was not enough data to assess if they represent a putative conserved boundary or recent intron gain or loss.

An alternative explanation for the presence of conserved introns at similar positions is intron sliding or slippage. Intron sliding is defined as the shift of an exon/intron position over time during evolution, such as through nucleotide insertions before the boundary [38,39]. Calculations have shown that changes of one to 15 nucleotides may occur; shifts of one nucleotide have been observed [39]. We calculated a maximum distance of seven amino acids, which is very close to the proposed maximum of intron slippage, supporting our claim of conserved boundaries in those regions.

High *CACTA* diversity existed already in the ancestor of monocotyledons and eudicotyledons

Our phylogenetic reconstruction clustered the transposases according to their exon number rather than by host species. This supports earlier studies, which compared intron gain and loss across several eukaryotic species and showed the evolutionary conservation of intron positions and their use as additional sources of phylogenetic information [40-42]. All clades contained a mixture of several host species, although Clade θ harbored only transposases from eudicotyledonous hosts. The monocotyledonous and eudicotyledonous hosts in all clades diverged approximately 120 to 340 million years ago [43]. This supports the existence of diversity among *CACTA* transposases already in the common ancestor of the monocotyledons and eudicotyledons.

The ancestral *CACTA* transposase likely had four exons

The number of exons in the transposases varies between species. Our analysis of boundaries between the transposases showed that 55 out of 73 exon/intron boundaries are conserved between 2 or more transposases. This raises the question of whether the ancestral transposase, which predated the divergence of the clades that we analyzed contained one exon and later gained additional exons or instead contained several exons and then lost them over time. A third alternative is a mixture of both mechanisms, in which exons are arbitrarily gained and lost. In most transposases, we annotated between two and six exons. The conservation of the boundaries in Regions I to III across several clades indicates a loss of introns in *CACTA* transposases rather than a gain.

Boundaries in Region I have the least conservation level among the boundaries analyzed. However, these boundaries were mapped on, or close to, the E of the DDE motif. Because this motif is considered to be highly conserved and from a common origin [22], the boundaries in Region I are very likely to have been generally conserved but lost in some transposases. Nevertheless, unique introns indicate that intron gain may occur, albeit at a low frequency. The putative intron gain in Clade γ is supported by its unique occurrence, whereas the conserved boundaries are found in Regions I to III and in several clades. This is in accordance with observations of ancestral introns in plants, fungi, and animals [44].

Taking these lines of evidence into account, we propose an ancestral *CACTA* transposase configuration with at least four exons. Subsequent and differential intron loss was a major force in *CACTA* transposase evolution. Our prediction is that the ancestor *CACTA* transposase with four exons predates the divergence of monocotyledons and eudicotyledons. Given the ancestry and abundance of DDE transposases, the *CACTA* transposases appear to follow the model of 'many introns early in eukaryotic evolution' [38,45,46].

Potential selection for intron gain

Against a background of general intron loss, we observed only one conspicuous case of intron gain, that of the first intron in Clade γ, where the intron is found within the entire clade. This clade contains *A. thaliana* and strawberry as hosts. Other introns were found outside Regions I to III, particularly in Clade θ, but are not present throughout an entire clade. These others are either remnants of an intron that was gained at the root of the clade, but then differentially lost in various families within the clade, or alternatively represent later insertions on the family level. Our dataset cannot resolve these alternatives. Moreover, the boundaries are based on models; a wrong prediction cannot be excluded. Due to the sparse number and weak support for introns with spotty distributions, we eliminated them from the analysis. Intron gain has been proposed to occur through the insertion of TEs and subsequent loss of TE mobility [33,47]. However, we did not identify TEs in *CACTA* transposase introns.

Interestingly, the putative gained intron in Clade γ represents the first intron, which is the one nearest the N-terminus. Studies in both eudicots and monocots suggest that first introns in particular have roles either as enhancers or in controlling the tissue specificity of expression [48-50]. Introns in *A. thaliana* have been shown to increase expression best when near the promoter [48] and to have the capacity for mediating differential expression patterns [51,52]. Therefore, intron gain at the first position in *A. thaliana* transposases may well have constituted an advantage. Although first introns have regulatory roles in monocots as well, we found no clade-wide examples of gain and retention of new transposase introns.

Intron loss in *CACTA* transposase was reverse transcriptase -mediated

Loss of introns in the analyzed transposase genes occurred in-frame, because putative functional ORFs have been identified. Therefore, intron loss in *CACTA* transposases most likely did not influence the coding capability of the transposases. We observed only small perturbations in the alignment where introns were lost in Region I, while Regions II and III show larger disturbances at positions of intron loss. The most commonly postulated means for intron loss are by reverse transcription of spliced transcripts, by direct genomic deletion, by intron removal as a result of double strand break (DSB) repair, and by exonization.

Exonization may occur if a donor splice site is mutated so that an intron is retained in the transcript [53,54]. This would lead to a fusion of the intron with its flanking exons and therefore the shifting of an annotated boundary in the MSA. Only unique boundaries could represent an intron lost by exonization. However, unique boundaries were annotated in highly similar blocks in the MSA, indicating no gain of sequence (Figure 1). If exonization has been responsible for intron loss, it would follow that *CACTA* transposases may undergo alternative splicing, similar to the *P* element in *Drosophila* or to *Ac/Ds* in maize. [33,34,55]. Intron loss by DSB repair [56] first requires a DSB, initiated either by excision of a mobile element such as a DNA transposon or by other means. However, no mobile elements have been identified in the transposase introns, making intron loss due to DSB repair unlikely. Evidence for a DSB initiated by other means was not found, but the DSB repair model cannot be excluded. Direct genomic deletion may lead to in-frame loss of introns if small direct repeats are present at the intron ends [25,57].

Intron loss by the action of reverse transcriptase (RT) is a frequently proposed model [58-61]. The mechanism comprises reverse transcription of processed or partially processed mRNA into cDNA and subsequent integration of the cDNA into the genome by homologous recombination [44,62,63]. This mechanism can lead either to loss of all introns, as suggested for gene *EP-1α* in the zooplankton *Oikopleura longicauda* [62], or to partial loss of introns as proposed in the *catalase 3* genes in *Z. mays* [63]. A modification of the RT model has been proposed to explain the partial loss of introns, in which enzymes that recognize and degrade aberrant DNA generate fragments from the cDNA [57]. These fragments then would recombine with genomic DNA. Alternatively, selective and precise in-frame loss of introns in the *str* gene family of *Caenorhabditis briggsae* and *C. elegans* was proposed to be due to a non-homologous recombination mechanism [64].

In the *CACTA* transposases, the phylogenetically close relationship of Clade α to Clades β and γ indicates a loss of all introns (Figure 2) as the simple RT-mediation model would predict. Similarly, in several clades transposases with one exon are grouped together with transposases containing several exons (Figure 2). Therefore, loss of all introns in a *CACTA* transposase was not a unique event; it has occurred several times in different clades. Moreover, Clade α consists of eighteen transposases from all five monocotyledonous hosts and the one transposase from soybean. This indicates no species specificity exists for transposases with one exon. Moreover, intron loss due to DSB repair, intron retention, or genomic deletion would target individual elements. In contrast, in RT-mediated intron loss, the reverse transcribed transposases could undergo homologous recombination with highly similar regions such as the DDE motif that is also found in a variety of other transposases. Plants, especially grasses, are known to have high numbers of retroelements, providing the potential for RT to interact with transcripts from *CACTA* transposases [65]. Taking these strands together, it appears that RT-mediation is the most likely pathway for intron loss in *CACTA* transposases and possibly in DNA transposon transposases as a whole.

Intron loss and gain in transposases and genes indicates transposases are ancient genomic components

Evolution of the *CACTA* transposase gene structure has parallels to that of the GDSL-lipase gene family [66]. By analysis of intron gain and loss across several land plants, it appears that the common ancestor of this gene family contained six exons. Through gain and loss of introns, different subfamilies arose, some containing unique introns. Intron loss in GDSL-lipase genes was prevalent in grasses, especially in sorghum. By contrast, in the widely distributed regulatory SnRK2 kinase family, monocots and eudicots are distinct regarding their patterns of intron retention, with the rice genes retaining more introns than those in *Arabidopsis* [67]. Most *CACTA* transposases without introns were found in sorghum, although this may merely represent sampling error. Independent loss of introns has been

reported as well for the *4f-rnp* genes in *Drosophila melanogaster* [68]. The similar trajectories followed by both different gene families and the *CACTA* transposases indicates that intron gain and loss in transposases has been driven by the same evolutionary mechanisms in TEs and in genes for various cellular functions. This is consonant with the view of transposable elements as ancient genomic components and not genome 'invaders' [69].

Conclusion

The presented analysis and comparison of exon/intron boundaries among 64 *CACTA* elements from monocotyledonous and eudicotyledonous hosts gives an insight into the dynamics of intron loss and gain in eukaryotic transposases in general and *CACTA* transposases in detail. Our results explain the observed variety in intron numbers among *CACTA* elements found in monocotyledonous and dicotyledonous and possibly further diverged hosts. The observed predominant loss of introns in *CACTA* transposases differs from previous studies in *Mariner*-like elements, indicating differences of intron gain and loss between DNA transposons. Our study strongly indicates a high variety among *CACTA* transposases before the divergence of monocotyledons and eudicotyledons hosts and provides a putative *CACTA* transposase configuration for the corresponding ancestor element. Our results support the view of transposable elements as genomic components and not as genome 'invaders'. However, to fully understand intron loss and gain in *CACTA* elements, or in DNA transposon in general, reliable transcription data will be required.

Materials and methods
Transposase selection
Transposase sequences from *O. sativa*, *T. aestivum*, *S. bicolor*, *Z. mays*, *A. thaliana*, *P. hybrida*, *F. vesca*, *M. domestica*, and *V. vinifera* were extracted from Repbase and PTREP, respectively, according to criteria described in the text. *CACTA* elements are described as *EnSpm*-like elements in Repbase while DTC in PTREP. *B. distachyon CACTA* consensus sequences were taken from [18] and annotated as described in the text.

Annotation of exon positions
For Repbase entries stored in the EMBL file format, we extracted the exon coordinates and transformed them from nucleotide positions into amino acid positions relative to the beginning of the predicted transposase protein. PTREP entries which stored protein sequences in the FASTA format were translated into DNA and aligned against the DNA consensus sequence of the corresponding *CACTA* element using dotter [70]. Despite the existence of multiple codons for each amino acid, exons could be visually recognized and annotated.

Multiple sequence alignments and GUIDANCE

To obtain the multiple sequence alignment and confidence scores the GUIDANCE web server (http://guidance.tau.ac.il, [71]) was used with following parameters: algorithm, GUIDANCE; number of bootstrap repeats, 100; multiple sequence alignment algorithm, MAFFT; advanced alignment options, maxiterate 1000; refinement strategy, genafpair. Perl scripts were written to extract and visualize data from GUIDANCE.

Generation of phylogenetic trees

All phylogenetic trees were calculated using RAxML-version 7.2.8 [32]. For the meaning of the used parameter and correct calling of RAxML, we referred to the RAxML manual. The PROTGAMMALGF protein substitution model was selected using the Perl script to identify the best protein substitution model provided on the RAxML website (http://sco.h-its.org/exelixis/web/software/raxml/index.html). Construction of the ML tree was made using following parameters: -m PROTGAMMALGF, -f d, -N 200. Bootstrap analysis was carried out using following parameters: -m PROTGAMMALGF, -f d, -x 54321, -N 1000. The consensus tree was computed using following parameters: -m PROTGAMMALGF, -J MR. Testing of outgroups was performed using following parameters: -f d -m PROTGAMMALGF -N 50 -o < outgroup>. Phylogenetic trees were prepared using FigTree (http://tree.bio.ed.ac.uk/software/figtree/) and TreeGraph [72].

Exon/intron boundary analysis

Various Perl scripts were written to analyze and visualize boundary data. All Perl programs can be obtained from the authors.

Additional files

Additional file 1: Table summarizing the analyzed transposases. Contains the names, length, and number of exons, host, and source for each analyzed transposase. Contains all annotated boundaries with positions on the original protein, on the trimmed MSA, its score and the residue.

Additional file 2: GUIDANCE results. Contains all files to recreate the analyzed MSA and consists of three files: msa_initial.fasta, the sequence alignment derived from GUIDANCE in FASTA format; msa_residueScores. txt, GUIDANCE scores for all residues; guidance output in HTML format.

Additional file 3: Best maximum likelihood tree for the 57 analyzed CACTA transposases. Describe s the best maximum likelihood tree out of 200 distinct, randomized, maximum parsimony trees for the 64 analyzed CACTA transposases. The tree has been mid-point rooted due to the lack of an available outgroup. Contains the 12 maximum likelihood trees in the Newick format which were used to check the robustness of the initial maximum likelihood tree. It can be opened using most modern phylogenetic programs.

Additional file 4: Distances between exon/intron boundaries within Region I. Contains a table with distances for all exon/intron boundaries within Region I depicted in Figure 1. The distances are given as residues on the MSA.

Additional file 5: Distances between exon/intron boundaries within Region II. Contains a table with distances for all exon/intron boundaries within Region II depicted in Figure 1. The distances are given as residues on the MSA.

Additional file 6: Distances between exon/intron boundaries within Region III. Contains a table with distances for all exon/intron boundaries within Region III depicted in Figure 1. The distances are given as residues on the MSA.

Additional file 7: Distances between all analyzed exon/intron boundaries. Contains a table with all distances between all analyzed exon/intron boundaries in the analyzed MSA. The distances are given as residues on the MSA.

Abbreviations

DSB: double-strand break; ML: maximum likelihood; MSA: multiple sequence alignment; ORF: open reading frame; RT: reverse transcriptase; TE: transposable element; TIR: terminal inverted repeat; TSD: target site duplication.

Competing interests

The authors declare that they have no competing interests.

Authors' contributions

JPB and TW conceived the study, JPB and AL performed the analyses, AHS contributed to the interpretation; JPB, AL, TW and AHS wrote the manuscript. All authors read and approved the final manuscript.

Authors' information

Co-senior authors: Thomas Wicker and Alan H Schulman.

Acknowledgements

JPB was supported by the SNF, Swiss National Science Foundation (Grant PBZHP3-143673). The authors acknowledge support from the Academy of Finland, Project 266430.
The authors thank the CSC - IT Center for Science Ltd., Finland, for use of their computer clusters.

Author details

[1]Institute of Biotechnology, Viikki Biocenter, University of Helsinki, PO Box 65, FIN-00014 Helsinki, Finland. [2]Institute of Plant Biology, University of Zurich, Zollikerstrasse 107, Zurich, Switzerland. [3]Biotechnology and Food Research, MTT Agrifood Research Finland, Myllytie 1, FIN-31600 Jokioinen, Finland. [4]Present address: Marie Bashir Institute for Infectious Diseases and Biosecurity, Charles Perkins Center, University of Sydney, Sydney NSW 2006, Australia.

References

1. Ueki N, Nishii I: **Idaten is a new cold-inducible transposon of** *Volvox carteri* **that can be used for tagging developmentally important genes.** *Genetics* 2008, **180:**1343–1353.
2. Pereira A, Cuypers H, Gierl A, Schwarz-Sommer Z, Saedler H: **Molecular analysis of the En/Spm transposable element system of** *Zea mays.* *EMBO J* 1986, **5:**835–841.
3. Inagaki Y, Hisatomi Y, Suzuki T, Kasahara K, Iida S: **Isolation of a Suppressor-mutator/Enhancer-like transposable element, Tpn1, from Japanese morning glory bearing variegated flowers.** *Plant Cell* 1994, **6:**375–383.
4. Snowden KC, Napoli CA: **Psl: a novel Spm-like transposable element from** *Petunia hybrida.* *Plant J* 1998, **14:**43–54.
5. Chopra S, Brendel V, Zhang J, Axtell JD, Peterson T: **Molecular characterization of a mutable pigmentation phenotype and isolation of the first active transposable element from** *Sorghum bicolor.* *Proc Natl Acad Sci USA* 1999, **96:**15330–15335.
6. Miura A, Yonebayashi S, Watanabe K, Toyama T, Shimada H, Kakutani T: **Mobilization of transposons by a mutation abolishing full DNA methylation in Arabidopsis.** *Nature* 2001, **411:**212–214.
7. Kapitonov VV, Jurka J: **Chapaev – a novel superfamily of DNA transposons.** *Repbase Reports* 2007, **7:**774–781.

8. Novick PA, Smith JD, Floumanhaft M, Ray DA, Boissinot S: The evolution and diversity of DNA transposons in the genome of the lizard *Anolis carolinensis*. *Genome Biol Evol* 2011, 3:1–14.

9. Langdon T, Jenkins G, Hasterok R, Jones RN, King IP: A high-copy-number CACTA family transposon in temperate grasses and cereals. *Genetics* 2003, 163:1097–1108.

10. Wicker T, Guyot R, Yahiaoui N, Keller B: CACTA transposons in Triticeae. A diverse family of high-copy repetitive elements. *Plant Physiol* 2003, 132:52–63.

11. Wicker T, Taudien S, Houben A, Keller B, Graner A, Platzer M, Stein N: A whole-genome snapshot of 454 sequences exposes the composition of the barley genome and provides evidence for parallel evolution of genome size in wheat and barley. *Plant J* 2009, 59:712–722.

12. Bennetzen JL: Transposable element contributions to plant gene and genome evolution. *Plant Mol Biol* 2000, 42:251–269.

13. Xu M, Brar HK, Grosic S, Palmer RG, Bhattacharyya MK: Excision of an active CACTA-like transposable element from DFR2 causes variegated flowers in soybean [*Glycine max* (L.) Merr.]. *Genetics* 2010, 184:53–63.

14. Zabala G, Vodkin L: Novel exon combinations generated by alternative splicing of gene fragments mobilized by a CACTA transposon in *Glycine max*. *BMC Plant Biol* 2007, 7:38.

15. Alix K, Joets J, Ryder CD, Moore J, Barker GC, Bailey JP, King GJ, Heslop-Harrison JSP: The CACTA transposon Bot1 played a major role in Brassica genome divergence and gene proliferation. *Plant J* 2008, 56:1030–1044.

16. Paterson AH, Bowers JE, Bruggmann R, Dubchak I, Grimwood J, Gundlach H, Haberer G, Hellsten U, Mitros T, Poliakov A, Schmutz J, Spannagl M, Tang H, Wang X, Wicker T, Bharti AK, Chapman J, Feltus FA, Gowik U, Grigoriev IV, Lyons E, Maher CA, Martis M, Narechania A, Otillar RP, Penning BW, Salamov AA, Wang Y, Zhang L, Carpita NC, *et al*: The Sorghum bicolor genome and the diversification of grasses. *Nature* 2009, 457:551–556.

17. Sabot F, Guyot R, Wicker T, Chantret N, Laubin B, Chalhoub B, Leroy P, Sourdille P, Bernard M: Updating of transposable element annotations from large wheat genomic sequences reveals diverse activities and gene associations. *Mol Genet Genomics* 2005, 274:119–130.

18. International Brachypodium Initiative: Genome sequencing and analysis of the model grass *Brachypodium distachyon*. *Nature* 2010, 463:763–768.

19. Nacken WK, Piotrowiak P, Saedler H, Sommer H: The transposable element Tam1 from *Antirrhinum majus* shows structural homology to the maize transposon En/Spm and has no sequence specificity of insertion. *Mol Gen Genet* 1991, 228:201–208.

20. Frey M, Reinecke J, Grant S, Saedler H, Gierl A: Excision of the En/Spm transposable element of *Zea mays* requires two element-encoded proteins. *EMBO J* 1990, 9:4037–4044.

21. Lewin B: *Genes VI*. New York: Oxford University Press, Inc; 1997.

22. Yuan Y-W, Wessler SR: The catalytic domain of all eukaryotic cut-and-paste transposase superfamilies. *Proc Natl Acad Sci USA* 2011, 108:7884–7889.

23. Greco R, Ouwerkerk PBF, Pereira A: Suppression of an atypically spliced rice CACTA transposon transcript in transgenic plants. *Genetics* 2005, 169:2383–2387.

24. Masson P, Strem M, Fedoroff N: The tnpA and tnpD gene products of the Spm element are required for transposition in tobacco. *Plant Cell* 1991, 3:73–85.

25. Rodríguez-Trelles F, Tarrío R, Ayala FJ: Origins and evolution of spliceosomal introns. *Annu Rev Genet* 2006, 40:47–76.

26. Churbanov A, Rogozin IB, Deogun JS, Ali H: Method of predicting splice sites based on signal interactions. *Biol Direct* 2006, 1:10.

27. Wicker T, Sabot F, Hua-Van A, Bennetzen JL, Capy P, Chalhoub B, Flavell A, Leroy P, Morgante M, Panaud O, Paux E, SanMiguel P, Schulman AH: A unified classification system for eukaryotic transposable elements. *Nat Rev Genet* 2007, 8:973–982.

28. Wicker T, Matthewsand DE, Beat K: TREP: a database for Triticeae repetitive elements. *Trends Plant Sci* 2002, 7:561–562.

29. Jurka J, Kapitonov VV, Pavlicek A, Klonowski P, Kohany O, Walichiewicz J: Repbase Update, a database of eukaryotic repetitive elements. *Cytogenet Genome Res* 2005, 110:462–467.

30. Talavera G, Castresana J: Improvement of phylogenies after removing divergent and ambiguously aligned blocks from protein sequence alignments. *Syst Biol* 2007, 56:564–577.

31. Penn O, Privman E, Landan G, Graur D, Pupko T: An alignment confidence score capturing robustness to guide tree uncertainty. *Mol Biol Evol* 2010, 27:1759–1767.

32. Stamatakis A: RAxML-VI-HPC: maximum likelihood-based phylogenetic analyses with thousands of taxa and mixed models. *Bioinformatics* 2006, 22:2688–2690.

33. Puruggganan M, Wessler S: The splicing of transposable elements and its role in intron evolution. *Genetica* 1992, 86:295–303.

34. Adams MD, Tarng RS, Rio DC: The alternative splicing factor PSI regulates P-element third intron splicing *in vivo*. *Genes Dev* 1997, 11:129–138.

35. Feschotte C, Wessler SR: Mariner-like transposases are widespread and diverse in flowering plants. *Proc Natl Acad Sci U S A* 2002, 99:280–285.

36. Schaack S, Gilbert C, Feschotte C: Promiscuous DNA: horizontal transfer of transposable elements and why it matters for eukaryotic evolution. *Trends Ecol Evol* 2010, 25:537–546.

37. Diao X, Freeling M, Lisch D: Horizontal transfer of a plant transposon. *PLoS Biol* 2006, 4:e5.

38. Stoltzfus A, Logsdon JM, Palmer JD, Doolittle WF: Intron 'sliding' and the diversity of intron positions. *Proc Natl Acad Sci U S A* 1997, 94:10739–10744.

39. Rogozin IB, Lyons-Weiler J, Koonin EV: Intron sliding in conserved gene families. *Trends Genet* 2000, 16:430–432.

40. Babenko VN, Rogozin IB, Mekhedov SL, Koonin EV: Prevalence of intron gain over intron loss in the evolution of paralogous gene families. *Nucleic Acids Res* 2004, 32:3724–3733.

41. Fedorov A, Merican AF, Gilbert W: Large-scale comparison of intron positions among animal, plant, and fungal genes. *Proc Natl Acad Sci USA* 2002, 99:16128–16133.

42. Venkatesh B, Ning Y, Brenner S: Late changes in spliceosomal introns define clades in vertebrate evolution. *Proc Natl Acad Sci USA* 1999, 96:10267–10271.

43. Wolfe KH, Gouy M, Yang YW, Sharp PM, Li WH: Date of the monocot-dicot divergence estimated from chloroplast DNA sequence data. *Proc Natl Acad Sci USA* 1989, 86:6201–6205.

44. Weiner AM, Deininger PL, Efstratiadis A: Nonviral retroposons: genes, pseudogenes, and transposable elements generated by the reverse flow of genetic information. *Annu Rev Biochem* 1986, 55:631–661.

45. Rogozin IB, Carmel L, Csuros M, Koonin EV: Origin and evolution of spliceosomal introns. *Biol Direct* 2012, 7:11.

46. Rogozin IB, Wolf YI, Sorokin AV, Mirkin BG, Koonin EV: Remarkable interkingdom conservation of intron positions and massive, lineage-specific intron loss and gain in eukaryotic evolution. *Curr Biol* 2003, 13:1512–1517.

47. Roy SW: The origin of recent introns: transposons? *Genome Biol* 2004, 5:251.

48. Rose AB: The effect of intron location on intron-mediated enhancement of gene expression in Arabidopsis. *Plant J* 2008, 40:744–751.

49. Mascarenhas D, Mettler IJ, Pierce DA, Lowe HW: Intron-mediated enhancement of heterologous gene expression in maize. *Plant Mol Biol* 1990, 15:913–920.

50. Snowden KC, Buchhholz WG, Hall TC: Intron position affects expression from the tpi promoter in rice. *Plant Mol Biol* 1996, 31:689–692.

51. Schauer SE, Philipp M, Baskar R, Gheyselinck J, Bolaños A, Curtis MD, Grossniklaus U: Intronic regulatory elements determine the divergent expression patterns of AGAMOUS-LIKE6 subfamily members in Arabidopsis. *Plant J* 2009, 59:987–1000.

52. Jeong Y-M, Mun J-H, Lee I, Woo JC, Hong CB, Kim S-G: Distinct roles of the first introns on the expression of Arabidopsis profilin gene family members. *Plant Physiol* 2006, 140:196–209.

53. Parma J, Christophe D, Pohl V, Vassart G: Structural organization of the 5′ region of the thyroglobulin gene. Evidence for intron loss and 'exonization' during evolution. *J Mol Biol* 1987, 196:769–779.

54. Catania F, Lynch M: Where do introns come from? *PLoS Biol* 2008, 6:e283.

55. Ner-Gaon H, Halachmi R, Savaldi-Goldstein S, Rubin E, Ophir R, Fluhr R: Intron retention is a major phenomenon in alternative splicing in Arabidopsis. *Plant J* 2004, 39:877–885.

56. Farlow A, Meduri E, Schlötterer C: DNA double-strand break repair and the evolution of intron density. *Trends Genet* 2011, 27:1–6.

57. Cho S, Jin S-W, Cohen A, Ellis RE: A phylogeny of caenorhabditis reveals frequent loss of introns during nematode evolution. *Genome Res* 2004, 14:1207–1220.

58. Cohen NE, Shen R, Carmel L: The role of reverse transcriptase in intron gain and loss mechanisms. *Mol Biol Evol* 2012, 29:179–186.

59. Mourier T, Jeffares DC: Eukaryotic intron loss. *Science* 2003, 300:1393.

60. Derr LK, Strathern JN: A role for reverse transcripts in gene conversion. *Nature* 1993, 361:170–173.

61. Fink GR: **Pseudogenes in yeast?** *Cell* 1987, **49**:5–6.
62. Wada H, Kobayashi M, Sato R, Satoh N, Miyasaka H, Shirayama Y: **Dynamic insertion-deletion of introns in deuterostome EF-1alpha genes.** *J Mol Evol* 2002, **54**:118–128.
63. Frugoli JA, McPeek MA, Thomas TL, McClung CR: **Intron loss and gain during evolution of the catalase gene family in angiosperms.** *Genetics* 1998, **149**:355–365.
64. Robertson HM: **Two large families of chemoreceptor genes in the nematodes *Caenorhabditis elegans* and *Caenorhabditis briggsae* reveal extensive gene duplication, diversification, movement, and intron loss.** *Genome Res* 1998, **8**:449–463.
65. Bennetzen JL: **The contributions of retroelements to plant genome organization, function and evolution.** *Trends Microbiol* 1996, **4**:347–353.
66. Volokita M, Rosilio-Brami T, Rivkin N, Zik M: **Combining comparative sequence and genomic data to ascertain phylogenetic relationships and explore the evolution of the large GDSL-lipase family in land plants.** *Mol Biol Evol* 2011, **28**:551–565.
67. Saha J, Chatterjee C, Sengupta A, Gupta K, Gupta B: **Genome-wide analysis and evolutionary study of sucrose non-fermenting 1-related protein kinase 2 (*SnRK2*) gene family members in Arabidopsis and Oryza.** *Comput Biol Chem* 2014, **49**:59–70.
68. Feiber AL, Rangarajan J, Vaughn JC: **The evolution of single-copy *Drosophila* nuclear 4f-rnp genes: spliceosomal intron losses create polymorphic alleles.** *J Mol Evol* 2002, **55**:401–413.
69. Schulman AH, Wicker T: **A field guide to transposable elements.** In *Plant Transposons and Genome Dynamics in Evolution*. Oxford, UK: Wiley-Blackwell; 2013:15–40.
70. Sonnhammer EL, Durbin R: **A dot-matrix program with dynamic threshold control suited for genomic DNA and protein sequence analysis.** *Gene* 1995, **167**:GC1–G10.
71. Penn O, Privman E, Ashkenazy H, Landan G, Graur D, Pupko T: **GUIDANCE: a web server for assessing alignment confidence scores.** *Nucleic Acids Res* 2010, **38**:W23–W28.
72. Stover BC, Muller KF: **TreeGraph 2: combining and visualizing evidence from different phylogenetic analyses.** *BMC Bioinformatics* 2010, **11**:7.

17

*Tc*1-like transposable elements in plant genomes

Yuan Liu[1,2] and Guojun Yang[1,2*]

Abstract

Background: The *Tc1/mariner* superfamily of transposable elements (TEs) is widespread in animal genomes. *Mariner*-like elements, which bear a DDD triad catalytic motif, have been identified in a wide range of flowering plant species. However, as the founding member of the superfamily, *Tc*1-like elements that bear a DD34E triad catalytic motif are only known to unikonts (animals, fungi, and Entamoeba).

Results: Here we report the identification of *Tc*1-like elements (TLEs) in plant genomes. These elements bear the four terminal nucleotides and the characteristic DD34E triad motif of *Tc*1 element. The two TLE families (*PpTc*1, *PpTc*2) identified in the moss (*Physcomitrella patens*) genome contain highly similar copies. Multiple copies of *PpTc*1 are actively transcribed and the transcripts encode intact full length transposase coding sequences. TLEs are also found in angiosperm genome sequence databases of rice (*Oryza sativa*), dwarf birch (*Betula nana*), cabbage (*Brassica rapa*), hemp (*Cannabis sativa*), barley (*Hordium valgare*), lettuce (*Lactuta sativa*), poplar (*Populus trichocarpa*), pear (*Pyrus x bretschneideri*), and wheat (*Triticum urartu*).

Conclusions: This study extends the occurrence of TLEs to the plant phylum. The elements in the moss genome have amplified recently and may still be capable of transposition. The TLEs are also present in angiosperm genomes, but apparently much less abundant than in moss.

Keywords: Transposable elements, Moss, *Tc1-mariner-IS630* superfamily, *Tc*1-like elements, *Mariner*-like elements, Plant genome, Evolution, Transposition activity

Background

Transposable elements (TEs) are a major component of most eukaryotic genomes. Their transposition in genomes may lead to increase in their copy numbers. TEs are classified into two categories (Class I and Class II) based on their mechanism for transposition. Class II elements are DNA transposons that adopt a 'cut-and-paste' approach catalyzed by enzymes called transposases. The elements of this class are further divided into superfamilies based on different types of transposases. All of the transposases of these elements bear a DDE/D triad motif, however, different superfamilies have distinct transposases and structural features such as the length of the duplicated target site sequences [1,2]. Despite the growing number of reported active TEs, the majority of transposable elements are not active [3,4]. These elements are important for the dynamic structure of genome during evolution [5,6]. The immobilized TEs can serve as raw genetic materials for genome tinkering [7-15]. Autonomous TEs encode and produce transposases for their mobilization. Non-autonomous elements have lost their ability to encode functional transposases and rely on other sources of transposases for transposition. An ultimate group of non-autonomous elements is miniature inverted-repeat transposable elements (MITEs). They are short elements and have high copy numbers [16-18].

Tc1-mariner-IS630 is a Class II TE superfamily first identified in nematode and insect genomes [19]. The superfamily was named after *Tc1* in *Caenorhabditis elegans* [20], and *mariner* in *Drosophila mauritiana* [21]. This superfamily is characterized by two terminal inverted repeats (TIRs) of typically 12 to 28 nt flanked by dinucleotide target site duplications (TSDs) of 'TA'. The transposases of this superfamily contain a triad catalytic motif consisted of two aspartic acid (D) residues and a glutamate residue (E) in *Tc*1-like elements (TLEs) or aspartic acid (DDD) in *Mariner*-like elements (MLEs) and *pogo*-like elements [22,23]. The pocket formed by these residues contains the metal ions needed in the DNA cleavage reaction during transposition [24]. Based on the number of residues

* Correspondence: gage.yang@utoronto.ca
[1]Department of Biology, University of Toronto at Mississauga, 3359 Mississauga Road, L5L 1C6 Mississauga, ON, Canada
[2]Cell and Systems Biology, University of Toronto, Toronto, Canada

between the second and third catalytic residues of the DDE/D motif, *Tc1/mariner* catalytic domains can be DD34E, DD34D, DD31-33D, DD35E, DD37D, DD37E, or DD39D, each defining a subgroup of the *Tc1/mariner* superfamily [18,22,25-27]. *Tc1/mariner* elements have been considered to be confined to animals until the recent identification of DD39D *mariner*-like elements and *pogo*-like elements in plants [18,22,23]. *Tc1*-like elements are the founding subgroup of the *Tc1/mariner* superfamily and they bear the DD34E triad catalytic motif [20]. Previous studies have identified TLEs in a variety of animals and fungi [23] as well as in the parasitic amoebozoa Entamoeba invadens [28]. However, to the best of our knowledge, there has been no report of TLEs outside the unikonts (animal, fungi, and amboebozoa) [29]. Previous studies have identified TLEs in a number of animal or fungal genomes, some have been demonstrated to be active, including *Tc1* and *Tc3* in *C. elegans* [20,30,31], *Minos* in *Drosophila hydei* [32], and *Impala* in fungus *Fusarium oxysporum* [33,34]. The reconstructed fish element *Sleeping Beauty* is also a TLE [35]. *Tc1*-like elements named *Hydargo* have been identified in Entamoeba parasites [28].

Here we report the identification of TLEs in plants. The two families of full-length TLEs in the moss (*Physcomitrella patens*) genome have multiple copies that contain an intact open reading frame (ORF). These ORFs are actively transcribed and presumably also translated into functional transposases in moss. TLEs were also found in the genome sequence databases of angiosperm plants.

Results

Tc1-like elements in moss

Mariner-like elements are widespread in plant genomes [18,36]. To investigate whether plant genomes contain TLEs, moss genome sequence databases were screened because mosses are among the first terrestrial plants. When the sequence of *Tc1* transposase was used as the query sequence for BLAST search against the moss (*Physcomitrella patens*) genome database that has a coverage of approximately 8.6X [37], 118 high scoring hits (e-value: <e-8) were obtained. Close inspection of the output revealed two groups of elements that have complete terminal inverted repeats (TIRs) with terminal 5′-CAGT … ACTG-3′ sequences flanked by TSDs of dinucleotide 'TA'. Both groups of elements contain open reading frames for transposases bearing a DD34E motif. These characteristics suggest that these two groups are TLEs and were designated as *PpTc1* and *PpTc2*. Neither of the two families has been previously described or annotated [37]. No similar elements or their transposase sequences were found in the genome of the spike moss *Selaginella moellendorffii*.

The full-length *PpTc1* elements are 1,584 bp long with TIRs of 33 bp. It has an ORF of 338 aa with two helix-turn-helix domains and a catalytic DD34E domain (Figure 1). A total of 85 copies were retrieved from the *P. patens* genome sequence database. Among them, 75 were full length bearing the intact ends with average sequence identity of 96.3%, and 52 of which were highly similar copies with >98% sequence identity, but there were no identical copies. Nine copies were found to carry an intact full-length ORF (338 aa). To gain insights into the insertion sites of *PpTc1* elements, it is important to inspect the sequences homologous to the flanking sequences of *PpTc1* insertion sites. Such sequences that do not bear the TE insertions are called related empty sites (RESs). The sequence signatures of the TE insertion sites on RESs may reflect historical transposition events. Among the 75 full length copies, RESs can be found for the flanking sequences of 42 copies with 14 of them in AT rich simple repeat flanking sequences (Additional file 1: Figure S1). Most of the 28 RESs that are not AT-rich simple repeats correspond to the sequences before insertion of elements, some (for example, that of scaffold 54) may have resulted from excision events and subsequent repairing.

The full-length *PpTc2* elements are 1,709 bp long, with TIRs of 33 bp (Figure 1). A total of 22 copies of *PpTc2* were retrieved from the genome database. The 20 full-length copies have an average sequence identity of 96.6%. *PpTc2* has eight copies bearing a full-length intact 338aa ORF. Among the 20 full-length copies, RESs can be found for the flanking sequences of three copies (Additional file 1: Figure S1). While the RES of scaffold 10 clearly represents a site before insertion of an element, that of scaffold 136 may have resulted from excision events and subsequent repairing of the excision sites. Interestingly, insertion of the *PpTc2* in scaffold 281 is accompanied by a duplication of a microsatellite unit at the insertion site. These RESs of *PpTc* insertions sites demonstrated the genomic changes caused by the activity of these elements during evolution.

Comparison of PpTc1 and PpTc2

The history of activities of these elements in the genome is an important part of the evolution of these elements. According to the molecular clock theory, the mutations accumulated in each copy of an element in a TE family can be used to infer the time of divergence from their ancestral element [38]. The sequences of the ancestral element of a TE family may be approximated to the consensus sequences of the TE family. Therefore, the elements produced at the same time frame can be expected to have similar levels of sequence divergence from the ancestral element. Based on the consensus sequences of *PpTc1* and *PpTc2*, the average sequence divergence score was calculated for each copy and the number of elements in a certain range of sequence divergence value

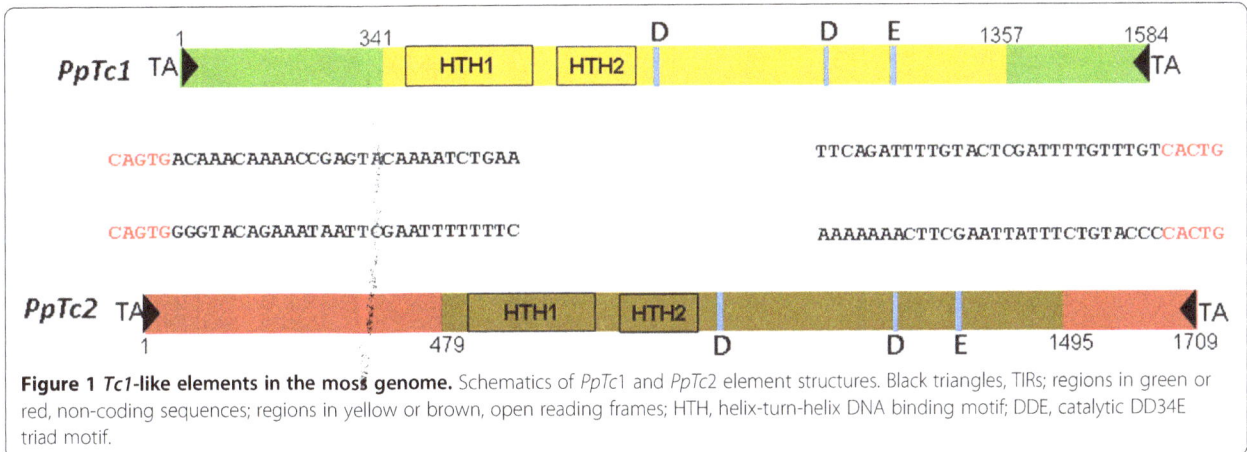

Figure 1 *Tc1*-like elements in the moss genome. Schematics of *PpTc*1 and *PpTc*2 element structures. Black triangles, TIRs; regions in green or red, non-coding sequences; regions in yellow or brown, open reading frames; HTH, helix-turn-helix DNA binding motif; DDE, catalytic DD34E triad motif.

was plotted against the sequence divergence range. The *PpTc*1 family has an average divergence value of 2.18 ± 0.08% with a significant peak at 1.5% sequence divergence (Figure 2), suggesting a recent burst of amplification events of this family occurred about 1.5 million years ago and the rate of amplification has since decreased according to a rate of 1% sequence divergence per million years. The *PpTc*2 family have an average sequence divergence value of 2.17 ± 0.20% with the most recent peak at about 1%, suggesting that *PpTc*2, similar to *PpTc*1, recently amplified about 1 million years ago. Interestingly, the *PpTc*2 dynamics is similar to the cycles of TE amplification described previously [39].

Although *PpTc*1 and *PpTc*2 bear identical extreme terminal sequences 'CAGT' (Figure 1), their internal regions do not bear detectable DNA sequence similarities. Even the transposase coding sequences do not share significant sequence similarities between the two elements. When the putative peptide sequences of the two transposases were aligned, they share 26% (89/338) sequence identify with 47% positive (161/338) (Figure 3A). These results suggest that the two elements shared a very distant common ancestor. However, the very similar intra-family sequence divergence levels of the two families suggest that they may have invaded and amplified in the moss genome at a similar time during evolution.

Since the crystal structures of *Mos*1 and the DNA binding domain of *Tc*3 were determined, the transposase structures of *PpTc*1 and *PpTc*2 can be predicted based on these templates [24,40]. Using Phyre2 web server, the transposase structure of *Mos*1 was used by the algorithm to model the transposases. The homologous models have 100% confidence with about 95% coverage of the query sequences, suggesting highly similar protein structures between these two proteins and to the *Mos*1 transposase (Figure 3B). Based on the structural features of *Mos*1, similar features were predicted on the models of *PpTc*1 and *PpTc*2 transposases. These models provide important starting information to understand the functionality of these transposases and their structural and functional deviations from other transposases in the *Tc1/mariner* superfamily.

Expression of PpTc1 in moss

The high intra-family sequence similarity in *PpTc*1 and *PpTc*2 and the presence of multiple copies of elements that contain intact transposase coding sequences indicate that they are potentially active. Expression of transposase is required for transposition activity, therefore it is important to determine whether *PpTc*1 and *PpTc*2 are

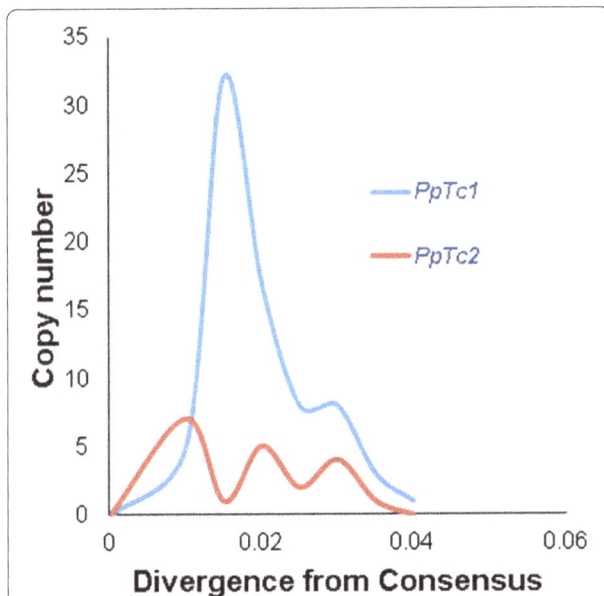

Figure 2 Sequence divergence of full-length elements of *PpTc*1 and *PpTc*2. Y-axis, number of elements; x-axis, level of sequence divergence from the consensus sequence of *PpTc*1 or *PpTc*2 family.

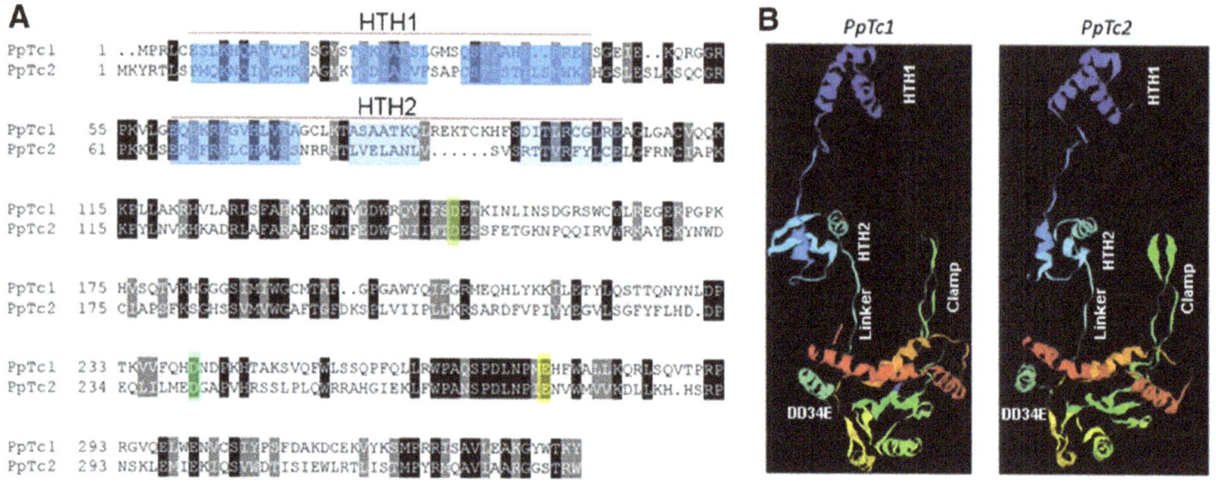

Figure 3 Comparison of the putative transposases of *PpTc*1 and *PpTc*2. **(A)** Alignment of peptide sequences. Colored residues: blue to cyan, α-helices of HTH motifs; green to yellow, DD34E triad motif. **(B)** Predicted three-dimensional ribbon models of transposases. Blue to red, N terminus to C terminus; HTH1 and HTH2, putative DNA binding (both) and dimerization (HTH1 only); clamp, loop structure potentially interacts with the linker of the other monomer in a transposase dimer; linker, potentially interacts with the clamp loop of the other monomer in a dimer; DD34E, catalytic active center.

actively transcribed. Extensive sequencing of the moss transcriptome has been previously performed and reported [41]. The expressed sequence tags derived from protonemal tissue and gametophores have been analyzed extensively and resulted in an assembled transcript database Pp0409 that contains 47,557 entries (www.cosmoss.org). Expressed sequence tag coverage of the genome assembly is 98% [37]. *PpTc* elements and the CDS of moss *actin*1 gene (*PpAct1*) were used to retrieve assembled transcripts from the database. Compared to the 17 transcripts from *PpAct*1, 68 assembled transcripts containing the nucleotide sequences of the ORF region were retrieved for *PpTc*1 and no transcript for *PpTc*2, suggesting that the level of transcripts of *PpTc*1 in moss cells is higher than the constitutive gene *actin*1. Each of these transcripts corresponds to a specific copy of *PpTc*1 element. Nine of the *PpTc*1 transcripts can be conceptually translated into a full-length intact transposase (Figure 4, Additional file 1: Table S1). Each of these transcripts bearing intact ORFs is derived from a specific copy of the nine genomic copies of *PpTc*1 bearing intact transposase coding sequences, suggesting that these elements are actively transcribed and yielded mature mRNA. The fact that no identical copies of *PpTc*1 were present in the genomic sequence database suggests an attenuated transposition activity after the peak amplification of the family around 1.5 million years ago. Since TE transcripts can be degraded by siRNA and their translation may be blocked by microRNAs, the *PpTc*1 transcripts were used to search against the small RNA databases [42-45]. However, no small RNA matching the coding sequences of *PpTc*1 transposase gene were retrieved, suggesting that the *PpTc*1 mRNAs are not

degraded or their translation blocked, therefore may be translated into transposase proteins. Because of the abundance of the transcripts of the transposase gene, it is possible that a post-translational mechanism such as over production inhibition demonstrated for animal *Tc1/mariner* elements may have led to the repression of its transposition [46,47]. When *PpTc*2 sequences were used to search against the assembled transcript database, no transcripts were retrieved. This suggests that the expression of the transposase genes of this family is probably repressed at the transcriptional levels.

Evolutionary relationship of transposases encoded by moss TLEs to those of animal and fungal TLEs

Since TLEs have been previously described only in animal and fungal genomes, the relationship of the moss TLEs to other TLEs will help to understand the propagation of TLEs in plant genomes. Even though there are only a few well characterized TLEs in literature, recent progress in whole genome sequencing produced TLE sequences in many different genomes. Using well characterized TLE transposase sequences including *Tc*1 (X01005), *Tc*3 (P34257.1), *Minos* (CAP09075.1), and *Impala* (AF282722), together with *PpTc*1 and *PpTc*2, we retrieved representative TLE sequences in different genomes from the nonredundant protein database of Genbank. The majority of these sequences were not classified therefore named as hypothetical proteins or unknown proteins. Notably, the TLE element in *Rhizopus delemar* was found to have at least 60 copies. After removal of redundancy of sequences belonging to the same family, together with *PpTc* elements, the sequences were aligned with the previously

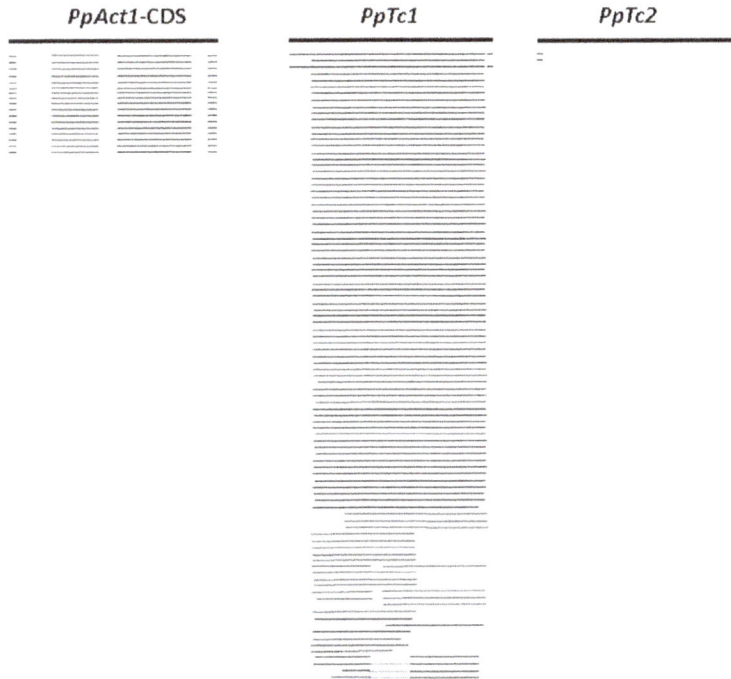

Figure 4 Transcripts from *PpTc* elements. Thick lines on top, query sequences; solid thin lines, matched regions between the queries and hits in the transcript database; dotted lines, unmatched regions reflecting intronic regions; the coding DNA sequence (CDS) of moss *actin*1 gene was used as a control.

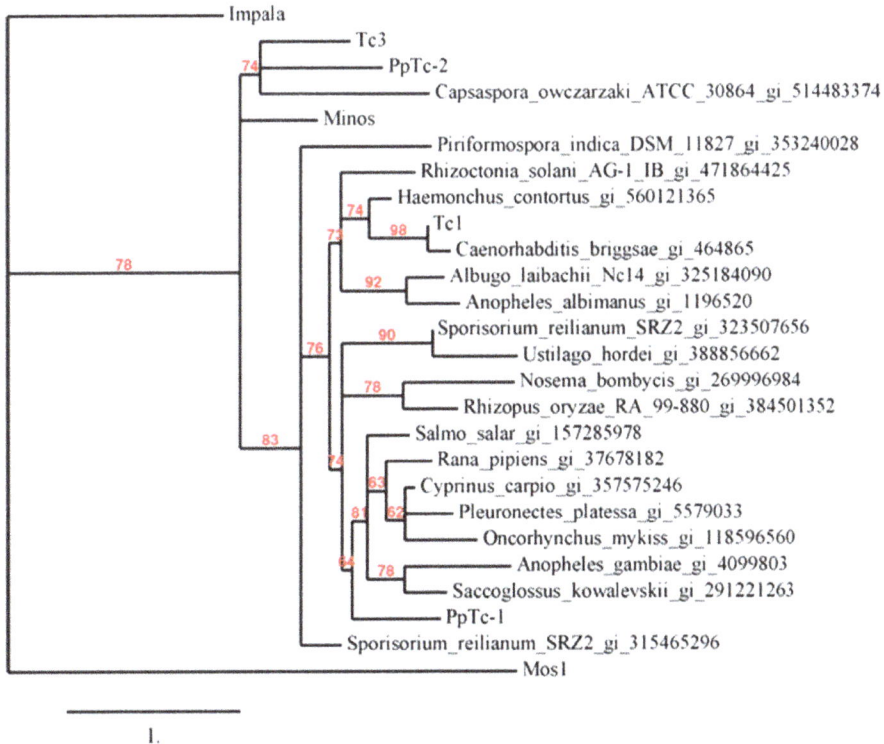

Figure 5 Phylogenetic relationship of transpoases of moss TLEs to those of animal and fungal TLEs. Names, species followed by GI numbers of each sequence; numbers on branches, percentage of bootstrap value of 1,000 reiterations.

described TLEs and a phylogenetic tree was constructed (Figure 5). Similar to that reported previously, the branches on the phylogenetic tree of these elements have relatively low bootstrap values (98% to 62%) [48]. Nevertheless, the topology of the previously analyzed elements such as *Tc1*, *Tc3*, *Impala*, and *Minos* is consistent with that shown in the previous report. *Impala* appeared to have branched off early from the rest of the TLEs. The rest of elements are grouped into two clades: *Tc1* clade and *Tc3* clade. The majority of these elements belong to the *Tc1* clade. The fact that the phylogenetic relationship among these elements is clearly incongruent with that of their host species may suggest ancestral polymorphism or long branch attraction [49], alternatively horizontal transfer of these elements among eukaryotic species may have also contributed to the observation [50,51]. The two moss elements belong to different clades with *PpTc1* in the *Tc1* and *PpTc2* in the *Tc3* clade, further suggesting that these two elements may have different origins.

TLEs in angiosperm genome sequence databases
To determine whether TLEs have proliferated throughout plant genomes, the predicted transposase sequences of *PpTc1* and *PpTc2* were used as query sequences to search against all other plant genomic sequences in the GenBank WGS and NR/NT databases using TBLASTN. Segments of *Tc1*-like transposase coding sequences were identified in nine angiosperm genomes including rice (*Oryza sativa*), dwarf birch (*Betula nana*), cabbage (*Brassica rapa*), hemp (*Cannabis sativa*), barley (*Hordium valgare*), lettuce (*Lactuta sativa*), poplar (*Populus Trichocarpa*), pear (*Pyrus x bretschneideri*), and wheat (*Triticum urartu*) (Table 1). The conserved regions including at least the second (aspartic acid) and the third (glutamic acid) residues of the DD34E catalytic motif were retrieved. Most of these elements are single copies and they are not uniform in size. While TLE in the database of *Oryza sativa* is a complete element with intact terminal sequences, the majority of the plant TLEs are fragmented and do not encode a complete transposase. When the regions between the second D and the E residues of the DD34E motifs were aligned, conserved motifs surrounding these two residues were revealed (Figure 6A and Additional file 1: Figure S2). The conserved motifs surrounding the E residues of these TLEs are apparently different from those surrounding the corresponding D residue of the MLEs such as *Mos1* (X78906), *Soymar1* (AF078934.1), and *Osmar5* (ACV32571.1). Among the sequenced plant genomes, the distribution of the species containing TLEs is apparently patchy (Figure 7). These results suggest that TLEs are also

Table 1 Plant Tc1-like transposases described in this study

Element	Organism	Accession	ORF start	ORF end	Complete DD34E triad?
Plant					Y: yes; N, no
PpTc1	*Physcomitrella patens*	ABEU01007491	7,186	8,199	Y
PpTc2	*Physcomitrella patens*	ABEU01006878	162,826	161,813	Y
OsTc1	*Oryza sativa Indica*	AAAA02041396	3,821	2,697	Y
BnTc1	*Betula nana*	CAOK01056615	1,484	1,978	N
BnTc2	*Betula nana*	CAOK01550459	168	1,214	Y
BnTc3	*Betula nana*	CAOK01014729	14,272	14,472	N
BnTc4	*Betula nana*	CAOK01486111	2	244	N
BrTc1	*Brassica rapa*	AENI01020305	162	572	N
BrTc2	*Brassica rapa*	AENI01036930	17	328	N
CsTc1	*Cannabis sativa*	AGQN01308320	302	517	N
HvTc1	*Hordium valgare*	CAJV010227559	1	1,684	Y
HvTc2	*Hordium valgare*	CAJV010272453	49	555	Y
HvTc3	*Hordium valgare*	CAJV012609061	1,716	2,114	N
HvTc4	*Hordium valgare*	CAJV011622646	1	222	N
LsTc1	*Lactuta sativa*	AFSA01593962	2	394	N
LsTc2	*Lactuta sativa*	AFSA01593962	87	485	N
PtTc1	*Populus trichocarpa*	AARH01030986	1	714	Y
PxbTc1	*Pyrus x bretschneideri*	AJSU01007483	3,055	3,606	Y
PxbTc2	*Pyrus x bretschneideri*	AJSU01007483	3,055	3,606	N
TuTc1	*Triticum urartu*	AOTI010070343	376	1,368	Y

All TLE elements bear the D34E of the DD34E catalytic motif.

A

```
Lemi1     RVLFVVDNGPAHPKIIEGL.........QNVELFFLPPNMTSKIQFCDAGII
Soymar1   TIFIQQDNARTHINPDDPEFVQAATQDGFDIRLMCQPPN.SPDFNVLDLGFF
Osmar5    TIWIQQDNARTHLPIDDAQFGVAVAQSGLDIRLVNQPPN.SPDMNCLDLGFF
Mos1      RVIFLHDNAPSHTARAVRDTLET.....LNWEVLPHAAY.SPDIAFSDYHLF
HvTc2     KIHIILDNSGYHCSQRVKDAALE.....KAIVLHYLPPY.SPNLNPIE.RLW
LsTc2     KIHIILDNSGYHCSQRVKDAALE.....KAIIHYLPPY.SPNLNPIE.RLW
Impala    GDIFMHDNASVHTARIVKALLEE.....LGVDLMTWPPY.SPDLNPIE.NLW
CsTc1     NSVVVMDNAAFHKRADIQELLEQ.....QGHKILWLPAY.SPDLNPIE.HMW
HvTc3     NSVMVMDNASFHKRKDIQDAIKD.....AGFILEYLPVY.SPDLNPIE.KKW
HvTc1     GAVIVMDNVSFHKRQDTQAAIQK.....AGFILEYLPTY.SPDMNPIE.HKW
LsTc1     GAVIVRDNVSFHKRQDIQAAIQK.....AGFILEYLPTY.SPDMNPIE.HKW
BrTc2     .LSVMQDNAPAHACENTMEEVRE.....RSIIPIDWPPN.SPDLNPIE.AVW
BnTc1     NLYFQQDNAPIHTSAKSARFVKQ.....NGLKMLLWPAN.SPDLNPIE.HIW
PxbTc2    NAVCMHDNARAHTAQVVDEYLHD.....VGIHKMEWPAR.SPDLNPIE.HAW
BnTc2     IDVIVLEDNAPCHSSKATCAARQN.....LGITSLKHPSN.SPDLNAIE.NLW
BnTc3     DALVVEDGASCHWSKQTNKGREK.....LHIVNLNHPPQ.SPNLNPVP.NVC
BnTc4     DILVVEDGAPCHTCKLAKEARSK.....LGIPSLIHPPS.SPDLNPIE.NVW
PpTc2     QLILMEDGAPVHRSSLPLQWRRA.....HGIEKLFWPAN.SPDLNPIE.NVW
HvTc4     ....QHDLAPAHSAKTTGKWFTD.....HGITVLNWPAN.SPDLNPVE.NLW
OsTc1     LFQLMHDNARPHTARVVRQTLAA.....ANINVLPWPAQ.SPDLNPIE.HAW
BrTc1     DFVLMHDNARCHTARVSRQFLRE.....KELRTMDWPAL.SPDLNPIE.HLW
PxbTc1    DFVLMHDNARCHTARVSRQFLRE.....KELRTMDWPAL.SPDLNPIE.HLW
PpTc1     KVVFQHDNDPKHTAKSVQFWLSS.....QPFQLLRWPAQ.SPDLNPME.HFW
PtTc1     GWMFQHDNDPKHTAKATKEWLKK.....KHIKVMEWPSQ.SPDLNPIE.TY.
Tc1       GFVFQQDNDPKHTSLHVRSWFQR.....RHVHLLDWPSQ.SPDLNPIE.HLW
TuTc1     NWIFQQDNDPKHSSILVRNWLAE.....NGVAVMQWPSQ.SPDLNPIE.HLW
```

B

```
Tc1       TACAGTGCTGGCCAAAAAGATATCCACTTT
PpTc1     TACAGTGACAAACAAAACCGAGTACAAAAT
PpTc2     TACAGTGGGGTACAGAAATAATTCGAATTT
OsTc1     TACACACATCAAAAGTGTCTAGGGATCAAC
HvTc1*    TACAGTTGATTTTTGGAGATAAGCACAACG
HvTc4     TATGGCATAAGCGGTATAAATTGATTTAAA
PxbTc-1   TACACTCGCGAGCAACCAAATCGACTCAAG
PxbTc2*   TACAGTAATAAAAATTTAAATTAAATGTTT
PtTc1*    TACAGTGGAGGAAATAATTATTTGACCCCT
TuTc-1    TACAGTGGTGGCCAAAAATTTAAGAACGAC
```

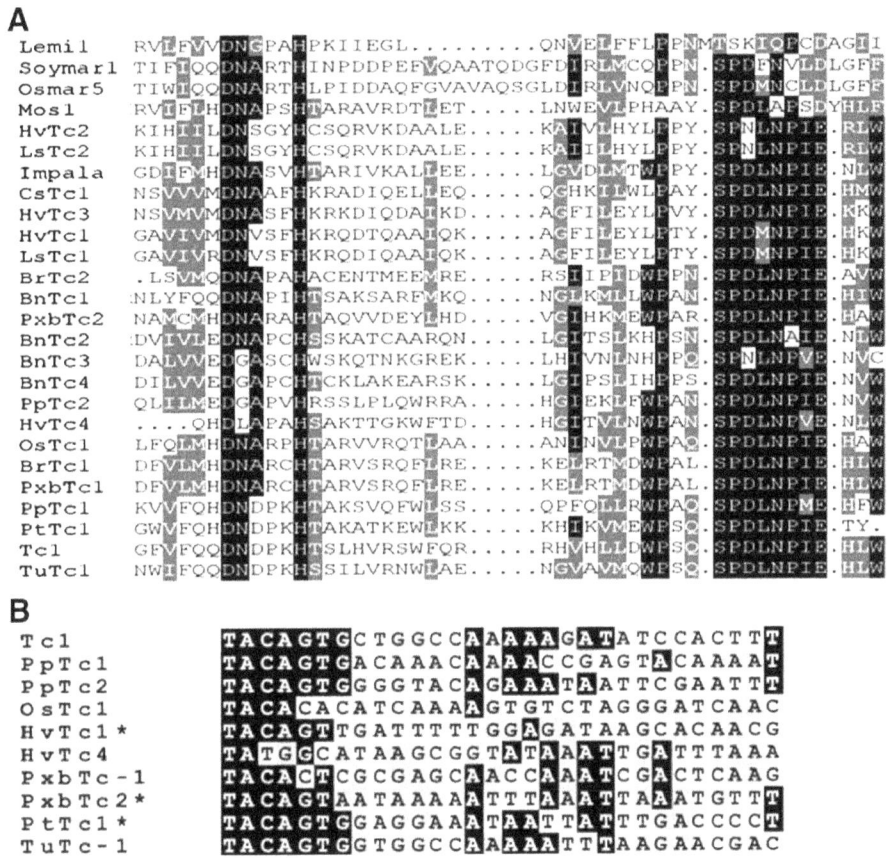

Figure 6 Sequence alignment of the catalytic motifs of transposases (A) and end sequences (B) of plant TLEs. (A) The regions containing the DDE/D catalytic motifs of the transposase sequences. Plant MLEs are shown at the bottom. (B) The terminal sequences of plant TLEs and *Tc1*. The degree of background shading indicates different levels of conservation of sequences. Asterisks indicate elements that only have one end present in a genomic contig. Abbreviation for species names: Os, *Oryza sativa*; Bn, *Betula nana*; Br, *Brassica rapa*); Cs, *Cannabis sativa*; Hv, *Hordium valgare*; Ls, *Lactuta sativa*; Pt, *Populus trichocarpa*; Pb, *Pyrus x bretschneideri*; Tu, *Triticum urartu*.

present in angiosperm genomes, but are much less abundant than in the moss genome.

Discussion

The identification of TLEs in plant genomes expanded our knowledge on the distribution and diversity of *Tc1/mariner* elements. Elements belonging to the *mariner*-subgroup have been found to be widespread in plant genomes [18]. TLEs, however, have not been previously reported in plants. In fact, *PpTc1* and *PpTc2* are the first *Tc1/mariner* elements described in moss. They not only expand the range of distribution of TLEs into plants, but also provide information for the development of TE-based tools for gene discovery in moss in the future.

PpTc elements have undergone a recent wave of proliferation. The results suggest that their transposition activities appear to have subsequently been contained in the current moss genome. Although most copies of *PpTc* elements have lost the capacity to encode a functional transposase due to mutations that interrupt the transposase

coding sequences, both families have members bearing full length intact transposase-coding genes and *PpTc1* elements are actively expressed in moss. These observations indicate that, even though the transposition activity of *PpTc1* may have been attenuated, it may still be modestly active. In addition, since the genome was sequenced with shot gun approaches, the reads for these repetitive sequences may have been misassembled. Therefore, it is possible that identical *PpTc1* sequences are present in the genome. The absence of transcripts from *PpTc2* may indicate a high level of repression of transposition. It remains mysterious how these elements are repressed. It is possible that, under certain environmental factors, these elements may become fully active in transposition. Alternatively, the activities of these elements may be restricted to certain tissues/organs or specific temporal stages during the life cycle of the plant. Further investigation on the repression of the transposition activities of both families will facilitate our understanding of the interaction between TEs and their host genomes.

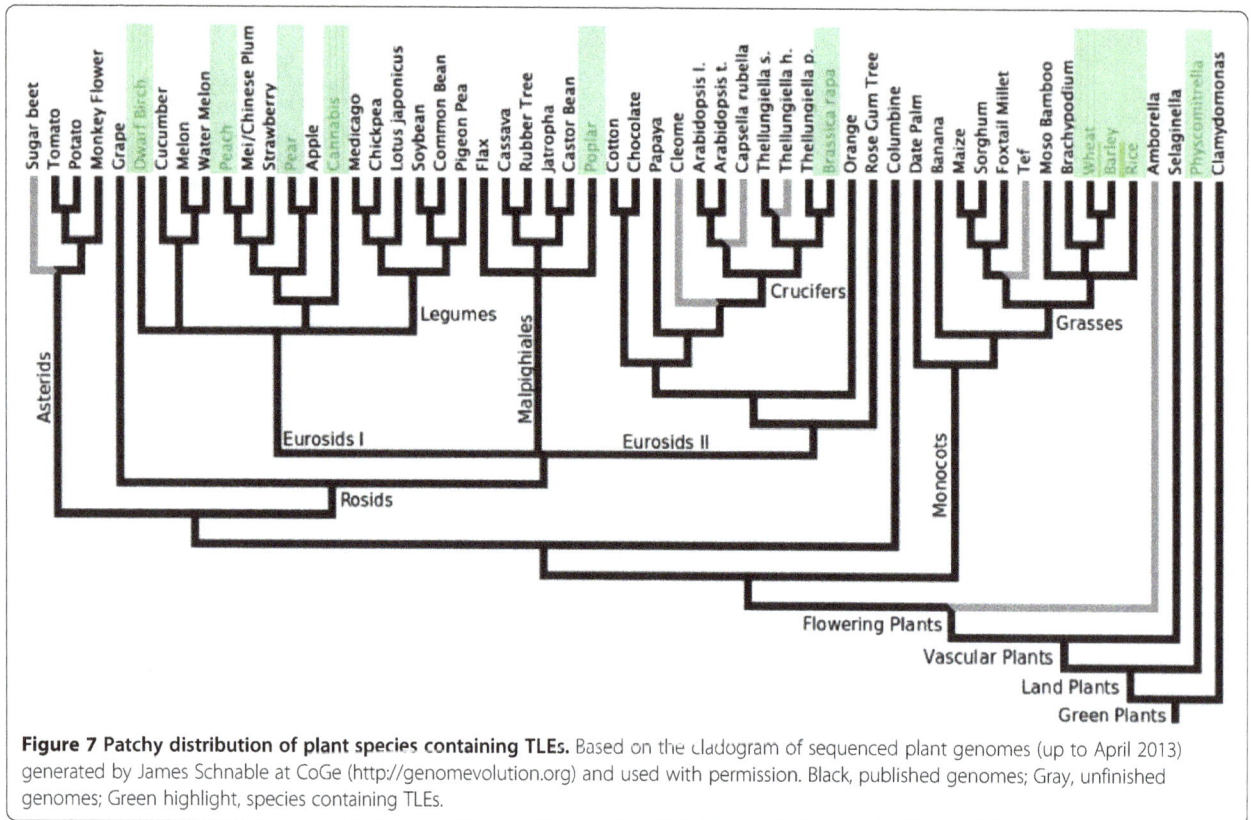

Figure 7 Patchy distribution of plant species containing TLEs. Based on the cladogram of sequenced plant genomes (up to April 2013) generated by James Schnable at CoGe (http://genomevolution.org) and used with permission. Black, published genomes; Gray, unfinished genomes; Green highlight, species containing TLEs.

Conclusions

TLEs are present in plant genomes. The two families of TLEs in the moss genome have recently amplified 1 to 2 million years ago. These families contain elements that are potentially capable of transposition but their transposition activities appear to have been attenuated. TLEs were also identified in the genome databases of angiosperm plants, suggesting their distribution in multiple plant orders. The results presented in this report further our understanding of the evolutionary history of *Tc1/mariner* elements and provide important information for future investigations into the interaction between TEs and host genomes.

Methods
Retrieval of moss *Tc1*-like elements

To identify transposons related to *Tc1*-like elements, the *Tc1* transposase peptide sequence was used as the query sequence to search against GenBank databases of *P. patens* genome with the default parameters. Each returned hit was retrieved and inspected for TIRs. Complete elements were searched against its host genome to obtain the members of its family. Nucleotide sequences of full-length TLE copies were retrieved with MITE Analysis Kit function MEMBER (http://labs.csb.utoronto.ca/yang/MAK/) [52,53]. Members of each family were retrieved with MAK with zero tolerance for end mismatches.

Characterization of moss TLEs

Alignments of all retrieved members in each *PpTc* family were obtained with CLUSTAL, and a consensus sequence was generated. The elements were conceptually translated and scanned for long ORFs with the APE program (http://biologylabs.utah.edu/jorgensen/wayned/ape/). HTH motifs were predicted with NPS webserver (http://npsa-pbil.ibcp.fr/cgi-bin/npsa_automat.pl?page=/NPSA/npsa_hth.html) and the conserved domain database at NCBI. The putative models of *PpTc1* and *PpTc2* were predicted with Phyre2 (http://www.sbg.bio.ic.ac.uk/phyre2/). Sequence alignments were performed with MUSCLE at the EBI webserver (http://www.ebi.ac.uk/Tools/msa/muscle/) and the phylogenetic tree was constructed with Phylogy.fr (www.phylogeny.fr) with 1,000 bootstrap reiterations.

Sequence divergence of *PpTc1* and *PpTc2* families

To calculate the average sequence divergence of a family, the consensus sequence of each family was constructed. The consensus sequence was used as the input for the Divergence function of MAK. Each divergence value is the complementary percentage of the similarity value in the pairwise alignment of a copy and the consensus sequence. The output contains the sequence divergence values for each member. The average divergence for each family was calculated. To plot the number of elements against divergence, values of individual divergence were grouped into bins of 0.5% and the number of elements in each bin was counted. The overall sequence similarity for a family is calculated as the complement of the average sequence divergence.

Expression analysis of *PpTc* families
Moss TLEs *PpTc*1 and *PpTc*2 (ABEU01007491 and ABEU01006878, respectively) were used to search against the assembled transcripts database Pp0409 on the moss genome browser (http://www.cosmoss.org/) [41]. Returned hits were inspected for a long ORF that encodes a transposase bearing a DD34E catalytic motif. The loci of transcripts were cross-referenced to the nucleotide BLAST hits to remove redundancy. The sequences were also used to search for moss small RNA databases [42-45].

Analyses of TLEs in other plant genome databases
Plant genome databases WGS and NR/NT were searched at NCBI using TBLASTN with the peptide sequences of the putative transposases of *PpTc*1 and *PpTc*2. Hits and their flanking sequences were retrieved to identify putative transposase or TIR sequences.

Abbreviations
MLE: *Mariner*-like element; TE: Transposable element; TIR: Terminal inverted repeat; TLE: *Tc1*-like element; TSD: Target site duplication.

Competing interests
The author(s) declare that they have no competing interests.

Authors' contributions
GY conceived the study. YL and GY designed and performed the analyses. YL and GY drafted the manuscript. YL and GY revised the manuscript. GY edited and finished the manuscript. Both authors read and approved the final manuscript.

Acknowledgements
We would like to thank Dr. Isam Fattash for assistance on the analyses of the expression of moss elements. Funded by the Natural Sciences and Engineering Research Council (NSERC) Discovery Grant (Canada) (RGPIN 371565 to GY, RGPIN-2014-04709), Canadian Foundation for Innovation (CFI24456 and IOF-12 to GY), Ontario Research Foundation (ORF24456 to GY), and the University of Toronto. The funding sources played no role in research design; the collection, analysis, and interpretation of data; the writing of the manuscript; or in the decision to submit the manuscript for publication.

References
1. Yuan YW, Wessler SR: The catalytic domain of all eukaryotic cut-and-paste transposase superfamilies. *Proc Natl Acad Sci U S A* 2011, 108:7884–7889.
2. Wicker T, Sabot F, Hua-Van A, Bennetzen JL, Capy P, Chalhoub B, Flavell A, Leroy P, Morgante M, Panaud O, Paux E, SanMiguel P, Schulman AH: A unified classification system for eukaryotic transposable elements. *Nat Rev Genet* 2007, 8:973–982.
3. Huang CR, Burns KH, Boeke JD: Active transposition in genomes. *Annu Rev Genet* 2012, 46:651–675.
4. Feschotte C, Pritham EJ: DNA transposons and the evolution of eukaryotic genomes. *Annu Rev Genet* 2007, 41:331–368.
5. Yang LX, Bennetzen JL: Distribution, diversity, evolution, and survival of Helitrons in the maize genome. *Proc Natl Acad Sci U S A* 2009, 106:19922–19927.
6. Du C, Fefelova N, Caronna J, He LM, Dooner HK: The polychromatic Helitron landscape of the maize genome. *Proc Natl Acad Sci U S A* 2009, 106:19916–19921.
7. Rebollo R, Romanish MT, Mager DL: Transposable elements: an abundant and natural source of regulatory sequences for host genes. *Annu Rev Genet* 2012, 46:21–42.
8. Jordan IK: Evolutionary tinkering with transposable elements. *Proc Natl Acad Sci U S A* 2006, 103:7941–7942.
9. Cowley M, Oakey RJ: Transposable elements re-wire and fine-tune the transcriptome. *PLoS Genet* 2013, 9:e1003234.
10. Mukamel Z, Tanay A: Hypomethylation marks enhancers within transposable elements. *Nat Genet* 2013, 45:717–718.
11. Feschotte C: Transposable elements and the evolution of regulatory networks. *Nat Rev Genet* 2008, 9:397–405.
12. Kraitshtein Z, Yaakov B, Khasdan V, Kashkush K: Genetic and epigenetic dynamics of a retrotransposon after allopolyploidization of wheat. *Genetics* 2010, 186:801–812.
13. Jiang N, Ferguson AA, Slotkin RK, Lisch D: Pack-Mutator-like transposable elements (Pack-MULEs) induce directional modification of genes through biased insertion and DNA acquisition. *Proc Natl Acad Sci U S A* 2011, 108:1537–1542.
14. Naito K, Zhang F, Tsukiyama T, Saito H, Hancock CN, Richardson AO, Okumoto Y, Tanisaka T, Wessler SR: Unexpected consequences of a sudden and massive transposon amplification on rice gene expression. *Nature* 2009, 461:1130–U1232.
15. Hollister JD, Smith LM, Guo YL, Ott F, Weigel D, Gaut BS: Transposable elements and small RNAs contribute to gene expression divergence between Arabidopsis thaliana and Arabidopsis lyrata. *Proc Natl Acad Sci U S A* 2011, 108:2322–2327.
16. Fattash I, Rooke R, Wong A, Hui C, Luu T, Bhardwaj P, Yang G: Miniature Inverted-repeat Transposable Elements (MITEs): discovery, distribution and activity. *Genome* 2013, 56:475–486.
17. Jiang N, Feschotte C, Zhang XY, Wessler SR: Using rice to understand the origin and amplification of miniature inverted repeat transposable elements (MITEs). *Curr Opin Plant Biol* 2004, 7:115–119.
18. Feschotte C, Wessler SR: Mariner-like transposases are widespread and diverse in flowering plants. *Proc Natl Acad Sci U S A* 2002, 99:280–285.
19. Plasterk RH, Izsvak Z, Ivics Z: Resident aliens: the Tc1/mariner superfamily of transposable elements. *Trends Genet* 1999, 15:326–332.
20. Eide D, Anderson P: Transposition of Tc1 in the Nematode Caenorhabditis-Elegans. *Proc Natl Acad Sci U S A* 1985, 82:1756–1760.
21. Jacobson JW, Medhora MM, Hartl DL: Molecular structure of a somatically unstable transposable element in Drosophila. *Proc Natl Acad Sci U S A* 1986, 83:8684–8688.
22. Feschotte C, Mouches C: Evidence that a family of miniature inverted-repeat transposable elements (MITEs) from the Arabidopsis thaliana genome has arisen from a pogo-like DNA transposon. *Mol Biol Evol* 2000, 17:730–737.
23. Robertson HM: Evolution of DNA transposons in eukaryotes. In *Mobile DNA II*. Edited by Craig NL, Craigie R, Gellert M, Lambowitz AM. Washington, DC: ASM Press; 2002:1093–1110.
24. Richardson JM, Colloms SD, Finnegan DJ, Walkinshaw MD: Molecular architecture of the Mos1 paired-end complex: the structural basis of DNA transposition in a eukaryote. *Cell* 2009, 138:1096–1108.
25. Doak TG, Doerder FP, Jahn CL, Herrick G: A proposed superfamily of transposase genes - transposon-like elements in ciliated protozoa and a common D35e motif. *Proc Natl Acad Sci U S A* 1994, 91:942–946.
26. Shao HG, Tu ZJ: Expanding the diversity of the IS630-Tc1-mariner superfamily: discovery of a unique DD37E transposon and reclassification of the DD37D and DD39D transposons. *Genetics* 2001, 159:1103–1115.
27. Negoua A, Rouault JD, Chakir M, Capy P: Internal deletions of transposable elements: the case of Lemi elements. *Genetica* 2013, 141:369–379.
28. Pritham EJ, Feschotte C, Wessler SR: Unexpected diversity and differential success of DNA transposons in four species of Entamoeba protozoans. *Mol Biol Evol* 2005, 22:1751–1763.

29. Keeling PJ, Burger G, Durnford DG, Lang BF, Lee RW, Pearlman RE, Roger AJ, Gray MW: The tree of eukaryotes. *Trends Ecol Evol* 2005, 20:670–676.

30. Collins J, Forbes E, Anderson P: The Tc3 family of transposable genetic elements in Caenorhabditis-elegans. *Genetics* 1989, 121:47–55.

31. Ruan KS, Emmons SW: Precise and imprecise somatic excision of the transposon Tc1 in the nematode C-elegans. *Nucleic Acids Res* 1987, 15:6875–6881.

32. Pavlopoulos A, Oehler S, Kapetanaki MG, Savakis C: The DNA transposon Minos as a tool for transgenesis and functional genomic analysis in vertebrates and invertebrates. *Genome Biol* 2007, Suppl 1:S2.

33. Carr PD, Tuckwell D, Hey PM, Simon L, D'Enfert C, Birch M, Oliver JD, Bromley MJ: The transposon impala is activated by low temperatures: use of a controlled transposition system to identify genes critical for viability of Aspergillus fumigatus. *Eukaryot Cell* 2010, 9:438–448.

34. Hua-Van A, Langin T, Daboussi MJ: Aberrant transposition of a Tc1-mariner element, impala, in the fungus Fusarium oxysporum. *Mol Genet Genomics* 2002, 267:79–87.

35. Ivics Z, Hackett PB, Plasterk RH, Izsvak Z: Molecular reconstruction of Sleeping beauty, a Tc1-like transposon from fish, and its transposition in human cells. *Cell* 1997, 91:501–510.

36. Jarvik T, Lark KG: Characterization of soymar1, a mariner element in soybean. *Genetics* 1998, 149:1569–1574.

37. Rensing SA, Lang D, Zimmer AD, Terry A, Salamov A, Shapiro H, Nishiyama T, Perroud PF, Lindquist EA, Kamisugi Y, Tanahashi T, Sakakibara K, Fujita T, Oishi K, Shin-I T, Kuroki Y, Toyoda A, Suzuki Y, Hashimoto S, Yamaguchi K, Sugano S, Kohara Y, Fujiyama A, Anterola A, Aoki S, Ashton N, Barbazu WB, Barker E, Bennetzen JL, Blankenship R, *et al*: The Physcomitrella genome reveals evolutionary insights into the conquest of land by plants. *Science* 2008, 319:64–69.

38. Kapitonov V, Jurka J: The age of Alu subfamilies. *J Mol Evol* 1996, 42:59–65.

39. Le Rouzic A, Boutin TS, Capy P: Long-term evolution of transposable elements. *Proc Natl Acad Sci U S A* 2007, 104:19375–19380.

40. van Pouderoyen G, Ketting RF, Perrakis A, Plasterk RHA, Sixma TK: Crystal structure of the specific DNA-binding domain of Tc3 transposase of C-elegans in complex with transposon DNA. *EMBO J* 1997, 16:6044–6054.

41. Lang D, Eisinger J, Reski R, Rensing SA: Representation and high-quality annotation of the Physcomitrella patens transcriptome demonstrates a high proportion of proteins involved in metabolism in mosses. *Plant Biol (Stuttg)* 2005, 7:238–250.

42. Griffiths-Jones S, Saini HK, Van Dongen S, Enright AJ: miRBase: tools for microRNA genomics. *Nucleic Acids Res* 2008, 36:D154–D158.

43. Fattash I, Voss B, Reski R, Hess WR, Frank W: Evidence for the rapid expansion of microRNA-mediated regulation in early land plant evolution. *BMC Plant Biol* 2007, 7:13.

44. Talmor-Neiman M, Stav R, Klipcan L, Buxdorf K, Baulcombe DC, Arazi T: Identification of trans-acting siRNAs in moss and an RNA-dependent RNA polymerase required for their biogenesis. *Plant J* 2006, 48:511–521.

45. Axtell MJ, Jan C, Rajagopalan R, Bartel DP: A two-hit trigger for siRNA biogenesis in plants. *Cell* 2006, 127:565–577.

46. Lohe AR, Hartl DL: Autoregulation of mariner transposase activity by overproduction and dominant-negative complementation. *Mol Biol Evol* 1996, 13:549–555.

47. Lampe DJ, Grant TE, Robertson HM: Factors affecting transposition of the Himar1 mariner transposon in vitro. *Genetics* 1998, 149:179–187.

48. Feschotte C, Swamy L, Wessler SR: Genome-wide analysis of mariner-like transposable elements in rice reveals complex relationships with stowaway miniature inverted repeat transposable elements (MITEs). *Genetics* 2003, 163:747–758.

49. Capy P, Anxolabehere D, Langin T: The strange phylogenies of transposable elements - are horizontal transfers the only explanation. *Trends Genet* 1994, 10:7–12.

50. Biedler JK, Sha HG, Tu ZJ: Evolution and horizontal transfer of a DD37E DNA transposon in mosquitoes. *Genetics* 2007, 177:2553–2558.

51. Robertson HM: The mariner transposable element is widespread in insects. *Nature* 1993, 362:241–245.

52. Yang G, Hall TC: MAK, a computational tool kit for automated MITE analysis. *Nucleic Acids Res* 2003, 31:3659–3665.

53. Janicki M, Rooke R, Yang G: Bioinformatics and genomic analysis of transposable elements in eukaryotic genomes. *Chromosome Res* 2011, 19:787–808.

Permissions

All chapters in this book were first published in MD, by BioMed Central; hereby published with permission under the Creative Commons Attribution License or equivalent. Every chapter published in this book has been scrutinized by our experts. Their significance has been extensively debated. The topics covered herein carry significant findings which will fuel the growth of the discipline. They may even be implemented as practical applications or may be referred to as a beginning point for another development.

The contributors of this book come from diverse backgrounds, making this book a truly international effort. This book will bring forth new frontiers with its revolutionizing research information and detailed analysis of the nascent developments around the world.

We would like to thank all the contributing authors for lending their expertise to make the book truly unique. They have played a crucial role in the development of this book. Without their invaluable contributions this book wouldn't have been possible. They have made vital efforts to compile up to date information on the varied aspects of this subject to make this book a valuable addition to the collection of many professionals and students.

This book was conceptualized with the vision of imparting up-to-date information and advanced data in this field. To ensure the same, a matchless editorial board was set up. Every individual on the board went through rigorous rounds of assessment to prove their worth. After which they invested a large part of their time researching and compiling the most relevant data for our readers.

The editorial board has been involved in producing this book since its inception. They have spent rigorous hours researching and exploring the diverse topics which have resulted in the successful publishing of this book. They have passed on their knowledge of decades through this book. To expedite this challenging task, the publisher supported the team at every step. A small team of assistant editors was also appointed to further simplify the editing procedure and attain best results for the readers.

Apart from the editorial board, the designing team has also invested a significant amount of their time in understanding the subject and creating the most relevant covers. They scrutinized every image to scout for the most suitable representation of the subject and create an appropriate cover for the book.

The publishing team has been an ardent support to the editorial, designing and production team. Their endless efforts to recruit the best for this project, has resulted in the accomplishment of this book. They are a veteran in the field of academics and their pool of knowledge is as vast as their experience in printing. Their expertise and guidance has proved useful at every step. Their uncompromising quality standards have made this book an exceptional effort. Their encouragement from time to time has been an inspiration for everyone.

The publisher and the editorial board hope that this book will prove to be a valuable piece of knowledge for researchers, students, practitioners and scholars across the globe.

List of Contributors

Francois Sabot
UMR DIADE IRD/UM2, 911 Avenue Agropolis
BP64503, F-34394 Montpellier Cedex 5, France

Min-Jin Han, Chu-Lin Xiong, Hong-Bo Zhang,
Meng-Qiang Zhang and Ze Zhang
School of Life Sciences, Chongqing University,
Chongqing 400044, China

Hua-Hao Zhang
College of Pharmacy and Life Science, Jiujiang
University, Jiujiang 332000, China

Anika Schmith, Thomas Spaller, Friedemann
Gaube and Thomas Winckler
Department of Pharmaceutical Biology, Institute of
Pharmacy, University of Jena, Semmelweisstrasse
10, 07743 Jena, Germany

Åsa Fransson
Department of Molecular Biology, Biomedical
Center, Swedish University of Agricultural Sciences,
Uppsala, Sweden
Present address: Aprea AB, Karolinska Institutet
Science Park, Nobels väg 3, 17175 Solna, Sweden.

Benjamin Boesler
Institute of Biology – Genetics, University of Kassel,
Kassel, Germany

Wolfgang Nellen
Institute of Biology – Genetics, University of Kassel,
Kassel, Germany
Present address: Department of Biology, Brawijaya
University, Jl. Veteran, Malang, East Java, Indonesia.

Sandeep Ojha and Christian Hammann
Department of Life Sciences and Chemistry,
Molecular Life Sciences Research Center, Jacobs
University Bremen, Bremen, Germany.

Fredrik Söderbom
Department of Cell and Molecular Biology,
Biomedical Center, Uppsala University, Uppsala,
Sweden

David M. Gilbert, Ashley E. Strother, Courtney E.
Burckhalter and C. Nathan Hancock
Department of Biology and Geology, University
of South Carolina Aiken, 471 University Parkway,
Aiken, SC 29801, USA

M. Catherine Bridges
Present Address: Department of Pathology and
Laboratory Medicine, Medical University of South
Carolina, Charleston, SC 29425, USA

James M. Burnette III
Present Address: College of Natural and
Agricultural Sciences, University of California
Riverside, Riverside, CA 92521, USA

Andrea L. Koenigsberg and Linden T. Hu
Department of Molecular Biology and Microbiology,
Tufts University Sackler School of Biomedical
Sciences, Boston, MA 02111, USA

Brian A. Klein
Department of Molecular Biology and Microbiology,
Tufts University Sackler School of Biomedical
Sciences, Boston, MA 02111, USA
Department of Microbiology, The Forsyth Institute,
Cambridge, MA 02142, USA

Tsute Chen, Jodie C. Scott and Margaret J. Duncan
Department of Microbiology, The Forsyth Institute,
Cambridge, MA 02142, USA

Arun Kannanganat, Kartikeya Joshi and Philip J.
Farabaugh
Department of Biological Sciences and Program in
Molecular and Cell Biology, University of Maryland
Baltimore County, Baltimore, MD 21250, USA

Susmitha Suresh
Department of Biological Sciences and Program in
Molecular and Cell Biology, University of Maryland
Baltimore County, Baltimore, MD 21250, USA
Present address: Division of Infectious Diseases,
Department of Internal Medicine, Stanford
University School of Medicine, Stanford, California
94305, USA

Arun Dakshinamurthy
Department of Biological Sciences and Program in Molecular and Cell Biology, University of Maryland Baltimore County, Baltimore, MD 21250, USA
Present address: Department of Nanosciences and Technology, Karunya University, Karunya Nagar, Coimbatore 641 114, Tamil Nadu, India

David J. Garfinkel and Hyo Won Ahn
Department of Biochemistry & Molecular Biology, University of Georgia, Athens, GA 30602, USA

Stefan Roffler, Fabrizio Menardo and Thomas Wicker
Institute of Plant Biology, University of Zürich, Zollikerstrasse 107, Zürich
CH-8008, Switzerland

Xiao Chen
Department of Molecular Biology, Princeton University, Princeton, NJ 08544, USA

Laura F. Landweber
Department of Ecology and Evolutionary Biology, Princeton University, Princeton, NJ 08544, USA

Shuang Jiang
Department of Horticulture, Zhejiang University, Hangzhou, Zhejiang 310058, China
Forest & Fruit Tree Institute, Shanghai Academy of Agricultural Sciences, Shanghai 201403, China

Danying Cai
Institute of Horticulture, Zhejiang Academy of Agricultural Sciences, Hangzhou, Zhejiang 310021, China.

Yongwang Sun and Yuanwen Teng
Department of Horticulture, Zhejiang University, Hangzhou, Zhejiang 310058, China
The Key Laboratory of Horticultural Plant Growth, Development and Quality Improvement, The Ministry of Agriculture of China, Hangzhou, Zhejiang 310058, China.
Zhejiang Provincial Key Laboratory of Horticultural Plant Integrative Biology, Hangzhou, Zhejiang 310058, China

Kimberly J. Skuster, Mark D. Urban and Stephen C. Ekker
Department of Biochemistry and Molecular Biology, Mayo Clinic, 200 1st St
SW, 1342C Guggenheim, Rochester, MN 55905, USA

Jun Ni
Department of Biochemistry and Molecular Biology, Mayo Clinic, 200 1st St SW, 1342C Guggenheim, Rochester, MN 55905, USA
Department of Chemical and Systems Biology, Stanford University School of Medicine, Stanford, CA 94305, USA

David Nelsen
Department of Biochemistry, Molecular Biology, and Biophysics, University of Minnesota, Minneapolis, MN 55455, USA

Kirk J. Wangensteen
Department of Biochemistry, Molecular Biology, and Biophysics, University of Minnesota, Minneapolis, MN 55455, USA
University of Pennsylvania, 9 Penn Tower, 3400 Spruce ST, Philadelphia, PA 19104, USA

Darius Balciunas
Department of Biochemistry, Molecular Biology, and Biophysics, University of Minnesota, Minneapolis, MN 55455, USA
Department of Biology, Temple University, Philadelphia, PA 19122, USA

Thomas Spaller, Eva Kling and Thomas Winckler
Institute of Pharmacy, Department of Pharmaceutical Biology, Friedrich
Schiller University Jena, Semmelweisstraße 10, Jena 07743, Germany

Gernot Glöckner
Institute for Biochemistry I, Medical Faculty, University of Cologne, Berlin, Germany.
Institute for Freshwater Ecology and Inland Fisheries, IGB, Berlin, Germany.

Falk Hillmann
Junior Research Group Evolution of Microbial Interaction, Leibniz Institute for Natural Product Research and Infection Biology—Hans Knöll Institute, Jena, Germany

Biju Vadakkemukadiyil Chellapan
Department of Computational Biology and Bioinformatics, University of Kerala, Karyavattom Campus, Karyavattom PO, Trivandrum, Kerala, India
Molecular Plant Pathology, Swammerdam Institute for Life Sciences, Faculty of Science, University of Amsterdam, P.O. Box 94215, 1090 Amsterdam, GE, the Netherlands.

Peter van Dam, Martijn Rep, Ben J. C. Cornelissen and Like Fokkens
Molecular Plant Pathology, Swammerdam Institute for Life Sciences, Faculty of Science, University of Amsterdam, P.O. Box 94215, 1090 Amsterdam, GE, The Netherlands.

Kirill Ustyantsev and Alexandr Blinov
Institute of Cytology and Genetics, Laboratory of Molecular Genetic Systems, Prospekt Lavrentyeva 10, 630090 Novosibirsk, Russia

Georgy Smyshlyaev
Structural and Computational Biology Unit, European Molecular Biology Laboratory, 69117 Heidelberg, Germany

Hao Yin, Xiao Wu, Dongqing Shi, Yangyang Chen, Kaijie Qi and Shaoling Zhang
Center of Pear Engineering Technology Research, College of Horticulture, Nanjing Agricultural University, Nanjing 210095, China
State Key Laboratory of Crop Genetics and Germplasm Enhancement, Nanjing Agricultural University, Nanjing, China

Zhengqiang Ma
State Key Laboratory of Crop Genetics and Germplasm Enhancement, Nanjing Agricultural University, Nanjing, China
College of Agricultural Sciences, Nanjing Agricultural University, Nanjing, China

Stefan Roffler and Thomas Wicker
Institute for Plant Biology, University of Zürich, Zollikerstrasse 107, CH-8008 Zürich, Switzerland

Ari Löytynoja
Institute of Biotechnology, Viikki Biocenter, University of Helsinki, PO Box 65, FIN-00014 Helsinki, Finland

Alan H Schulman
Institute of Biotechnology, Viikki Biocenter, University of Helsinki, PO Box 65, FIN-00014 Helsinki, Finland.
Biotechnology and Food Research, MTT Agrifood Research Finland, Myllytie 1, FIN-31600 Jokioinen, Finland

Jan P Buchmann
Institute of Biotechnology, Viikki Biocenter, University of Helsinki, PO Box 65, FIN-00014 Helsinki, Finland.
Present address: Marie Bashir Institute for Infectious Diseases and Biosecurity, Charles Perkins Center, University of Sydney, Sydney NSW 2006, Australia

Thomas Wicker
Institute of Plant Biology, University of Zurich, Zollikerstrasse 107, Zurich, Switzerland

Yuan Liu and Guojun Yang
Department of Biology, University of Toronto at Mississauga, 3359 Mississauga Road, L5L 1C6 Mississauga, ON, Canada
Cell and Systems Biology, University of Toronto, Toronto, Canada

Index

www.ingramcontent.com/pod-product-compliance
Lightning Source LLC
Chambersburg PA
CBHW082044190326
41458CB00010B/3451